DUALITY IN OPTIMIZATION AND VARIATIONAL INEQUALITIES

Optimization Theory and Applications
A series of books and monographs on the theory and applications of optimization.
Edited by K.-H. Elster † and F. Giannessi, University of Pisa, Italy

Volume 1

Stochastic Linear Programming Algorithms: A Comparison Based on a Model Management System
J. Mayer

Volume 2

Duality in Optimization and Variational Inequalities
C.J. Goh and X.Q. Yang

DUALITY IN OPTIMIZATION AND VARIATIONAL INEQUALITIES

C.J.Goh

Department of Mathematics, Auckland University
and
Department of Mathematics and Statistics,
University Of Western Australia

and

X.Q. Yang

Department of Applied Mathematics,
The Hong Kong Polytechnic University

CRC Press
Taylor & Francis Group
Boca Raton London New York

CRC Press is an imprint of the
Taylor & Francis Group, an **informa** business
A TAYLOR & FRANCIS BOOK

CRC Press
Taylor & Francis Group
6000 Broken Sound Parkway NW, Suite 300
Boca Raton, FL 33487-2742

First issued in paperback 2019

Publisher's Note
This book has been prepared from camera-ready copy provided by the authors.

ISBN-13: 978-0-415-27479-1 (hbk)
ISBN-13: 978-0-367-39615-2 (pbk)

British Library Cataloguing in Publication Data
A catalogue record for this book is available from the British Library

Library of Congress Cataloging in Publication Data
A catalog record has been applied for.

Visit the Taylor & Francis Web site at
http://www.taylorandfrancis.com

and the CRC Press Web site at
http://www.crcpress.com

Dedicated to

April and Linus

Zoe and Lucy

CONTENTS

FIGURE CAPTIONS

PREFACE

The concept of duality is ubiquitous in mathematics. It appears in group theory, number theory, mathematical physics, control theory, stochastic processes, financial engineering, the list goes on. The primary objective of this book is to focus on duality in optimization, and its generalization in variational inequalities. This is a domain that can sometimes be considered totally esoteric and suitable only to the purest of pure mathematicians, but at other times can be used in practical ways for solving engineering and business problems. Needless to say, the concept of optimization is paramount in just about every area in physical/natural/social/financial science and engineering. Companies use it to save money, engineers use it to design superior products, governments use it to plan the economy, biologists use it to study brain functions, and even mathematicians have used it to win the Nobel prize in economics! While there is an abundance of books that specialize in optimization, the great majority of them are concerned with furnishing methods, algorithms and applications. Some advanced monographs dwell more in the theory, but there are very few books in the literature that are entirely devoted to discussing duality. Two notable exceptions are found in [R2] and [W1], which both are devoted to a generalized Lagrangian model based on conjugate duality. Some justifications for the existence of this book are as follows.

There are, in the main, two approaches to the study of duality in optimization models. Classically, the primal optimization problem is first defined and then the duality theory is constructed and developed from there. At a more abstract level, the primal and dual optimization problems are derived from an abstract Lagrangian function $L : \mathcal{X} \times \mathcal{Y} \to \mathbb{R}$ in the following manner:

$$\text{(primal problem)} \qquad \min_{\mathbf{x}\in\mathcal{X}} f(\mathbf{x}) \qquad \text{where} \quad f(\mathbf{x}) = \sup_{\mathbf{y}\in\mathcal{Y}} L(\mathbf{x},\mathbf{y}),$$

$$\text{(dual problem)} \qquad \max_{\mathbf{y}\in\mathcal{Y}} g(\mathbf{y}) \qquad \text{where} \quad g(\mathbf{y}) = \inf_{\mathbf{x}\in\mathcal{X}} L(\mathbf{x},\mathbf{y}),$$

where $\mathcal{X}$ and $\mathcal{Y}$ are appropriate domains defined in some primal and dual spaces. This approach to duality is founded on conjugate duality and invariably a convexity assumption is made. This approach also places strong emphasis on the saddle point property and minimax theorems (see Section 1.6). There is much merit in pursuing duality in this abstract way. Apart from its complete symmetry, it can be reduced to, as special cases, many other duality results in optimization. Nevertheless, we have chosen not to pursue duality in the abstract Lagrangian way, the main reason being that the two books [R2] and [W1] have already done an excellent job with such an approach.

Furthermore, this book is intended to cover specialized duality results ranging from network programming, to vector optimization, to vector variational inequalities, and some of these are yet to be considered as special cases of the abstract Lagrangian approach.

Lastly, convexity is imperative in the abstract approach. The conventional wisdom is that when the convexity assumption is not satisfied, then it is difficult to derive necessary and sufficient optimality conditions apart from those trivial ones based on definitions. In Chapter Five, we discuss strong duality results and equivalent optimality conditions in the absence of convexity. This is made possible only if we are prepared to depart from the Lagrangian model.

The first exposure to duality that most undergraduates encounter is likely to be the duality in linear programming (LP). From our experience in teaching this subject to undergraduates, the probability of the following question, or its variant, recurring each year is close to one: "OK, this duality stuff is all very nice, but so what? Does it help me to solve the problem?" Our answer to this question is not usually well accepted by the average undergraduate who is typically obsessed with doing sums. In fact, it would usually take several years of work in optimization theory before one would appreciate the assertion that many methods and ideas in optimization are practically driven by duality. *To help users of optimization to understand and appreciate this assertion is thus the main aim of this book.*

Most readers would understand duality as:

• A pair of optimization problems, the primal and the dual, both share the same optimal cost, and the solution of one yields the solution to the other. Examples of this are found in Fenchel Duality Theorem, the Linear Programming Duality, the Lagrangian Duality Theorem, and vector optimization.

In a broader sense, duality relates two systems, or geometric entities, or sets, or functions, or objects, in one of the following ways:

• Theorems of alternatives: Given two systems or sets, if one is non-empty, the other one must be empty, and vice-versa. Examples of this are found in the Farkas Lemma, the Painted network/index theorem, the feasible flow and the feasible potential theorems.

• Duality via transformation: A set of transformation rules that allow one object/set/function to be transformed into another. A further transformation under the same rules returns the original object. For example, planar graph duality, cone duality, and conjugate functions.

• Properties of one system have a one-to-one correspondence with the properties of another. For example, cycles and cuts, match and block, flow and potential, controllability and observability in control theory, and the inverse of an one-to-one map.

Readers will no doubt discover other ways in which duality can occur. It is clear that all these manifestations are centered around some kind of symmetry. In fact duality is effectively the study of a form of symmetry.

By and large, this book is intended for readers with a prior knowledge of optimization, and who have already gained some exposure to computational algorithms for solving optimization problems. Our contention is to emphasize and explain duality as clearly as possible from a theoretical perspective. To avoid getting entangled in unnecessarily abstract notions, just about all results in this book are presented for finite dimensional vector spaces such as $\mathbb{R}^n$. Most results in this book can be easily extended to abstract spaces, albeit with added notational complexity.

This book is organized in nine chapters. Loosely speaking the first four chapters are of relatively light weight while the last five are of relatively heavy weight. The light weight chapters are used to remind the readers of some foundational material and to facilitate the understanding of duality in a clear and concise manner.

Chapter One contains the standard mathematical preliminaries that advanced readers may skip. Chapter Two is a collection of duality results in network optimization, which in our opinion is probably the simplest way of understanding duality as there is usually some kind of visualizable geometry involved. Chapter Three is concerned with duality in linear programming and associated linear systems. Linear programming is likely to be the first encounter of most readers with duality. The reason why we choose to put this chapter after that of network optimization is due to several reasons. First, it is easier to understand duality in the network context. Second, the topics on painted index theory and monotropic optimization are generalizations of the network problem, yet these contain significant overlaps with linear programming (LP) and readers would have been better off understanding LP duality first. Chapter Four is concerned with traditional duality results in nonlinear programming.

Chapters Five to Nine contain fairly recent research materials, many of which have yet to appear in book form. Chapter Five attempts to break away from the traditional Lagrangian duality theory and proposes new approaches to the duality of nonconvex nonlinear systems. Chapter Six is about duality in variational inequalities, a subject that includes optimization as a special case. From Chapter Seven onward, multicriteria consideration dominates, and we shall begin with an introduction to multicriteria optimization and vector variational inequalities in Chapter Seven. Chapter Eight is devoted to the study of duality in multicriteria optimization, a topic of great theoretical interest. The last chapter contains fairly recent research results in the duality of vector variational inequalities.

We are indebted to Bruce Craven, Gaven Martin, Alistair Mees, Elijah Polak, Alex Rubinov and Kok Lay Teo for their valuable advice, comments, and feedback on earlier drafts of this book. The final draft has been improved by the watchful eyes of Xuexiang Huang, Shengjie Li and Xinmin Yang who assisted in the proofreading. Les Jennings' help in our struggle with TEX has made it possible for us to meet the publication deadline. We thank Frances Horrocks of Taylor and Francis for coordinating the final effort in getting this book to press. This book could not have been written without the earlier inspiring work of Terry Rockafellar, who has also provided us with his encouragement, and his kind permission in allowing us to extract materials from his book. We are in particular grateful to Franco Giannessi for his continuous encouragement and detailed comments during the preparation of this book. We acknowledge that our research activities have been continuously supported by several grants from the Australian Research Council and the Research Grants Council of Hong Kong.

CHAPTER 1

MATHEMATICAL PRELIMINARIES

The material in this book spans a fairly wide range of mathematics. It would be foolhardy to attempt to cover all the requisite knowledge and theories and hence only that relatively small subset of fundamental material is presented here so as to make the book as self-contained as possible. The purpose of this chapter is not to facilitate comprehensive understanding by a novice but rather to ensure the notations and basic concepts introduced are consistent and to remind the reader of certain fundamental results. The reader is therefore assumed to have an elementary grounding in linear algebra, real analysis, calculus, and have done at least a first course in linear and nonlinear programming. Some exposure to graph theory, vector optimization and variational inequalities would certainly be an added advantage.

Nothing in this chapter is new of course. It is inevitable that materials are often imported, in an abridged form or otherwise, from other sources, but with notation suitably modified to conform to the standard notation of this book. We have tried to present the results complete with proof most of the time, although occasionally some results of a more involved nature are presented without proof. Whenever in doubt, the readers should refer to the following classics. Much of the materials in this chapter are also extracted from these references.

- Real Analysis: [R4], [R5].
- Convex Analysis: [B3], [ET1], [R1], [HL1], [vT1].
- Linear and Nonlinear Programming: [L1], [BS1], [C1], [C2].
- Graph Theory and Network Optimization: [R3], [KH1], [AMO1], [H3].
- Vector Optimization: [SNT1], [L2], [J1].
- Variational Inequalities: [N1].

1.1 Basic Notation and a List of Symbols

For the benefit of the reader, we have compiled a (possibly incomplete) list of notations and symbols used throughout the book, so that one may refer to this section for a vaguely-familiar-but-can't-remember symbol. This section is not intended to be carefully read and digested at the first sitting, but to be browsed through and its existence noted.
In general,

• a *vector* is a column vector, unless otherwise specified, and is typically represented by a lower case Roman or Greek alphabet in bold face, e.g., $\mathbf{x}$ or $\boldsymbol{\xi}$. The superscript of a vector represents its ordering in a sequence of vectors, e.g., $\mathbf{x}^i$ is the ith vector in a sequence of vectors. The transpose of a vector $\mathbf{x}$ is denoted by $\mathbf{x}^\top$. The ith component of the vector $\mathbf{x}$ is typically denoted by x_i or $[\mathbf{x}]_i$. The unit vector $\mathbf{e}^i$ has entries given by $[\mathbf{e}^i]_j = \delta_{ij}$, where δ_{ij} is the Kronecker delta tensor with $\delta_{ij} = 1$ if $i = j$ and $\delta_{ij} = 0$ otherwise.

• A *matrix* is typically represented by an upper case Roman or Greek alphabet in bold face, e.g., $\mathbf{A}$ or $\boldsymbol{\Lambda}$; the superscript of a matrix represents its ordering in a sequence of matrices, e.g., $\mathbf{A}^i$ is the ith matrix in a sequence of matrices. The (i, j) entry of a matrix $\mathbf{A}$ is typically represented by A_{ij} or $[\mathbf{A}]_{ij}$. The *transpose, rank, determinant, trace, kernel* of a matrix $\mathbf{A}$ are denoted respectively by $\mathbf{A}^\top, \text{rank}(\mathbf{A}), \det(\mathbf{A}), \text{tr}(\mathbf{A}), \ker(\mathbf{A})$. The identity matrix is denoted by $\mathbf{I}$. The kth column of $\mathbf{A}$ is denoted by $\mathbf{a}^k$, and the kth row of $\mathbf{A}$ is denoted by $\mathbf{a}_k$.

• A scalar is typically represented by a lower case Roman or Greek alphabet; A superscripted scalar represents its ordering in a sequence of scalars.

• A set is typically represented by an upper case Roman or Greek alphabet in calligraphic font, e.g., $\mathcal{A}$ or $\mathcal{N}$; if a set $\mathcal{A}$ is discrete, $|\mathcal{A}|$ denotes its cardinality.

• An abstract space or subspace, is typically represented by an upper case Roman or Greek alphabet in script font, e.g., $\mathscr{S}$ or $\mathscr{B}$; $\dim(\mathscr{S})$ denotes the dimension of the spasce $\mathscr{S}$.

• A scalar-valued function is represented by a lower case Roman or Greek alphabet, e.g., $f(\mathbf{x})$ is a scalar-valued function of the vector variable $\mathbf{x}$, a vector-valued function is represented by a lower case Roman or Greek alphabet in bold face, e.g., $\mathbf{f}(\mathbf{x})$.

• $\mathbb{R}$ is the set of real numbers;
$\bar{\mathbb{R}} = \mathbb{R} \cup \{\infty\}$;
$\mathbb{R}^n = \mathbb{R} \times \mathbb{R} \times \cdots \times \mathbb{R}$ is the Euclidean vector space of dimension n;
$\mathbb{R}^n_+ = \{\boldsymbol{\xi} \in \mathbb{R}^n |\ \xi_j \geq \mathbf{0},\ j = 1, 2, \cdots, n\}$ is the closed positive orthant of $\mathbb{R}^n$; and int $\mathbb{R}^n_+ = \{\boldsymbol{\xi} \in \mathbb{R}^n |\ \xi_j > 0,\ j = 1, 2, \cdots, n\}$ is the interior of $\mathbb{R}^n_+$.
$\bar{\mathbb{R}}^n = \mathbb{R}^n \cup \{\infty\}$ where ∞ is the imaginary point which has ∞ for every component.

Given a real/scalar-valued function f,

• $\text{dom}(f)$ is the effective domain of the function f;

- Range(f) is the range of f;
- $\nabla f(\mathbf{x}^0)$ or $\left.\frac{\partial f}{\partial \mathbf{x}}\right|_{\mathbf{x}^0}$ is the gradient of f with respect to $\mathbf{x}$ evaluated at $\mathbf{x}^0$, and is a row vector;
- $\partial f(\mathbf{x}^0)$ is the convex subdifferential of f at $\mathbf{x}^0$;
- $\partial^\circ f(\mathbf{x}^0)$ is the Clarke subdifferential of f at $\mathbf{x}^0$;
- $f'_+(\xi)$ and $f'_-(\xi)$ denote the right and left derivative of a scalar function f of a scalar variable ξ;
- $f'(\mathbf{x}; \mathbf{d})$ is the directional derivative of f at $\mathbf{x}$ in the direction of $\mathbf{d}$;
- f^* is the Fenchel transform or conjugate of f;
- f^{**} is the bi-Fenchel transform or biconjugate of f.

Given a vector-valued function $\mathbf{f}$,

- $\nabla \mathbf{f}(\mathbf{x}^0)$ or $\left.\frac{\partial \mathbf{f}}{\partial \mathbf{x}}\right|_{\mathbf{x}^0}$ is the Jacobian of $\mathbf{f}$ with respect to $\mathbf{x}$ evaluated at $\mathbf{x}^0$, and is a matrix;
- $\partial \mathbf{f}(\mathbf{x}^0)$ is the set of all type I subgradients of $\mathbf{f}$ at $\mathbf{x}^0$;
- $\partial_w \mathbf{f}(\mathbf{x}^0)$ is the set of all type I weak subgradients of $\mathbf{f}$ at $\mathbf{x}^0$;
- $\partial_s \mathbf{f}(\mathbf{x}^0)$ is the set of all type I strong subgradients of $\mathbf{f}$ at $\mathbf{x}^0$;
- $\partial_{\mathbf{e}} \mathbf{f}(\mathbf{x}^0)$ is the set of all type II subgradients of $\mathbf{f}$ at $\mathbf{x}^0$;
- $\partial_{w\mathbf{e}} \mathbf{f}(\mathbf{x}^0)$ is the set of all type II weak subgradients of $\mathbf{f}$ at $\mathbf{x}^0$;
- $\partial_{s\mathbf{e}} \mathbf{f}(\mathbf{x}^0)$ is the set of all type II strong subgradients of $\mathbf{f}$ at $\mathbf{x}^0$;
- $\mathbf{f}^*$ is the type I conjugate or type I Fenchel transform of $\mathbf{f}$;
- $\mathbf{f}^*_w$ is the type I weak conjugate or type I weak Fenchel transform of $\mathbf{f}$;
- $\mathbf{f}^{**}$ is the type I biconjugate or type I bi-Fenchel transform of $\mathbf{f}$;
- $\mathbf{f}^{**}_w$ is the type I weak biconjugate or type I weak bi-Fenchel transform of $\mathbf{f}$;
- $\mathbf{f}^*_{\mathbf{e}}$ is the type II conjugate or type II Fenchel transform of $\mathbf{f}$;
- $\mathbf{f}^*_{w\mathbf{e}}$ is the type II weak conjugate or type II weak Fenchel transform of $\mathbf{f}$;
- $\mathbf{f}^{**}_{\mathbf{e}}$ is the type II biconjugate or type II bi-Fenchel transform of $\mathbf{f}$;
- $\mathbf{f}^{**}_{w\mathbf{e}}$ is the type II weak biconjugate or type II weak bi-Fenchel transform of $\mathbf{f}$.

- A graph or network $\mathcal{G} = (\mathcal{N}, \mathcal{A})$ is typically a directed graph (or digraph), unless otherwise specified. The following symbols are reserved for special meanings in network optimization (see Section 1.4 and Chapter Two):
  - $\mathcal{N}$ is the node set of $\mathcal{G}$;
  - $\mathcal{A}$ is the arc set of $\mathcal{G}$;

- $\mathbf{E}$ usually denotes the node-arc incidence matrix of $\mathcal{G}$;
- $\mathcal{T}$ usually denotes a tree;
- $\mathcal{F}$ usually denotes a forest;
- $\mathcal{P} : \mathcal{M} \to \mathcal{M}'$ usually denotes a path joining some node in $\mathcal{M}$ to another node in $\mathcal{M}'$;

  $\mathbf{e}_{\mathcal{P}} \in \mathbb{R}^{|\mathcal{A}|}$ denotes the incidence vector for the path $\mathcal{P}$;
  $\mathcal{P}^+$ is the set of arcs having the same orientation as $\mathcal{P}$;
  $\mathcal{P}^-$ is the set of arcs having the opposite orientation as $\mathcal{P}$;
- $\mathcal{Q} : \mathcal{M} \downarrow \mathcal{M}'$ usually denotes a cut separating the node sets $\mathcal{M}$ and $\mathcal{M}'$;

  $\mathbf{e}_{\mathcal{Q}} \in \mathbb{R}^{|\mathcal{A}|}$ denotes the incidence vector for the cut $\mathcal{Q}$;
  $\mathcal{Q}^+$ is the set of arcs having the same orientation as $\mathcal{Q}$;
  $\mathcal{Q}^-$ is the set of arcs having the opposite orientation as $\mathcal{Q}$;
- $\mathbf{e}_{\mathcal{S}} \in \mathbb{R}^{|\mathcal{N}|}$ denotes the incidence vector for a set of nodes $\mathcal{S}$;
- c_j denotes the unit cost of differential on an arc j;
- d_j denotes the unit cost of flow on an arc j;
- c_j^+ and c_j^- denote the upper and lower flow capacities of an arc j respectively;

  $c_{\mathcal{Q}}^+$ and $c_{\mathcal{Q}}^-$ denote the upper and lower flow capacities of a cut $\mathcal{Q}$ respectively;
- d_j^+ and d_j^- denote the upper and lower spans of an arc j respectively;

  $d_{\mathcal{P}}^+$ and $d_{\mathcal{P}}^-$ denote the upper and lower spans of a path $\mathcal{P}$ respectively;
- $\mathbf{x} \in \mathbb{R}^{|\mathcal{A}|}$ usually denotes the flow vector on the digraph $\mathcal{G}$;
- $\mathbf{y} \in \mathbb{R}^{|\mathcal{N}|}$ usually denotes the divergence vector on the digraph $\mathcal{G}$;
- $\mathbf{u} \in \mathbb{R}^{|\mathcal{N}|}$ usually denotes the node potential vector on the digraph $\mathcal{G}$;
- $\mathbf{v} \in \mathbb{R}^{|\mathcal{A}|}$ usually denotes the tension (or differential) vector on the digraph $\mathcal{G}$;
- $\mathbf{b} \in \mathbb{R}^{|\mathcal{N}|}$ usually denotes the flow requirement vector on the digraph $\mathcal{G}$.

In monotropic programming, the notations used are very similar to that used for network optimization, since the former is in fact a powerful generalization of the latter. As the descriptions of the terms are not as straightforward as that for the network case, we do not list them here.

- Given $\boldsymbol{\xi}, \boldsymbol{\eta} \in \mathbb{R}^p$, the ordering relationships induced by the cone $\mathbb{R}^p_+$ ($\leq_{\mathbf{R}^p_+}, \leq_{\mathbf{R}^p_+\setminus\{\mathbf{0}\}}, \not\leq_{\mathbf{R}^p_+\setminus\{\mathbf{0}\}}, \leq_{\text{int } \mathbf{R}^p_+}, \not\leq_{\text{int } \mathbf{R}^p_+}$) are defined as:

$$\boldsymbol{\xi} \leq_{\mathbf{R}^p_+} \boldsymbol{\eta} \iff \boldsymbol{\eta} - \boldsymbol{\xi} \in \mathbb{R}^p_+;$$

$$\boldsymbol{\xi} \leq_{\mathbf{R}^p_+\setminus\{\mathbf{0}\}} \boldsymbol{\eta} \iff \boldsymbol{\eta} - \boldsymbol{\xi} \in \mathbb{R}^p_+ \setminus \{\mathbf{0}\};$$

$$\boldsymbol{\xi} \not\leq_{\mathbf{R}^p_+\setminus\{\mathbf{0}\}} \boldsymbol{\eta} \iff \boldsymbol{\eta} - \boldsymbol{\xi} \notin \mathbb{R}^p_+ \setminus \{\mathbf{0}\};$$

$$\boldsymbol{\xi} \leq_{\text{int } \mathbf{R}^p_+} \boldsymbol{\eta} \iff \boldsymbol{\eta} - \boldsymbol{\xi} \in \text{int } \mathbb{R}^p_+;$$

$$\boldsymbol{\xi} \not\leq_{\text{int } \mathbf{R}^p_+} \boldsymbol{\eta} \iff \boldsymbol{\eta} - \boldsymbol{\xi} \notin \text{int } \mathbb{R}^p_+.$$

The subscript associated with the ordering relationships in an Euclidean vector space denotes the dimension of the vector space in which the ordering is defined. The ordering relationships ($\geq_{\mathbf{R}^p_+}, \geq_{\mathbf{R}^p_+\setminus\{\mathbf{0}\}}, \not\geq_{\mathbf{R}^p_+\setminus\{\mathbf{0}\}}, \geq_{\text{int } \mathbf{R}^p_+}, \not\leq_{\text{int } \mathbf{R}^p_+}$) are defined similarly. In Chapters one to six, a simpler notation without the subscript is used.

By default, $<$, $\leq$, $\leqq$, $>$, $\geq$, $\geqq$ represent $\leq_{\text{int } \mathbf{R}^n_+}$, $\leq_{\mathbf{R}^n_+\setminus\{\mathbf{0}\}}$, $\leq_{\mathbf{R}^n_+}$, $\geq_{\text{int } \mathbf{R}^n_+}$, $\geq_{\mathbf{R}^n_+\setminus\{\mathbf{0}\}}$, $\geq_{\mathbf{R}^n_+}$ respectively, where n is the dimension of the underlying Euclidean space.

• Given two subsets $\mathcal{A}$ and $\mathcal{B}$ of $\mathbb{R}^p$, the following define the ordering relationships on sets induced by the cone $\mathbb{R}^p_+$:

$$\mathcal{A} \leq_{\mathbf{R}^p_+} \mathcal{B} \Longleftrightarrow \boldsymbol{\xi} \leq_{\mathbf{R}^p_+} \boldsymbol{\eta}, \ \forall \boldsymbol{\xi} \in \mathcal{A}, \boldsymbol{\eta} \in \mathcal{B};$$
$$\mathcal{A} \leq_{\mathbf{R}^p_+\setminus\{\mathbf{0}\}} \mathcal{B} \Longleftrightarrow \boldsymbol{\xi} \leq_{\mathbf{R}^p_+\setminus\{\mathbf{0}\}} \boldsymbol{\eta}, \ \forall \boldsymbol{\xi} \in \mathcal{A}, \boldsymbol{\eta} \in \mathcal{B};$$
$$\mathcal{A} \not\leq_{\mathbf{R}^p_+\setminus\{\mathbf{0}\}} \mathcal{B} \Longleftrightarrow \boldsymbol{\xi} \not\leq_{\mathbf{R}^p_+\setminus\{\mathbf{0}\}} \boldsymbol{\eta}, \ \forall \boldsymbol{\xi} \in \mathcal{A}, \boldsymbol{\eta} \in \mathcal{B};$$
$$\mathcal{A} \leq_{\text{int } \mathbf{R}^p_+} \mathcal{B} \Longleftrightarrow \boldsymbol{\xi} \leq_{\text{int } \mathbf{R}^p_+} \boldsymbol{\eta}, \ \forall \boldsymbol{\xi} \in \mathcal{A}, \boldsymbol{\eta} \in \mathcal{B};$$
$$\mathcal{A} \not\leq_{\text{int } \mathbf{R}^p_+} \mathcal{B} \Longleftrightarrow \boldsymbol{\xi} \not\leq_{\text{int } \mathbf{R}^p_+} \boldsymbol{\eta}, \ \forall \boldsymbol{\xi} \in \mathcal{A}, \boldsymbol{\eta} \in \mathcal{B}.$$

• $\min_{\mathbf{R}^p_+\setminus\{\mathbf{0}\}} \mathcal{Y}$ is the set of all minimal points, or efficient points, of the set $\mathcal{Y}$;

• $\min_{\text{int } \mathbf{R}^p_+} \mathcal{Y}$ is the set of all weakly minimal points, or weakly efficient points, of the set $\mathcal{Y}$;

• $\max_{\mathbf{R}^p_+\setminus\{\mathbf{0}\}} \mathcal{Y}$ is the set of all maximal points of the set $\mathcal{Y}$;

• $\max_{\text{int } \mathbf{R}^p_+} \mathcal{Y}$ is the set of all weakly maximal points of the set $\mathcal{Y}$.

Given a vector-valued function $\mathbf{f}$ and a set $\mathcal{X} \subset \text{dom}(\mathbf{f})$,

• $\min_{\mathbf{R}^p_+\setminus\{\mathbf{0}\}} \mathbf{f}(\mathcal{X})$ is the set of all minimal points, or efficient points, of the set $\mathbf{f}(\mathcal{X})$, and $\text{argmin}_{\mathbf{R}^p_+\setminus\{\mathbf{0}\}}(\mathbf{f}, \mathcal{X})$ is the set of all minimal solutions or efficient solutions to the corresponding multicriteria optimization problem.

• $\min_{\text{int } \mathbf{R}^p_+} \mathbf{f}(\mathcal{X})$ is the set of all weakly minimal points, or weakly efficient points, of the set $\mathbf{f}(\mathcal{X})$, and $\text{argmin}_{\text{int } \mathbf{R}^p_+}(\mathbf{f}, \mathcal{X})$ is the set of all weakly minimal solutions or weakly efficient solutions to the corresponding multicriteria optimization problem.

1.2 Elements of Convex Analysis

All the results in this book are presented in the context of the Euclidean vector space, although these results can mostly be generalized to abstract spaces like Banach space or Hausdorff space. As this section is not intended to be a first course in real and convex analysis, the readers may find the presentation a bit dry with the seemingly endless stream of definitions-theorems-proofs. If in doubt or if a better understanding is required, the reader is urged to supplement the reading of this section with more comprehensive texts listed in the beginning of this chapter.

Definition 1.2.1 Let $\mathbb{R}$ be the set of real numbers. Then,
$\bar{\mathbb{R}} = \mathbb{R} \cup \{\infty\}$;
$\mathbb{R}^n = \mathbb{R} \times \mathbb{R} \times \cdots \times \mathbb{R}$ is the Euclidean vector space of dimension n;
$\mathbb{R}^n_+ = \{\boldsymbol{\xi} \in \mathbb{R}^n | \ \xi_j \geq \mathbf{0}, \ j = 1, 2, \cdots, n\}$ is the closed positive orthant of $\mathbb{R}^n$;
int $\mathbb{R}^n_+ = \{\boldsymbol{\xi} \in \mathbb{R}^n | \ \xi_j > 0, \ j = 1, 2, \cdots, n\}$.

Definition 1.2.2 Let $\mathcal{X}$ and $\mathcal{Y}$ be subsets of $\mathbb{R}^n$, $\lambda, \mu \in \mathbb{R}$, $\mathbf{x} \in \mathbb{R}^n$. Then
$\mu\mathcal{X} = \{\mu\mathbf{x} \mid \mathbf{x} \in \mathcal{X}\}$;
$\mathcal{X} + \mathcal{Y} = \{\mathbf{x} + \mathbf{y} \mid \mathbf{x} \in \mathcal{X}, \ \ \mathbf{y} \in \mathcal{Y}\}$;
$\mathcal{X} \cap \mathcal{Y} = \{\mathbf{x} \mid \mathbf{x} \in \mathcal{X}$, and $\mathbf{x} \in \mathcal{Y}\}$;
$\mathcal{X} \cup \mathcal{Y} = \{\mathbf{x} \mid \mathbf{x} \in \mathcal{X}$, or $\mathbf{x} \in \mathcal{Y}\}$;
$\|\mathbf{x}\|_p = [\sum_{i=1}^n |x_i|^p]^{\frac{1}{p}}$;
by convention, $\|\cdot\|$ denotes the usual Hilbert-Schmidt norm $\|\cdot\|_2$.

Definition 1.2.3 Let $\mathcal{X} \subset \mathbb{R}^n$, $\mathbf{x} \in \mathbb{R}^n$, $\delta > 0$, then
(*Open ball*) $\mathcal{B}(\mathbf{x}, \delta) = \{\mathbf{y} \in \mathbb{R}^n \mid \|\mathbf{y} - \mathbf{x}\| < \delta\}$.
(*Interior point*) $\mathbf{x}$ is said to be an interior point of $\mathcal{X}$ if there exists some $\delta > 0$ such that $\mathcal{B}(\mathbf{x}, \delta) \subset \mathcal{X}$.
(*Interior*) int($\mathcal{X}$) = set of all interior points of $\mathcal{X}$, i.e., int($\mathcal{X}$) = $\{\mathbf{x} \in \mathcal{X} \mid \exists\delta > 0$ such that $\mathcal{B}(\mathbf{x}, \delta) \subset \mathcal{X}\}$.
A sequence of vectors $\{\mathbf{x}^i, i = 1, 2, \cdots\}$ is said to *converge* to the limit point $\bar{\mathbf{x}}$ if $\|\mathbf{x}^i - \bar{\mathbf{x}}\| \to 0$ as $i \to \infty$.
(*Closure*) $\bar{\mathcal{X}} = \mathcal{X} \cup \{$ limits of all convergent sequences in $\mathcal{X}\}$.
A set $\mathcal{X}$ is said to be *closed* if $\mathcal{X} = \bar{\mathcal{X}}$.
A point $\mathbf{x} \in \mathbb{R}^n$ is said to be a *boundary point* of the set $\mathcal{X}$ if there exists some $\delta > 0$ such that $\mathcal{B}(\mathbf{x}, \delta) \cap \mathcal{X} \neq \emptyset$ and $\mathcal{B}(\mathbf{x}, \delta) \cap (\mathbb{R}^n \setminus \mathcal{X}) \neq \emptyset$.
(*Boundary*) $\partial\mathcal{X}$ = set of all boundary points. It is clear that $\partial\mathcal{X} = \bar{\mathcal{X}} \setminus \text{int}(\mathcal{X})$.
A set $\mathcal{X} \in \mathbb{R}^n$ is *bounded* if there exists $r > 0$ such that $\mathcal{X} \subset \mathcal{B}(\mathbf{0}, r)$.
A *compact* set in $\mathbb{R}^n$ is a closed and bounded set.

Definition 1.2.4
(i) The point $\mathbf{x} \in \mathbb{R}^n$ is said to be a *convex combination* of two points $\mathbf{x}^1, \mathbf{x}^2 \in \mathbb{R}^n$ if $\mathbf{x} = \alpha\mathbf{x}^1 + (1 - \alpha)\mathbf{x}^2$, for some $\alpha \in \mathbb{R}, 0 \leq \alpha \leq 1$.
(ii) The point $\mathbf{x} \in \mathbb{R}^n$ is said to be a *convex combination* of m points $\mathbf{x}^1, \mathbf{x}^2, \cdots \mathbf{x}^m \in \mathbb{R}^n$ if $\mathbf{x} = \sum_{i=1}^m \alpha_i\mathbf{x}^i$, for $\alpha_i \geq 0$, $\sum_{i=1}^m \alpha_i = 1$.
(iii) The point $\mathbf{x} \in \mathbb{R}^n$ is said to be an *affine combination* of two points $\mathbf{x}^1, \mathbf{x}^2 \in \mathbb{R}^n$ if $\mathbf{x} = \alpha\mathbf{x}^1 + (1 - \alpha)\mathbf{x}^2$, for some $\alpha \in \mathbb{R}$. Note that α is unrestricted in sign and magnitude here, as opposed to (i).
(iv) The point $\mathbf{x} \in \mathbb{R}^n$ is said to be an *affine combination* of m points $\mathbf{x}^1, \mathbf{x}^2, \cdots \mathbf{x}^m \in \mathbb{R}^n$ if $\mathbf{x} = \sum_{i=1}^m \alpha_i\mathbf{x}^i$, for $\sum_{i=1}^m \alpha_i = 1$.

Definition 1.2.5 A set $\mathcal{X} \subset \mathbb{R}^n$ is said to be *convex* if for every $\mathbf{x}^1, \mathbf{x}^2 \in \mathcal{X}$ and every real number $\alpha \in \mathbb{R}, 0 \leq \alpha \leq 1$, the point $\alpha\mathbf{x}^1 + (1 - \alpha)\mathbf{x}^2 \in \mathcal{X}$. In other words, $\mathcal{X}$ is convex if the convex combination of every pair of points in $\mathcal{X}$ lies in $\mathcal{X}$.

Theorem 1.2.1
(i) If $\mathcal{X}$ is convex, then the set $\beta\mathcal{X} = \{\mathbf{x} \mid \mathbf{x} = \beta\mathbf{y}, \ \ \mathbf{y} \in \mathcal{X}\}$ for some $\beta \in \mathbb{R}$ is convex.

(ii) If $\mathcal{X}$ and $\mathcal{Y}$ are convex, then the set $\mathcal{X}+\mathcal{Y}=\{\mathbf{z} \mid \mathbf{z}=\mathbf{x}+\mathbf{y}, \mathbf{x} \in \mathcal{X}, \mathbf{y} \in \mathcal{Y}\}$ is convex.
(iii) If $\mathcal{X}$ and $\mathcal{Y}$ are convex, then $\mathcal{X} \cap \mathcal{Y}$ is convex.

Proof: (i) Given $\mathbf{x}^1, \mathbf{x}^2 \in \mathcal{X}$ and $\alpha \in \mathbb{R}$, then for $\alpha \in [0,1]$, we have

$$\alpha\beta\mathbf{x}^1 + (1-\alpha)\beta\mathbf{x}^2 = \beta(\alpha\mathbf{x}^1 + (1-\alpha)\mathbf{x}^2) \in \beta\mathcal{X}.$$

(ii) Given $\mathbf{z}^1 = \mathbf{x}^1 + \mathbf{y}^1$ and $\mathbf{z}^2 = \mathbf{x}^2 + \mathbf{y}^2$, where $\mathbf{x}^1, \mathbf{x}^2 \in \mathcal{X}$, $\mathbf{y}^1, \mathbf{y}^2 \in \mathcal{Y}$, then for $\alpha \in [0,1]$, we have

$$\begin{aligned}\alpha\mathbf{z}^1 + (1-\alpha)\mathbf{z}^2 &= \alpha(\mathbf{x}^1 + \mathbf{y}^1) + (1-\alpha)(\mathbf{x}^2 + \mathbf{y}^2) \\ &= \alpha\mathbf{x}^1 + (1-\alpha)\mathbf{x}^2 + \alpha\mathbf{y}^1 + (1-\alpha)\mathbf{y}^2 \in \mathcal{X} + \mathcal{Y}.\end{aligned}$$

(iii) Trivial. ∎

Corollary 1.2.2 If $\mathcal{X}^i, i = 1, 2, \cdots, m$ are convex sets in $\mathbb{R}^n$, then $\cap_{i=1}^m \mathcal{X}^i$ is convex. Note that m can be taken to infinity if necessary.

Theorem 1.2.3 $\mathcal{X} \subset \mathbb{R}^n$ is convex if and only if every convex combination of points in $\mathcal{X}$ belongs to $\mathcal{X}$.

Proof:
(Sufficiency) Follows trivially from the definition.

(Necessity) The assertion is trivially true if we take the convex combination of $k = 2$ points in $\mathcal{X}$. Assume the assertion is true for k and proceed by induction on k. Then for $\lambda_{k+1} < 1$,

$$\sum_{i=1}^{k+1} \lambda_i \mathbf{x}^i = \sum_{i=1}^{k} \lambda_i \mathbf{x}^i + \lambda_{k+1}\mathbf{x}^{k+1} = (1-\lambda_{k+1}) \sum_{i=1}^{k} \frac{\lambda_i}{1-\lambda_{k+1}} \mathbf{x}^i + \lambda_{k+1}\mathbf{x}^{k+1}.$$

Let

$$\mu_i = \frac{\lambda_i}{1-\lambda_{k+1}} \geq 0$$

and hence

$$\sum_{i=1}^{k} \mu_i = \frac{1}{1-\lambda_{k+1}} \sum_{i=1}^{k} \lambda_i = 1.$$

Since $\mathbf{y} = \sum_{i=1}^{k} \mu_i \mathbf{x}^i \in \mathcal{X}$, by the induction hypothesis, we have

$$\sum_{i=1}^{k+1} \lambda_i \mathbf{x}^i = (1-\lambda_{k+1})\mathbf{y} + \lambda_{k+1}\mathbf{x}^{k+1} \in \mathcal{X}.$$

∎

Definition 1.2.6 Let $\mathcal{X} \subset \mathbb{R}^n$. The *convex hull* of $\mathcal{X}$, denoted co$(\mathcal{X})$, is the intersection of all convex sets containing $\mathcal{X}$ (i.e., convex supersets of $\mathcal{X}$).

Remark 1.2.1 Clearly

- $\mathcal{X} \subseteq \text{co}(\mathcal{X})$;
- $\text{co}(\mathcal{X}) = \mathcal{X}$ if and only if $\mathcal{X}$ is convex;
- $\text{co}(\mathcal{X})$ is the smallest convex set containing $\mathcal{X}$.

Theorem 1.2.4 $\text{co}(\mathcal{X})$ is the set of all convex combinations of points in $\mathcal{X}$.

Proof: Let the set $\Gamma(\mathcal{X})$ of all convex combinations of points in $\mathcal{X}$ be defined by:

$$\Gamma(\mathcal{X}) = \{\mathbf{x} = \sum_{i=1}^{k} \lambda_i \mathbf{x}^i \mid \mathbf{x}^i \in \mathcal{X}, \lambda_i \geq 0, \sum_{i=1}^{k} \lambda_i = 1\}.$$

If $\mathbf{x} \in \Gamma(\mathcal{X})$, then $\mathbf{x} = \sum_{i=1}^{k} \lambda_i \mathbf{x}^i$ with $\mathbf{x}^i \in \mathcal{X} \subset \text{co}(\mathcal{X})$. But $\text{co}(\mathcal{X})$ is convex by Corollary 1.2.2 (with $m = \infty$), hence $\mathbf{x} \in \text{co}(\mathcal{X})$. This implies that $\Gamma(\mathcal{X}) \subseteq \text{co}(\mathcal{X})$.

To show that $\text{co}(\mathcal{X}) \subseteq \Gamma(\mathcal{X})$, we need only to show that $\Gamma(\mathcal{X})$ is a convex set that contains $\mathcal{X}$. Take

$$\mathbf{y} = \sum_{i=1}^{\ell} \lambda_i \mathbf{y}^i, \quad \lambda_i \geq 0, \quad \sum_{i=1}^{\ell} \lambda_i = 1, \quad \mathbf{y}^i \in \mathcal{X},$$

$$\mathbf{z} = \sum_{j=1}^{k} \mu_j \mathbf{z}^j, \quad \mu_j \geq 0, \quad \sum_{j=1}^{k} \mu_j = 1, \quad \mathbf{z}^j \in \mathcal{X},$$

and $\gamma \in [0,1]$. Then

$$\gamma \mathbf{y} + (1-\gamma)\mathbf{z} = \sum_{i=1}^{\ell} \gamma \lambda_i \mathbf{y}^i + \sum_{j=1}^{k} (1-\gamma)\mu_j \mathbf{z}^j,$$

where $\gamma\lambda_i \geq 0, \quad (1-\gamma)\mu_j \geq 0$, and

$$\sum_{i=1}^{\ell} \gamma \lambda_i + \sum_{j=1}^{k} (1-\gamma)\mu_j = \gamma + (1-\gamma) = 1.$$

Consequently $\gamma \mathbf{y} + (1-\gamma)\mathbf{z} \in \Gamma(\mathcal{X})$ and hence $\Gamma(\mathcal{X})$ is convex. Furthermore $\mathcal{X} \subseteq \Gamma(\mathcal{X})$ since every $\mathbf{x} \in \mathcal{X}$ can be written as $1\mathbf{x}$. ∎

Definition 1.2.7 A set $\mathcal{X} \subset \mathbb{R}^n$ is called an *affine set* (or *a linear variety*) if given $\mathbf{x}^1, \mathbf{x}^2 \in \mathcal{X}$, all affine combinations of $\mathbf{x}^1$ and $\mathbf{x}^2$ also belong to $\mathcal{X}$, i.e. $\alpha\mathbf{x}^1 + (1-\alpha)\mathbf{x}^2 \in \mathcal{X}$.

Theorem 1.2.5 Every affine set $\mathcal{X} \subseteq \mathbb{R}^n$ is a translate of a subspace.

Proof: We need only to show that the set $\mathcal{Y} = \mathcal{X} - \mathbf{x}$ for some $\mathbf{x}$ is a subspace. Choose any point $\mathbf{x} \in \mathcal{X}$, then clearly $\mathcal{Y}$ contains the origin. Pick any $\mathbf{y}^1, \mathbf{y}^2 \in \mathcal{Y}$, then by definition, there exist $\mathbf{x}^1, \mathbf{x}^2 \in \mathcal{X}$ such that $\mathbf{y}^1 = \mathbf{x}^1 - \mathbf{x}$ and $\mathbf{y}^2 = \mathbf{x}^2 - \mathbf{x}$. But $\mathbf{y}^1 + \mathbf{y}^2 = \mathbf{x}^1 - \mathbf{x} + \mathbf{x}^2 - \mathbf{x} = (\mathbf{x}^1 + \mathbf{x}^2 - \mathbf{x}) - \mathbf{x} \in \mathcal{Y}$, since $(\mathbf{x}^1 + \mathbf{x}^2 - \mathbf{x}) \in \mathcal{X}$. ∎

Definition 1.2.8 Let $\mathcal{X} \subset \mathbb{R}^n$. The *affine hull* of $\mathcal{X}$, denoted $\text{aff}(\mathcal{X})$, is the intersection of all affine sets containing $\mathcal{X}$ (i.e., all affine supersets of $\mathcal{X}$).

Remark 1.2.2 Clearly

- $\mathcal{X} \subseteq \text{aff}(\mathcal{X})$;
- $\text{aff}(\mathcal{X}) = \mathcal{X}$ if and only if $\mathcal{X}$ is affine;
- $\text{aff}(\mathcal{X})$ is the smallest affine set containing $\mathcal{X}$.

Theorem 1.2.6 $\text{aff}(\mathcal{X})$ is the set of all affine combinations of points in $\mathcal{X}$.

Proof: Let the set $\Gamma(\mathcal{X})$ of all affine combinations of points in $\mathcal{X}$ be defined by:

$$\Gamma(\mathcal{X}) = \{\mathbf{x} = \sum_{i=1}^{k} \lambda_i \mathbf{x}^i \mid \mathbf{x}^i \in \mathcal{X},\ \sum_{i=1}^{k} \lambda_i = 1\}.$$

If $\mathbf{x} \in \Gamma(\mathcal{X})$, then $\mathbf{x} = \sum_{i=1}^{k} \lambda_i \mathbf{x}^i$ with $\mathbf{x}^i \in \mathcal{X} \subset \text{aff}(\mathcal{X})$. But $\text{aff}(\mathcal{X})$ is affine, hence $\mathbf{x} \in \text{aff}(\mathcal{X})$. This implies that $\Gamma(\mathcal{X}) \subseteq \text{aff}(\mathcal{X})$.

To show that $\text{aff}(\mathcal{X}) \subseteq \Gamma(\mathcal{X})$, we need only to show that $\Gamma(\mathcal{X})$ is an affine set that contains $\mathcal{X}$. Take

$$\mathbf{y} = \sum_{i=1}^{\ell} \lambda_i \mathbf{y}^i, \qquad \sum_{i=1}^{\ell} \lambda_i = 1, \quad \mathbf{y}^i \in \mathcal{X},$$
$$\mathbf{z} = \sum_{j=1}^{k} \mu_j \mathbf{z}^j, \qquad \sum_{j=1}^{k} \mu_j = 1, \quad \mathbf{z}^j \in \mathcal{X},$$

and $\gamma \in \mathbb{R}$. Then

$$\gamma \mathbf{y} + (1-\gamma)\mathbf{z} = \sum_{i=1}^{\ell} \gamma \lambda_i \mathbf{y}^i + \sum_{j=1}^{k} (1-\gamma)\mu_j \mathbf{z}^j$$

where

$$\sum_{i=1}^{\ell} \gamma \lambda_i + \sum_{j=1}^{k} (1-\gamma)\mu_j = \gamma + (1-\gamma) = 1.$$

Consequently $\gamma \mathbf{y} + (1-\gamma)\mathbf{z} \in \Gamma(\mathcal{X})$ and hence $\Gamma(\mathcal{X})$ is affine. Furthermore $\mathcal{X} \subseteq \Gamma(\mathcal{X})$ since every $\mathbf{x} \in \mathcal{X}$ can be written as $1\mathbf{x}$. ∎

Definition 1.2.9 The *dimension* $\dim(\mathcal{X})$ of an affine set $\mathcal{X}$ is the dimension of the unique subspace that is translated from $\mathcal{X}$. The dimension of an arbitrary set $\mathcal{X}$ is defined to be $\dim(\text{aff}(\mathcal{X}))$, i.e., the dimension of the affine hull of $\mathcal{X}$.

Definition 1.2.10 The *relative interior* of a set $\mathcal{X}$ is defined by

$$\mathrm{ri}(\mathcal{X}) = \{\mathbf{x} \in \mathcal{X} \mid \mathcal{B}(\mathbf{x}, \delta) \cap \mathrm{aff}(\mathcal{X}) \subseteq \mathcal{X} \quad \text{for some } \delta > 0\}.$$

The *relative boundary* of $\mathcal{X}$ is defined by $\mathrm{rebd}(\mathcal{X}) = \bar{\mathcal{X}} \setminus \mathrm{ri}(\mathcal{X})$.

Definition 1.2.11 The points $\mathbf{x}^1, \mathbf{x}^2, \cdots, \mathbf{x}^k$ are said to be *affinely independent* if $\dim(\mathrm{aff}\{\mathbf{x}^1, \mathbf{x}^2, \cdots, \mathbf{x}^k\}) = k-1$. Alternatively, this is equivalent to the $k-1$ vectors $\mathbf{x}^i - \mathbf{x}^l$, $i \neq l$ being linearly independent for any given l.

Definition 1.2.12 A *hyperplane* in $\mathbb{R}^n$ is an $n-1$ dimensional (or co-dimension 1) affine set.

Theorem 1.2.7 Let $\mathbf{a} \in \mathbb{R}^n$, and $b \in \mathbb{R}$. The set $\mathcal{H} = \{\mathbf{x} \in \mathbb{R}^n : \mathbf{a}^\top \mathbf{x} = b\}$ is a hyperplane in $\mathbb{R}^n$. Conversely, let $\mathcal{H}$ be a hyperplane in $\mathbb{R}^n$. Then there exists a vector $\mathbf{a} \in \mathbb{R}^n$ and a scalar $b \in \mathbb{R}$ such that $\mathcal{H} = \{\mathbf{x} \in \mathbb{R}^n \mid \mathbf{a}^\top \mathbf{x} = b\}$.

Proof: It is easy to check that $\mathcal{H}$ is an affine set. Let $\mathbf{x}^1$ be any vector in $\mathcal{H}$. Translating $\mathcal{H}$ by $-\mathbf{x}^1$ gives the set $\mathcal{H}' = \mathcal{H} - \mathbf{x}^1$. $\mathcal{H}'$ contains the origin and consists of all vectors in $\mathbb{R}^n$ orthogonal to $\mathbf{a}$, hence it is an $n-1$ dimensional subspace.

Conversely, let $\mathbf{x}^1 \in \mathcal{H}$ and translating by $-\mathbf{x}^1$ gives the set $\mathcal{H}' = \mathcal{H} - \mathbf{x}^1$ which is an $n-1$ dimensional subspace. Let $\mathbf{a}$ be a nontrivial vector orthogonal to this subspace, i.e., $\mathbf{a} \in (\mathcal{H}')^\perp$, so that $\mathcal{H}' = \{\mathbf{x} : \mathbf{a}^\top \mathbf{x} = 0\}$. Let $b = \mathbf{a}^\top \mathbf{x}^1$. clearly $\mathbf{x}^2 \in \mathcal{H} \Rightarrow \mathbf{x}^2 - \mathbf{x}^1 \in \mathcal{H}'$, and hence $\mathbf{a}^\top \mathbf{x}^1 - \mathbf{a}^\top \mathbf{x}^2 = 0$ which implies that $\mathbf{a}^\top \mathbf{x}^2 = b$. Thus $\mathcal{H} \subseteq \{\mathbf{x} : \mathbf{a}^\top \mathbf{x} = b\}$. Since $\mathcal{H}$ is, by definition, of dimension $n-1$ and $\{\mathbf{x} : \mathbf{a}^\top \mathbf{x} = b\}$ is of dimension $n-1$ from above, by uniqueness these two sets must be equal. ∎

Definition 1.2.13 Let $\mathbf{a} \in \mathbb{R}^n$, and $b \in \mathbb{R}$. The hyperplane $\mathcal{H} = \{\mathbf{x} \in \mathbb{R}^n : \mathbf{a}^\top \mathbf{x} = b\}$ divides $\mathbb{R}^n$ into the *positive closed half space* $\mathcal{H}^+$ and *negative closed half space* $\mathcal{H}^-$, where

$$\begin{aligned}\mathcal{H}^+ &= \{\mathbf{x} \in \mathbb{R}^n : \mathbf{a}^\top \mathbf{x} \geq b\}\\ \mathcal{H}^- &= \{\mathbf{x} \in \mathbb{R}^n : \mathbf{a}^\top \mathbf{x} \leq b\}.\end{aligned}$$

Closed half spaces are convex sets.

Definition 1.2.14 A *polyhedral set* or *polyhedron* is the intersection of a *finite* number of closed half spaces. A *polytope* is a nonempty bounded polyhedron. Note that in some books, (notably [L1]), the roles of polyhedrons and polytopes are exchanged, (i.e., polytopes need not be bounded while polyhedrons are bounded).

By Corollary 1.2.2, polytopes and polyhedrons are convex sets. Furthermore, one can think of them as the feasible space (the set of all points satisfying the constraints) of a set of linear inequality constraints:

$$\begin{aligned}(\mathbf{a}^1)^\top \mathbf{x} &\leq b^1,\\ (\mathbf{a}^2)^\top \mathbf{x} &\leq b^2,\\ &\vdots\\ (\mathbf{a}^m)^\top \mathbf{x} &\leq b^m.\end{aligned}$$

Definition 1.2.15 A point $\mathbf{x}$ in a convex set $\mathcal{X}$ is called an *extreme point* of $\mathcal{X}$ if there are no two distinct points $\mathbf{x}^1$ and $\mathbf{x}^2$ in $\mathcal{X}$ such that $\mathbf{x}$ can be expressed as a convex combination of $\mathbf{x}^1$ and $\mathbf{x}^2$, i.e. $\mathbf{x} = \lambda\mathbf{x}^1 + (1-\lambda)\mathbf{x}^2$ for some $\lambda \in (0,1)$. Alternatively, $\mathbf{x}$ is an extreme point of $\mathcal{X}$ if it is not an interior point of any line segment in $\mathcal{X}$.

Definition 1.2.16 A hyperplane $\mathcal{H} = \{\mathbf{x} \in \mathbb{R}^n \mid \mathbf{a}^\top\mathbf{x} = b\}$ is said to *separate* two nonempty subsets $\mathcal{X}$ and $\mathcal{Y}$ of $\mathbb{R}^n$ if

$$\forall \mathbf{x} \in \mathcal{X},\ \mathbf{y} \in \mathcal{Y},\ \mathbf{a}^\top\mathbf{x} \leq b \leq \mathbf{a}^\top\mathbf{y}.$$

The separation is said to be *strict* if

$$\forall \mathbf{x} \in \mathcal{X},\ \mathbf{y} \in \mathcal{Y},\ \mathbf{a}^\top\mathbf{x} < b < \mathbf{a}^\top\mathbf{y}.$$

Definition 1.2.17 Let $\mathcal{X} \subset \mathbb{R}^n$ and $\mathbf{x}^0 \in \mathcal{X}$. The hyperplane $\mathcal{H} = \{\mathbf{x} \in \mathbb{R}^n \mid \mathbf{a}^\top\mathbf{x} = b\}$ is said to *support* $\mathcal{X}$ at $\mathbf{x}^0$ if $\mathbf{a}^\top\mathbf{x} \leq b\ \forall \mathbf{x} \in \mathcal{X}$ and $\mathbf{a}^\top\mathbf{x}^0 = b$. $\mathcal{H}$ is called a *supporting hyperplane* of the set $\mathcal{X}$.

Theorem 1.2.8 (Separation Hyperplane Theorem) Let $\mathcal{X} \subset \mathbb{R}^n$ be convex and let $\boldsymbol{\zeta} \notin \bar{\mathcal{X}}$. Then there exists a hyperplane that strictly separates $\mathcal{X}$ and $\boldsymbol{\zeta}$.

Proof: First, we show that there exists a point $\mathbf{x}^0 \in \bar{\mathcal{X}}$ such that $0 < \delta = \|\boldsymbol{\zeta} - \mathbf{x}^0\| = \inf_{\mathbf{x}\in\mathcal{X}} \|\boldsymbol{\zeta} - \mathbf{x}\|$. Let $f(\mathbf{x}) = \|\boldsymbol{\zeta} - \mathbf{x}\|$ and let $\delta = \inf_{\mathbf{x}\in\mathcal{X}} f(\mathbf{x})$. Then $\bar{\mathcal{B}}(\zeta, 2\delta) \cap \bar{\mathcal{X}}$ is compact and nonempty. Since f is continuous, it must attain its minimum at some point $\mathbf{x}^0 \in \bar{\mathcal{X}}$. Note that $\mathbf{x}^0$ may lie in $\mathrm{ri}(\mathcal{X})$, but certainly if $\boldsymbol{\zeta} \notin \mathcal{X}$ then $\mathbf{x}^0 \in \partial\mathcal{X}$.

Next, consider the hyperplane:

$$\mathcal{H} = \left\{\mathbf{y} \in \mathbb{R}^n \mid (\boldsymbol{\zeta} - \mathbf{x}^0)^\top \left(\mathbf{y} - \frac{1}{2}(\boldsymbol{\zeta} + \mathbf{x}^0)\right) = 0\right\}.$$

If $\mathbf{y} = \boldsymbol{\zeta}$, then

$$(\boldsymbol{\zeta} - \mathbf{x}^0)^\top(\boldsymbol{\zeta} - \frac{1}{2}(\boldsymbol{\zeta} + \mathbf{x}^0)) = \frac{1}{2}(\boldsymbol{\zeta} - \mathbf{x}^0)^\top(\boldsymbol{\zeta} - \mathbf{x}^0) = \frac{1}{2}\delta^2 > 0.$$

On the other hand, given any $\mathbf{x} \in \mathcal{X}$ and $\lambda \in [0,1]$, then $\lambda\mathbf{x} + (1-\lambda)\mathbf{x}^0 \in \mathcal{X}$ by the convexity of $\mathcal{X}$. Since $\mathbf{x}^0$ is the closest point of $\boldsymbol{\zeta}$ to the set $\mathcal{X}$,

$$\|\lambda\mathbf{x} + (1-\lambda)\mathbf{x}^0 - \boldsymbol{\zeta}\|^2 \geq \|\mathbf{x}^0 - \boldsymbol{\zeta}\|^2,$$

or

$$\lambda^2\|\mathbf{x}^0 - \mathbf{x}\|^2 + 2\lambda(\mathbf{x} - \mathbf{x}^0)^\top(\mathbf{x}^0 - \boldsymbol{\zeta}) \geq 0.$$

In the limit as $\lambda \downarrow 0$, we have

$$(\mathbf{x} - \mathbf{x}^0)^\top(\mathbf{x}^0 - \boldsymbol{\zeta}) \geq 0. \tag{1.2.1}$$

If $\mathbf{y} = \mathbf{x} \in \mathcal{X}$, then

$$\begin{aligned}(\zeta - \mathbf{x}^0)^\top \left(\mathbf{x} - \frac{1}{2}(\mathbf{x}^0 + \zeta)\right) &= (\zeta - \mathbf{x}^0)^\top \left(\mathbf{x} - \mathbf{x}^0 + \frac{1}{2}(\mathbf{x}^0 - \zeta)\right) \\ &= (\zeta - \mathbf{x}^0)^\top (\mathbf{x} - \mathbf{x}^0) - \frac{1}{2}\|\zeta - \mathbf{x}^0\|^2 \\ &= -(\mathbf{x} - \mathbf{x}^0)^\top (\mathbf{x}^0 - \zeta) - \frac{1}{2}\delta^2 < 0 \quad \text{by (1.2.1).}\end{aligned}$$

Hence $\mathcal{H}$ separates $\mathcal{X}$ and the point ζ strictly. ∎

Corollary 1.2.9 Let $\mathcal{X} \subset \mathbb{R}^n$ be closed and convex. Then $\mathcal{X}$ is the intersection of all the closed half spaces containing $\mathcal{X}$.

Proof: Since $\mathcal{X} \neq \mathbb{R}^n$, there exists a point $\mathbf{q} \notin \mathcal{X}$ and by Theorem 1.2.8, there exists a closed half space containing $\mathcal{X}$ but not $\mathbf{q}$. Let $\mathcal{Y}$ be the intersection of all such closed half spaces, each being generated by a point not in $\mathcal{X}$. Clearly $\mathcal{X} \subseteq \mathcal{Y}$. But $\mathbf{q} \notin \mathcal{X} \Rightarrow \mathbf{q} \notin \mathcal{Y}\ \forall \mathbf{q} \notin \mathcal{X}$. In other words, $\mathbf{q} \in \mathcal{Y} \Rightarrow \mathbf{q} \in \mathcal{X}$ or $\mathcal{Y} \subseteq \mathcal{X}$. ∎

Theorem 1.2.10 Let $\mathcal{X}^1$ and $\mathcal{X}^2$ be disjoint closed and convex sets in $\mathbb{R}^n$ of which at least one is compact. There exists a hyperplane separating $\mathcal{X}^1$ and $\mathcal{X}^2$ strictly.

Proof: If $\mathcal{X}^1$ is convex, so is $-\mathcal{X}^1$ by Theorem 1.2.1(i). Then $\mathcal{X}^2 - \mathcal{X}^1$ is also convex by Theorem 1.2.3 (i) and (ii) and is also closed. Since $\mathcal{X}^1$ and $\mathcal{X}^2$ are disjoint, $\mathcal{X}^2 \cap \mathcal{X}^1 = \emptyset$ and hence $\mathbf{0} \notin \mathcal{X}^2 - \mathcal{X}^1$. By Theorem 1.2.8, there exists $\mathbf{a} \in \mathbb{R}^n$ and b (with the assumption that $b > 0$ without loss of generality) such that $\mathbf{x} \in \mathcal{X}^2 - \mathcal{X}^1 \Rightarrow \mathbf{a}^\top \mathbf{x} > b$.

For all $\mathbf{x}^1 \in \mathcal{X}^1$ and $\mathbf{x}^2 \in \mathcal{X}^2$, we have $\mathbf{x}^2 - \mathbf{x}^1 \in \mathcal{X}^2 - \mathcal{X}^1$, hence $\mathbf{a}^\top(\mathbf{x}^2 - \mathbf{x}^1) > b$. Consequently,

$$\mathbf{a}^\top \mathbf{x}^2 > \mathbf{a}^\top \mathbf{x}^1 + b \quad \forall \mathbf{x}^1 \in \mathcal{X}^1, \mathbf{x}^2 \in \mathcal{X}^2,$$

or

$$\inf_{\mathbf{x}^2 \in \mathcal{X}^2} \mathbf{a}^\top \mathbf{x}^2 \geq \sup_{\mathbf{x}^1 \in \mathcal{X}^1} \mathbf{a}^\top \mathbf{x}^1 + b > \sup_{\mathbf{x}^1 \in \mathcal{X}^1} \mathbf{a}^\top \mathbf{x}^1 \quad \text{since } b > 0.$$

Let γ be such that

$$\inf_{\mathbf{x}^2 \in \mathcal{X}^2} \mathbf{a}^\top \mathbf{x}^2 > \gamma > \sup_{\mathbf{x}^1 \in \mathcal{X}^1} \mathbf{a}^\top \mathbf{x}^1.$$

Then the hyperplane $\mathcal{H} = \{\mathbf{y} \mid \mathbf{a}^\top \mathbf{y} - \gamma = 0\}$ separates $\mathcal{X}^1$ and $\mathcal{X}^2$ in a strict sense. ∎

Note that Theorem 1.2.10 is not true in general without the compactness assumption.

Theorem 1.2.11 (Supporting Hyperplane Theorem) If $\mathcal{X} \subset \mathbb{R}^n$ is closed and convex, then there exists a supporting hyperplane at each boundary point in $\partial\mathcal{X}$.

Proof: Let $\mathbf{d}$ be a direction pointing to the exterior of $\mathcal{X}$ from some boundary point $\mathbf{x}^0 \in \partial\mathcal{X}$. Consider the sequence of points $\boldsymbol{\zeta}^i = \mathbf{x}^0 + \epsilon_i\mathbf{d} \notin \mathcal{X}$ as $\epsilon_i \downarrow 0$ through a decreasing sequence $\{\epsilon_i,\ i = 1, 2, \cdots\}$. For each point $\boldsymbol{\zeta}^i$, there exists a set of separating hyperplane $\{\mathbf{y} \in \mathbb{R}^n \mid (\boldsymbol{\beta}^i)^\top\mathbf{y} - \gamma_i = 0\}$ that separates $\boldsymbol{\zeta}^i$ from $\mathcal{X}$ in a non-strict sense, i.e.,

$$(\boldsymbol{\beta}^i)^\top\boldsymbol{\zeta}^i - \gamma_i \leq 0 \quad \text{and} \quad (\boldsymbol{\beta}^i)^\top\mathbf{x} - \gamma_i \geq 0 \quad \forall\mathbf{x} \in \mathcal{X}. \tag{1.2.2}$$

Let $\mathcal{L}_i$ be the set of normalized $\boldsymbol{\beta}^i$, i.e., $\|\boldsymbol{\beta}^i\| = 1$, then
(i) the set $\mathcal{L}_i$ is bounded for each i, since $\|\boldsymbol{\beta}^i\| = 1$;
(ii) the set $\mathcal{L}_i$ is closed for each i, since equality is included in (1.2.2);
(iii) the sequence $\mathcal{L}_i$, $i = 1, 2, \cdots$ is monotone, since $\epsilon_i < \epsilon_j \Rightarrow \mathcal{L}_i \subset \mathcal{L}_j$.
Since the sequence $\{\mathcal{L}_i\}$ is closed, bounded and monotone, by the Bolzano Weierstrass Theorem [R5], there exists a non-empty limit $\mathcal{L}_\infty$, i.e., there exists (at least) one supporting hyperplane at $\mathbf{x}^0$. ∎

Lemma 1.2.12 Let $\mathcal{X} \subseteq \mathbb{R}^n$ be convex, $\mathcal{H}$ be a supporting hyperplane of $\mathcal{X}$, and $\mathcal{M} = \mathcal{X} \cap \mathcal{H}$. Then every extreme point of $\mathcal{M}$ is also an extreme point of $\mathcal{X}$.

Proof: Assume that $\mathcal{H}$ takes the form $\mathcal{H} = \{\mathbf{y} \mid \mathbf{a}^\top\mathbf{y} = b\}$ where $\mathbf{a}^\top\mathbf{x} \geq b,\ \forall\mathbf{x} \in \mathcal{X}$. Suppose $\mathbf{x}^0 \in \mathcal{M}$ is not an extreme point of $\mathcal{X}$. Since $\mathbf{x}^0 \in \mathcal{X}$, there exist $\mathbf{x}^1 \in \mathcal{X}$ and $\mathbf{x}^2 \in \mathcal{X}$, $\lambda \in (0, 1)$ such that $\mathbf{x}^0 = \lambda\mathbf{x}^1 + (1 - \lambda)\mathbf{x}^2$. Since $\mathbf{x}^0 \in \mathcal{M}$, we have $b = \mathbf{a}^\top\mathbf{x}^0 = \lambda\mathbf{a}^\top\mathbf{x}^1 + (1 - \lambda)\mathbf{a}^\top\mathbf{x}^2$. This, and the fact that $\mathbf{a}^\top\mathbf{x}^1 \geq b$ and $\mathbf{a}^\top\mathbf{x}^2 \geq b$, imply that $\mathbf{a}^\top\mathbf{x}^1 = \mathbf{a}^\top\mathbf{x}^2 = b$ and therefore $\mathbf{x}^1 \in \mathcal{M}$ and $\mathbf{x}^2 \in \mathcal{M}$. Hence $\mathbf{x}^0$ is not an extreme point of $\mathcal{M}$. ∎

Theorem 1.2.13 A compact convex set $\mathcal{X} \subseteq \mathbb{R}^n$ is the convex hull of all its extreme points.

Proof: We proceed by induction on n. Clearly the assertion holds in $\mathbb{R}^n$ for $n = 1$. Next assume it holds in $\mathbb{R}^{n-1}$. Let $\mathcal{Y}$ be the convex hull of all the extreme points of $\mathcal{X}$. Clearly $\mathcal{Y} \subseteq \mathcal{X}$. We need to establish the converse $\mathcal{X} \subseteq \mathcal{Y}$. Suppose there exists $\mathbf{x} \in \mathcal{X}$ with $\mathbf{x} \notin \mathcal{Y}$, then Theorem 1.2.8 asserts that there exists $\mathbf{a} \neq \mathbf{0}$ such that $\mathbf{a}^\top\mathbf{x} < \inf_{\mathbf{y}\in\mathcal{Y}} \mathbf{a}^\top\mathbf{y}$. Let $\alpha = \min_{\mathbf{x}\in\mathcal{X}} \mathbf{a}^\top\mathbf{x} < \infty$. Since $\mathcal{X}$ is compact and $\mathbf{a}^\top\mathbf{x}$ is a continuous function, the Weierstrass Theorem (see Remark 1.5.1) asserts that there exists $\mathbf{x}^0 \in \mathcal{X}$ such that $\mathbf{a}^\top\mathbf{x}^0 = \alpha$, and consequently $\mathcal{H} = \{\mathbf{y} \in \mathbb{R}^n \mid \mathbf{a}^\top\mathbf{y} = \alpha\}$ is a supporting hyperplane for $\mathcal{X}$. Let $\mathcal{M} = \mathcal{X} \cap \mathcal{H}$, then $\mathcal{M}$ is a compact convex set in $\mathbb{R}^{n-1}$ and is non-empty since it contains the point $\mathbf{x}^0$.

By the induction hypothesis, $\mathcal{M}$ is the convex hull of all its extreme points, and there exists at least one such extreme point. By Lemma 1.2.12, the extreme points of $\mathcal{M}$ are also the extreme points of $\mathcal{X}$. Thus we have found extreme points of $\mathcal{X}$ that do not belong to $\mathcal{Y}$, a contradiction. ∎

Corollary 1.2.14 A polytope is the convex hull of all its extreme points.

We shall now turn our attention to functions.

Definition 1.2.18 Let $f : \mathbb{R}^n \to \bar{\mathbb{R}}$ be an *extended real-valued function.* The *(effective) domain* of f is defined as

$$\mathcal{X} = \text{dom}(f) = \{\mathbf{x} \in \mathbb{R}^n : \ |f(\mathbf{x})| < +\infty\}$$

The *range* of f is $f(\mathcal{X}) \subseteq \mathbb{R}$.

Remark 1.2.3 If $f : \mathcal{X} \to \mathbb{R}$ defined on a set $\mathcal{X} \subseteq \mathbb{R}^n$ associates with each point of $\mathcal{X}$ a real number $f(\mathbf{x})$, then define

$$F(\mathbf{x}) = \begin{cases} f(\mathbf{x}) & \text{if } \mathbf{x} \in \mathcal{X} \\ +\infty & \text{if } \mathbf{x} \in \mathbb{R}^n \setminus \mathcal{X}. \end{cases}$$

Then $dom(F) = \mathcal{X}$. So the set $\mathcal{X}$ is called the domain of f.

Consequently the study of the constrained optimization problem $\min\{f(\mathbf{x}) \mid \mathbf{x} \in \mathcal{X}\} = \min \ f(\mathcal{X})$ is equivalent to the study of the unconstrained problem $\min\{F(\mathbf{x}) \mid \mathbf{x} \in \mathbb{R}^n\}$. The min, or strictly speaking, infimum is ∞ if the feasible set $\mathcal{X}$ is empty.

Definition 1.2.19 (Continuous and semicontinuous functions)

• A real-valued function $f : \mathcal{X} \to \mathbb{R}$ is *upper semicontinuous (u.s.c.)* at $\mathbf{x} \in \mathcal{X}$ if, for any given $\epsilon > 0$, there exists a $\delta > 0$ such that

$$\mathbf{y} \in \mathcal{X} \text{ and } \|\mathbf{y} - \mathbf{x}\| < \delta \Rightarrow f(\mathbf{y}) - f(\mathbf{x}) < \epsilon.$$

f is an *upper semicontinuous function* if f is upper semicontinuous at every point $\mathbf{x} \in \mathcal{X}$. A vector-valued function $\mathbf{f} : \mathcal{X} \to \mathbb{R}^m$ is *upper semicontinuous* at $\mathbf{x} \in \mathcal{X}$ if every component of $\mathbf{f}$ is upper semicontinuous at $\mathbf{x}$. $\mathbf{f}$ is an upper semicontinuous (vector-valued) function if every component of $\mathbf{f}$ is an upper semicontinuous function.

• A real-valued function $f : \mathcal{X} \to \mathbb{R}$ is *lower semicontinuous (l.s.c.)* at $\mathbf{x} \in \mathcal{X}$ if, for any given $\epsilon > 0$, there exists a $\delta > 0$ such that

$$\mathbf{y} \in \mathcal{X} \text{ and } \|\mathbf{y} - \mathbf{x}\| < \delta \Rightarrow f(\mathbf{y}) - f(\mathbf{x}) > -\epsilon.$$

f is a *lower semicontinuous function* if f is lower semicontinuous at every point $\mathbf{x} \in \mathcal{X}$. A vector-valued function $\mathbf{f} : \mathcal{X} \to \mathbb{R}^m$ is *lower semicontinuous* at $\mathbf{x} \in \mathcal{X}$ if every component of $\mathbf{f}$ is lower semicontinuous at $\mathbf{x}$. $\mathbf{f}$ is a lower semicontinuous (vector-valued) function if every component of $\mathbf{f}$ is a lower semicontinuous function.

• A real-valued function $f : \mathcal{X} \to \mathbb{R}$ is *continuous* at $\mathbf{x} \in \mathcal{X}$ if, for any given $\epsilon > 0$, there exists a $\delta > 0$ such that

$$\mathbf{y} \in \mathcal{X} \text{ and } \|\mathbf{y} - \mathbf{x}\| < \delta \Rightarrow |f(\mathbf{y}) - f(\mathbf{x})| < \epsilon.$$

f is a *continuous function* if f is continuous at every point $\mathbf{x} \in \mathcal{X}$. A vector-valued function $\mathbf{f} : \mathcal{X} \to \mathbb{R}^m$ is *continuous* at $\mathbf{x} \in \mathcal{X}$ if every component of $\mathbf{f}$ is continuous at $\mathbf{x}$. $\mathbf{f}$ is a continuous (vector-valued) function if every component of $\mathbf{f}$ is a continuous function.

• Note that a function that is simultaneously upper and lower semicontinuous is continuous. A function f is u.s.c. (resp., l.s.c.) if and only if $-f$ is l.s.c. (resp. u.s.c.).

Definition 1.2.20 Let $\mathcal{X} \subseteq \mathbb{R}^n$ be a nonempty set, and $f : \mathcal{X} \to \mathbb{R}$.

• f is said to be *differentiable* at some $\mathbf{x} \in \text{int}(\mathcal{X})$ if there exists a (row) vector $\nabla f(\mathbf{x})$ such that

$$f(\mathbf{y}) = f(\mathbf{x}) + \nabla f(\mathbf{x})(\mathbf{y} - \mathbf{x}) + g(\mathbf{y} - \mathbf{x})\|\mathbf{y} - \mathbf{x}\|,$$

where g is a function satisfying

$$\lim_{\mathbf{y} \to \mathbf{x}} g(\mathbf{y} - \mathbf{x}) = 0. \tag{1.2.3}$$

$\nabla f(\mathbf{x})$, or sometimes written as $\left.\frac{\partial f}{\partial \mathbf{x}}\right|_{\mathbf{x}}$ or $f'(\mathbf{x})$ is called the *gradient* of f at $\mathbf{x}$, and it has components $[\nabla f(\mathbf{x})]_j = \left.\dfrac{\partial f}{\partial x_j}\right|_{\mathbf{x}}$, $j = 1, 2, \cdots, n$.

• f is said to be *twice differentiable* at some $\mathbf{x} \in \text{int}(\mathcal{X})$ if there exists a matrix $H(\mathbf{x})$ such that

$$f(\mathbf{y}) = f(\mathbf{x}) + \nabla f(\mathbf{x})(\mathbf{y} - \mathbf{x}) + \frac{1}{2}(\mathbf{y} - \mathbf{x})^\top H(\mathbf{x})(\mathbf{y} - \mathbf{x}) + g(\mathbf{y} - \mathbf{x})\|\mathbf{y} - \mathbf{x}\|^2,$$

where g satisfies (1.2.3). $H(\mathbf{x})$ is called the *Hessian* of f evaluated at $\mathbf{x}$, and has components $[H(\mathbf{x})]_{ij} = \dfrac{\partial^2 f(\mathbf{x})}{\partial x_i \partial x_j}$, $i, j = 1, 2, \cdots, n$.

• f is said to be differentiable/twice differentiable if f is differentiable/twice differentiable at every interior point of its domain.

• A vector-valued function $\mathbf{f}$ is said to be differentiable/twice differentiable if every component of $\mathbf{f}$ is differentiable/twice differentiable.

Definition 1.2.21 The *epigraph* of a function $f : \mathbb{R}^n \to \bar{\mathbb{R}}$ is the set $\text{epi}(f) = \{(\mathbf{x}, y) \in \mathbb{R}^{n+1} \mid \mathbf{x} \in \mathbb{R}^n,\ y \in \mathbb{R}, f(\mathbf{x}) \leq y\}$.

Definition 1.2.22 A function $f : \mathbb{R}^n \to \bar{\mathbb{R}}$ is said to be *convex* if its epigraph $\text{epi}(f)$ is a convex subset of $\mathbb{R}^{n+1}$. If, furthermore, $\text{dom}(f) \neq \emptyset$ and $f(\mathbf{x}) > -\infty\ \forall \mathbf{x} \in \mathbb{R}^n$, then f is a *proper convex function*. f is said to be *concave* if $-f$ is convex. Alternatively, a more conventional definition is as follows. Let the set $\mathcal{X}$ be convex and $f : \mathcal{X} \to \mathbb{R}$. f is convex on $\mathcal{X}$ if given any $\mathbf{x}, \mathbf{y} \in \mathcal{X}$ and $\lambda \in [0, 1]$, then

$$f(\lambda\mathbf{x} + (1 - \lambda)\mathbf{y}) \leq \lambda f(\mathbf{x}) + (1 - \lambda)f(\mathbf{y}).$$

This is equivalent to saying that the function F as defined in Remark 1.2.3 is convex. Moreover, f is said to be *strictly convex* if given any $\mathbf{x}, \mathbf{y} \in \mathcal{X}, x \neq y$ and $\lambda \in (0,1)$,

$$f(\lambda \mathbf{x} + (1-\lambda)\mathbf{y}) < \lambda f(\mathbf{x}) + (1-\lambda) f(\mathbf{y}).$$

Remark 1.2.4 If $f : \mathcal{X} \to \mathbb{R}$ is convex, then it is well known that (see, for example, [R1]):

• f is continuous in ri$(\mathcal{X})$;

• if f is differentiable on an open set containing $\mathcal{X}$, then f is convex on $\mathcal{X}$ if and only if

$$f(\mathbf{y}) \geq f(\mathbf{x}) + \nabla f(\mathbf{x})(\mathbf{y} - \mathbf{x}) \quad \forall \mathbf{x}, \mathbf{y} \in \mathbb{R}^n;$$

• if f is twice differentiable on an open set containing $\mathcal{X}$, then f is convex on $\mathcal{X}$ if and only if the Hessian of f is positive semidefinite on $\mathcal{X}$. If the Hessian of f is positive definite, then f is strictly convex on $\mathcal{X}$. In general, the converse is not true.

Definition 1.2.23 A function $f : \mathcal{X} \to \mathbb{R}$ is said to be *quasiconvex* if for $\lambda \in [0,1]$ and $\mathbf{x}, \mathbf{y} \in \mathcal{X}$,

$$f(\lambda \mathbf{x} + (1-\lambda)\mathbf{y}) \leq \max\{f(\mathbf{x}), f(\mathbf{y})\};$$

and *strictly quasiconvex* if

$$\lambda \in (0,1), \quad f(\mathbf{x}) \neq f(\mathbf{y}) \Rightarrow f(\lambda \mathbf{x} + (1-\lambda)\mathbf{y}) < \max\{f(\mathbf{x}), f(\mathbf{y})\};$$

and *strongly quasiconvex* if

$$\lambda \in (0,1), \quad \mathbf{x} \neq \mathbf{y} \Rightarrow f(\lambda \mathbf{x} + (1-\lambda)\mathbf{y}) < \max\{f(\mathbf{x}), f(\mathbf{y})\}.$$

Definition 1.2.24 A function $f : \mathcal{X} \to \mathbb{R}$ is said to be *pseudoconvex* if for

$$\mathbf{x},\ \mathbf{y} \in \mathcal{X}, \mathbf{x} \neq \mathbf{y}\ ,\ \nabla f(\mathbf{x})(\mathbf{y} - \mathbf{x}) \geq 0 \Rightarrow f(\mathbf{y}) \geq f(\mathbf{x});$$

and *strictly pseudoconvex* if

$$\nabla f(\mathbf{x})(\mathbf{y} - \mathbf{x}) > 0 \Rightarrow f(\mathbf{y}) > f(\mathbf{x}).$$

Definition 1.2.25 A proper convex function $f : \mathbb{R}^n \to \bar{\mathbb{R}}$ is said to be *closed* if epi(f) is a closed set in $\mathbb{R}^{n+1}$.

Definition 1.2.26 The *closure* $\bar{f}$ of a function $f : \mathbb{R}^n \to \bar{\mathbb{R}}$ is defined to be

$$\bar{f}(\mathbf{x}) = \sup_{F \in A(f)} F(\mathbf{x}) \quad \forall \mathbf{x} \in \mathbb{R}^n,$$

where $A(f)$ is the family of all affine functions F defined on $\mathbb{R}^n$ such that $F(\mathbf{x}) \leq f(\mathbf{x})\ \forall \mathbf{x} \in \mathbb{R}^n$. $A(f)$ is called the *Affine Minorant* of f.

Remark 1.2.5 It can be shown (see [vT1]) that, a proper convex function f is closed if and only if it is lower semicontinuous, if and only if $f = \bar{f}$.

Lemma 1.2.15 Let $\mathcal{F}$ be a family of convex functions on $\mathbb{R}^n$, then,

$$f(\mathbf{x}) = \sup_{g \in \mathcal{F}} g(\mathbf{x}) \quad \forall \mathbf{x} \in \mathbb{R}^n,$$

is a convex function.

Proof: $\text{epi}(f) = \cap_{g \in \mathcal{F}} \text{epi}(g)$ is convex by Corollary 1.2.2. ■

Definition 1.2.27 Let $f : \mathbb{R}^n \to \bar{\mathbb{R}}$ be a convex function, and let $\mathbf{x} \in \text{dom}(f)$. The vector $\mathbf{z} \in \mathbb{R}^n$ is said to be a *subgradient* of f at $\mathbf{x}$ if

$$\mathbf{z}^\top(\mathbf{y} - \mathbf{x}) \leq f(\mathbf{y}) - f(\mathbf{x}) \quad \forall \mathbf{y} \in \mathbb{R}^n.$$

The set

$$\partial f(\mathbf{x}) = \{\mathbf{z} \in \mathbb{R}^n \mid \mathbf{z}^\top(\mathbf{y} - \mathbf{x}) \leq f(\mathbf{y}) - f(\mathbf{x}) \ \forall \mathbf{y} \in \mathbb{R}^n\}$$

is called the *subdifferential* of f at $\mathbf{x}$. If $\mathbf{x} \notin \text{dom}(f)$, then $\partial f(\mathbf{x}) = \emptyset$ by convention.

Remark 1.2.6 It can be easily established that for convex functions,

- $\partial f(\mathbf{x})$ is closed and convex;
- $\partial f(\mathbf{x})$ is a singleton if and only if f is differentiable at $\mathbf{x}$. In this case $\partial f(\mathbf{x}) = \{\nabla f(\mathbf{x})\}$.

Definition 1.2.28 Let $f : \mathbb{R}^n \to \bar{\mathbb{R}}$, $\mathbf{x} \in \text{dom}(f)$ and $\mathbf{d} \in \mathbb{R}^n$. The directional derivative of f at $\mathbf{x}$ in the direction of $\mathbf{d}$, if it exists, is defined by

$$f'(\mathbf{x}, \mathbf{d}) = \lim_{t \downarrow 0} \frac{f(\mathbf{x} + t\mathbf{d}) - f(\mathbf{x})}{t}.$$

Theorem 1.2.16 Let $f : \mathbb{R}^n \to \bar{\mathbb{R}}$ be a convex function, and let $\mathbf{x} \in \text{dom}(f)$, $\mathbf{y} \in \mathbb{R}^n$. Then $\mathbf{y} \in \partial f(\mathbf{x})$ if and only if $\mathbf{y}^\top\mathbf{d} \leq f'(\mathbf{x}, \mathbf{d})$ $\forall \mathbf{d} \in \mathbb{R}^n$.

Proof: (Necessity) If $\mathbf{y} \in \partial f(\mathbf{x})$, let $t > 0$. Then from Definition 1.2.27,

$$f(\mathbf{x} + t\mathbf{d}) - f(\mathbf{x}) \geq \mathbf{y}^\top(\mathbf{x} + t\mathbf{d} - \mathbf{x}) \quad \text{or}$$

$$\mathbf{y}^\top\mathbf{d} \leq \frac{f(\mathbf{x} + t\mathbf{d}) - f(\mathbf{x})}{t} \quad \forall t > 0.$$

The conclusion follows by letting $t \downarrow 0$.

(Sufficiency) If $\mathbf{y}^\top \mathbf{d} \le f'(\mathbf{x}, \mathbf{d})$, then

$$\begin{aligned} \mathbf{y}^\top \mathbf{d} &\le \lim_{t \downarrow 0} \frac{f(\mathbf{x} + t\mathbf{d}) - f(\mathbf{x})}{t} \quad \forall \mathbf{x} \in \mathbb{R}^n \\ &\le \lim_{t \downarrow 0} \frac{tf(\mathbf{x} + \mathbf{d}) + (1 - t)f(\mathbf{x}) - f(\mathbf{x})}{t} \qquad \text{by convexity of } f \\ &= f(\mathbf{x} + \mathbf{d}) - f(\mathbf{x}), \end{aligned}$$

hence $\mathbf{y} \in \partial f(\mathbf{x})$. ∎

Definition 1.2.29 Let $f : \mathbb{R}^n \longrightarrow \mathbb{R}$ be a locally Lipschitz function. The *Clarke subgradient* of f at $\mathbf{x} \in \mathbb{R}^n$ is defined by

$$\partial^\circ f(\mathbf{x}) = \{\mathbf{z} \in \mathbf{R}^n : f^\circ(\mathbf{x}; \mathbf{y}) \ge \mathbf{z}^\top \mathbf{y}, \forall \mathbf{y} \in \mathbf{R}^n\},$$

where

$$f^\circ(\mathbf{x}; \mathbf{y}) = \limsup_{\mathbf{x}' \to \mathbf{x}, s \downarrow 0} \frac{f(\mathbf{x}' + s\mathbf{y}) - f(\mathbf{x}')}{s},$$

is the *Clarke generalized directional derivative* of f at $\mathbf{x}$ in the direction $\mathbf{y}$.

f is said to be *subdifferentially regular* at $\mathbf{x}$ if the directional derivative $f'(\mathbf{x}; \mathbf{y})$ exists and

$$f'(\mathbf{x}; \mathbf{y}) = f^\circ(\mathbf{x}; \mathbf{y}), \quad \forall \mathbf{y} \in \mathbf{R}^n.$$

The following properties of $f^\circ(\mathbf{x}; \mathbf{y})$ are stated without proof, see [C5].

Theorem 1.2.17
(i) $f^\circ(\mathbf{x}; \mathbf{y}) = \max_{\mathbf{z} \in \partial^\circ f(\mathbf{x})} \mathbf{z}^\top \mathbf{y}$.
(ii) Let $f(\mathbf{x}) = G(g(\mathbf{x}))$ where $g : \mathbb{R}^n \to \mathbb{R}^m, G : \mathbb{R}^m \to \mathbb{R}$. Let

$$P(\mathbf{x}) = \text{co}\{\mathbf{p} \in \mathbb{R}^n : \mathbf{p} = [\mathbf{h}^1, \cdots, \mathbf{h}^m] \cdot \mathbf{W}, \mathbf{h}^i \in \partial^\circ g_i(\mathbf{x}), \mathbf{W} \in \partial^\circ G(f(\mathbf{x}))\}.$$

Then

$$\partial^\circ f(\mathbf{x}) \subseteq P(\mathbf{x}).$$

1.3 Fenchel Transform and Conjugate Duality

The theory of conjugate duality [F2] plays a very important role in convex analysis. It is aesthetically appealing as a nice symmetry prevails in many of the results. An entire book [R2] has been devoted to the subject. This contains many more results than what this section can afford to discuss, together with an extensive range of useful applications. Again, while the results in this section are presented for finite dimensional vector spaces, they can be generalized to infinite dimensional versions in abstract spaces.

Definition 1.3.1 (Fenchel Transform) Given $f : \mathbb{R}^n \to \bar{\mathbb{R}}$, the *Fenchel transform (or conjugate, or polar)* of f is the function $f^* : \mathbb{R}^n \to \bar{\mathbb{R}}$ defined as:

$$f^*(\mathbf{y}) = \sup_{\mathbf{x}\in\mathbb{R}^n} \{\mathbf{y}^\top \mathbf{x} - f(\mathbf{x})\}.$$

The *biconjugate (or bipolar)* $f^{**} : \mathbb{R}^n \to \bar{\mathbb{R}}$ of f is the conjugate of f^*.

Remark 1.3.1 One can interpret the meaning of the Fenchel transform in the following way. Given a $\mathbf{y} \in \mathbb{R}^n$ and an $\alpha \in \mathbb{R}$, one defines a linear function $\ell : \mathbb{R}^n \to \bar{\mathbb{R}}$ by $\ell(\mathbf{x}) = \mathbf{y}^\top \mathbf{x} - \alpha$. One desires that the linear function ℓ belongs to $A(f)$, the family of affine minorant (see Definition 1.2.26) of f. This means that $\ell(\mathbf{x}) = \mathbf{y}^\top \mathbf{x} - \alpha \leq f(\mathbf{x}) \quad \forall \mathbf{x} \in \mathbb{R}^n$, or $\alpha \geq \mathbf{y}^\top \mathbf{x} - f(\mathbf{x})$. This property is accomplished by choosing $\alpha \geq \sup_{\mathbf{x}\in\mathbb{R}^n}\{\mathbf{y}^\top \mathbf{x} - f(\mathbf{x})\} = f^*(\mathbf{y})$. Another economic interpretation goes as follows: if x_i represents the quantity held of each of the ith of n commodities, $f(\mathbf{x})$ is the cost of producing them, and y_i is the price of selling the ith commodity, then $\mathbf{y}^\top \mathbf{x} - f(\mathbf{x})$ is the net profit from producing and selling the commodities, and $f^*(\mathbf{y})$ is the maximum possible profit for the given price $\mathbf{y}$.

Remark 1.3.2 Note that the function f does not need to be convex for its Fenchel transform to exist. Although more often than not we talk only about the Fenchel transform of convex functions, and f^* is often called the *convex conjugate* of f.

Lemma 1.3.1 (Fenchel's or Young's Inequality)

$$f(\mathbf{x}) + f^*(\mathbf{y}) \geq \mathbf{y}^\top \mathbf{x} \qquad \forall \mathbf{x}, \mathbf{y} \in \mathbb{R}^n.$$

Proof: Follows directly from Definition 1.3.1. ∎

Example 1.3.1 Let $\mathcal{X} \subset \mathbb{R}^n$. The indicator function $\delta_{\mathcal{X}} : \mathbb{R}^n \to \{0, \infty\}$ of $\mathcal{X}$ is defined by

$$\delta_{\mathcal{X}}(\mathbf{x}) = \begin{cases} 0 & \text{if } \mathbf{x} \in \mathcal{X} \\ \infty & \text{otherwise.} \end{cases}$$

Then the Fenchel transform of $\delta_{\mathcal{X}}$ is given by the support function of $\mathcal{X}$ defined by

$$\delta^*_{\mathcal{X}}(\mathbf{x}) = \sup_{\mathbf{y}\in\mathcal{X}} \mathbf{y}^\top \mathbf{x}.$$

Theorem 1.3.2 Let $f : \mathbb{R}^n \to \bar{\mathbb{R}}$. Then the Fenchel transform $f^* : \mathbb{R}^n \to \bar{\mathbb{R}}$ is convex on $\mathbb{R}^n$.

Proof: For each $\mathbf{x} \in \mathbb{R}^n$, the mapping $\mathbf{y} \to \mathbf{y}^\top \mathbf{x} - f(\mathbf{x})$ is linear (and convex) in $\mathbf{y}$. Let $F = \{\mathbf{y}^\top \mathbf{x} - f(\mathbf{x}) \mid \mathbf{x} \in \mathbb{R}^n\}$ be the family of functions which are linear in $\mathbf{y}$. Then, by Lemma 1.2.15,

$$f^*(\mathbf{y}) = \sup_{\mathbf{x} \in \mathbb{R}^n} \{\mathbf{y}^\top \mathbf{x} - f(\mathbf{x})\}$$

is convex. ∎

Theorem 1.3.3 (Properties of Fenchel transform) Let $f : \mathbb{R}^n \to \bar{\mathbb{R}}$.
(i) If $f_1(\mathbf{x}) = f(\mathbf{x} + \mathbf{x}^0) \quad \forall \mathbf{x} \in \mathbb{R}^n$, then $f_1^*(\mathbf{y}) = f^*(\mathbf{y}) - \mathbf{y}^\top \mathbf{x}^0 \quad \forall \mathbf{y} \in \mathbb{R}^n$.
(ii) If $\beta \in \mathbb{R}$ and $f_2(\mathbf{x}) = f(\mathbf{x}) + \beta \quad \forall \mathbf{x} \in \mathbb{R}^n$, then $f_2^*(\mathbf{y}) = f^*(\mathbf{y}) - \beta \quad \forall \mathbf{y} \in \mathbb{R}^n$.
(iii) If $\beta \neq 0$ and $f_3(\mathbf{x}) = f(\beta \mathbf{x}) \quad \forall \mathbf{x} \in \mathbb{R}^n$, then $f_3^*(\mathbf{y}) = f^*(\mathbf{y}/\beta) \quad \forall \mathbf{y} \in \mathbb{R}^n$.
(iv) If $\beta > 0$ and $f_4(\mathbf{x}) = \beta f(\mathbf{x}) \quad \forall \mathbf{x} \in \mathbb{R}^n$, then $f_4^*(\mathbf{y}) = \beta f^*(\mathbf{y}/\beta) \quad \forall \mathbf{y} \in \mathbb{R}^n$.

Proof:
(i)

$$\begin{aligned} f_1^*(\mathbf{y}) &= \sup\{\mathbf{y}^\top \mathbf{x} - f(\mathbf{x} + \mathbf{x}^0) \mid \mathbf{x} \in \mathbb{R}^n\} \\ &= \sup\{\mathbf{y}^\top (\mathbf{x} - \mathbf{x}^0) - f(\mathbf{x}) \mid \mathbf{x} \in \mathbb{R}^n\} \\ &= \sup\{\mathbf{y}^\top \mathbf{x} - f(\mathbf{x}) \mid \mathbf{x} \in \mathbb{R}^n\} - \mathbf{y}^\top \mathbf{x}^0 \\ &= f^*(\mathbf{y}) - \mathbf{y}^\top \mathbf{x}^0. \end{aligned}$$

(ii)

$$\begin{aligned} f_2^*(\mathbf{y}) &= \sup\{\mathbf{y}^\top \mathbf{x} - f(\mathbf{x}) - \beta \mid \mathbf{x} \in \mathbb{R}^n\} \\ &= \sup\{\mathbf{y}^\top \mathbf{x} - f(\mathbf{x}) \mid \mathbf{x} \in \mathbb{R}^n\} - \beta \\ &= f^*(\mathbf{y}) - \beta. \end{aligned}$$

(iii)

$$\begin{aligned} f_3^*(\mathbf{y}) &= \sup\{\mathbf{y}^\top \mathbf{x} - f(\beta \mathbf{x}) \mid \mathbf{x} \in \mathbb{R}^n\} \\ &= \sup\{\mathbf{y}^\top (\mathbf{x}/\beta) - f(\mathbf{x}) \mid \mathbf{x} \in \mathbb{R}^n\} \quad \text{since } \beta\mathbb{R}^n = \mathbb{R}^n \\ &= \sup\{(\mathbf{y}/\beta)^\top \mathbf{x} - f(\mathbf{x}) \mid \mathbf{x} \in \mathbb{R}^n\} \\ &= f^*(\mathbf{y}/\beta). \end{aligned}$$

(iv)

$$\begin{aligned} f_4^*(\mathbf{y}) &= \sup\{\mathbf{y}^\top \mathbf{x} - \beta f(\mathbf{x}) \mid \mathbf{x} \in \mathbb{R}^n\} \\ &= \beta \sup\{(\mathbf{y}/\beta)^\top \mathbf{x} - f(\mathbf{x}) \mid \mathbf{x} \in \mathbb{R}^n\} \quad \text{since } \beta > 0 \\ &= \beta f^*(\mathbf{y}/\beta). \end{aligned}$$

∎

Theorem 1.3.4 If f is a proper convex function, then $f^{**} = \bar{f}$.

Proof: Young's inequality (Lemma 1.3.1) asserts that, for a given $\mathbf{x}$,

$$\mathbf{y}^\top \mathbf{x} - f^*(\mathbf{y}) \leq f(\mathbf{x}) \quad \forall \mathbf{y} \in \mathbb{R}^n,$$

From Definition 1.2.26, $\bar{f}(\mathbf{x}) = \sup_{F \in A(f)} F(\mathbf{x})$ where $A(f)$ is the affine minorant, it follows that

$$\mathbf{y}^\top \mathbf{x} - f^*(\mathbf{y}) \leq \bar{f}(\mathbf{x}) \quad \forall \mathbf{y}, \mathbf{x} \in \mathbb{R}^n.$$

In particular,

$$f^{**}(\mathbf{x}) = \sup_{\mathbf{y} \in \mathbf{R}^n} \{\mathbf{y}^\top \mathbf{x} - f^*(\mathbf{y})\} \leq \bar{f}(\mathbf{x}) \quad \forall \mathbf{x} \in \mathbb{R}^n.$$

On the other hand, if for some $\mathbf{y} \in \mathbb{R}^n$ and $b \in \mathbb{R}$ such that

$$\mathbf{y}^\top \mathbf{x} - b \leq f(\mathbf{x}) \quad \forall \mathbf{x} \in \mathbb{R}^n,$$

then

$$\sup_{\mathbf{x} \in \mathbf{R}^n} \{\mathbf{x}^\top \mathbf{y} - f(\mathbf{x})\} = f^*(\mathbf{y}) \leq b.$$

Consequently,

$$\mathbf{x}^\top \mathbf{y} - b \leq \mathbf{x}^\top \mathbf{y} - f^*(\mathbf{y}) \leq \sup_{\mathbf{y} \in \mathbf{R}^n} \{\mathbf{x}^\top \mathbf{y} - f^*(\mathbf{y})\} = f^{**}(\mathbf{x}) \quad \forall \mathbf{x} \in \mathbb{R}^n.$$

By Definition 1.2.26, $\bar{f}(x)$ is the supremum over all $\mathbf{y}$ and b such that $\mathbf{x}^\top \mathbf{y} - b \leq f(\mathbf{x})$, therefore, $\bar{f}(\mathbf{x}) \leq f^{**}(\mathbf{x})$. ∎

Corollary 1.3.5 $f^{**} = f$ if f is closed.

Theorem 1.3.6 Let $f : \mathbb{R}^n \to \bar{\mathbb{R}}$ be convex, and let $\mathbf{x}, \mathbf{y} \in \mathbb{R}^n$. Then

$$\mathbf{y} \in \partial f(\mathbf{x}) \text{ if and only if } f^*(\mathbf{y}) = \mathbf{y}^\top \mathbf{x} - f(\mathbf{x}).$$

Proof:

$$\begin{aligned} f^*(\mathbf{y}) = \mathbf{y}^\top \mathbf{x} - f(\mathbf{x}) &\Leftrightarrow \mathbf{y}^\top \mathbf{x} - f(\mathbf{x}) = \sup_{\mathbf{x}'} \{\mathbf{y}^\top \mathbf{x}' - f(\mathbf{x}')\} \\ &\Leftrightarrow \mathbf{y}^\top (\mathbf{x}' - \mathbf{x}) \leq f(\mathbf{x}') - f(\mathbf{x}) \quad \forall \mathbf{x}' \in \mathbb{R}^n \\ &\Leftrightarrow \mathbf{y} \in \partial f(\mathbf{x}). \end{aligned}$$

∎

Corollary 1.3.7 Let $f : \mathbb{R}^n \to \bar{\mathbb{R}}$ be closed and convex, and let $\mathbf{x}, \mathbf{y} \in \mathbb{R}^n$. Then $\mathbf{y} \in \partial f(\mathbf{x})$ if and only if $\mathbf{x} \in \partial f^*(\mathbf{y})$.

Proof:

$$\begin{aligned} \mathbf{y} \in \partial f(\mathbf{x}) &\Leftrightarrow f^*(\mathbf{y}) = \mathbf{y}^\top \mathbf{x} - f(\mathbf{x}) \quad \text{by Theorem 1.3.6} \\ &\Leftrightarrow f(\mathbf{x}) = \mathbf{y}^\top \mathbf{x} - f^*(\mathbf{y}) \\ &\Leftrightarrow f^{**}(\mathbf{x}) = \mathbf{y}^\top \mathbf{x} - f^*(\mathbf{y}) \quad \text{by Corollary 1.3.5} \\ &\Leftrightarrow \mathbf{x} \in \partial f^*(\mathbf{y}). \end{aligned}$$

■

Definition 1.3.2 (Concave functions and their Fenchel transforms) A function $g : \mathbb{R}^n \to \bar{\mathbb{R}}$ is said to be *proper concave* if $-g$ is proper convex. The Fenchel transform of a proper concave function g is defined as

$$g_*(\mathbf{y}) = \inf_{\mathbf{x} \in \mathbf{R}^n} \{\mathbf{y}^\top \mathbf{x} - g(\mathbf{x})\}.$$

The following counterpart of Lemma 1.3.1 follows directly from Definition 1.3.2.

Lemma 1.3.8 (Young's Inequality for concave functions) Let $g : \mathbb{R}^n \to \bar{\mathbb{R}}$ be a *proper concave* function, and let g_* be its Fenchel transform. Then

$$g(\mathbf{x}) + g_*(\mathbf{y}) \le \mathbf{y}^\top \mathbf{x} \qquad \forall \mathbf{x}, \mathbf{y} \in \mathbb{R}^n.$$

Theorem 1.3.9 (Fenchel Duality Theorem) Let $f : \mathbb{R}^n \to \bar{\mathbb{R}}$ be a proper convex function, and $g : \mathbb{R}^n \to \bar{\mathbb{R}}$ be a proper concave function where $\text{dom}(f) \cap \text{dom}(g) \neq \emptyset$. Then

$$\inf_{\mathbf{x} \in \mathbf{R}^n} \{f(\mathbf{x}) - g(\mathbf{x})\} = \sup_{\mathbf{y} \in \mathbf{R}^n} \{g_*(\mathbf{y}) - f^*(\mathbf{y})\}.$$

Proof: From Lemma 1.3.1 and Lemma 1.3.8, we have, for all $\mathbf{x}, \mathbf{y} \in \mathbb{R}^n$,

$$f(\mathbf{x}) + f^*(\mathbf{y}) \ge \mathbf{x}^\top \mathbf{y} \ge g(\mathbf{x}) + g_*(\mathbf{y}),$$

and hence

$$f(\mathbf{x}) - g(\mathbf{x}) \ge g_*(\mathbf{y}) - f^*(\mathbf{y}) \quad \forall \mathbf{x}, \mathbf{y} \in \mathbb{R}^n.$$

This yields the following weak duality result:

$$\inf_{\mathbf{x}} \{f(\mathbf{x}) - g(\mathbf{x})\} \ge \sup_{\mathbf{y}} \{g_*(\mathbf{y}) - f^*(\mathbf{y})\}. \tag{1.3.1}$$

Let $\xi = \inf_{\mathbf{x}} \{f(\mathbf{x}) - g(\mathbf{x})\}$, and

$$\begin{aligned} \mathcal{E}_1 &= \text{epi}(f) = \{(\mathbf{x}, a) \in \mathbb{R}^{n+1} \mid f(\mathbf{x}) \le a\}. \\ \mathcal{E}_2 &= \{(\mathbf{x}, a) \in \mathbb{R}^{n+1} \mid a \le g(\mathbf{x}) + \xi\}. \end{aligned}$$

Since $f(\mathbf{x}) - g(\mathbf{x}) \geq \xi \quad \forall \mathbf{x} \in \mathbb{R}^n$, the two closed sets $\mathcal{E}_1$ and $\mathcal{E}_2$ are convex sets that are disjoint except possibly for a common point, hence by a weaker version of Theorem 1.2.10 there exists a non-vertical hyperplane that separates the two sets in a non-strict sense, i.e., there exists $(\mathbf{y}, \alpha) \in \mathbb{R}^{n+1}, \alpha \neq 0, \beta \in \mathbb{R}$ such that

$$\mathbf{y}^\top \mathbf{x} + \alpha f(\mathbf{x}) \geq \beta \geq \mathbf{y}^\top \mathbf{x} + \alpha(g(\mathbf{x}) + \xi) \quad \forall \mathbf{x} \in \mathbb{R}^n. \tag{1.3.2}$$

Let $\gamma = -\beta/\alpha$ and $\mathbf{z} = -\mathbf{y}/\alpha$, then (1.3.2) can be expressed as

$$f(\mathbf{x}) \geq \mathbf{z}^\top \mathbf{x} - \gamma \geq g(\mathbf{x}) + \xi \qquad \forall \mathbf{x} \in \mathbb{R}^n. \tag{1.3.3}$$

From the first inequality of (1.3.3),

$$\mathbf{z}^\top \mathbf{x} - f(\mathbf{x}) \leq \gamma \quad \forall \mathbf{x} \in \mathbb{R}^n.$$

In particular,

$$\sup_{\mathbf{x}} \{\mathbf{z}^\top \mathbf{x} - f(\mathbf{x})\} = f^*(\mathbf{z}) \leq \gamma. \tag{1.3.4}$$

From the second inequality of (1.3.3),

$$\inf_{\mathbf{x}} \{\mathbf{z}^\top \mathbf{x} - g(\mathbf{x})\} = g_*(\mathbf{z}) \geq \gamma + \xi. \tag{1.3.5}$$

Subtracting (1.3.4) from (1.3.5), we have, together with (1.3.1),

$$\xi \leq g_*(\mathbf{z}) - f^*(\mathbf{z}) \leq \sup_{\mathbf{z}} \{g_*(\mathbf{z}) - f^*(\mathbf{z})\} \leq \xi,$$

or

$$\sup_{\mathbf{z} \in \mathbf{R}^n} \{g_*(\mathbf{z}) - f^*(\mathbf{z})\} = \xi = \inf_{\mathbf{x} \in \mathbf{R}^n} \{f(\mathbf{x}) - g(\mathbf{x})\}.$$

∎

1.4 Elements of Graph Theory

The occurrence of networks or graphs in modern mathematics is ubiquitous. The Königsberg seven bridge problem was first addressed as a graph problem by Euler back in the eighteen century. Now we have little choice but to live amongst computer and telecommunication networks, utility (water and electricity) distribution networks, transportation and railway systems. In commerce and industries, we have flexible manufacturing systems, international trading and arbitrage models in finance, which are essentially network models.

Most network models are deterministic in nature, but the study of stochastic and queueing networks have also become extremely important and popular. Needless to say, the study of networks reaps enormous benefits in numerous practical applications, although it is not the intention of this book to dwell on practicalities. These applications can be found in many excellent operations research text books, with [AMO1] and [KH1] dedicated to both algorithms and applications.

The terms *graphs* and *networks* appear to appeal differently to different people, depending on one's affiliation with pure mathematics, applied mathematics or engineering. We make no such distinction however, and will use both terms interchangeably, although more often than not, a network is invariably considered as a directed graph, as defined below.

Definition 1.4.1 A *graph* $\mathcal{G} = (\mathcal{N}, \mathcal{A})$ is made up of a (finite) node set $\mathcal{N}$ and a (finite) arc set $\mathcal{A}$. Members of $\mathcal{N}$ are called *nodes (or vertices)* and members of $\mathcal{A}$ are called *arcs (or edges)*. The graph is said to be *proper* if $|\mathcal{N}| \geq 2$ and $|\mathcal{A}| \geq 1$. If an arc is associated with a direction (as indicated by an arrow on that arc), it is called a *directed arc*. If all arcs of the graph are directed, then the graph is called a *directed graph (or a digraph* in short). Each arc of $\mathcal{A}$ is *incident* to two end points, and if an arc $j \in \mathcal{A}$ is also directed, then it is sometimes represented as an ordered pair of nodes $(\alpha(j), \omega(j))$, where $\alpha(j)$ is the *source node* of j and $\omega(j)$ is the *sink node* of j. In this manner, the arc set $\mathcal{A}$ can be perceived as a subset of $\mathcal{N} \times \mathcal{N}$. Two nodes are said to be *adjacent* to each other if they are joined by some arc. The digraph is *complete* if $\mathcal{A} = \mathcal{N} \times \mathcal{N}$.

Definition 1.4.2 The *node-arc incidence matrix* $\mathbf{E}$ of a digraph $\mathcal{G} = (\mathcal{N}, \mathcal{A})$ is defined as an $|\mathcal{N}| \times |\mathcal{A}|$ integer-valued matrix with entries given by:

$$E_{ij} = \begin{cases} 1 & \text{if } \alpha(j) = i, \\ -1 & \text{if } \omega(j) = i, \\ 0 & \text{otherwise.} \end{cases}$$

Lemma 1.4.1 $\mathbf{1}^\top \mathbf{E} = \mathbf{0}^\top$ where $\mathbf{1}^\top = (1, 1, \cdots, 1)$.

Proof: Each column of $\mathbf{E}$ has a 1 and a -1. ∎

Definition 1.4.3 A *path* $\mathcal{P}$ in a digraph $\mathcal{G} = (\mathcal{N}, \mathcal{A})$ is an alternating finite sequence of nodes and arcs $\{i_0, j_1, i_1, j_2, \cdots, j_K, i_K\}$ where $K \geq 1$, $i_k \in \mathcal{N}$ and an arc $j_k \in \mathcal{A}$

is either $j_k = (i_k, i_{k-1})$ (whence it is traversed negatively) or $j_k = (i_{k-1}, i_k)$ (whence it is traversed positively). The *source node and sink node* of $\mathcal{P}$ are i_0 and i_K respectively, and the notation $\mathcal{P} : i_0 \to i_K$ is used to denote that the path joins the source node i_0 to the sink node i_K. The *length* of the path is the number of arcs in it. If some arc appears more than once in the path, either positively or negatively, the path is said to be one with *multiplicities*. An *elementary* path is a path without multiplicities and where no node occurs more than once in the path. In a path without multiplicities, the set of all positively traversed arcs is denoted by $\mathcal{P}^+$ while the set of all negatively traversed arcs is denoted by $\mathcal{P}^-$. The *arc-path incidence vector* $\mathbf{e}_\mathcal{P} \in \mathbb{R}^{|\mathcal{A}|}$ is defined by

$$[\mathbf{e}_\mathcal{P}]_j = \begin{cases} 1 & \text{if } j \in \mathcal{P}^+, \\ -1 & \text{if } j \in \mathcal{P}^-, \\ 0 & \text{otherwise.} \end{cases}$$

Definition 1.4.4 A *cycle* is a special path $\mathcal{P} = \{i_0, j_1, i_1, j_2, \cdots, j_K, i_K\}$ where $i_0 = i_K$. An *elementary cycle* is one without multiplicity and that, apart from $i_0 = i_k$, each node of $\mathcal{P}$ occurs only once. The *orientation* of the cycle is identified with the order of appearance of the arcs and nodes.

Theorem 1.4.2 If $\mathcal{P} = \{i_0, j_1, i_1, j_1, \cdots, j_K, i_K\}$ is an elementary path or cycle of a proper digraph $\mathcal{G} = (\mathcal{N}, \mathcal{A})$, then

$$\mathbf{E}\mathbf{e}_\mathcal{P} = \mathbf{e}^{i_0} - \mathbf{e}^{i_K},$$

where $\mathbf{e}^i \in \mathbb{R}^{|\mathcal{N}|}$ is a vector with $[\mathbf{e}^i]_j = \delta_{ij}$ and $\mathbf{E}$ is the node-arc incidence matrix of $\mathcal{G}$.

Proof: Let E^j be the jth column of $\mathbf{E}$, then if

$$j_k = (i_{k-1}, i_k), \text{ then } [\mathbf{e}_\mathcal{P}]_{j_k} = 1, \text{ hence, } [\mathbf{e}_\mathcal{P}]_{j_k} E^{j_k} = \mathbf{e}^{i_{k-1}} - \mathbf{e}^{i_k}, \quad \text{otherwise;}$$
$$j_k = (i_k, i_{k-1}), \text{ then } [\mathbf{e}_\mathcal{P}]_{j_k} = -1, \text{ hence, } [\mathbf{e}_\mathcal{P}]_{j_k} E^{j_k} = \mathbf{e}^{i_{k-1}} - \mathbf{e}^{i_k}.$$

Thus in all cases

$$\mathbf{E}\mathbf{e}_\mathcal{P} = \sum_{j \in \mathcal{A}} E^j [\mathbf{e}_\mathcal{P}]_j = \mathbf{e}^{i_0} - \mathbf{e}^{i_1} + \mathbf{e}^{i_1} - \mathbf{e}^{i_2} + \cdots + \mathbf{e}^{i_{K-1}} - \mathbf{e}^{i_K} = \mathbf{e}^{i_0} - \mathbf{e}^{i_K}.$$

∎

Corollary 1.4.3 If $\mathcal{P}$ is a cycle, then $\mathbf{E}\mathbf{e}_\mathcal{P} = \mathbf{0}$, and therefore the columns of $\mathbf{E}$ corresponding to this cycle are linearly dependent.

Definition 1.4.5 A graph $\mathcal{G}' = (\mathcal{N}', \mathcal{A}')$ is a *subgraph* of the graph $\mathcal{G} = (\mathcal{N}, \mathcal{A})$ if $\mathcal{N}' \subseteq \mathcal{N}$ and $\mathcal{A}' \subseteq \mathcal{A}$. $\mathcal{G}' = (\mathcal{N}', \mathcal{A}')$ is a *spanning subgraph* of $\mathcal{G} = (\mathcal{N}, \mathcal{A})$ if $\mathcal{N}' = \mathcal{N}$.

Definition 1.4.6 A graph $\mathcal{G} = (\mathcal{N}, \mathcal{A})$ is said to be *connected* if given any pair of nodes (i, i'), a path $\mathcal{P} : i \to i'$ connecting the nodes can be found. A component of

$\mathcal{G}$ is a connected subgraph $\mathcal{G}_1$ of $\mathcal{G}$ such that there is no other connected subgraph of $\mathcal{G}$ which contains $\mathcal{G}_1$ strictly.

Definition 1.4.7 A proper graph $\mathcal{G} = (\mathcal{N}, \mathcal{A})$ is *acyclic* if no cycle can be found in $\mathcal{G}$. An acyclic graph is called a *forest*. A connected forest is called a *tree*. A *spanning tree* of $\mathcal{T}$ is a spanning subgraph of $\mathcal{G}$ which is also a tree. The *cotree* $\mathcal{T}^*$ of a spanning tree $\mathcal{T}$ is another spanning subgraph containing exactly those arcs which are not in $\mathcal{T}$.

Theorem 1.4.4 Let $\mathcal{G} = (\mathcal{N}, \mathcal{A})$ be a proper graph and $\mathcal{T} = (\mathcal{N}', \mathcal{A}')$ be a subgraph of $\mathcal{G}$. Then the following are equivalent:
(i) $\mathcal{T}$ is a tree, i.e., acyclic and connected.
(ii) Given any pair of nodes (i, i') in $\mathcal{N}'$, there exists an unique path in $\mathcal{T}$ joining i to i'.
(iii) $\mathcal{T}$ is connected and $|\mathcal{N}'| = |\mathcal{A}'| + 1$.
(iv) $\mathcal{T}$ is acyclic and $|\mathcal{N}'| = |\mathcal{A}'| + 1$.

Proof:
(i)⇒(ii): If not, then a cycle can be formed in $\mathcal{T}$ by concatenating any two paths joining i and i', contradicting the fact that $\mathcal{T}$ is acyclic.

(ii)⇒(iii): If there exists a unique path in $\mathcal{T}$ joining any pair of nodes, then clearly $\mathcal{T}$ is connected. We prove $|\mathcal{N}'| = |\mathcal{A}'| + 1$ by induction. This is clearly true for a proper graph with two nodes and one arc. Assume that it is true for any tree with less than m nodes. If $\mathcal{T}$ has m nodes then removal of an arc will disconnect the graph into two components, $\mathcal{T}_1 = (\mathcal{N}_1, \mathcal{A}_1)$ and $\mathcal{T}_2 = (\mathcal{N}_2, \mathcal{A}_2)$. Since each of $\mathcal{T}_1$ and $\mathcal{T}_2$ has less than m nodes,

$$|\mathcal{N}_1| = |\mathcal{A}_1| + 1 \text{ and } |\mathcal{N}_2| = |\mathcal{A}_2| + 1. \tag{1.4.1}$$

Thus

$$|\mathcal{N}'| = |\mathcal{N}_1| + |\mathcal{N}_2|, \tag{1.4.2}$$

and

$$|\mathcal{A}'| = |\mathcal{A}_1| + |\mathcal{A}_2| + 1 \tag{1.4.3}$$

(the last one being that of the removed arc). The conclusion follows from combining (1.4.1) to (1.4.3).

(iii)⇒(iv): If $\mathcal{T}$ is not acyclic, then a cycle $\mathcal{P}$ with m nodes and m arcs exists. For any other node not lying on this cycle, it must be part of a path that joins to one of the nodes on the cycle by the connectedness assumption. Call this a *tentacle*. Each tentacle will have the same number of nodes (not counting the last one in the cycle) and arcs. Adding up, $\mathcal{T}$ will have $|\mathcal{N}'| = |\mathcal{A}'|$, a contradiction.

(iv)⇒(i): The acyclic part is directly implied. If $\mathcal{T}$ is not connected, then there exist $K > 1$ components, each being a tree with one less arcs than nodes. Adding up, $\mathcal{T}$ will have $|\mathcal{N}'| = |\mathcal{A}'| + K$ nodes, a contradiction.

∎

Remark 1.4.1 The concept of path has a dual concept called *cut*. It will become clear in Chapter Two that *the duality between path and cut plays a central role behind many network optimization problems.*

Definition 1.4.8 A *cut* $\mathcal{Q}$ in a graph $\mathcal{G} = (\mathcal{N}, \mathcal{A})$ is a set of arcs whose removal results in a disconnected graph. The cut is said to be *elementary* if the removal of arcs in this cut increases the number of components by exactly one. If $\mathcal{G}$ is a digraph, a cut is conveniently expressed as $\mathcal{Q} = [\mathcal{S}, \mathcal{N} \setminus \mathcal{S}]$ where $\mathcal{S} \subseteq \mathcal{N}$. This represents the set of arcs which straddle the node set $\mathcal{S}$ and its complement $\mathcal{N} \setminus \mathcal{S}$. The set of arcs in $\mathcal{Q}$ can be partitioned into two disjoint subsets:

$$\mathcal{Q}^+ = \{j \in \mathcal{A} \mid \alpha(j) \in \mathcal{S}, \omega(j) \in \mathcal{N} \setminus \mathcal{S}\} \quad \text{(the set of forward arcs)},$$
$$\mathcal{Q}^- = \{j \in \mathcal{A} \mid \omega(j) \in \mathcal{S}, \alpha(j) \in \mathcal{N} \setminus \mathcal{S}\} \quad \text{(the set of reverse arcs)}.$$

If $\mathcal{S} = \emptyset$ or $\mathcal{N}$, then the resulting cut $\mathcal{Q} = [\mathcal{S}, \mathcal{N} \setminus \mathcal{S}]$ is an *empty cut.* The *cut incidence vector* $\mathbf{e}_{\mathcal{Q}} \in \mathbb{R}^{|\mathcal{A}|}$ is defined as:

$$[\mathbf{e}_{\mathcal{Q}}]_j = \begin{cases} 1 & \text{if } j \in \mathcal{Q}^+, \\ -1 & \text{if } j \in \mathcal{Q}^-, \\ 0 & \text{otherwise.} \end{cases}$$

The *set indicator vector* $\mathbf{e}_{\mathcal{S}} \in \mathbb{R}^{|\mathcal{N}|}$ for the node set $\mathcal{S}$ that generates the cut is defined as:

$$[\mathbf{e}_{\mathcal{S}}]_i = \begin{cases} 1 & \text{if } i \in \mathcal{S}, \\ 0 & \text{otherwise.} \end{cases}$$

Definition 1.4.9 A node is *an end point* if it has only one arc incident on it.

Remark 1.4.2 The following facts are intuitively obvious, and rigorous proofs can be easily established following the proof of Theorem 1.4.4.

- A tree has at least two end points.
- The deletion of an end point of a tree and the arc incident on it results in another tree.
- Every connected graph has a spanning tree.
- There exists a unique path in a tree that joins the two end points of a cotree arc.

The following result establishes the correspondence of a spanning tree to a basis of the node arc incidence matrix $\mathbf{E}$.

Theorem 1.4.5 Let $\mathcal{G} = (\mathcal{N}, \mathcal{A})$ be a proper connected digraph with a node arc incidence matrix $\mathbf{E}$. A set of columns $\{E^j \mid j \in \mathcal{A}'\}$ with $|\mathcal{A}'| = |\mathcal{N}| - 1$ is a basis for the column space of $\mathbf{E}$ if and only if $\mathcal{T} = (\mathcal{N}, \mathcal{A}')$ is a spanning tree for $\mathcal{G}$.

Proof: (sufficiency) We first prove that the set of columns $\{E^j \mid j \in \mathcal{A}'\}$ corresponding to a spanning tree $\mathcal{T}$ of $\mathcal{G}$ is (i)linearly independent and (ii)spans the column space of $\mathbf{E}$. This establishes that $\{E^j \mid j \in \mathcal{A}'\}$ is a basis.

(i) Since $\mathcal{T}$ is a spanning tree, $|\mathcal{A}'| = |\mathcal{N}| - 1$. If the set $\{E^j,\ j \in \mathcal{A}'\}$ is linearly dependent, then there exists $\lambda_j,\ j \in \mathcal{A}'$, not all zero, such that

$$\sum_{j \in \mathcal{A}'} \lambda_j E^j = \mathbf{0}. \tag{1.4.4}$$

Let t be an end point of $\mathcal{T}$ and let μ be the arc in $\mathcal{T}$ incident on t. Since t is an end point, $[\mathbf{E}]_{t\mu} = 1$, or -1, but $[\mathbf{E}]_{tj} = 0\ \forall j \neq \mu$. (1.4.4) thus implies that $\lambda_\mu = 0$. The deletion of the node t and the arc μ results in another tree. If we apply the above argument recursively, then eventually $\lambda_j = 0\ \forall j \in \mathcal{A}'$, a contradiction.

(ii) Any arc k in $\mathcal{G}$ but not in $\mathcal{T}$ is a cotree arc. Let $\mathcal{P}$ be the unique path in $\mathcal{T}$ joining the two end points of k, then k together with $\mathcal{P}$ forms a cycle. By Corollary 1.4.3, the columns of $\mathbf{E}$ corresponding to all the arcs in the cycle are linearly dependent, implying that E^k can be expressed as a linear combination of the columns of $\mathbf{E}$ corresponding to the tree arcs. Consequently $\{E^j\ |j \in \mathcal{A}'\}$ spans the column space of $\mathbf{E}$.

(necessity) If $\{E^j\ |j \in \mathcal{A}'\}$ with $|\mathcal{A}'| = |\mathcal{N}| - 1$ is a basis for $\mathbf{E}$, it must be linearly independent. By Corollary 1.4.3, no cycle can be found in $\mathcal{T} = (\mathcal{N}, \mathcal{A}')$. By part (iv) of Theorem 1.4.4, $\mathcal{T}$ is acyclic and has one less arcs than nodes, hence it must be a tree. Since it contains all the nodes in $\mathcal{G}$, it is also a spanning tree. ∎

Corollary 1.4.6 Let $\mathbf{E}$ be the incidence matrix of a connected graph. Then rank($\mathbf{E}$) $= |\mathcal{N}| - 1$.

Proof: Follows from Lemma 1.4.1 and Theorem 1.4.5. ∎

Note that the property rank($\mathbf{E}$) $= |\mathcal{N}| - 1$ is known as the *rank one deficiency* of $\mathbf{E}$.

1.5 Elements of Optimization and Variational Inequalities

The reader of this book should already be familiar with the concept of optimization. We now present a very general overview of the subject and a brief introduction to variational inequalities in order to lay the foundation for subsequent duality analyses.

The need to order vectors arises frequently in this book. A formal discussion of this topic will be given in Chapter Seven. For the purpose of the first six chapters, we adopt a set of informal vector ordering relationships: $<, \leq, \leqq, >, \geq, \geqq$. Given two vectors $\boldsymbol{\xi}$ and $\boldsymbol{\eta}$,

• $\boldsymbol{\xi} > \boldsymbol{\eta}$ means that every component of $\boldsymbol{\xi}$ is strictly greater than the corresponding component of $\boldsymbol{\eta}$;

• $\boldsymbol{\xi} \geq \boldsymbol{\eta}$ means that every component of $\boldsymbol{\xi}$ is greater than or equal to the corresponding component of $\boldsymbol{\eta}$, with at least one component strictly greater;

• $\boldsymbol{\xi} \geqq \boldsymbol{\eta}$ means that every component of $\boldsymbol{\xi}$ is greater than or equal to the corresponding component of $\boldsymbol{\eta}$.

The most general optimization problem is succinctly stated as follows:

$$\text{(Prototype Optimization Problem)} \qquad \min_{\mathbf{x} \in \mathcal{S}} f(\mathbf{x}) \qquad \text{subject to} \quad \mathbf{x} \in \mathcal{X}. \quad (1.5.1)$$

Sometimes the above is succinctly stated as:

$$\min f(\mathcal{X}).$$

Some elaboration of the various terminology is warranted here.

• $\mathbf{x}$ is called the *variable* of the optimization problem, and is usually a member of some space $\mathcal{S}$ which could be continuous or discrete, finite or infinite dimensional. We are mainly concerned with duality in continuous spaces in this book. Duality in discrete spaces is quite a lot more subtle, and is perhaps not as clean as the continuous case. Nevertheless some discussion of discrete optimization is included in Chapter Five. The space $\mathcal{S}$ can be an infinite dimensional Banach space or Hausdorff space, although more often it is just the finite dimensional vector space $\mathbb{R}^n$. To quote from a well-known applied mathematician, *one can indulge in any fancy space one likes, but if one has to compute a solution, then everything is back to* $\mathbb{R}^n$. To avoid an unnecessarily abstract notation that may obscure the underlying ideas, we have chosen to discuss duality mainly in the context of $\mathbb{R}^n$, although it should be noted once again that most results are generalizable to abstract spaces with only little modification required.

• The function $f : \mathcal{S} \to \text{range}(f)$ is usually called the *cost function, or functional* if $\mathcal{S}$ is infinite dimensional, of the optimization problem. Optimization is mostly concerned with real-valued functions. Usually, the cost function f is real-valued, and hence range $(f) \subseteq \mathbb{R}$, or possibly $\bar{\mathbb{R}}$. *Scalar optimization* is often referred to as just optimization by default. In the more general context, f could be a vector-valued function, in which case, range $f \subseteq \mathbb{R}^m$ with $m > 1$, and the optimization problem is called a *multicriteria optimization, or vector optimization, or multiobjective optimization.* Multicriteria optimization will be dealt with in the latter half of this book, as it entails a more complex interpretation of optimality.

• By convention, optimization usually means minimization by default, hence min is used. In the event when certain closedness or compactness assumptions cannot be satisfied, we often replace min by inf to denote finding the greatest lower bound (or infimum) instead of the minimum, which by convention the latter is attainable.

In the case where the minimum is attainable, inf and min are used interchangeably. Any optimization problem concerns with maximization can be easily converted to a minimizing one by simply changing the sign of the objective function.

• The set $\mathcal{X}$ is a subset of the space $\mathcal{S}$, and is called the *feasible set (or constraint set)* of the optimization problem. If $\mathcal{X} = \mathcal{S}$, then the problem is *unconstrained*, and is *constrained* otherwise. Usually, $\mathcal{X}$ is nonempty, otherwise the problem is called *infeasible* and the minimum cost is assigned ∞ by convention. Often, $\mathcal{X}$ is assumed to be connected with a nonempty interior. Sometimes, the set $\mathcal{X}$ may have additional properties such as compactness or convexity, in which cases, there are many stronger results pertaining to the solution when f is nice. Often the feasible set $\mathcal{X}$ is expressed as a set of points that satisfy a number of *constraints*, e.g.,

$$\mathcal{X} = \{\mathbf{x} \in \mathcal{S} \mid \mathbf{g}(\mathbf{x}) = \mathbf{0}, \;\; \mathbf{h}(\mathbf{x}) \leqq \mathbf{0}, \;\; \boldsymbol{\ell} \leqq \mathbf{x} \leqq \mathbf{u}\}.$$

Here the functions $\mathbf{g}$ and $\mathbf{h}$ are defined on the same domain as the cost function f and can either be scalar valued or vector-valued, or non-existent. Each component of $\mathbf{g}$ is called an *equality constraint* and each component of $\mathbf{h}$ is called an *inequality constraint*. The special set of inequality constraints given as $\boldsymbol{\ell} \leqq \mathbf{x} \leqq \mathbf{u}$ are called *simple bounds*.

• If the infimum in (1.5.1) is attainable by some feasible point, then the minimum cost is denoted by $\min f(\mathcal{X}) = f(\mathbf{x}^*)$ where $\mathbf{x}^*$ is understood to be the optimal solution to the problem. Strictly speaking, this should be a globally optimal solution, although often $\mathbf{x}^*$ is expected to be only locally optimal.

• Some special cases deserve special attention. If $f, \mathbf{g}$ and $\mathbf{h}$ are all linear, then the problem is called a *linear programming problem*. This will be dealt with in some length in Chapter Three. Otherwise if one or more of $f, \mathbf{g}$ or $\mathbf{h}$ is nonlinear, the problem is called a *nonlinear programming problem*. If $\mathbf{g}$ and $\mathbf{h}$ are linear and f is convex and separable, i.e., $f(\mathbf{x}) = \sum_j f_j(x_j)$, then the problem is called a *monotropic programming problem*, and this will also be dealt with in Chapter Three. If f is convex and $\mathcal{X}$ is convex, we have a *convex optimization problem*. Much stronger results can be obtained for this convex case than otherwise, and these will be presented in Chapter Four. If the problem is defined on some special structure, e.g., a graph or network, then ad-hoc titles like network optimization are ascribed to the problem. Many beautiful and geometrically meaningful duality results are found in network optimization, and the whole of Chapter Two is devoted to this special case.

• Optimization problems are sometimes made complicated by other considerations. If $\mathcal{S}$ is an infinite dimensional space, then the problem is called *infinite dimensional*. Optimal control problems [C2] and calculus of variations [GF1], [E1] are special classes of infinite dimensional optimization problem. If the cost and constraint functions are perturbed by noise or random factors, then the problem is called a *stochastic* optimization problem [KW1]. Interesting duality results prevail in both optimal control problems as well as stochastic programming problems. Unfortunately, these are not within the scope of this book, and the readers are referred to [H1] and [R8] for details.

Definition 1.5.1 Let $\mathcal{X} \subseteq \mathbb{R}^n$ and $f : \mathcal{X} \to \mathbb{R}$ be a real-valued function.

• A point $\mathbf{x}^* \in \mathcal{X}$ is said to be a *local minimum* of (1.5.1) if there exists $\epsilon > 0$ such that $\mathbf{x} \in \mathcal{X}$, and $\|\mathbf{x} - \mathbf{x}^*\| < \epsilon \Rightarrow f(\mathbf{x}) \geq f(\mathbf{x}^*)$.

• A point $\mathbf{x}^* \in \mathcal{X}$ is said to be a *local maximum* of (1.5.1) if there exists $\epsilon > 0$ such that $\mathbf{x} \in \mathcal{X}$, and $\|\mathbf{x} - \mathbf{x}^*\| < \epsilon \Rightarrow f(\mathbf{x}) \leq f(\mathbf{x}^*)$.

• A point $\mathbf{x}^* \in \mathcal{X}$ is said to be a *global minimum* of (1.5.1) if $f(\mathbf{x}^*) \leq f(\mathbf{x})\ \forall \mathbf{x} \in \mathcal{X}$.

• A point $\mathbf{x}^* \in \mathcal{X}$ is said to be a *global maximum* of (1.5.1) if $f(\mathbf{x}^*) \geq f(\mathbf{x})\ \forall \mathbf{x} \in \mathcal{X}$.

Remark 1.5.1 Let $\mathcal{X} \subseteq \mathbb{R}^n$ be a nonempty compact set, and $f : \mathcal{X} \to \mathbb{R}$. The following facts are well known (Weierstrass Theorem, see [R5]):

• If f is lower semicontinuous, then f attains a global minimum over $\mathcal{X}$.

• If f is upper semicontinuous, then f attains a global maximum over $\mathcal{X}$.

• If f is continuous, then f attains both a global minimum and a global maximum over $\mathcal{X}$.

The following properties for the minima of convex functions can be established.

Theorem 1.5.1 Let f be a closed and convex function on $\mathbb{R}^n$. Then
(i) a local minimum of f is also a global minimum,
(ii) the set of all global minima is convex, and
(iii) if f is strictly convex, then the global minimum is unique.

Proof: (i) If f has a local minimum at $\mathbf{x}^*$, then $f(\mathbf{x}^*) \leq f(\mathbf{x})\ \forall \mathbf{x} \in \mathcal{B}(\mathbf{x}^*, \epsilon)$ for some $\epsilon > 0$. For any $\mathbf{x} \in \mathbb{R}^n$ and a sufficiently small $\lambda > 0$, $\lambda\mathbf{x} + (1-\lambda)\mathbf{x}^* \in \mathcal{B}(\mathbf{x}^*, \epsilon)$, hence $f(\lambda\mathbf{x} + (1-\lambda)\mathbf{x}^*) \geq f(\mathbf{x}^*)$. Since f is convex, this means that

$$f(\mathbf{x}^*) \leq f(\lambda\mathbf{x} + (1-\lambda)\mathbf{x}^*) \leq \lambda f(\mathbf{x}) + (1-\lambda) f(\mathbf{x}^*).$$

From which we have $\lambda f(\mathbf{x}) - \lambda f(\mathbf{x}^*) \geq 0$ and since $\lambda > 0$, the conclusion that $\mathbf{x}^*$ is a global minimum follows.

(ii) If $\mathbf{x}^1$ and $\mathbf{x}^2$ are (global) minima, $f(\mathbf{x}^1) = f(\mathbf{x}^2)$, then, for $0 < \lambda < 1$,

$$f(\lambda\mathbf{x}^1 + (1-\lambda)\mathbf{x}^2) \leq \lambda f(\mathbf{x}^1) + (1-\lambda) f(\mathbf{x}^2) = f(\mathbf{x}^2). \tag{1.5.2}$$

But

$$f(\lambda\mathbf{x}^1 + (1-\lambda)\mathbf{x}^2) \geq f(\mathbf{x}^2) \tag{1.5.3}$$

since $\mathbf{x}^2$ is a global minimum. (1.5.2) and (1.5.3) together imply that $\lambda\mathbf{x}^1 + (1-\lambda)\mathbf{x}^2$ is a global minimum.

(iii) If f is strictly convex, and $\mathbf{x}^1 \neq \mathbf{x}^2$ are global minima with $f(\mathbf{x}^1) = f(\mathbf{x}^2)$, then

$$f(\lambda\mathbf{x}^1 + (1-\lambda)\mathbf{x}^2) < \lambda f(\mathbf{x}^1) + (1-\lambda) f(\mathbf{x}^2) = f(\mathbf{x}^2).$$

This contradicts (1.5.3), so $\mathbf{x}^1 = \mathbf{x}^2$. ∎

In particular, we have the following characterization of a global minimum for a convex function in terms of the subdifferential.

Theorem 1.5.2 Let $f : \mathbb{R}^n \to \bar{\mathbb{R}}$ be a convex function, and let $\mathbf{x}^* \in \text{dom}(f)$. Then $\mathbf{x}^*$ is the (global) minimum of f if and only if $0 \in \partial f(\mathbf{x}^*)$.

Proof: (Necessity) If $0 \in \partial f(\mathbf{x}^*)$, then by Definition 1.2.30, $f(\mathbf{x}) \geq f(\mathbf{x}^*)\ \forall \mathbf{x} \in \mathbb{R}^n$, and hence $\mathbf{x}^*$ is a (global) minimum.
(Sufficiency) If f is globally minimized at $\mathbf{x}^*$, then $f(\mathbf{x}) - f(\mathbf{x}^*) \geq 0 = \mathbf{0}^\top(\mathbf{x} - \mathbf{x}^*)$, implying that $\mathbf{0} \in \partial f(\mathbf{x}^*)$.

■

We now turn our attention to variational inequalities, henceforth abbreviated as VI. In some sense, VI can often be perceived to be a generalization of optimization. The topic was first studied in the context of solving partial differential equations in infinite dimensional spaces [KS1], but it has since found many applications in optimization and general equilibrium models in finite dimensional spaces.

Prototype Variational Inequality Problem (VI) Let $\mathcal{X} \subseteq \mathbb{R}^n$ be a convex set and $\mathbf{f} : \mathbb{R}^n \to \mathbb{R}^n$ be a vector-valued function.

$$\text{Find a point} \quad \mathbf{x}^* \in \mathcal{X} \text{ such that } \mathbf{f}(\mathbf{x}^*)^\top(\mathbf{x} - \mathbf{x}^*) \geq 0 \quad \forall \mathbf{x} \in \mathcal{X}. \tag{1.5.4}$$

The problem VI as stated in (1.5.4) reduces to several well-known problems under further assumptions on the function $\mathbf{f}$ and the set $\mathcal{X}$. These are discussed briefly in turn as follows, readers are referred to [N1] for details.

• **Solution of simultaneous equations.** Let $\mathcal{X} = \mathbb{R}^n$, then $\mathbf{x}^*$ solves $\mathbf{f}(\mathbf{x}) = \mathbf{0}$ if and only if $\mathbf{x}^*$ solves the VI (1.5.4).

• **Fixed point problems.** Let $\mathcal{X}$ be a closed and convex set in $\mathbb{R}^n$, then for any $\gamma > 0$, $\mathbf{x}^*$ satisfies the fixed point relationship

$$\mathbf{x}^* = \text{Pr}_{\mathcal{X}}(\mathbf{x}^* - \gamma \mathbf{f}(\mathbf{x}^*))$$

if and only if $\mathbf{x}^*$ solves the VI (1.5.4), where the projection operator $\text{Pr}_{\mathcal{X}}$ is defined as

$$\text{Pr}_{\mathcal{X}}(\mathbf{x}) = \text{argmin}_{\mathbf{y} \in \mathcal{X}} \|\mathbf{x} - \mathbf{y}\|.$$

• **Complementarity problems.** Let $\mathcal{X} = \mathbb{R}^n_+$, then $\mathbf{x}^*$ solves the complementarity problem:

$$\mathbf{f}(\mathbf{x}^*) \geqq \mathbf{0}, \quad \mathbf{x}^* \geqq \mathbf{0}, \quad \mathbf{f}(\mathbf{x}^*)^\top \mathbf{x}^* = 0$$

if and only if $\mathbf{x}^*$ solves the VI (1.5.4).

Our primary interest, however, lies in the context of VI as a generalization of optimization problems. Consider the prototype optimization as stated in (1.5.1). To avoid confusing f with $\mathbf{f}$, we shall restate the optimization problem using a different symbol for the cost function as follows:

$$\text{Optimization problem} \qquad \min_{\mathbf{x} \in \mathcal{X}} \ g(\mathbf{x}) \tag{1.5.5}$$

where g is assumed to be a differentiable function. The following result is well-known, and supplement the results to be discussed in Chapter Four.

Theorem 1.5.3 (Optimality conditions for optimization) Assuming that the feasible set $\mathcal{X}$ is closed and convex. A necessary condition for $\mathbf{x}^*$ to be a local minimum of the optimization problem (1.5.5) is that $\mathbf{x}^*$ solves the following VI,

$$\nabla g(\mathbf{x}^*)(\mathbf{x}-\mathbf{x}^*) \geq 0 \quad \forall \mathbf{x} \in \mathcal{X}. \tag{1.5.6}$$

If in addition, g is convex, then $\mathbf{x}^*$ is a global minimum of the optimization problem (1.5.5) if and only if $\mathbf{x}^*$ solves the VI (1.5.6).

Proof: Let $0 < t \ll 1$, then

$$g(\mathbf{x}^* + t(\mathbf{x}-\mathbf{x}^*)) = g(\mathbf{x}^*) + t\nabla g(\mathbf{x}^*)(\mathbf{x}-\mathbf{x}^*) + o(t^2).$$

Clearly for $\mathbf{x}^*$ to be a local minimum, the VI (1.5.6) must hold.

If g is convex, then

$$g(\mathbf{x}) \geq g(\mathbf{x}^*) + \nabla g(\mathbf{x}^*)(\mathbf{x}-\mathbf{x}^*) \quad \forall \mathbf{x} \in \mathcal{X}.$$

If, in addition, (1.5.6) holds, then

$$g(\mathbf{x}) \geq g(\mathbf{x}^*) \quad \forall \mathbf{x} \in \mathcal{X}.$$

∎

¿From Theorem 1.5.3 and the definition of VI in (1.5.4), it is clear that if we can find a convex function g such that $\nabla g = \mathbf{f}^\top$, then the optimization problem (1.5.5) and the VI problem (1.5.6) are equivalent. This is a concept well known in Physics, where we conveniently borrow the following terminology.

Definition 1.5.2 A vector function $\mathbf{f} : \mathbb{R}^n \to \mathbb{R}^n$ is said to be *integrable* if $\nabla \mathbf{f}$ is symmetric, i.e.,

$$\frac{\partial f_i}{\partial x_j} = \frac{\partial f_j}{\partial x_i} \quad \forall i, j.$$

The following result is a well-known extension of Greens Theorem in higher dimension (see [OR1]).

Theorem 1.5.4 If $\mathbf{f} : \mathbb{R}^n \to \mathbb{R}^n$ is integrable, then there exists a (scalar-valued) function $g : \mathbb{R}^n \to \mathbb{R}$ denoted by

$$g(\mathbf{x}) = \oint^{\mathbf{x}} \mathbf{f}(\mathbf{y})d\mathbf{y}$$

such that

$$\nabla g(\mathbf{x}) = \mathbf{f}(\mathbf{x})^\top. \tag{1.5.7}$$

Since it is well-known that a function is convex if and only if its Hessian is positive semi-definite, we have the following equivalence result.

Theorem 1.5.5 Assuming that the Jacobian of the function f as in (1.5.4) is symmetric and positive semidefinite, then there exists a convex function $g(\mathbf{x})$ such that (1.5.7) holds, and $\mathbf{x}^*$ solves the VI (1.5.4) if and only if $\mathbf{x}^*$ solves the optimization problem (1.5.5).

There exists an extensive literature on the analysis and computational methods of VIs. As we are only interested in the duality aspect of the subject, the readers are referred to [N1] for a comprehensive treatment of these results.

1.6 Weak/Strong Duality, Duality Gap and Gap Function

This section is intended to be a very general discussion of some fundamental concepts in duality. As such it is not meant to be elaborate and exhaustive. In the framework of optimization, for every optimization problem (referred to as the primal problem), one can often identify a corresponding dual optimization problem, such that the primal and the dual are intimately related to each other. Dual optimization problems occur in various different forms, but in general they take the following structure.

Definition 1.6.1 (Prototype Primal-Dual Optimization problems)

$$\text{(Prototype Primal Optimization Problem)} \quad \inf f(\mathbf{x}) \quad \text{subject to} \, \mathbf{x} \in \mathcal{X}, \quad (1.6.1)$$
$$\text{(Prototype Dual Optimization Problem)} \quad \sup g(\mathbf{y}) \quad \text{subject to} \, \mathbf{y} \in \mathcal{Y}, \quad (1.6.2)$$

where $\mathbf{x} \in \mathcal{S}$ and $\mathbf{y} \in \mathcal{T}$ are the primal and dual variables respectively, $\mathcal{S}$ and $\mathcal{T}$ (as the dual space of $\mathcal{S}$) are the underlying spaces on which the primal and dual are defined respectively, $\mathcal{X}$ and $\mathcal{Y}$ are the primal and dual feasible sets respectively, f and g are the primal and dual cost functions respectively. Strictly speaking, the cost function of the dual should be referred to as the "objective function" since it is intended to be maximized. For convenience, however, we shall loosely call it "cost" although the reader should be reminded of its intended meaning. Note that we have not assumed that the minimum primal cost or the maximum dual cost are attainable, hence inf and sup are used. Note that sometimes the dual problem of the dual problem becomes the original primal problem. In this case we refer to the primal and dual pair as *symmetric*.

In duality theory, the primal and dual problems are structured in such a way that a number of properties relating the two problems hold, often under certain convexity assumptions. The most well known ones pertain to the so-called weak and strong duality.

Definition 1.6.2 (Weak Duality)

$$\text{(Prototype Weak Duality)} \qquad f(\mathbf{x}) \geq g(\mathbf{y}) \quad \forall \mathbf{x} \in \mathcal{X},\ \forall \mathbf{y} \in \mathcal{Y}, \tag{1.6.3}$$

or, equivalently,

$$\inf_{\mathbf{x}\in\mathcal{X}} f(\mathbf{x}) \geq \sup_{\mathbf{y}\in\mathcal{Y}} g(\mathbf{y}), \qquad \text{or} \qquad \inf f(\mathcal{X}) \geq \sup g(\mathcal{Y}).$$

The difference between the infimum primal cost and supremum dual cost is called the *duality gap.*

Definition 1.6.3 (Duality Gap) $\quad \Delta = \inf f(\mathcal{X}) - \sup g(\mathcal{Y}).$

Under further assumptions that usually involve convexity, the duality gap can become zero, i.e., $\Delta = 0$. A theorem that asserts that the duality gap is zero is called a *duality theorem.* However a zero duality gap does not necessarily imply that there exists feasible primal and dual solutions with the same primal and dual cost, because the infimum in the primal or the supremum in the dual may not be attainable by feasible solutions. If, under further assumptions, there exists $\mathbf{x}^* \in \mathcal{X}$ and $\mathbf{y}^* \in \mathcal{Y}$ such that

$$f(\mathbf{x}^*) = \inf f(\mathcal{X}) = \min f(\mathcal{X})$$
$$g(\mathbf{y}^*) = \sup g(\mathcal{Y}) = \max g(\mathcal{Y})$$

then, we say that *the optimal solutions are realizable.* If the duality gap is zero and the optimal solutions are realizable, then we have,

Definition 1.6.4 (Strong Duality)

$$\text{(Prototype Strong Duality)} \qquad \min f(\mathcal{X}) = \max g(\mathcal{Y}), \tag{1.6.4}$$

where the use of min and max implicitly implies that the infimum in the primal and the supremum in the dual are both attainable.

Once weak duality is established, the recipe for establishing strong duality usually takes the following step. *Find a primal feasible solution* $\mathbf{x}$ *and a dual feasible solution* $\mathbf{y}$ *such that* $f(\mathbf{x}) = g(\mathbf{y})$. The following result then ensures that strong duality follows as a consequence.

Theorem 1.6.1 Assume that weak duality holds for the primal and dual pair. If there exists a primal feasible $\mathbf{x}^*$, and a dual feasible $\mathbf{y}^*$ such that $f(\mathbf{x}^*) = g(\mathbf{y}^*)$, then $\mathbf{x}^*$ solves the primal and $\mathbf{y}^*$ solves the dual.

Proof: If $\mathbf{x}^*$ does not solve the primal, then there exists some $\mathbf{x}^0 \in \mathcal{X}$ such that $f(\mathbf{x}^0) < f(\mathbf{x}^*) = g(\mathbf{y}^*)$. This contradicts the weak duality. Similarly, if $\mathbf{y}^*$ does not solve the dual, then there exists some $\mathbf{y}^0 \in \mathcal{Y}$ such that $g(\mathbf{y}^0) > g(\mathbf{y}^*) = f(\mathbf{x}^*)$. This contradicts the weak duality. ∎

Abstract Lagrangian Duality

Duality results are manifested in many different forms, perhaps the best known and most important one is the Lagrangian form which will be studied in Chapter Four. This form of duality typically begins with a primal problem and then the Lagrangian and the Lagrangian dual problem are defined accordingly. Here we shall briefly discuss an abstract version that is discussed in great detail in [W1] and [R2]. In this case, the primal and dual costs are induced, under suitable conditions, by a certain real-valued abstract Lagrangian function: $L : \mathcal{S} \times \mathcal{T} \to \mathbb{R}$ such that

$$f(\mathbf{x}) = \sup_{\mathbf{y} \in \mathcal{Y}} L(\mathbf{x}, \mathbf{y})$$

$$g(\mathbf{y}) = \inf_{\mathbf{x} \in \mathcal{X}} L(\mathbf{x}, \mathbf{y})$$

The following are specialized duality theorems pertaining to the Lagrangian form:

• **Saddle Point Theorem.** Under suitable conditions, there exists a saddle point $(\mathbf{x}^*, \mathbf{y}^*) \in \mathcal{X} \times \mathcal{Y}$ such that for all $(\mathbf{x}, \mathbf{y}) \in \mathcal{X} \times \mathcal{Y}$,

$$L(\mathbf{x}^*, \mathbf{y}) \leq L(\mathbf{x}^*, \mathbf{y}^*) \leq L(\mathbf{x}, \mathbf{y}^*).$$

• **Minimax Theorem.**

$$\sup_{\mathbf{y} \in \mathcal{Y}} \inf_{\mathbf{x} \in \mathcal{X}} L(\mathbf{x}, \mathbf{y}) = \inf_{\mathbf{x} \in \mathcal{X}} \sup_{\mathbf{y} \in \mathcal{Y}} L(\mathbf{x}, \mathbf{y})$$

As we are concerned with duality in a broader sense, our discussion of Lagrangian duality is confined to a restricted form in Chapter Four. For a detailed treatment of generalized Lagrangian duality, the readers are referred to [MP1], [W1] and [R2].

Duality in variational inequalities

Because of notational complexity, we shall not pursue a general discussion of the duality in variational inequalities here. This requires an extended definition of variational inequalities presented in the previous section, and hence entails a more complex analysis. Essentially, for a given primal VI (in extended form), there exists another dual counterpart VI such that, under suitable assumptions on the underlying functions and sets, the solution of the primal VI and the solution of the dual VI are related. The details of this are deferred to Chapter Six.

Gap functions

The notion of a gap function is of strong theoretical interest in the study of duality for optimization as well as variational inequalities problems, usually under a certain convexity assumption. There are both primal and dual versions of gap functions, and of course they are related to each other. Essentially, a (primal) gap function γ for an optimization or VI problem is a scalar-valued function of the primal variable with the following defining properties:

- Property 1. $\gamma(\mathbf{x}) \geq 0 \quad \forall \mathbf{x}$;
- Property 2. $\gamma(\mathbf{x}^*) = 0$ if and only if $\mathbf{x}^*$ solves the problem.

As it will become clear when the details are discussed in latter Chapters, *Property 1 is essentially a restatement of weak duality, and Property 2 is a restatement of strong duality.* By virtue of these defining properties, it is clear that the optimization problem or VI problem can be recast as the following unconstrained problem, where the optimal cost is known a priori to be zero.

$$\text{(Equivalent optimization problem)} \qquad \min\ \gamma(\mathbf{x}).$$

However, gap functions are in general not differentiable, unless additional conditions are imposed. Furthermore, gap functions are usually not explicitly computable since it often requires the solution of an auxiliary optimization problem. Thus it is probably harder to solve the above problem than to solve the original optimization or VI problem. There are, however, useful computational methods which in fact solve the equivalent problem after some suitable transformation. As computational issues are beyond the scope of this book, the reader is referred to [H2] and [F4] for details.

1.7 Motivations for Studying Duality

There had been a number of questions raised in the course of writing this book. What warrants another theoretical book in optimization, especially one that specializes in a rather narrow area of the subject? Shouldn't there be more books on providing efficient computational methods and algorithms, or better still, just powerful software packages for solving complex optimization problems? Is a book on duality of any relevance to other than pure mathematicians? Does the understanding of duality make any difference to the problem solving ability of scientists, economists and engineers? Isn't it sufficient that the practising engineers/scientists/economists master the use of standard optimization software to solve their problems effectively? What is the advantage of looking at the dual problem instead of the primal one? Our answers to all of the above questions are not likely to be overwhelmingly convincing to the wider readers outside the mathematics community. Nevertheless we shall attempt to provide as much justification and motivation as possible in this section.

We believe that there are four main reasons for the study of duality.

- It has a tremendous aesthetic appeal.
- It deepens the theoretical understanding of optimization and variational inequalities.

• It provides the insights for devising effective computational methods and algorithms.

• It furnishes a meaningful intepretation to many physical, economics and engineering problems.

We shall discuss these in turns.

First and foremost, from a purely mathematical point of view duality is a supremely beautiful example of how complex pairs of systems or problems can be brought to fit together in a perfect jigsaw puzzle. The word *dual* has its origin in the Latin word *duo*, which means "two". It takes two to trade, to mate, to make babies, to quarrel. Men are made with two legs, two arms, two eyes, two ears. Computers work with Boolean algebra and we routinely make decision of a binary nature. There is day and night, ying and yang, good and evil, ... If one were asked what is the most significant number in mathematics, apart perhaps from the trivial zero and one, one would justifiably argue that it is two. This assertion comes about perhaps not so much in the magnitude of the number two, but more because the number two, as embodied in the concept of duality, is an aesthetic manifestation of symmetry. A nice and simple example of this is found in the Fenchel duality theorem (Theorem 1.3.9).

Second, duality is the backbone of optimization theory. One cannot really claim to have understood optimization completely until one has a firm grip on the underlying duality theory. Whilst it is true that many practitioners of optimization have spent a large part of their career solving optimization problems without really understanding duality, many have also fallen into the occasional trap of believing in the wrong optimal solution. There are important theoretical issues in optimization such as if the optimality condition is only a necessary condition, and if so, is the problem convex, etc. Unfortunately such issues are often of lesser worries to practitioners who are merely concerned with finding a solution. The fact is that many real world physical/engineering/economics problems are non-convex. The lack of strong duality in non-convex problems means that conventional methods based on the Lagrangian theory will often fail to find the true optimal solution. Another example is found in the study of convergence issues where duality plays an important role, such as the cutting plane methods in nonlinear programming. As a matter of fact, the application of the cutting plane method to a problem of minimising a linear cost function subject to a convex set constraint can be viewed as an extension of duality in linear programming. This is clearly the case if the convex set is viewed as an intersection of all the half-spaces that contains it. The problem can therefore be thought of as an (infinite) linear programming problem, see [L1] for detail.

The third reason for studying duality is that a lot of computational methods would not have worked without duality. Furthermore, the understanding of duality provides the fuel needed for the invention of new algorithms. Note that duality in optimization is usually associated with problems subject to one or more constraints. Computational methods are usually derived from studying the optimality condition for the problem. For the sake of simplifying our present discussion, let us assume that the underlying problem is convex, and thus the optimality condition is both necessary and sufficient, and hence the duality gap is zero. Problems that do not satisfy the convexity assumption are much more complex and in general lacking in effective computational methods.

Constrained optimization problems are usually solved iteratively. An algorithm typically starts with a phase one, which entails the finding of an initial (primal) feasible solution. Once this is obtained, a phase two is entered into whereby the feasibility of the intermediate solutions are maintained, and the algorithm works towards satisfying a certain optimality condition. The optimal solution is declared found when the optimality condition is satisfied. This approach is referred to as the *primal method*, in the sense that at each step of the computation, *primal feasibility* is always maintained, while one works towards *primal optimality*. The best known example of this approach is the primal simplex method for solving linear programs, to be discussed in Chapter Three. Some readers are probably aware that while nothing is explicitly said about duality in such a primal method, tacitly duality plays a crucial role. This is because *primal optimality equates to dual feasibility*, and *primal feasibility equates to dual optimality*. This concept will be elaborated in Chapter Three. Thus it can be said that a primal method maintains *primal feasibility* while working towards *dual feasibility*. An optimal solution is reached when both primal feasibility and dual feasibility are attained.

In a totally symmetric way, one can do it the other way round, i.e., one maintains dual feasibility (primal optimality) and works towards primal feasibility (dual optimality). The latter approach forms the basis of the so-called *dual method*. The dual simplex method in linear programming is a well-known example. There are practical advantages to using the dual method. This happens when finding an initial feasible solution is not quite as easy as finding an infeasible solution that satisfies the primal optimality (or dual feasibility) condition, or when maintaining primal feasibility is not quite as easy as maintaining dual feasibility. Just as in the primal method, an optimal solution is reached when both primal feasibility and dual feasibility are attained.

Then there is a third option: one neither maintains primal feasibility nor dual feasibility but switches between a *primal phase* and a *dual phase* iteratively until both primal feasibility and dual feasibility are attained. In the primal phase, the primal variable is being adjusted towards primal feasibility, while in the dual phase, the dual variable is being adjusted towards dual feasibility. The obvious advantage of such a method is that there is no need for a phase one (for either finding an initial feasible solution as in a primal method, or finding an initial solution that is not feasible but satisfying the optimality condition as in a dual method). The best known example for this approach is perhaps the out-of-kilter method for network flow problems (see Chapter Two).

Throughout the course of this book, a recurring theme is the discussion of the *dual problem* to a primal problem. Needless to say, *the dual method for solving a primal optimization problem is intimately related to the dual optimization problem*, and conversely, *the dual method for solving the dual optimization problem is intimately related to the primal optimization problem.*

In terms of computational efficiency, however, there is little evidence to suggest that which of the primal method, the dual method, or the primal-dual method is the more superior one. The actual performance often depends on the circumstances and the characteristics of the underlying problem. The detail of these comparisons is beyond the scope of this book. It must be stressed again, however, that the understanding of duality is of fundamental importance in devising any efficient and effective computational algorithm for solving optimization problems.

Finally, duality provides a meaningful interpretation to many physical, economic and engineering problems, which in turns enhances the understanding of the underlying optimal solution. This fourth reason for studying duality probably means more to practitioners in the real world, as it often helps to convince them the necessity of studying duality.

Meaningful interpretations of of duality abound in many real world problems, in particular linear ones. In the study of network problems, the concept of *flow* is closely related to the concept of *potential* (see Section 2.3), and the concept of *path* is closely related to the concept of *cut* (see Section 2.1). In the study of linear programming, duality is first encountered in the concept of *relative cost or shadow price* (see Section 3.1). In economics, duality is central in the price equilibrium model and consumer/producer behavioral models (see [C8]). Mechanical engineers are concerned with *mechanical networks of rods, beams and springs* (see [R3]). Electrical engineers study equilibria models in electrical circuits, and hydraulic engineers study equilibria models in hydraulic systems. In calculus of variations and optimal control theory, the concept of *an adjoint variable or co-state* is also related to duality. A large number of optimization problems also have corresponding dual problems with a meaningful physical interpretation. Most undergraduates in optimization are first introduced to duality by the classical diet problem due to Stigler (see [C1], or Section 3.2). The primal problem is interpreted as a cost minimizing problem for composing a diet with constraints on the nutrient content. Its corresponding dual is interpreted as a revenue maximizing problem for a drug company that synthesizes pills which provide the same nutrient content. Other well known practical examples with meaningful duals are found in:

- The max-flow problem (see Section 2.4);
- The max-match problem (see Section 2.5);
- The max-tension problem (see Section 2.5);
- The assignment problem (see Section 2.5);
- The transportation problem (see Chapter 5 of [L1]).

CHAPTER 2

DUALITY IN NETWORK OPTIMIZATION

This chapter is entirely devoted to the study of duality in systems with a predominantly combinatorial structure, specifically, that of a graph. Almost all, except the latter half of the first section, is based on material taken from the book "Network Flows and Monotropic Optimization" by R.T. Rockafellar [R3]. Both authors have had the fortunate opportunity of learning much about duality from this book when we were asked to teach it to undergraduate students at the honours level. We have since been convinced that the study of network problems is the best, if not the easiest, way of understanding duality, both from a geometric as well as an algebraic perspective.

The study of duality in the context of networks also reaps extra benefits because many of the theoretical results are often obtained by constructive proofs, i.e., the proof also precipitates a solution to its dual, if the dual exists. Furthermore, the proof is often concise and elegant, and the notions of strong and weak duality can also be "visualized" in geometrical forms.

While this chapter only intends to highlight those theoretical results in [R3] relevant to duality, it is still nevertheless a highly nontrivial task, as much of [R3] is in fact about duality. Many of the algorithmic developments (except when the algorithm forms part of a constructive proof) and fine details such as convergence and refinement of network and monotropic optimization algorithms are either bypassed or mentioned briefly. By doing so, some clarity has had to make way for brevity. Without doubt, [R3] is a must-read if the reader really wishes to master the subject, this chapter merely whets the appetite. The extension of network optimization to monotropic optimization in [R3] brings out many beautiful generalizations of duality results. However, we have chosen to cover these material in Chapter Three instead in view of the significant overlap with linear programming. For readers who are already familiar with linear programming, it would be expedient to go straight on to Sections 3.5 and 3.6 upon the completion of this chapter.

As Rockafellar has pointed out in [R3], the theory of graphs and networks has long suffered from a lack of uniformity in terminology and notation. It is not uncommon for the same word or term to mean different things in different sources, sometimes with the opposite meaning. In fact, the terminology and notation used in graph theory are arguably the most non-uniform among all areas of mathematics. We offer no apology if the terminology used in this book differs from what the reader is already familiar with, as it is intended to highlight the duality aspects of graph and network optimization. We had tried, in the first instance, to adopt a set of notations significantly different from that of [R3]. After a while, this effort proved to be futile for we came to the conclusion that the concept of duality is most apparent when explained with the notation adopted by [R3]. Any other notation would only obscure the symmetry of duality. For example, the unit cost of arc flow d_j and the spans d_j^+, d_j^- all use the alphabet d, and these have corresponding dual notations in c_j (unit cost of differential) and c_j^+, c_j^- (flow capacities). This is no coincidence, and it all has to do with duality. Similarly, many other pairs of symbols tend to appear in tandem, e.g., $\mathcal{P}$ (path) versus $\mathcal{Q}$ (cut); $\mathbf{x}$ (flow) versus $\mathbf{y}$ (divergence); $\mathbf{u}$ (potential) versus $\mathbf{v}$ (differential), etc.

2.1 Duality of Paths and Cuts and Minimum Spanning Trees

The duality between paths and cuts can be most clearly seen geometrically in the case of a planar graph. For the present discussion, it suffices to consider only undirected graphs, but the concept is applicable to digraphs as well. By definition, *a planar graph is one which can be embedded in a plane without any arcs crossing each other.* In particular, we are interested in a planar graph which cannot be disconnected by the removal of a single arc. An example of this is given in Figure 2.1.1.

In the (undirected) graph $\mathcal{G}$ depicted in Figure 2.1.1, we call a region bounded by two or more adjoining arcs a *face* of $\mathcal{G}$. Each arc of $\mathcal{G}$ then borders two different faces due to the connectedness assumption. A dual graph $\mathcal{G}^*$ can be constructed by regarding each face as a node of $\mathcal{G}^*$, and an arc exists between two nodes of $\mathcal{G}^*$ if and only if the two corresponding faces share a common border arc in $\mathcal{G}$. It is easy to see that $\mathcal{G}^*$ is also a planar graph which cannot be disconnected by the removal of an arc. Furthermore, the dual $\mathcal{G}^{**}$ of $\mathcal{G}^*$ is clearly $\mathcal{G}$. More importantly, the duality between a cycle (as a special path) and a cut is most apparent from the following observation:

- *There is a one-to-one correspondence between cycles in $\mathcal{G}$ and cuts in $\mathcal{G}^*$, and there is a one-to-one correspondence between cuts in $\mathcal{G}$ and cycles in $\mathcal{G}^*$. The arcs in a cycle of $\mathcal{G}$ constitute a cut in $\mathcal{G}^*$, and vice versa.*

This fact unfortunately does not have a great deal of practical applications, as this particular form of graph is rather restrictive. Nevertheless, once the duality between paths and cuts is firmly set in one's mind, it becomes very helpful in appreciating the working mechanism behind a large number of network flow and differential problems to be discussed in this chapter.

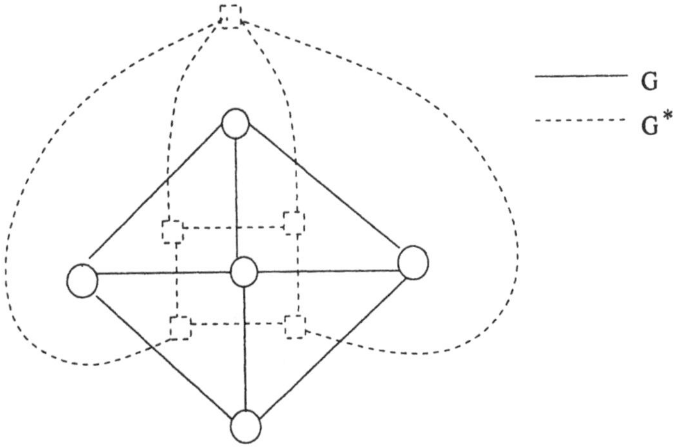

FIGURE 2.1.1 A planar graph and its dual

Minimum Spanning Tree

The duality of paths and cuts in a (undirected) graph can be further understood by studying the problem of finding an *Minimum Spanning Tree (MST)*. This is one of those combinatorial problems which admits surprisingly simple methods of solution. This of course is driven by the underlying duality between paths and cuts. The problem is stated as:

Definition 2.1.1 (Minimum Spanning Tree (MST) Problem) Given an *undirected* graph $\mathcal{G} = (\mathcal{N}, \mathcal{A})$, and a cost d_j associated with each arc $j \in \mathcal{A}$, find a spanning tree $\mathcal{T} = (\mathcal{N}, \mathcal{A}')$ that has a minimum total cost, i.e., find $\mathcal{A}' \subset \mathcal{A}$ such that $\sum_{j \in \mathcal{A}'} d_j$ is minimized.

Note that the problem is stated for an undirected graph. There is also a version of the MST problem for digraphs: *Find an MST on a digraph with all paths directed away from a given root node.* This, however, is a much more difficult problem than the undirected case. The undirected MST problem occurs in a large number of applications, some are obvious and others are not. These range from networking of computers or water pipelines, to cluster analysis, and minimizing the risk of espionage. For a comprehensive discussion of these and other applications, see [AMO1].

Recall, from Chapter One, that a spanning tree of a graph $\mathcal{G}$ is a spanning subgraph of $\mathcal{G}$ that is connected and acyclic. We refer to the arcs in the tree as the *tree arcs*, and those not in the tree as the *cotree arcs*. Recall also that:

• The spanning tree $\mathcal{T}$ contains a unique path joining any two nodes, and thus for a cotree arc (k,l), the arc (k,l) together with the unique path from k to l forms a unique cycle.

• Deleting a tree arc from a spanning tree disconnects the node set into two disjoint node subsets, the arcs from the underlying graph $\mathcal{G}$ which straddle the two node subsets constitute a cut. On the other hand, there exists a unique path joining the two end points of a cotree arc. From these observation, we have the following two optimality conditions that aptly manifest the duality between paths and cuts.

Theorem 2.1.1 (Cut optimality condition of an MST) A spanning tree $\mathcal{T}$ is an MST if and only if : for every tree arc $(i,j) \in \mathcal{T}, d_{ij} \leq d_{kl}$ for every (k,l) contained in the cut formed by deleting (i,j) from $\mathcal{T}$.

Proof: If $d_{ij} > d_{kl}$, then replacing (i,j) by (k,l) results in a tree with less cost. Conversely, if $\mathcal{T}$ satisfies the optimality condition, we show that it must be optimal. Let's say the MST is $\mathcal{T}_0$ and $\mathcal{T}_0 \neq \mathcal{T}$. Thus $\mathcal{T}$ contains an arc (i,j) which is not in $\mathcal{T}_0$. Deleting (i,j) from $\mathcal{T}$ gives rise to a cut $[\mathcal{S}, \mathcal{N} \setminus \mathcal{S}]$. Adding (i,j) to $\mathcal{T}_0$ creates a cycle which must contain another arc $(k,l) \in \mathcal{T}_0$ in $[\mathcal{S}, \mathcal{N} \setminus \mathcal{S}]$. Since $\mathcal{T}$ satisfies the optimality condition $d_{ij} \leq d_{kl}$. Now $\mathcal{T}_0$ is an MST, so $d_{ij} \geq d_{kl}$, otherwise replacing (k,l) by (i,j) will result in a tree with less cost. Hence $d_{ij} = d_{kl}$. Replacing $(i,j) \in \mathcal{T}$ by (k,l) (note: the cost is unchanged) results in another spanning tree with one more common arc with $\mathcal{T}_0$. By repeating this argument, we can transform $\mathcal{T}$ into $\mathcal{T}_0$. Since the cost does not change in each transformation, $\mathcal{T}$ is an MST. ∎

Theorem 2.1.2 (Path optimality condition of an MST) A spanning tree $\mathcal{T}$ is an MST if and only if : for every cotree arc $(k,l) \notin \mathcal{T}, d_{ij} \leq d_{kl}$ for every arc (i,j) contained in the path in $\mathcal{T}$ connecting node k and l.

Proof: We only need to show that the path optimality condition is equivalent to the cut optimality condition. Let (i,j) be a tree arc in $\mathcal{T}$, and let $\mathcal{S}$ and $\bar{\mathcal{S}}$ be the two disjoint node sets obtained from deleting (i,j) from $\mathcal{T}$. Consider any cotree arc $(k,l) \in [\mathcal{S}, \mathcal{N} \setminus \mathcal{S}]$. Since $\mathcal{T}$ contains a unique path joining k to l, and since (i,j) is the only arc of $\mathcal{T}$ in the cut, (i,j) must be in this path. The path optimality condition asserts that $d_{ij} \leq d_{kl}$. Since this is true for every cotree arc in $[\mathcal{S}, \mathcal{N} \setminus \mathcal{S}]$, $\mathcal{T}$ satisfies the cut optimality condition. Similarly it can be shown that the cut condition implies the path condition, and hence the two are equivalent. ∎

The path optimality condition and cut optimality condition lead to trivial algorithms for solving the MST problem. We either start with an arbitrary spanning tree, and test the path optimality condition against every cotree arc. If the condition is satisfied, it is an MST. Otherwise find the cotree arc that fails the optimality condition. The two end points of this cotree arc are joined by a unique path of tree arcs. In the spanning tree we replace the cotree arc by the tree arc in that path which has the lower cost. This yields a new spanning tree with a lower cost. Repeat this procedure until the optimality condition is satisfied. On the other hand, we can also test the cut optimality condition against every tree arc, and if it is not satisfied, we replace the tree arc in the cut by some other cotree arc in the cut with a lower cost. These conceptual algorithms, however, are not very efficient. We shall now discuss two other popular algorithms for solving MST problems. Both algorithms, like the great majority of algorithms for solving network flow problems, grow a tree or a forest until the tree/forest becomes a spanning tree, i.e., until all the nodes of the graph $\mathcal{G}$ are covered. The mechanism that drives the growth of the tree/forest is dependent on the optimality condition used. In the case of Kruskal's algorithm, we grow a forest; and in the case of Prim's algorithm, we grow a tree.

Algorithm 2.1.1 (Kruskal's algorithm)

Initialization: Rank all the arcs in ascending order of their costs, breaking ties arbitrarily. Form a subgraph $\mathcal{F} = (\mathcal{N}', \mathcal{A}')$ with the single cheapest arc, together with its end points.

Do until $\mathcal{F}$ becomes a spanning tree:

Check if adding the next cheapest arc a to $\mathcal{F}$ creates a cycle, if so, discard this arc. Otherwise add this arc together with the other end points to $\mathcal{F}$, i.e., $\mathcal{A}' \leftarrow \mathcal{A}' \cup \{a\}$, $\mathcal{N}' \leftarrow \mathcal{N}' \cup \{$ both end points of $a\}$.

End ∎

Theorem 2.1.3 Kruskal's algorithm yields the MST.

Proof: When a (cotree) arc is discarded, it is so because otherwise it would have formed a cycle with arcs already in $\mathcal{F}$. The cost of this (discarded) arc is higher than every other arc in that cycle, since we examine them in increasing order of cost. Therefore at termination, $\mathcal{F}$ satisfies the path optimality condition. ∎

While Kruskal's algorithm exploits the path optimality condition, the cut optimality condition leads to the Prim's algorithm.

Algorithm 2.1.2 (Prim's algorithm)

Initialization: Form a subgraph $\mathcal{T} = (\mathcal{N}', \mathcal{A}')$, with $\mathcal{N}'$ to contain a single arbitrary node, $\mathcal{A}' \leftarrow \emptyset$. Let the cut $[\mathcal{N}', \mathcal{N} \setminus \mathcal{N}']$.

Do until $\mathcal{T}$ becomes an MST:

Append the arc a with the least cost amongst all arcs in the cut $[\mathcal{N}', \mathcal{N} \setminus \mathcal{N}']$ to $\mathcal{T}$, i.e., $\mathcal{A}' \leftarrow \mathcal{A}' \cup \{a\}$, $\mathcal{N}' \leftarrow \mathcal{N}' \cup \{$the other end point of a not in $\mathcal{N}'\}$.

End ∎

Theorem 2.1.4. Prim's algorithm yields the MST.

Proof: It follows directly from construction that at each iteration, the appended arc has the least cost in the cut resulting from deleting that arc from the spanning tree, hence the cut optimality condition is also satisfied at termination. ∎

2.2 Duality in a Painted Network

The previous section illustrates the apparent duality between paths and cuts in undirected graphs. The path-cut duality for digraphs shall now be discussed in the context of the painted network theory, which is without exaggeration the heart of the great majority of network optimization methods. This elegant theory is first published by Minty [M3] in 1960, and the earliest theory was concerned with the case of finding a path between two nodes, and used only three colors. Rockafellar in [R3] extended this theory to the case of two disjoint subsets of nodes, and used four colors instead. This simple extension turns out to allow a much wider class of problems to be treated as special cases of the painted path/cut problem.

Definition 2.2.1 (Painting of a digraph) A directed graph $\mathcal{G} = (\mathcal{N}, \mathcal{A})$ is called a *painted network* if each arc in $\mathcal{A}$ is painted by one of the four colors: `Green`, `Red`, `White`, and `Black`.

The painting is supposed to reflect the physical significance of the arcs. One can think of the network as a road system with the arcs representing roads and the nodes representing origins and destinations of the commuters. If it helps to remember, a `Green` road allows traffic to flow in either direction, a `Red` road is out of bound in either direction, i.e., no go both ways, a `White` road is a one-way street in the direction of the arc, and a `Black` road is a one-way street in the reverse direction of the arc. Note that a painted network does not necessarily have arcs painted in all four colors.

Given a painted network $\mathcal{G} = (\mathcal{N}, \mathcal{A})$ and two disjoint node subsets $\mathcal{N}^+$ and $\mathcal{N}^-$, two fundamental problems, each being a dual to the other, are defined as follows:

Definition 2.2.2 (Painted Path Problem) Find a path $\mathcal{P}$ joining some node in $\mathcal{N}^+$ to some node in $\mathcal{N}^-$ (denoted by $\mathcal{P} : \mathcal{N}^+ \to \mathcal{N}^-$) such that every arc in $\mathcal{P}^+$ is painted `Green` or `White`, and every arc in $\mathcal{P}^-$ is painted `Green` or `Black`. Any path satisfying this property is said to be *compatible* with the painting.

Definition 2.2.3 (Painted Cut Problem) Find a cut $\mathcal{Q}$ separating all the nodes in $\mathcal{N}^+$ to all the nodes in $\mathcal{N}^-$ (denoted by $\mathcal{Q} : \mathcal{N}^+ \downarrow \mathcal{N}^-$) such that every arc in $\mathcal{Q}^+$ is painted `Red` or `Black`, and every arc in $\mathcal{Q}^-$ is painted `Red` or `White`. Any cut satisfying this property is said to be *compatible* with the painting.

It turns out that there is no loss in generality in restricting the painted path/cut to an elementary path/cut. The duality of paths and cuts is again highlighted in the following theorem, which is a sort of "theorem of alternative" pertaining to discrete systems.

Theorem 2.2.1 (Painted Network Theorem) Exactly one of the painted path problem or the painted cut problem has a solution.

Proof: This is done by construction using the painted network algorithm. The idea here is to grow a tree (in the case when $\mathcal{N}^+$ is a single node), or a forest (in the case when $\mathcal{N}^+$ has two or more nodes) that is rooted from node(s) in $\mathcal{N}^+$. At some intermediate step of the algorithm, we have a tree/forest $\mathcal{F} = (\mathcal{S}, \mathcal{A})$ where $\mathcal{N}^+ \subset \mathcal{S}$. The node set $\mathcal{S}$ yields a cut $\mathcal{Q} = [\mathcal{S}, \mathcal{N} \setminus \mathcal{S}]$. To grow the tree, we look for an arc in $\mathcal{Q}^+$ which are either `White` or `Green`, or an arc in $\mathcal{Q}^-$ which are either `Black` or `Green`. If no such arc can be found, $\mathcal{Q}$ is a painted cut. Otherwise append the arc and its other end point into the tree/forest $\mathcal{F}$. This is continued until $\mathcal{F}$ reaches some node in $\mathcal{N}^-$, whence the painted path is found. Since the network $\mathcal{G}$ only has a finite number of arcs, this procedure must terminate after at most $|\mathcal{N}|$ steps, and must exit with either a painted path or a painted cut. ∎

The algorithm for finding either a painted path or a painted cut is now formalized as follows:

Algorithm 2.2.1 (Painted Network Algorithm)

Initialization: Let $\mathcal{F} = (\mathcal{S}, \mathcal{B})$ where $\mathcal{S} \leftarrow \mathcal{N}^+, \mathcal{B} \leftarrow \emptyset$.

Do until $\mathcal{S} \cap \mathcal{N}^- \neq \emptyset$ (whence, the painted path is obtained from the unique path in $\mathcal{F}$ joining some node in $\mathcal{N}^+$ to some node in $\mathcal{N}^-$).

Let the cut $\mathcal{Q} = [\mathcal{S}, \mathcal{N} \setminus \mathcal{S}]$.

If all arcs in $\mathcal{Q}^+$ are painted `Black` or `Red`, or all arcs in $\mathcal{Q}^-$ are painted `White` or `Red`, the painted cut is then given by $\mathcal{Q}$.

Else there exists a `Green` or `White` arc $a \in Q^+$, or a `Green` or `Black` arc $a \in Q^-$. Then $\mathcal{B} \leftarrow \mathcal{B} \cup \{a\}$, $\mathcal{S} \leftarrow \mathcal{S} \cup \{$ the other end point of $a\}$.

End

End ∎

The painted path/cut problem has a surprisingly wide range of applications. Some of the more obvious ones are checking the connectedness of graphs and finding flow augmenting paths (to be discussed in the next section). By choosing the painting scheme according to the flow or differential characteristics, the painted network algorithm is often used as a subroutine to a large number of network algorithms, some of these will be discussed in the latter sections of this chapter.

2.3 Flow, Divergence, Potential and Differential

The concept of flow in networks is ubiquitous. The applications range from the obvious ones such as current in electrical circuits, water in pipelines, telephone messages in telecommunication networks, and data flow in local area (computer) networks, to the not-so-obvious ones such as balancing political representations in a council, international trading, stocks and options trading. For a comprehensive coverage of such applications, see [AMO1]. On the other hand, while the concept of potential is not as pervasive, its role in electrical circuits has been around for a very long time. As it will become obvious shortly,

- *flow and potential are dual to each other.*

As we are dealing with orientable quantities, henceforth all reference to a network will be that of a directed graph $\mathcal{G} = (\mathcal{N}, \mathcal{A})$. Let the node-arc incidence matrix of $\mathcal{G}$ be $\mathbf{E}$. We shall study the properties of flow and potentials, together with their derivatives, divergence and differential in a general setting in this section, before going into details in subsequent sections. In an abstract sense, we have:

Definition 2.3.1 A *flow* (vector) $\mathbf{x}$ is defined to be a function $\mathbf{x} : \mathcal{A} \to \mathbb{R}^{|\mathcal{A}|}$. The flow x_j in an arc j is the jth component of $\mathbf{x}$ and is sometimes known as the *flux* of arc j. Whether the term "flow" pertains to that of a network or that of a particular arc shall be determined by the context of the discussion. The derivative (not in a calculus sense!) of the flow $\mathbf{x}$ defined on $\mathcal{G}$ is the *divergence* vector $\mathbf{y}$ which is obtained from $\mathbf{x}$ by a linear mapping:

$$\mathbf{y} = \nabla \mathbf{x} = \mathbf{E}\mathbf{x}. \tag{2.3.1}$$

The component of $\mathbf{y}$ can be computed to be

$$y_i = \sum_{i \in \alpha(j)} x_j - \sum_{i \in \omega(j)} x_j.$$

The divergence $\mathbf{y}$ of $\mathcal{G}$ is therefore a (composite) mapping from the set of nodes $\mathcal{N}$ to $\mathbb{R}^{|\mathcal{N}|}$. Physically, the divergence of a node y_i is the net outflow from the node

i. If $y_i > 0$, the node i is said to be a *source* or *supply node*; if $y_i < 0$, the node i is said to be a *sink* or *demand node*; and if $y_i = 0$, the node i is said to be a *transshipment* node where the flow is *conserved*.

The following result is intuitively obvious. Basically it says, supply equals demand.

Theorem 2.3.1 (Total Divergence Principle)

$$\mathbf{1}^{\top}\mathbf{y} = \sum_{i \in \mathcal{N}} y_i = 0.$$

Proof: Follows from Lemma 1.4.1 and the definition of $\mathbf{y}$. ∎

A special kind of flow plays an important role in the duality of flow and potential.

Definition 2.3.2 A *circulation* $\mathbf{x}$ is a special flow on $\mathcal{G}$ for which the corresponding divergence $\mathbf{y} = \mathbf{E}\mathbf{x} = \mathbf{0}$. The set of all circulations is clearly the null-space of the linear mapping defined by the incidence matrix $\mathbf{E}$.

Remark 2.3.1 Just as it is more appropriate to talk about the *duality of cycle and cut* (instead of path and cut), it will become apparent later that it is often more appropriate to talk about the *duality of circulation and potential* (instead of flow and potential). This can be done without loss of generality, because for a given flow in a given network, there exists a corresponding augmented network where the flow is a circulation. This is easily achieved by creating an additional super node, which is joined to all the source nodes and all the sink nodes by additional arcs. By making the flow on the arcs joining the super node to the source nodes as the net outflow of the source nodes, and the flow on the arcs joining the sink nodes to the super node as the net inflow to the sink nodes, the resulting flow in the augmented network results in a circulation.

We shall now define the concept of *potential* formally.

Definition 2.3.3 A *potential* (vector) $\mathbf{u}$ is defined to be a function: $\mathbf{u} : \mathcal{N} \to \mathbb{R}^{|\mathcal{N}|}$. The *potential* u_i of a node i is the ith component of $\mathbf{u}$. The derivative (not in a calculus sense!) of the potential $\mathbf{u}$ defined on $\mathcal{G}$ is the *differential* $\mathbf{v}$ (and often known also as a *tension*) which is obtained from $\mathbf{u}$ by the linear mapping:

$$\mathbf{v} = \Delta\mathbf{u} = -\mathbf{E}^{\top}\mathbf{u}. \tag{2.3.2}$$

The differential $\mathbf{v}$ of $\mathcal{G}$ is therefore a (composite) mapping from the set of arcs $\mathcal{A}$ to $\mathbb{R}^{|\mathcal{A}|}$. Physically, the differential or tension v_j of an arc j is the potential difference between its sink node and its source node, i.e., $v_j = u_{\omega(j)} - u_{\alpha(j)}$.

The set of all differentials, called the *differential space* $\mathcal{D}$ is also a linear subspace of $\mathbb{R}^{|\mathcal{A}|}$ since it is spanned by the rows of the incidence matrix, which contains the null vector (as the sum of all rows). The following relationship between flow, divergence, potential and differential follows from the definitions of divergence and differential:

$$\mathbf{v}^\top \mathbf{x} = -\mathbf{u}^\top \mathbf{y} = -\mathbf{u}^\top \mathbf{E}\mathbf{x}. \tag{2.3.3}$$

Consequently, if $\mathbf{x}$ is a circulation ($\mathbf{y} = \mathbf{0}$), then

$$\mathbf{v}^\top \mathbf{x} = 0. \tag{2.3.4}$$

If it is still not obvious by now, flow/divergence and potential/differential form a pair of *dual linear systems*, as:

• The linear mapping $\mathbf{u} \mapsto -\mathbf{E}^\top \mathbf{u}$ is the negative adjoint of the linear mapping $\mathbf{x} \mapsto \mathbf{E}\mathbf{x}$.

• The circulation space and the differential space are orthogonally complementary to each other, i.e.,

$$\mathcal{D} = \mathcal{C}^\perp = \{\mathbf{v} \in \mathbb{R}^{|\mathcal{A}|} \mid \mathbf{v}^\top \mathbf{x} = 0 \ \forall \mathbf{x} \in \mathcal{C}\},$$
$$\mathcal{C} = \mathcal{D}^\perp = \{\mathbf{x} \in \mathbb{R}^{|\mathcal{A}|} \mid \mathbf{v}^\top \mathbf{x} = 0 \ \forall \mathbf{v} \in \mathcal{D}\}.$$

Tucker Representations of the circulation and the differential spaces.

As defined before, a circulation $\mathbf{x}$ for a network $\mathcal{G} = (\mathcal{N}, \mathcal{A})$ is such that

$$\mathbf{E}\mathbf{x} = \sum_{j \in \mathcal{A}} x_j E^j = \mathbf{0}, \tag{2.3.5}$$

where E^j is the jth column of $\mathbf{E}$. To simplify the notation slightly, we assume that $\mathcal{G}$ is connected, although this assumption can be removed easily, in which case we need to talk about forests instead. Furthermore, we assume that $\mathcal{G}$ is not a tree. as otherwise (2.3.5) implies that $\mathbf{x} = \mathbf{0}$. From Corollary 1.4.6, $\mathbf{E}$ is rank one deficient, and so the set of vectors $\{E^j, j \in \mathcal{A}\}$ must be linearly dependent, as asserted in (2.3.5). Consequently, some of the flow variables can be expressed uniquely as a function of the others. (Some familiarity with linear programming, in particular, (3.1.3) of Chapter Three will facilitate the understanding of this assertion.) In particular, there is some (non-unique) subset $\mathcal{F} \subset \mathcal{G}$, such that $\forall j \in \mathcal{F}$, the flow in arc j is uniquely determined by the flow in arcs in $\mathcal{A} \setminus \mathcal{F}$:

$$x_j = \sum_{k \in \mathcal{A} \setminus \mathcal{F}} a_{ij} x_k, \qquad \forall j \in \mathcal{F}. \tag{2.3.6}$$

The latter representation is called a *Tucker representation* [T1] of the circulation space $\mathcal{C}$. If flux values for the arcs in $\mathcal{A} \setminus \mathcal{F}$ are chosen arbitrarily, then the flux values of arcs in $\mathcal{F}$ are uniquely determined by (2.3.6) in order for $\mathbf{x}$ to be a circulation. Recall from Theorem 1.4.5 that the columns of $\mathbf{E}$ corresponding to a spanning tree of $\mathcal{G}$ is a basis for the remaining columns of $\mathbf{E}$. The following theorem is essentially a restatement of this fact.

Theorem 2.3.2 An arc set $\mathcal{F} \subset \mathcal{A}$ corresponds to a Tucker representation of the circulation space $\mathcal{C}$ if and only if $\mathcal{F}$ is a spanning tree for $\mathcal{G}$.

Proof: (2.3.5) can be rewritten as

$$\sum_{j \in \mathcal{F}} x_j E^j = - \sum_{k \in \mathcal{A} \setminus \mathcal{F}} x_k E^k. \tag{2.3.7}$$

The set $\mathcal{F}$ corresponds to a Tucker representation if and only if given any arbitrary values of x_k, $k \in \mathcal{A} \setminus \mathcal{F}$, the values of $x_j, j \in \mathcal{F}$ are uniquely determined from (2.3.7). This is equivalent to saying that the set of vectors $\{E^j, j \in \mathcal{F}\}$ forms a basis for the column space of $\mathbf{E}$, i.e., $\mathcal{F}$ is a spanning tree. ■

From (2.3.4) and (2.3.6), we have, if $\mathbf{x} \in \mathcal{C}$ and $\mathbf{v} \in \mathcal{D}$,

$$\begin{aligned} 0 &= \mathbf{x}^{\mathsf{T}} \mathbf{v} \\ &= \sum_{j \in \mathcal{F}} x_j v_j + \sum_{k \in \mathcal{A} \setminus \mathcal{F}} x_k v_k \\ &= \sum_{k \in \mathcal{A} \setminus \mathcal{F}} \left[\sum_{j \in \mathcal{F}} v_j a_{jk} + v_k \right] x_k. \end{aligned}$$

Since this holds for arbitrary x_k, $k \in \mathcal{A} \setminus \mathcal{F}$, we have the following Tucker representation of the differential space $\mathcal{D}$:

$$v_k = - \sum_{k \in \mathcal{F}} v_j a_{jk} \quad \forall k \in \mathcal{A} \setminus \mathcal{F}. \tag{2.3.8}$$

It is therefore clear from (2.3.6) and (2.3.8) that:

- *The Tucker representations of $\mathcal{C}$ and $\mathcal{D}$ occur in dual pairs in a one-to-one correspondence with a spanning tree of $\mathcal{G}$.*

To wrap up this section, we present two fundamental problems of network optimization. Each of these includes a number of other network optimization problems as special cases, although the solution methods for these special cases are often tailored to exploit the structure of the special case involved. Detailed discussions of these two general problems will only take place in Section 2.8. Their statement here is intended to facilitate easy reference from subsequent sections. Under suitable convexity and continuity assumptions about the underlying cost functions and capacity/span intervals, these two problems can be shown to be dual to each other, the detail of which will be discussed in Section 2.9.

Given a digraph $\mathcal{G} = (\mathcal{N}, \mathcal{A})$ with an incidence matrix $\mathbf{E}$, we associate:

- with each arc $j \in \mathcal{A}$, two intervals $C_j \subset \mathbb{R}$ and $D_j \subset \mathbb{R}$ and two functions $f_j : C_j \to \mathbb{R}$ and $g_j : D_j \to \mathbb{R}$;

• with each node $i \in \mathcal{N}$, two intervals $C_i \subset \mathbb{R}$ and $D_i \subset \mathbb{R}$ and two functions $f_i : C_i \to \mathbb{R}$ and $g_i : D_i \to \mathbb{R}$.

Definition 2.3.4 (Optimal Flow Problem)

$$\begin{aligned} \min_{\mathbf{x}\in\mathbf{R}^{|\mathcal{A}|},\mathbf{y}\in\mathbf{R}^{|\mathcal{N}|}} \quad & \sum_{j\in\mathcal{A}} f_j(x_j) + \sum_{i\in\mathcal{N}} f_i(y_i) \\ \text{subject to} \quad & \mathbf{y} = \nabla\mathbf{x} = \mathbf{E}\mathbf{x}, \\ & x_j \in C_j \ \forall j \in \mathcal{A}, \\ & y_i \in C_i \ \forall i \in \mathcal{N}. \end{aligned}$$

Definition 2.3.5 (Optimal Potential Problem)

$$\begin{aligned} \max_{\mathbf{v}\in\mathbf{R}^{|\mathcal{A}|},\mathbf{u}\in\mathbf{R}^{|\mathcal{N}|}} \quad & -\sum_{j\in\mathcal{A}} g_j(v_j) - \sum_{i\in\mathcal{N}} g_i(u_i) \\ \text{subject to} \quad & \mathbf{v} = \Delta\mathbf{u} = -\mathbf{E}^\top\mathbf{u}, \\ & v_j \in D_j \ \forall j \in \mathcal{A}, \\ & u_i \in D_i \ \forall i \in \mathcal{N}. \end{aligned}$$

2.4 Duality of Max Flow Min Cut

The topic of this section is arguably the first and best-known network optimization model. The max flow problem was first posed and solved by Ford and Fulkerson [FF1] in 1956. This paved the way for many more elegant and practical network optimization models.

The max flow problem is easy to explain. Given a network $\mathcal{G} = (\mathcal{N}, \mathcal{A})$ with incidence matrix $\mathbf{E}$ and two special nodes, a source s and a sink t, what is the maximum amount of flow, or flux, that one can push from s to t? Of course this question will only make sense if the flows on arcs are bounded by some capacities. This has been briefly mentioned in the previous section, and will be formalized in this section. But before that, it will help to gain some insight from the following.

Remark 2.4.1 (An intuitive idea) Before we discuss the problem in its entire generality and in a rigorous manner, it is instructive to understand the underlying idea in an intuitive, albeit handwaving, way. Think of two nodes, a source and a sink. The flow of goods from the source to the sink is only through a single path consisting of several arcs arranged in series, with each arc having a capacity. The amount of flow through this path is clearly restricted by the capacities of the arcs. In particular, the maximum amount of flow through this path must be constrained by the capacity of the arc with the least capacity. This is somewhat analogous to the old saying that a chain is only as strong as its weakest link. The removal of the arc with the least capacity will disconnect the (single path) network into two components, and these components are separated by a cut which consists of a single arc. The capacity of flow through this cut is given by the capacity of the removed arc, and this has the smallest capacity of any cut separating the source and sink node. Thus *the max flow from the source to the sink is equal to the capacity of the cut with the smallest capacity and which separates the source from the sink.*

With the above idea in mind, it remains to generalize this simple idea into a formal mathematical model for the rest of this section.

Definition 2.4.1 The (flow) *capacity* for a network $\mathcal{G} = (\mathcal{N}, \mathcal{A})$ is a set-valued function $\mathbf{C} : \mathcal{A} \to 2^{\bar{\mathbf{R}}^{|\mathcal{A}|}}$. For an arc j, its capacity is a nonempty interval $C_j = [c_j^-, c_j^+]$, where $\forall j, c_j^- \leq c_j^+, c_j^+ > -\infty, c_j^- < +\infty$. Note that is possible for $c_j^+ = +\infty$ or $c_j^- = -\infty$. A flow $\mathbf{x}$ on $\mathcal{G}$ is *feasible with respect to capacities* if $\forall j, x_j \in C_j$.

Given a node subset $\mathcal{S} \subset \mathcal{N}$, the corresponding cut $\mathcal{Q} = [\mathcal{S}, \mathcal{N} \setminus \mathcal{S}]$ also has a capacity, which represents the lower and upper bounds of flux across the cut.

Definition 2.4.2 The capacity interval associated with the cut $\mathcal{Q}$ is given by $C_\mathcal{Q} = [c_\mathcal{Q}^-, c_\mathcal{Q}^+]$ where the *upper capacity* $c_\mathcal{Q}^+$ of $\mathcal{Q}$, and the *lower capacity* $c_\mathcal{Q}^-$ of $\mathcal{Q}$, are given by:

$$c_\mathcal{Q}^+ = \sum_{j \in \mathcal{Q}^+} c_j^+ - \sum_{j \in \mathcal{Q}^-} c_j^-,$$

$$c_\mathcal{Q}^- = \sum_{j \in \mathcal{Q}^+} c_j^- - \sum_{j \in \mathcal{Q}^-} c_j^+.$$

Definition 2.4.3 Given a flow $\mathbf{x}$ on $\mathcal{G}$ and a node set $\mathcal{S} \subseteq \mathcal{N}$, the divergence of $\mathbf{x}$ from $\mathcal{S}$ is defined by $y_\mathcal{S} = \sum_{i \in \mathcal{S}} y_i$ where $\mathbf{y} = \nabla \mathbf{x} = \mathbf{E}\mathbf{x}$.

Theorem 2.4.1 (Divergence Principle) The divergence of $\mathbf{x}$ from $\mathcal{S}$ is equal to the flux of $\mathbf{x}$ across the cut $\mathcal{Q} = [\mathcal{S}, \mathcal{N} \setminus \mathcal{S}]$, where the latter is defined by

$$\text{flux of } \mathbf{x} \text{ across } \mathcal{Q} = \mathbf{e}_\mathcal{Q}^T \mathbf{x} = \sum_{j \in \mathcal{Q}^+} x_j - \sum_{j \in \mathcal{Q}^-} x_j,$$

where $\mathbf{e}_\mathcal{Q}$ is the cut-incidence vector of $\mathcal{Q}$. (Note that the total divergence principle of Theorem 2.3.1 is a special case of the divergence principle where $\mathcal{S} = \mathcal{N}$ and $\mathcal{Q}$ is an empty cut.)

Proof: Both quantities are equal to $\sum_{i \in \mathcal{S}} \sum_{j \in \mathcal{A}} [\mathbf{E}]_{ij} x_j$. ∎

Intuitively, the flux across a cut $\mathcal{Q} = [\mathcal{S}, \mathcal{N} \setminus \mathcal{S}]$ must be bounded above and below by the capacity of the cut. This is formalized as follows:

Theorem 2.4.2 If x is feasible, then $c_{\mathcal{Q}}^- \leq \mathbf{e}_{\mathcal{Q}}^T \mathbf{x} \leq c_{\mathcal{Q}}^+$.

Proof: If $j \in \mathcal{Q}^+$, then $c_j^- \leq x_j \leq c_j^+$; otherwise if $j \in \mathcal{Q}^-$, then $-c_j^+ \leq -x_j \leq -c_j^-$. Summing over all $j \in \mathcal{Q}$ yields the result. ∎

The original max flow min cut problem was formulated for a single source node and a single sink node, and where all the lower capacities of flow are zero. A much more general version of the max-flow problem due to Rockafellar [R3] requires only a marginal increase in complexity, yet turns out to be very useful later when applied to other related problems.

Definition 2.4.4 (Max Flow Problem) Given two disjoint node sets $\mathcal{N}^+$ and $\mathcal{N}^-$ of $\mathcal{G}$, maximize the flux from $\mathcal{N}^+$ to $\mathcal{N}^-$ over all feasible flows (i.e., $x_j \in C_j\ \forall j$) such that $y_i = 0\ \forall i \notin \mathcal{N}^+ \cup \mathcal{N}^-$. Alternatively, this is equivalent to maximizing $y_{\mathcal{N}^+} = -y_{\mathcal{N}^-}$.

The dual of the max flow problem is the min cut problem:

Definition 2.4.5 (Min Cut Problem) Given two disjoint node sets $\mathcal{N}^+$ and $\mathcal{N}^-$ of $\mathcal{G}$, minimize $c_{\mathcal{Q}}^+$ over all cuts $\mathcal{Q}$ separating $\mathcal{N}^+$ from $\mathcal{N}^-$.

Remark 2.4.2 Note that the max flow problem is a special case of the optimal flow problem defined in Section 2.3, and is thus a continuous optimization problem. The cost function to be minimized in this case is just $y_{\mathcal{N}^-} = \sum_{i \in \mathcal{N}^-} y_i$ and there is no upper or lower bound on the divergence. On the other hand, the min cut problem is purely combinatorial in nature.

Theorem 2.4.3 (Weak Duality) If $\mathbf{x}$ is a feasible flow satisfying the assumption of Definition 2.4.4, and $\mathcal{Q}$ is any cut separating $\mathcal{N}^+$ from $\mathcal{N}^-$, then

$$\text{flux of } \mathbf{x} \text{ from } \mathcal{N}^+ \text{ to } \mathcal{N}^- \leq c_{\mathcal{Q}}^+.$$

Proof: If $\mathcal{Q} = [\mathcal{S}, \mathcal{N} \setminus \mathcal{S}]$ separates $\mathcal{N}^+$ from $\mathcal{N}^-$, then $\mathcal{N}^+ \subset \mathcal{S}$ and $\mathcal{S} \cap \mathcal{N}^- = \emptyset$. Since $y_i = 0\ \forall i \in \mathcal{S} \setminus \mathcal{N}^+$, therefore the flux of $\mathbf{x}$ from $\mathcal{N}^+$ to $\mathcal{N}^-$, $y_{\mathcal{N}^+} = y_{\mathcal{S}}$, is equal to the flux across the cut $\mathcal{Q}$. The proof thus follows from the second inequality of Theorem 2.4.2. ∎

Before presenting the strong duality result, we shall discuss a special case of the painted path problem, the solution of which will be used as a subroutine for solving the max flow problem.

Definition 2.4.6 (Flow Augmenting Path) Given a flow $\mathbf{x}$ that is feasible with respect to the capacity $\mathbf{C}$, and two disjoint node sets $\mathcal{N}^+$ and $\mathcal{N}^-$ of a network $\mathcal{G}$, a *flow augmenting path* $\mathcal{P} : \mathcal{N}^+ \to \mathcal{N}^-$ is such that $x_j < c_j^+\ j \in \mathcal{P}^+$, and $x_j > c_j^-\ j \in \mathcal{P}^-$.

Clearly, if $\mathcal{P}$ is a flow augmenting path, then the flux from $\mathcal{N}^+$ to $\mathcal{N}^-$ can be increased further by pushing more flow along $\mathcal{P}$. The maximal allowable increase in flow along $\mathcal{P}$ is given by

$$\alpha = \min \begin{cases} c_j^+ - x_j, & j \in \mathcal{P}^+, \\ x_j - c_j^-, & j \in \mathcal{P}^-. \end{cases} \tag{2.4.1}$$

Algorithm 2.4.1 (Finding a flow augmenting path)

Initialization: Given a flow $\mathbf{x}$ that is feasible with respect to the (flow) capacity $\mathbf{C}$, paint the arcs according to the following:

$$\begin{aligned} \texttt{Green} \text{ if } c_j^- &< x_j < c_j^+ \\ \texttt{Red} \text{ if } c_j^- &= x_j = c_j^+ \\ \texttt{White} \text{ if } c_j^- &= x_j < c_j^+ \\ \texttt{Black} \text{ if } c_j^- &< x_j = c_j^+ \end{aligned}$$

Do: Apply the Painted Network Algorithm 2.2.1. If a painted cut $\mathcal{Q}$ is found, then no flow augmenting path exists. Otherwise return a flow augmenting path P as the painted path.
End ■

The basic mechanics of solving the max flow problem is simple: one keeps finding flow augmenting paths until it is no longer possible.

Algorithm 2.4.2 (Max Flow (Ford and Fulkerson) Algorithm)

Initialization: Find an initial feasible flow.

Do until no flow augmenting path can be found:

Apply Algorithm 2.4.1 to find a flow augmenting path $\mathcal{P}$. Change the flow by: $\mathbf{x} \leftarrow \mathbf{x} + \alpha \mathbf{e}_P$ where α is computed from (2.4.1).

End ■

In some ill-posed optimization problem, the max flow algorithm may continue to run forever, even though the flux from $\mathcal{N}^+$ to $\mathcal{N}^-$ increases at every iteration by ever diminishing amounts. An example of such a situation is given by [R3]. It turns out that a fairly reasonable assumption which is easily satisfied in practice will guarantee the max flow algorithm to terminate in a finite number of iterations.

Definition 2.4.7 A set of real numbers are said to be *commensurable* if they are all multiples of a certain quantum δ.

Theorem 2.4.4 Assume that the arc flow capacities c_j^+, c_j^- and the initial arc flow x_j, $\forall j \in \mathcal{A}$ are commensurable (with respect to a quantum δ), and that there exists a cut separating $\mathcal{N}^+$ and $\mathcal{N}^-$ with a finite upper capacity. Then the max flow algorithm will result in an integral solution (i.e., a solution that is a multiple of the quantum δ) in a finite number of iterations.

Proof: If the assumptions of the theorem are satisfied, then the flow increment α computed in the flow augmenting path algorithm 2.4.1 must also be a multiple of some quantum δ, and therefore the increment is at least δ. Since there is at least one cut separating $\mathcal{N}^+$ and $\mathcal{N}^-$, with finite capacity, the flow increment cannot take place forever, hence the algorithm must eventually terminate after a finite number of steps. The final flow is a sum of a series of quantities integral with respect to δ, and is therefore integral. ∎

Not only does the max flow Algorithm 2.4.2 find the max flow, it also finds the min cut as a by-product. This is formalized in the following celebrated theorem due to Ford and Fulkerson [FF1].

Theorem 2.4.5 (Strong Duality - Max Flow Min Cut Theorem) Assume that the commensurability assumption of Theorem 2.4.4 holds, and that there is at least one flow that satisfies the constraints of the max flow problem, and that there exists a cut separating $\mathcal{N}^+$ and $\mathcal{N}^-$ with a finite upper capacity. Then

$$\text{max in the max flow problem} = \text{capacity of the min cut.}$$

Proof: The commensurability assumption of Theorem 2.4.4 guarantees that Algorithm 2.4.2 terminates in a finite number of steps, where the application of the painted network algorithm results in a painted cut $\mathcal{Q}$ and a flow $\mathbf{x}$. From the definition of the painting scheme in Algorithm 2.4.1, we have

$$\begin{aligned} x_j &= c_j^+ \ \forall j \in \mathcal{Q}^+, \\ x_j &= c_j^- \ \forall j \in \mathcal{Q}^-. \end{aligned}$$

The flux across the cut, which is also the flux from $\mathcal{N}^+$ to $\mathcal{N}^-$ (see the proof of Theorem 2.4.3) is given by

$$\begin{aligned} y_{\mathcal{N}^+} &= \mathbf{e}_{\mathcal{Q}}^\top \mathbf{x} \\ &= \sum_{j \in \mathcal{Q}^+} x_j - \sum_{j \in \mathcal{Q}^-} x_j \\ &= \sum_{j \in \mathcal{Q}^+} c_j^+ - \sum_{j \in \mathcal{Q}^-} c_j^- = c_{\mathcal{Q}}^+. \end{aligned}$$

Since there exists at least a cut with a finite upper capacity by assumption of the theorem, $c_{\mathcal{Q}}^+ < \infty$. By weak duality (Theorem 2.4.3), and the argument put forth in Theorem 1.6.1, the flow $\mathbf{x}$ must therefore be the max flow, and the cut $\mathcal{Q}$ must be the min cut. ∎

2.5 Duality of Max Match Min Block

A special case of the max flow min cut duality that is made possible by the integral characteristics of the max flow solution is found in two other optimization problems, which are perhaps less well known. Both of these problems are combinatorial in nature, yet admits surprisingly efficient methods of solution. Needless to say, the max match problem and the min blocking problem are dual to each other. This duality has a number of important applications, the most prominent of which is found in the Hungarian algorithm for solving the assignment problem. This problem is commonly found in most Operations Research textbooks, although very few of them actually relate the algorithm to the max match min block duality.

We begin by defining the concept of *match and block*. Given two nonempty finite sets $\mathcal{K}$ and $\mathcal{L}$ and a subset $\mathcal{H} \subset \mathcal{K} \times \mathcal{L}$. An element $(k, l) \in \mathcal{H}$ is said to be *compatible*, while those in $\mathcal{K} \times \mathcal{L}$ but not in $\mathcal{H}$ are said to be *incompatible*.

Definition 2.5.1 A *match* $\mathcal{M}$ is a set of compatible pairs

$$\{(k_1, \ell_1), (k_1, \ell_1), \cdots, (k_{|\mathcal{M}|}, \ell_{|\mathcal{M}|})\} \subset \mathcal{H}$$

such that $k_1, k_2, \cdots, k_{|\mathcal{M}|}$ are all distinct, and $\ell_1, \ell_2, \cdots, \ell_{|\mathcal{M}|}$ are all distinct. Such a set can be thought of as a one-to-one correspondence between a subset of $\mathcal{K}$ and a subset of $\mathcal{L}$.

Definition 2.5.2 A *block* $\mathcal{B}$ is a subset of $\mathcal{K} \cup \mathcal{L}$ such that for every $(k, l) \in \mathcal{H}$, either $k \in \mathcal{B}$, or $\ell \in \mathcal{B}$, or both.

The two combinatorial problems that are dual to each other are stated as follows:

Definition 2.5.3 (Max Matching Problem) Find a match $\mathcal{M}$ that maximizes $|\mathcal{M}|$.

Definition 2.5.4 (Min Blocking Problem) Find a block $\mathcal{B}$ that minimizes $|\mathcal{B}|$.

Before we state and prove the main duality theorem, we shall describe how the max match problem can be reduced to a special case of the max flow problem, and hence can be solved efficiently as such. First, a bipartite graph $\mathcal{G} = (\mathcal{N}, \mathcal{A})$ can be formed with two node sets, each corresponds to $\mathcal{K}$ and $\mathcal{L}$, see Figure 2.5.1. An arc joins some node $k \in \mathcal{K}$ to some other node $\ell \in \mathcal{L}$ if and only if (k, l) is an element of $\mathcal{H}$. Let the flow capacity of each of these arcs be $[0, \infty]$. Note that the transformation to max flow problem will work equally well if we set the capacity of arc (k, ℓ) to be $[0, 1]$, or in fact $[0, \zeta]$ where ζ is any integer greater than 1. But to facilitate an easy proof of the main duality theorem, infinite upper capacity is chosen. In addition two special nodes are added to the bipartite graph. A source node s is joined by $|\mathcal{K}|$ arcs to each of the nodes in $\mathcal{K}$, and $|\mathcal{L}|$ arcs join the nodes in $\mathcal{L}$ to a sink node t. The capacity of each of these new arcs is $[0, 1]$.

It is not difficult to see that the solution to the max flow problem of pushing as much flow as possible from s to t in this case corresponds to a max match. If this is not already obvious, some explanation is given as follows. From Theorem 2.4.5, the max flow solution must be integral. Thus if an arc (k, ℓ) has any flow at all, it must have exactly one unit of flow, since k only allows a maximum of one unit of flow into it, and ℓ only allows a maximum of one unit of flow out of it. (k, ℓ) thus corresponds to a match, and the max flow corresponds to the maximum number of match.

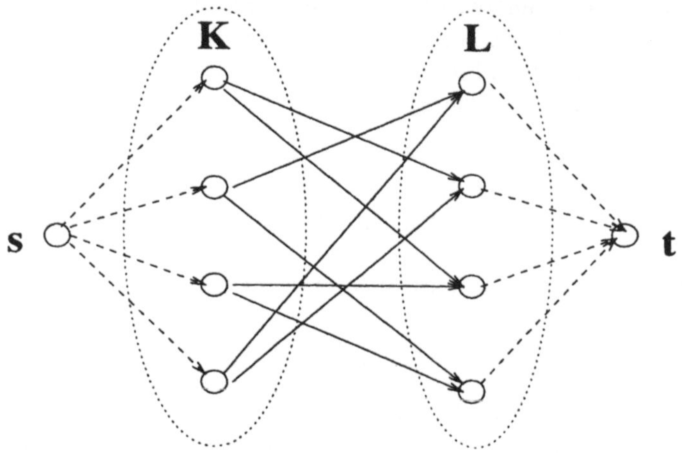

FIGURE 2.5.1 A graph representation of the matching problem

By the max flow min cut duality, the max flow also gives rise to a min cut. If we can establish that a cut corresponds to a block, then the duality between max matching and min blocking follows. This is the crux of the following main duality theorem. The proof here is due to Rockafellar [R3] and is much more elegant than the conventional one (see, e.g., [K3]) based on an elaborate matrix analysis.

Theorem 2.5.1 (Strong duality - König-Egerváry Theorem)

max $|M|$ in the matching problem = min $|B|$ in the blocking problem.

Proof: Since there is at least one cut, for example, $Q = [\{s\}, \mathcal{N} \setminus \{s\}]$ where $c_Q^+ = |\mathcal{K}| < \infty$, the min cut must have a finite capacity, so we need only to examine cuts with finite upper capacities. Such a cut $\mathcal{Q}$ cannot contain any arc in $\mathcal{Q}^+$ joining a node in $\mathcal{K}$ and a node in $\mathcal{L}$ as it will incur an infinite capacity. Thus it must be in the form as depicted in Figure 2.5.2. If this cut is given by $Q = [\mathcal{S}, \mathcal{N} \setminus \mathcal{S}]$, then it is not difficult to observe that

$$c_Q^+ = |\mathcal{K} \cap \bar{\mathcal{S}}| + |\mathcal{L} \cap \mathcal{S}|, \tag{2.5.1}$$

where we denote $\bar{\mathcal{S}} = \mathcal{N} \setminus \mathcal{S}$ for convenience.

If this is not obvious, note that

$$\mathcal{H} \cap \mathcal{Q}^+ = \emptyset, \tag{2.5.2}$$

otherwise the capacity of $\mathcal{Q}$ is ∞. Note also that

$$\mathcal{H} \cap \mathcal{Q}^- \neq \emptyset.$$

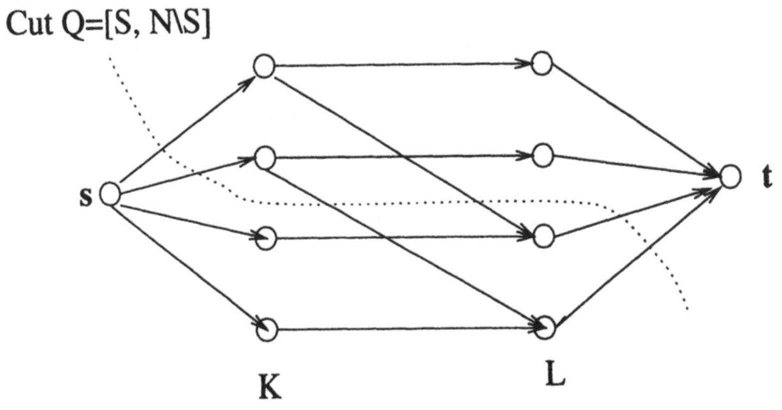

FIGURE 2.5.2 Proof of König-Egerváry Theorem

Consider four sets of nodes: $\mathcal{K} \cap \mathcal{S}, \mathcal{K} \cap \bar{\mathcal{S}}, \mathcal{L} \cap \mathcal{S}, \mathcal{L} \cap \bar{\mathcal{S}}$. All arcs in $\mathcal{H} \subseteq \mathcal{K} \times \mathcal{L}$ belong to one of the four groups:

$$\begin{aligned}
&(\mathcal{K} \cap \mathcal{S}) \times (\mathcal{L} \cap \mathcal{S}) \text{ which is not part of } \mathcal{Q}, \\
&(\mathcal{K} \cap \mathcal{S}) \times (\mathcal{L} \cap \bar{\mathcal{S}}) \subseteq \mathcal{Q}^+, \\
&(\mathcal{K} \cap \bar{\mathcal{S}}) \times (\mathcal{L} \cap \mathcal{S}) \subseteq \mathcal{Q}^-, \\
&(\mathcal{K} \cap \bar{\mathcal{S}}) \times (\mathcal{L} \cap \bar{\mathcal{S}}) \text{ which is not part of } \mathcal{Q}.
\end{aligned}$$

Since $\mathcal{Q}$ is the disjoint union of 4 sets:

$$\mathcal{Q}^- = (\mathcal{K} \cap \bar{\mathcal{S}}) \times (\mathcal{L} \cap \mathcal{S}), \tag{2.5.3}$$

$$\mathcal{Q}^+ = (\{s\} \times (\mathcal{K} \cap \bar{\mathcal{S}})) \cup ((\mathcal{L} \cap \mathcal{S}) \times \{t\}) \cup ((\mathcal{K} \cap \mathcal{S}) \times (\mathcal{L} \cap \bar{\mathcal{S}})). \tag{2.5.4}$$

The last set on the right hand side of (2.5.4) is empty by virtue of (2.5.2), and the upper capacity of each arc in the first two sets of $\mathcal{Q}^+$ is exactly one, and the lower capacity of each arc in $\mathcal{Q}^-$ is 0. It follows therefore that (2.5.1) must hold.

This cut thus corresponds one-to-one with the block $\mathcal{B} = (\mathcal{K} \cap \bar{\mathcal{S}}) \cup (\mathcal{L} \cap \mathcal{S})$. In particular, the min cut also corresponds to a min block. ■

Remark 2.5.1 The matching and blocking problem can be represented by a $|\mathcal{K}| \times |\mathcal{L}|$ matrix **A**, called the *compatibility matrix*, where $[\mathbf{A}]_{k\ell} = 0$ if $(k, \ell) \in \mathcal{H}$, and 1, or in fact any other non-zero number, otherwise. A match in this case corresponds to a special set of so-called *noncollinear zeros*, where each row or column of **A** contains exactly one of these special zeros. A block in this case corresponds to a set of horizontal or vertical lines that covers all the zeros of **A**. The duality theorem of

König-Egerváry can be interpreted in an easy-to-understand manner: *the maximum number of noncollinear zeros equals the minimum number of (horizontal or vertical) lines that cover all the zeros.* ∎

A direct application of Remark 2.5.1 is found in the (optimal) assignment problem. An assignment problem is a special case of the matching problem that concerns a bipartite graph $\mathcal{G} = (\mathcal{K} \cup \mathcal{L}, \mathcal{K} \times \mathcal{L})$ where $\mathcal{K} \cap \mathcal{L} = \emptyset$ and $|\mathcal{K}| = |\mathcal{L}| = n$. Each arc $(k, \ell) \in \mathcal{K} \times \mathcal{L}$ represents an assignment of k to ℓ with a cost $d_{k\ell} \geq 0$. The objective is to assign each node in $\mathcal{K}$ to exactly one other node in $\mathcal{L}$ such that the total cost of assignment is minimized. This can be formulated as the following $\{0, 1\}$ linear programming problem, where the variable $x_{k\ell} = 1$ if k is assigned to ℓ, and 0 otherwise. This is also closely related to a special case of linear programming called the transportation problem. In fact the transportation problem can be considered to be an extension of the assignment problem, otherwise known as the multi-assignment problem.

(Assignment Problem)

$$\min \sum_{i=1}^{n} \sum_{j=1}^{n} d_{ij} x_{ij}$$

$$\sum_{i=1}^{n} x_{ij} = 1 \quad \forall j,$$

$$\sum_{j=1}^{n} x_{ij} = 1 \quad \forall i,$$

$$x_{ij} \in \{0, 1\} \quad \forall i, j.$$

Unlike most integer programming problems which are often difficult to solve, this special discrete 0-1 programming problem admits a surprisingly simple solution method called the *Hungarian algorithm*, which was invented by Kuhn [K2] in honor of the Hungarian mathematician Egerváry. The Hungarian Algorithm can be found in just about all standard Operations Research textbooks. However, the idea behind the Hungarian algorithm is rarely explained in standard texts. The following attempts to give one (of possibly many) interpretation of the underlying mechanism. [R3] contains another interpretation of the Hungarian algorithm in terms of a primal-dual formulation.

It is well-known that adding or subtracting a constant to the cost function of an optimization problem does not change the optimum solution. The idea then is to modify the cost matrix d_{ij} by subtracting constants from the cost function, until some of the d_{ij}'s become zero, while others remain greater than zero. This can be achieved by noting that subtracting a constant α_i from row i, $i = 1, 2, \cdots, n$, and subtracting a constant β_j from column j, $j = 1, 2, \cdots, n$ results in the following cost function:

$$\begin{aligned}\sum_{i=1}^{n}\sum_{j=1}^{n}(d_{ij}-\alpha_i-\beta_j)x_{ij} &= \sum_{i=1}^{n}\sum_{j=1}^{n}d_{ij}x_{ij}-\sum_{i=1}^{n}\alpha_i\sum_{j=1}^{n}x_{ij}-\sum_{j=1}^{n}\beta_j\sum_{i=1}^{n}x_{ij}\\ &= \sum_{i=1}^{n}\sum_{j=1}^{n}d_{ij}x_{ij}-\sum_{i=1}^{n}\alpha_i-\sum_{j=1}^{n}\beta_j.\end{aligned}$$

This amounts to subtracting two constants $\sum_{i=1}^{n}\alpha_i$ and $\sum_{j=1}^{n}\beta_j$ from the original cost function. If α_i is the least cost entry in row i and β_j is the least cost entry in column j, then the modified cost matrix will contain at least n zeros, not all of which are necessarily noncollinear. The modified cost matrix can then be regarded as a compatibility matrix as defined in Remark 2.5.1. A match in this case corresponds to several pairs of zero cost matches. If the number of zero cost matches is n, i.e, there exists a set of n noncollinear zeros or, by duality, the minimum number of covering lines is n, the assignment problem is solved. Otherwise cover all the zeros using the minimum number of lines, look for the least uncovered cost, and then use it to modify the cost matrix again. This is repeated until a maximal match of cardinality n is found.

Remark 2.5.2 Some familiarity with linear programming (LP) is required to appreciate what is to be said here. Note that if the cost d_{ij} of the assignment problems are all integer, and if the 0-1 integer constraint of the assignment problem is replaced by a non-negativity constraint, then the solution to the relaxed non-discrete optimization problem will end up having the same (integer) solution as the 0-1 optimization problem. Consequently, there is a meaningful interpretation of the dual to the assignment problem as a continuous LP:

$$\begin{aligned}\max_{u_i,v_j} \quad & \sum_{j=1}^{n}v_j-\sum_{i=1}^{n}u_i\\ \text{such that} \quad & v_j-u_i\le d_{ij} \quad \forall i,j.\end{aligned}$$

One can think of the dual problem again as one of assigning n jobs to n applicants in a company. The applicant j that is assigned to job i will be paid a subsidy u_i in addition to an assignment fee of d_{ij} by the government. The company will then pay a payroll tax of v_j to the government for each applicant j. Then the objective function of the dual problem amounts to maximizing the net difference between the total tax collected by the government and the total subsidies paid to the company. The constraint of the dual says that for each assignment, the payroll tax imposed must be less than or equal to the subsidy plus the assignment fee.

Remark 2.5.3 Other applications of the max match min block duality can be found in many other combinatorial optimization problems, such as the bottleneck optimization problem, the connection problem, the matching problem with multiplicities, and the seating problem, see [R3] for details.

2.6 Duality of Max Tension Min Path

This section is justifiably the dual of Section 2.4. The max flow min cut duality is concerned with flow, cut, flux across cuts and capacity intervals. The max tension min path duality is concerned with differential, path, spread relative to paths, and span intervals. The symmetry between the two sets of concepts is so striking that it is tempting to replace the occurrence of terms in Section 2.4 by their dual counterparts to create this section. The min path problem has a number of important applications, notably that of knapsack problems and PERT, or *Project Evaluation and Review Techniques*, in addition to the perennial shortest path problem. These well known models can be found in any standard OR texts and hence will not be repeated here.

As in the case of max flow min cut, it is instructive to present the crux of the problem in a simple idea first.

An intuitive idea. Consider two beads, a source and a sink. The beads are joined by several strings arranged in parallel, each string having a physical length. Each string can be considered a path that joins the two beads. These strings together constitute a cut (and the only one) separating the source and the sink, since the removal of all these strings disconnects the two beads. The minimum distance from the source bead to the sink bead is given by the shortest string. On the other hand, if we try to pull the two beads apart, we cannot pull them apart indefinitely without breaking some string. The *maximum separation* that we can pull them apart is equal to the shortest string length (or path length), which is also the *minimum distance* between the beads.

The above idea shall now be generalized a bit further, albeit still in a handwaving fashion. Given two nodes s and t in a network $\mathcal{G} = (\mathcal{N}, \mathcal{A})$, find the path that joins s to t with the shortest length. One can think of the length of a path as the total physical length of the series of strings joining two given beads in a network of beads joined by inextensible strings. In such a scenario, there is a trivial way of solving the min path problem following the above intuitive idea, if one bothers to make up such a physical network of beads and strings. One simply lets the network hang from the bead s. The shortest path joining s to t is given by the series of taut strings hanging vertically between s and t. It is no coincidence that the length of the shortest path is also the maximum vertical difference (or maximum separation) in the heights of s and t under such an arrangement, and that one cannot pull s and t apart more than this distance.

The rest of this section is the formalization of this idea in a rigorous and entirely general manner.

We begin by assuming that some potential $\mathbf{u}$ for a digraph $\mathcal{G} = (\mathcal{N}, \mathcal{A})$ has been assigned. This potential induced a differential $\mathbf{v} = -\mathbf{E}^\top \mathbf{u}$ where $\mathbf{E}$ is the node-arc incidence matrix. Note that unlike in [R3], we refer to tension and differential interchangeably.

Definition 2.6.1 The *span interval* is a set-valued function $\mathbf{D} : \mathcal{A} \mapsto 2^{\mathbf{R}^{|\mathcal{A}|}}$. For an arc j, its span interval is a nonempty interval $D_j = [d_j^-, d_j^+]$, where $\forall j, d_j^- \leq$

$d_j^+, d_j^+ > -\infty, d_j^- < +\infty$. Note that it is possible for $d_j^+ = +\infty$ or $d_j^- = -\infty$. A differential $\mathbf{v}$ on $\mathcal{G}$ is *feasible with respect to spans* if $v_j \in D_j, \forall j \in \mathcal{A}$.

Definition 2.6.2 The *upper and lower span* of the path $\mathcal{P}$ are defined by:

$$d_{\mathcal{P}}^+ = \sum_{j \in \mathcal{P}^+} d_j^+ - \sum_{j \in \mathcal{P}^-} d_j^-,$$

$$d_{\mathcal{P}}^- = \sum_{j \in \mathcal{P}^+} d_j^- - \sum_{j \in \mathcal{P}^-} d_j^+.$$

The upper and lower span can be interpreted as a sort of distance when $\mathcal{P}$ is traversed in the forward and reverse direction respectively. Although in a more general context, a span can take the meaning of several other things. Note that the span interval can be compared to the capacity interval for flow, and the upper and lower span of a path can be compared to the upper and lower capacity of a cut. Let $\mathbf{e}_{\mathcal{P}}$ be the incidence vector of the path $\mathcal{P}$. The following notion can be considered as a dual counterpart to the concept of flux across a cut.

Definition 2.6.3 The spread of $\mathbf{v}$ relative to the path $\mathcal{P}$ is defined by:

$$\text{spread of } \mathbf{v} \text{ relative to } \mathcal{P} = \mathbf{e}_{\mathcal{P}}^{\top}\mathbf{v} = \sum_{j \in \mathcal{P}^+} v_j - \sum_{j \in \mathcal{P}^-} v_j.$$

Intuitively, the spread of a feasible $\mathbf{v}$ relative to the path $\mathcal{P}$ must be bounded above and below by the span of the path. This is formalized as follows (compare this to Theorem 2.4.2):

Theorem 2.6.1 If v is feasible, then for any path $\mathcal{P}$,

$$d_{\mathcal{P}}^- \leq \mathbf{e}_{\mathcal{P}}^{\top}\mathbf{v} \leq d_{\mathcal{P}}^+.$$

Proof: If $j \in \mathcal{P}^+$, then $d_j^- \leq v_j \leq d_j^+$; otherwise if $j \in \mathcal{P}^-$, then $-d_j^+ \leq -v_j \leq -d_j^-$. Summing over all $j \in \mathcal{P}$ yields the result. ∎

Theorem 2.6.2 (Integration Rule) Given a path $\mathcal{P} : s \to t$ and a potential $\mathbf{u}$,

$$u_t = u_s + \text{ spread of } \mathbf{v} \text{ relative to } \mathcal{P}.$$

Proof: We have

$$\begin{aligned}
\text{spread of } \mathbf{v} \text{ relative to } \mathcal{P} &= \mathbf{e}_{\mathcal{P}}^{\top}\mathbf{v} \\
&= -\mathbf{e}_{\mathcal{P}}^{\top}\mathbf{E}^{\top}\mathbf{u} = -(\mathbf{E}\mathbf{e}_{\mathcal{P}})^{\top}\mathbf{u} \\
&= -(\mathbf{e}^s - \mathbf{e}^t)^{\top}\mathbf{u} \qquad \text{from Theorem 1.4.2} \\
&= u_t - u_s.
\end{aligned}$$

∎

Remark 2.6.1 The spread of $\mathbf{v}$ relative to $\mathcal{P}$ thus depends only on the end points of $\mathcal{P}$, and not on the actual path taken. This is analogous to the well known fact in calculus that the line integral of a conservative field is dependent only on the end points. We call $u_t - u_s$ the spread of $\mathbf{u}$ from s to t, and this is the same as the spread of any path from s to t. If $\mathcal{P}$ is a cycle, then the spread of v relative to $\mathcal{P}$ is 0.

A formal generalization of the shortest path problem is as follows:

Definition 2.6.4 (Min Path Problem) Given two disjoint node sets $\mathcal{N}^+$ and $\mathcal{N}^-$ of $\mathcal{G}$, minimize $d^+_{\mathcal{P}}$ over all paths $\mathcal{P} : \mathcal{N}^+ \to \mathcal{N}^-$.

Remark 2.6.2 In the event of a disconnected network, where $\mathcal{N}^+$ is disconnected from $\mathcal{N}^-$ and hence there exist no paths between them, additional arcs with $d^+_j = \infty$ or $d^-_j = -\infty$ can be used to connect the graph artificially. The min path in this case will have $d^+_{\mathcal{P}} = \infty$. This is consistent with the convention that the minimum cost for an infeasible optimization problem is ∞. After the network is artificially connected, obviously there exists a *cut* $\mathcal{Q} : \mathcal{N}^+ \downarrow \mathcal{N}^-$ *of so-called unlimited span*, i.e., $d^+_j = \infty \; \forall j \in \mathcal{Q}^+$, $d^-_j = -\infty \; \forall j \in \mathcal{Q}^-$.

Remark 2.6.3 If a path with a finite upper span $d^+_{\mathcal{P}}$ exists, it is the solution of an associated painted network problem with the following painting scheme:

$$\begin{aligned} \texttt{Green} \text{ if } d^+_j < \infty,\; d^-_j > -\infty, \\ \texttt{Red} \text{ if } d^+_j = \infty,\; d^-_j = -\infty, \\ \texttt{White} \text{ if } d^+_j < \infty,\; d^-_j = -\infty, \\ \texttt{Black} \text{ if } d^+_j = \infty,\; d^-_j > -\infty. \end{aligned}$$

Definition 2.6.5 (Max Tension Problem) Given two disjoint node sets $\mathcal{N}^+$ and $\mathcal{N}^-$ of $\mathcal{G}$, find a potential $\mathbf{u}$ where u_i is constant on $\mathcal{N}^+$ and (a different) constant on $\mathcal{N}^-$, such that the spread of $\mathbf{u}$ from $\mathcal{N}^+$ to $\mathcal{N}^-$ is maximized, subject to the constraint that the resulting differential $\mathbf{v} = \Delta\mathbf{u}$ is feasible with respect to spans.

Theorem 2.6.3 (Weak Duality) If $\mathbf{u}$ is a potential satisfying the conditions of the max tension problem, and $\mathcal{P} : \mathcal{N}^+ \to \mathcal{N}^-$ is a path joining some node in $\mathcal{N}^+$ to some node in $\mathcal{N}^-$, then:

$$\text{spread of } \mathbf{u} \text{ from } \mathcal{N}^+ \text{ to } \mathcal{N}^- \leq d^+_{\mathcal{P}}.$$

Proof: Follows from the second inequality of Theorem 2.6.1 and the proof of Theorem 2.6.2. ∎

Theorem 2.6.4 (Strong Duality - Max Tension Min Path Theorem) Assuming that there exists at least one potential satisfying the conditions of the max tension problem, and that there does not exist a cut $\mathcal{Q} : \mathcal{N}^+ \downarrow \mathcal{N}^-$ of unlimited span, then

$$\text{max in max tension problem} = \text{min in min path problem.}$$

Proof: This is done by constructing a path $\mathcal{P} : \mathcal{N}^+ \to \mathcal{N}^-$ using the following min path algorithm. If a path can be found such that $d^+_{\mathcal{P}}$ is equal to the spread of a $\mathbf{u}$ from $\mathcal{N}^+$ to $\mathcal{N}^-$, then by the weak duality theorem 2.6.3, $\mathcal{P}$ must be the solution to the min path problem and $\mathbf{u}$ must be the solution to the max tension problem. ∎

It remains to present the min path algorithm. This has the same flavor as many network algorithms in that a tree (or forest if $\mathcal{N}^+$ is a set) is grown from its root in $\mathcal{N}^+$ until it reaches some node in $\mathcal{N}^-$. At each stage of the growth, an arc that stands for the shortest extension away from the tree is added to the tree.

If a feasible potential exists, it can be assumed, without loss of generality, that $d_j^- \le 0 \le d_j^+$ for all arc j, so that an initial feasible potential is given by $\mathbf{u} = \mathbf{0}$.

Algorithm 2.6.1 (Min Path Algorithm)

Initialization. Let $\mathbf{u} = \mathbf{0}$ be the initial feasible potential. Let $\mathcal{S} = \mathcal{N}^+, \mathcal{A}' = \emptyset, \mathcal{T} = (\mathcal{S}, \mathcal{A}'), \mathcal{Q} = [\mathcal{S}, \mathcal{N} \setminus \mathcal{S}]$, $w_i = 0\ \forall i \in \mathcal{S}$.

Do until $\mathcal{S} \cap \mathcal{N}^- \neq \emptyset$:

Compute

$$\beta = \min \begin{cases} w_{\alpha(j)} + d_j^+ & \text{if } j \in \mathcal{Q}^+; \\ w_{\omega(j)} - d_j^- & \text{if } j \in \mathcal{Q}^-. \end{cases} \tag{2.6.1}$$

If $\beta = \infty$, $\mathcal{Q}$ is a cut of unlimited span, stop.
Else take the minimizing arc j in (2.6.1),
$\mathcal{A}' \leftarrow \mathcal{A}' \cup \{j\}$;
$i \leftarrow \omega(j)$ if $j \in \mathcal{Q}^+, i \leftarrow \alpha(j)$ if $j \in \mathcal{Q}^-$;
$\mathcal{S} \leftarrow \mathcal{S} \cup \{i\}$;
$w_i \leftarrow \beta$
End

End

Upon exit, let $w_k = \beta\ \forall k \notin \mathcal{S}$, and let $t \in \mathcal{S} \cap \mathcal{N}^-$. The potential $\mathbf{u} = \mathbf{w}$ solves the max tension problem, and the path $\mathcal{P} : \mathcal{N}^+ \to t$ solves the min path problem, with $d^+_{\mathcal{P}} = w_t$. ∎

Remark 2.6.4 Note that the above min path algorithm is intended to facilitate understanding and is not designed for computational efficiency. A much more efficient version of the min path algorithm due to Dijkstra can be found in standard Operations Research texts such as [T2].

2.7 Duality of Feasible Flow and Feasible Potential

The ultimate goal of this chapter is to establish the duality between the optimal flow problem and the optimal potential problem as first defined in Section 2.3. However, before the optimality issue is discussed, it is necessary to establish if a feasible solution exists at all. This section is entirely devoted to studying the feasibility issue. This effort is particularly rewarding for it demonstrates many elegant applications of the painted network theory. As before, the driving force behind the theory is the duality between path/flow and cut/potential.

Consider a digraph $\mathcal{G} = (\mathcal{N}, \mathcal{A})$ with a node arc incidence matrix $\mathbf{E}$, a flow requirement vector $\mathbf{b} \in \mathbb{R}^{|\mathcal{N}|}$, a flow capacity interval $C_j = [c_j^-, c_j^+]$ and a span interval $D_j = [d_j^-, d_j^+]$ for each arc $j \in \mathcal{A}$. The two feasibility problems are defined as follows:

Definition 2.7.1 (Feasible Flow) A feasible flow $\mathbf{x} \in \mathbb{R}^{|\mathcal{A}|}$ is such that

$$x_j \in C_j \ \forall j \in \mathcal{A} \tag{2.7.1}$$

$$\text{and} \quad \mathbf{E}\mathbf{x} = \mathbf{b}. \tag{2.7.2}$$

Definition 2.7.2 (Feasible Potential) A feasible potential $\mathbf{u} \in \mathbb{R}^{|\mathcal{N}|}$ is such that

$$v_j \in D_j \ \forall j \in \mathcal{A} \tag{2.7.3}$$

$$\text{where} \quad \mathbf{v} = -\mathbf{E}^\top \mathbf{u}. \tag{2.7.4}$$

Trivially, the feasible flow problem seeks to find a feasible flow, and the feasible potential problem seeks to find a feasible potential. These two problems can be considered as dual to each other.

We shall devote the first half of this section to studying the feasible flow problem.

Definition 2.7.3 Given a set $\mathcal{S} \subseteq \mathcal{N}$, the *net supply* of the set $\mathcal{S}$ is given by $b_\mathcal{S} = \sum_{i \in \mathcal{S}} b_i$.

Theorem 2.7.1 (Feasible Flow Theorem) The feasible flow problem has a solution $\mathbf{x}$ if and only if $b_\mathcal{N} = 0$ and $b_\mathcal{S} \leq c_\mathcal{Q}^+$ for all cuts $\mathcal{Q} = [\mathcal{S}, \mathcal{N} \setminus \mathcal{S}]$.

Proof: (Necessity) If $b_\mathcal{N} = \mathbf{1}^\top \mathbf{b} \neq 0$, then (2.7.2) can never be satisfied, since $\mathbf{1}^\top \mathbf{E} = \mathbf{0}^\top$. Furthermore, $b_\mathcal{S}$ is the same as the flux of $\mathbf{x}$ across the cut $\mathcal{Q} = [\mathcal{S}, \mathcal{N} \setminus \mathcal{S}]$ which must be less than or equal to $c_\mathcal{Q}^+$ by the second inequality of Theorem 2.4.2. (Sufficiency) The proof of sufficiency is by construction through the following feasible flow algorithm. ∎

Algorithm 2.7.1 (Feasible Flow Algorithm)

Initialization. Find a flow $\mathbf{x}$ such that $x_j \in C_j\ \forall j$ (this is easy!) and compute the divergence $\mathbf{y} = \nabla\mathbf{x} = \mathbf{E}\mathbf{x}$.

Do until $\mathcal{N}^+ = \mathcal{N}^- = \emptyset$, return $\mathbf{x}$ as the feasible flow, stop.

Let $\mathcal{N}^+ = \{i \in \mathcal{N} \mid b_i > y_i\}$ (set of nodes with surplus) and $\mathcal{N}^- = \{i \in \mathcal{N} \mid b_i < y_i\}$ (set of nodes with deficit), and apply the painted network Algorithm 2.2.1 after painting all the arcs according to the scheme:

$$\begin{aligned}
\texttt{Green} \text{ if } c_j^- &< x_j < c_j^+, \\
\texttt{Red} \text{ if } c_j^- &= x_j = c_j^+, \\
\texttt{White} \text{ if } c_j^- &= x_j < c_j^+, \\
\texttt{Black} \text{ if } c_j^- &< x_j = c_j^+.
\end{aligned}$$

If a painted cut $\mathcal{Q} = [\mathcal{S}, \mathcal{N} \setminus \mathcal{S}]$ is found, then $b_\mathcal{S} > c_\mathcal{Q}^+$, no feasible flow can be found, stop.

Else a painted path $\mathcal{P}$ is found. Compute

$$\eta = \min \begin{cases} c_j^+ - x_j & \text{for arc } j \in \mathcal{P}^+, \\ x_j - c_j^- & \text{for arc } j \in \mathcal{P}^-, \\ b_i - y_i & \text{for the starting node of } \mathcal{P} \text{ in } \mathcal{N}^+, \\ y_i - b_i & \text{for the ending node of } \mathcal{P} \text{ in } \mathcal{N}^-. \end{cases}$$

Change the flow by: $\mathbf{x} \leftarrow \mathbf{x} + \eta \mathbf{e}_\mathcal{P}$.

End ∎

End

Remark 2.7.1 One can think of the feasible flow Theorem 2.7.1 as a theorem of alternative: either there exists a feasible flow, or there exists a cut $\mathcal{Q} = [\mathcal{S}, \mathcal{N} \setminus \mathcal{S}]$ such that $b_\mathcal{S} > c_\mathcal{Q}^+$, but not both.

Remark 2.7.2 Assuming that the flow capacity intervals are all finite, then at each step the flow increment η must be positive and finite. If we further assume that c_j^+, c_j^-, and b_i are all commensurable, then η must be integral, and hence the non-zero difference of $|b_i - y_i|$ for each node in $\mathcal{N}^+$ and $\mathcal{N}^-$ decreases to zero by non-diminishing amounts. The algorithm either exits with a feasible flow when $\mathcal{N}^+ = \mathcal{N}^- = \emptyset$ or returns a cut that violates the condition of the feasible flow Theorem 2.7.1. In either case, the number of steps needed to exit from Algorithm 2.7.1 must be finite.

There are two conditions for $\mathbf{x}$ to be a feasible flow. In the feasible flow algorithm, one starts with a flow that satisfies the capacity constraint (2.7.1), and then works towards satisfying the conservation constraint (2.7.2). In a symmetrical manner, one can also start with a flow that satisfies the conservation constraint

(2.7.2), and then works towards satisfying the capacity constraint (2.7.1). To do so, we change the flow by a circulation so that the conservation constraint is preserved.

Algorithm 2.7.2 (Flow Rectification Algorithm)

Initialization. Find a flow $\mathbf{x}$ such that $\mathbf{Ex} = \mathbf{b}$. (This is easy!)

Do until $\mathcal{A}^+ = \mathcal{A}^- = \emptyset$, return $\mathbf{x}$ as the feasible flow, stop.

Let $\mathcal{A}^+ = \{j \in \mathcal{A} \mid x_j > c_j^+\}$ (set of arcs with too much flow) and $\mathcal{A}^- = \{j \in \mathcal{A} \mid x_j < c_j^-\}$ (set of arcs with too little flow). Choose some arc $k \in \mathcal{A}^+ \cup \mathcal{A}^-$.

> **If** $k \in \mathcal{A}^+$ let $\mathcal{N}^+ \leftarrow \{\alpha(k)\}, \mathcal{N}^- \leftarrow \{\omega(k)\}$;
> **Else** $k \in \mathcal{A}^-$ let $\mathcal{N}^- \leftarrow \{\alpha(k)\}, \mathcal{N}^+ \leftarrow \{\omega(k)\}$.
> **End** ∎

Leave out the arc k from the digraph and apply the painted network Algorithm 2.2.1 after painting all the arcs according to the following scheme:

$$\begin{aligned} \texttt{Green} \text{ if } c_j^- &< x_j < c_j^+, \\ \texttt{Red} \text{ if } c_j^- &= x_j = c_j^+, \\ \texttt{White} \text{ if } x_j &\le c_j^-,\ x_j < c_j^+, \\ \texttt{Black} \text{ if } c_j^- &< x_j,\ c_j^+ \le x_j. \end{aligned}$$

If a painted cut $\mathcal{Q} = [\mathcal{S}, \mathcal{N} \setminus \mathcal{S}]$ is found, then $b_{\mathcal{S}} > c_{\mathcal{Q}}^+$, no feasible flow can be found, stop.

Else take the painted path $\mathcal{P}$ and compute:

$$\eta = \min \begin{cases} c_j^+ - x_j & \text{for arc } j \in \mathcal{P}^+, \\ x_j - c_j^- & \text{for arc } j \in \mathcal{P}^-, \\ c_j^- - x_k & \text{if } k \in \mathcal{A}^-, \\ x_k - c_j^+ & \text{if } k \in \mathcal{A}^+. \end{cases}$$

Change the flow by: $x_k \leftarrow x_k - \eta$ if $k \in \mathcal{A}^+$, $x_k \leftarrow x_k + \eta$ if $k \in \mathcal{A}^-$, and

$$\mathbf{x} \leftarrow \mathbf{x} + \eta \mathbf{e}_{\mathcal{P}}.$$

End

End

Remark 2.7.3 The justification of finite termination here is similar to that of Remark 2.7.2. The flow increment η must be positive and finite in each iteration under the assumption that c_j^+, c_j^-, and b_i are all commensurable. In which case η must be integral, and hence the non-zero difference of $x_j - c_j^+$ for each arc in $\mathcal{A}^+$ and $c_j^- - x_j$ for each arc in $\mathcal{A}^-$ will be reduced to zero by non-diminishing amounts.

The algorithm either exits with a feasible flow when $\mathcal{A}^+ = \mathcal{A}^- = \emptyset$ or returns a cut that violates the condition of the feasible flow Theorem 2.7.1.

We now turn to the dual problem of finding a feasible potential. As in the case of feasible flow, a necessary and sufficient condition for a feasible potential to exist is as follows:

Theorem 2.7.2 (Feasible Potential Theorem) The feasible potential problem has a solution $\mathbf{u}$ if and only if $d^+_{\mathcal{P}} \geq 0$ for all elementary cycles $\mathcal{P}$.

Proof: (Necessity) The spread of $\mathbf{v} = \Delta\mathbf{u}$ is 0 around any circuit $\mathcal{P}$, and this (i.e., 0) must be less than or equal to $d^+_{\mathcal{P}}$ by Theorem 2.6.1.
(Sufficiency) The proof of sufficiency is by construction through the following feasible potential algorithm. ■

Algorithm 2.7.3 (Feasible Potential Algorithm)

Initialization. Assign an arbitrary potential $\mathbf{u}$ and compute its induced differential $\mathbf{v} = \Delta\mathbf{x} = -\mathbf{E}^{\mathsf{T}}\mathbf{u}$.

Do: until $\rho_i = 0\ \forall i \in \mathcal{N}$, return $\mathbf{u}$ as the feasible flow, stop.

Let $\delta^+_j = d^+_j - v_j$ and $\delta^-_j = d^-_j - v_j$. For all $i \in \mathcal{N}$, compute

$$\rho_i = \min \begin{cases} \delta^+_j & \text{if } \alpha(j) = i, \\ -\delta^-_j & \text{if } \omega(j) = i, \\ 0 \end{cases}$$

Let $k \in \mathcal{N}$ be such that $\rho_k < 0, \mathcal{N}^+ \leftarrow \{k\}, \mathcal{N}^- \leftarrow \{k\}$. Except when traversing the first arc out of k, let $\delta^+_j \leftarrow \max\{0, \delta^+_j\}$ and $\delta^-_j \leftarrow \min\{0, \delta^-_j\}$. Apply the (modified) min path Algorithm 2.6.1 using the span interval $[\delta^-_j, \delta^+_j]\ \forall j$ to find a cycle from k back to itself.

If k is reached a second time, thus completing a cycle $\mathcal{P}$ with $\beta < 0$, then $d^+_{\mathcal{P}} < 0$ and no feasible potential can be found, stop.

Else stop (the min path algorithm) as soon as a value $\beta \geq 0$ has been reached. Set $w_i = 0$ for all nodes not in the tree. $\mathbf{u} \leftarrow \mathbf{u} + \mathbf{w}$.

End

End ■

Remark 2.7.4 The justification of finite termination here again requires the commensurability of d^+_j, d^-_j, and the starting potential. When the potential is changed, the new potential will be such that $\rho_k = 0$ and that for all other nodes, the corresponding ρ_i is no less than the old value. Thus feasibility is improved by a finite number of quanta each time. The algorithm either exits with a feasible potential when $\rho_i \geq 0\ \forall i$, or returns a cycle that violates the condition of the feasible potential Theorem 2.7.2.

Remark 2.7.5 Just as the case of feasible flow, the feasible potential theorem can be regarded as a theorem of alternatives: either there exists a cycle $\mathcal{P}$ such that $d^+_{\mathcal{P}} < 0$, or there exists a feasible potential, but not both.

Another algorithm for finding a feasible potential that can be thought of as a dual to the flow rectification Algorithm 2.7.2 is as follows:

Algorithm 2.7.4 (Tension Rectification Algorithm)

Initialization. Assign an arbitrary potential $\mathbf{u}$ and compute its induced differential $\mathbf{v} = \Delta\mathbf{x} = -\mathbf{E}^\top\mathbf{u}$.

Do until $\mathcal{A}^+ = \mathcal{A}^- = \emptyset$, return $\mathbf{u}$ as the feasible potential, stop.

Let

$$\mathcal{A}^- = \{j \in \mathcal{A} \mid v_j > d_j^+\} \text{ (set of arcs with too much tension)}$$

and

$$\mathcal{A}^+ = \{j \in \mathcal{A} \mid v_j < d_j^-\} \text{ (set of arcs with too little tension).}$$

Choose some arc $k \in \mathcal{A}^+ \cup \mathcal{A}^-$, and paint the network according to the scheme:

$$\begin{aligned} \texttt{Green if } & d_j^- = v_j = d_j^+, \\ \texttt{Red if } & d_j^- < v_j < d_j^+, \\ \texttt{White if } & v_j \geq d_j^+,\ v_j > d_j^-, \\ \texttt{Black if } & v_j \leq d_j^-,\ v_j < d_j^+. \end{aligned}$$

Apply the painted network Algorithm 2.2.1 to find a cycle containing k.

If a painted cycle $\mathcal{P}$ is found, then $d^+_{\mathcal{P}} < 0$, no feasible potential can be found, stop.

Else take the painted cut $\mathcal{Q}$ (which contains k) and compute

$$\eta = \min \begin{cases} d_j^+ - v_j & \text{for arc } j \in \mathcal{Q}^+, \\ v_j - d_j^- & \text{for arc } j \in \mathcal{Q}^-, \\ d_k^- - v_k & \text{if } k \in \mathcal{A}^+, \\ v_k - d_k^+ & \text{if } k \in \mathcal{A}^-. \end{cases}$$

Change the potential by: $\mathbf{u} \leftarrow \mathbf{u} + \eta\mathbf{e}_{\mathcal{N}\setminus\mathcal{S}}$.

End

End ∎

Remark 2.7.6 In each iteration of the tension rectification Algorithm 2.7.4, the new potential improves by a finite number of quanta over the previous one since the new differential v_j is no further away from the span interval D_j than the old differential. In particular, v_k is closer to its span interval by the positive amount η.

2.8 Duality of Linear Optimal Flow and Optimal Potential

We now turn to a fairly important class of network optimization models which includes all the network models, i.e., max flow, shortest path and assignment problems that we have discussed so far as special cases. Some of the better known special cases of network optimization models that are not mentioned yet include the transportation problem, the warehousing problem, critical path analysis, facility location planning, just to name a few. This class of problems apparently accounts for 70% of all linear programming applications in the real world (see [AMO1]). Under some appropriate assumptions on their cost and capacity/span intervals, the linear cost optimal flow problem and the linear cost optimal potential problem are also dual to each other.

Consider a digraph $\mathcal{G} = (\mathcal{N}, \mathcal{A})$ with a node arc incidence matrix $\mathbf{E}$, a flow requirement vector $\mathbf{b} \in \mathbb{R}^{|\mathcal{N}|}$, a flow capacity interval $C_j = [c_j^-, c_j^+]$ and a span interval $D_j = [d_j^-, d_j^+]$ for each arc $j \in \mathcal{A}$. Let $\mathbf{d}, \mathbf{c} \in \mathbb{R}^{|\mathcal{A}|}$ and $\mathbf{b} \in \mathbb{R}^{|\mathcal{N}|}, \mathbf{1}^\top \mathbf{b} = 0$. Unless otherwise stated specifically, all cuts and paths/cycles are assumed to be elementary.

The two optimization problems are defined as follows:

Definition 2.8.1 (Linear Optimal Flow Problem)

$$\min_{\mathbf{x}} \quad \mathbf{d}^\top \mathbf{x}$$

$$\text{subject to} \quad x_j \in C_j = [c_j^-, c_j^+], \ \forall j \in \mathcal{A} \tag{2.8.1}$$

$$\text{and} \quad \mathbf{Ex} = \mathbf{b}. \tag{2.8.2}$$

Definition 2.8.2 (Linear Optimal Potential Problem)

$$\max_{\mathbf{u},\mathbf{v}} \quad -\mathbf{b}^\top \mathbf{u} - \mathbf{c}^\top \mathbf{v}$$

$$\text{subject to} \quad v_j \in D_j = [d_j^-, d_j^+], \ \forall j \in \mathcal{A} \tag{2.8.3}$$

$$\text{and} \quad \mathbf{v} = -\mathbf{E}^\top \mathbf{u}. \tag{2.8.4}$$

Clearly both problems are special cases of linear programming. What makes them important and interesting to study is that because of the special network structure, these problems can be solved much more efficiently than the usual linear programs, usually by several orders of magnitude in computation time. One may be tempted to think that these two problems are dual to each other, but as they stand this is not quite true yet. At the end of this section, special assumptions on the cost and capacity/span intervals will be imposed to establish the duality.

We have addressed the issue of feasibility in the preceding sections. For optimization, an intimately related issue is whether the optimal solution is finite. As the reader will soon discover, these issues are dual to each other in the sense that

• *feasibility of the primal corresponds to finiteness of the dual optimal, and that finiteness of the primal optimal corresponds to feasibility of the dual.*

Definition 2.8.3 In the linear optimal flow problem, we say that a cycle $\mathcal{P}$ is *unbalanced* if

$$\mathbf{d}^\top \mathbf{e}_{\mathcal{P}} = \sum_{j \in \mathcal{P}^+} d_j - \sum_{j \in \mathcal{P}^-} d_j < 0$$
$$\text{and } c_j^+ = \infty \; \forall\, j \in \mathcal{P}^+, c_j^- = -\infty \; \forall\, j \in \mathcal{P}^-.$$

Note that the qualification $\mathbf{d}^\top \mathbf{e}_{\mathcal{P}} < 0$ looks rather similar to the upper span of the cycle in the feasible potential problem, although "span" has no meaning in a flow problem. Recall that the feasible potential problem has a feasible solution if and only if there is no elementary cycle with the upper span $d_{\mathcal{P}}^+ < 0$. The striking resemblance here is no coincidence. In due course it will become clear that feasibility of the dual and finiteness of the primal are closely related. This observation is also applicable to the notion of an unbalanced cut to be defined as follows:

Definition 2.8.4 In the linear optimal potential problem, we say that an elementary cut $\mathcal{Q} = [\mathcal{S}, \mathcal{N} \setminus \mathcal{S}]$ is *unbalanced* if

$$\mathbf{b}^\top \mathbf{e}_{\mathcal{N}\setminus\mathcal{S}} + \mathbf{c}^\top \mathbf{e}_{\mathcal{Q}} = -\sum_{i \in \mathcal{S}} b_i + \sum_{j \in \mathcal{Q}^+} c_j - \sum_{j \in \mathcal{Q}^-} c_j < 0,$$
$$\text{and } d_j^+ = \infty \; \forall\, j \in \mathcal{Q}^+, d_j^- = -\infty \; \forall\, j \in \mathcal{Q}^-.$$

In the above, note that $0 = b_{\mathcal{N}} = b_{\mathcal{S}} + b_{\mathcal{N}\setminus\mathcal{S}}$, so that $b_{\mathcal{N}\setminus\mathcal{S}} = -b_{\mathcal{S}}$. We shall study the optimal flow problem and the optimal potential problem in turn before proving an important duality result.

Theorem 2.8.1 (Existence for linear optimal flow) Assume that the linear optimal flow problem has at least one feasible solution. A finite optimal solution exists if and only if no elementary cycle is unbalanced.

Proof: (Necessity) If an unbalanced cycle exists, an infinite amount of flow can be sent around this cycle without violating the conservation constraint (2.8.2). This incurs an infinitely negative cost, inf $= -\infty$.
(Sufficiency) A specialized proof based on the theory of extreme flow can be found in [R3]. We shall defer the proof of sufficiency to the next section where the existence of the more general problem of convex optimization is proven in a unified framework. ∎

Definition 2.8.5 (Kilter curve for optimal flow) For each arc j, the *Kilter Curve* for arc j is the set

$$\Gamma_j = \left\{(x_j, v_j) \in \mathbb{R}^2 \mid c_j^- \leq x_j \leq c_j^+; v_j \leq d_j \text{ if } x_j < c_j^+;\; v_j \geq d_j \text{ if } x_j > c_j^- \right\}.$$

Note that the above conditions imply that $v_j = d_j$ if $c_j^- < x_j < c_j^+$, $v_j \leq d_j$ if $c_j^- = x_j$, and $v_j \geq d_j$ if $c_j^+ = x_j$. This curve is depicted in Figure 2.8.1 in the case where both c_j^- and c_j^+ are finite.

Theorem 2.8.2 (Optimality condition for linear optimal flow) $\mathbf{x}$ solves the linear optimal flow problem if and only if there exists a potential $\mathbf{u}$ such that $(x_j, v_j) \in \Gamma_j \; \forall j \in \mathcal{A}$, where $\mathbf{v} = \Delta \mathbf{u}$.

Proof: (Sufficiency) Assume that $\mathbf{x}, \mathbf{u}$ satisfy the optimality conditions. Given any other feasible flow $\mathbf{x}'$, then

$$\begin{aligned} &\text{either } c_j^- < x_j, x_j' < c_j^+ \text{ then } v_j - d_j = 0; \\ &\text{or } c_j^+ = x_j \geq x_j' \text{ then } v_j \geq d_j; \\ &\text{or } c_j^- = x_j \leq x_j' \text{ then } v_j \leq d_j. \end{aligned}$$

In all three cases $(d_j - v_j) \cdot (x_j' - x_j) \geq 0$. Summing over all j and using the identity (2.3.3) gives:

$$\mathbf{d}^\top (\mathbf{x}' - \mathbf{x}) \geq \mathbf{v}^\top (\mathbf{x}' - \mathbf{x}) = -\mathbf{u}^\top (\mathbf{b} - \mathbf{b}) = 0,$$

or

$$\mathbf{d}^\top \mathbf{x}' \geq \mathbf{d}^\top \mathbf{x}.$$

Hence $\mathbf{x}$ must be optimal.
(Necessity) Application of the following optimal flow algorithm either finds a better solution, i.e., one with a lower cost, or finds an optimal $\mathbf{x}$ and an optimal $\mathbf{u}$ that satisfies the kilter condition. ∎

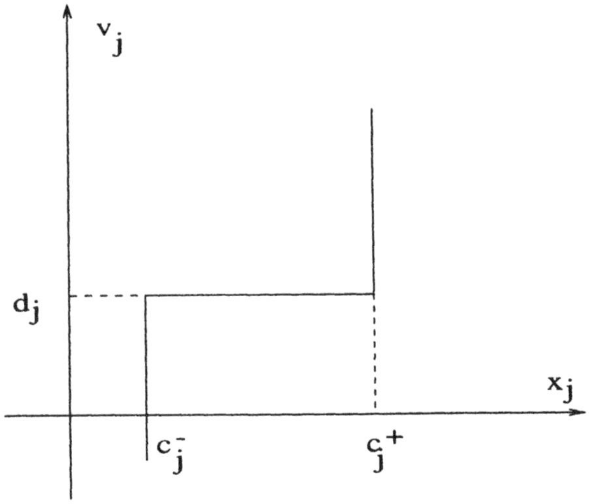

FIGURE 2.8.1 Kilter curve for arc j in optimal flow

Remark 2.8.1 We often refer to the optimality condition in Theorem 2.8.2 as the *Kilter Condition.* This has an interesting and intuitively obvious economic interpretation. If x_j is the flow of goods on arc j and u_i is the price of goods at node i, then $(d_j - v_j)x_j$ is the net cost in sending the goods on arc j. To minimize total cost, this quantity should be as small as possible for all arcs, as such if $(d_j - v_j) < 0$, more goods should be sent; and if $(d_j - v_j) > 0$, less goods should be sent. If $(d_j - v_j) = 0$, then it doesn't matter as long as the amount sent is feasible with respect to capacity.

Algorithm 2.8.1 (Optimal Flow Algorithm)

Initialization. Find a feasible flow $\mathbf{x}$.

Do: Let the (flow dependent) span interval be defined by:

$$D_j = [d_j^-, d_j^+] = \begin{cases} [d_j, d_j] & \text{if } c_j^- < x_j < c_j^+ \\ (-\infty, d_j] & \text{if } c_j^- = x_j < c_j^+ \\ [d_j, \infty) & \text{if } c_j^- < x_j = c_j^+ \\ (-\infty, \infty) & \text{if } c_j^- = x_j = c_j^+ \end{cases}$$

Apply either Algorithm 2.7.3 or Algorithm 2.7.4 to find a cycle $\mathcal{P}$ with $d_{\mathcal{P}}^+ < 0$ or a feasible potential.

If a feasible potential to the above span interval is found, return $\mathbf{x}$ as the optimal solution, stop.

Else no feasible potential is found, but a cycle $\mathcal{P}$ with $d_{\mathcal{P}}^+ < 0$ is found instead. Let

$$\eta = \min \begin{cases} c_j^+ - x_j & \text{for arc } j \in \mathcal{P}^+, \\ x_j - c_j^- & \text{for arc } j \in \mathcal{P}^-. \end{cases}$$

If $\eta = \infty$, $\mathcal{P}$ is an unbalanced cycle, the solution is unbounded, stop.
Else $\eta > 0$ is finite. Change the flow by: $\mathbf{x} \leftarrow \mathbf{x} + \eta \mathbf{e}_{\mathcal{P}}$.
End

End

End ∎

Remark 2.8.2 If a feasible potential is found, the kilter condition is satisfied and by Theorem 2.8.2, the solution is optimal. If a cycle $\mathcal{P}$ with $d_{\mathcal{P}}^+ < 0$ is found, then by definition of the span interval, the flow on this cycle must be such that $x_j < c_j^+ \ \forall\, j \in \mathcal{P}^+$, $x_j > c_j^- \ \forall\, j \in \mathcal{P}^-$, and consequently $\eta > 0$. The new solution has a cost $\mathbf{d}^\top(\mathbf{x} + \eta \mathbf{e}_{\mathcal{P}}) = \mathbf{d}^\top \mathbf{x} + \eta \mathbf{d}^\top \mathbf{e}_{\mathcal{P}}$. Since $\mathbf{d}^\top \mathbf{e}_{\mathcal{P}} = d_{\mathcal{P}}^+ < 0$, the new solution has a strictly lower cost. If $\eta = \infty$, by Theorem 2.8.1, the optimal solution is unbounded. Otherwise η is finite and if the usual commensurability assumption is satisfied, η will not be ever-diminishing, this process must terminate in a finite solution that gives the minimum flow.

We now turn to the problem of optimal potential. Because of duality, the analysis is very similar. The following result can be considered to be a dual counterpart to Theorem 2.8.1.

Theorem 2.8.3 (Existence for linear optimal potential) Assuming that the linear optimal potential problem has at least one feasible solution. A finite optimal solution exists if and only if no cut is unbalanced.

Proof: (Necessity) If an unbalanced cut $\mathcal{Q} = [\mathcal{S}, \mathcal{N} \setminus \mathcal{S}]$ exists, then we can change a feasible potential $\mathbf{u}$ to $\mathbf{u} + t\mathbf{e}_{\mathcal{N}\setminus\mathcal{S}}$ which will remain feasible for any large t, since the differential in $\mathcal{S}$ and in $\mathcal{N} \setminus \mathcal{S}$ are unchanged and the span interval across the cut can accommodate any arbitrarily large increase in differential. Furthermore, the new potential will have a higher cost for any positive t. This thus incurs an infinitely large increase in cost, sup $= \infty$.
(Sufficiency) As in the case of flow, we shall defer the proof of sufficiency to Section 2.9 as a special case of the unified proof. ∎

Definition 2.8.6 (Kilter curve for optimal potential) For each arc j, the *Kilter Curve* for arc j is the set

$$\Gamma_j = \left\{(x_j, v_j) \in \mathbb{R}^2 \mid d_j^- \le v_j \le d_j^+; x_j \le c_j \text{ if } v_j < d_j^+;\ x_j \ge c_j \text{ if } v_j > d_j^-\ \right\}.$$

Note that the above conditions imply that $x_j = c_j$ if $d_j^- < v_j < d_j^+$, $x_j \le c_j$ if $d_j^- = d_j$, and $x_j \ge c_j$ if $d_j^+ = v_j$. This curve is depicted in Figure 2.8.2 in the case where both d_j^- and d_j^+ are finite. Note the symmetry with the Kilter Curve for optimal flow in Figure 2.8.1.

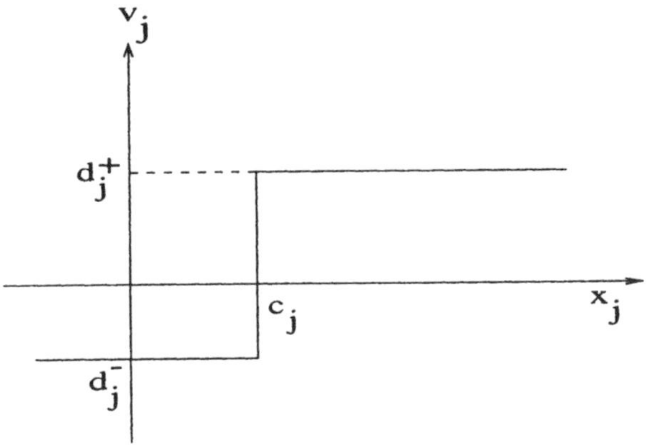

FIGURE 2.8.2 Kilter curve for arc j in optimal potential

Theorem 2.8.4 (Optimality condition for linear optimal potential) $\mathbf{u}$ solves the linear optimal potential problem if and only if there exists a flow $\mathbf{x}$ such that $\nabla\mathbf{x} = \mathbf{E}\mathbf{x} = \mathbf{b}$ and $(x_j, v_j) \in \Gamma_j\ \forall j \in \mathcal{A}$, where $\mathbf{v} = \Delta\mathbf{u}$.

Proof: (Sufficiency) Assume that $\mathbf{x}, \mathbf{u}$ satisfy the optimality conditions. Given any other feasible potential $\mathbf{u}'$ (with $\Delta\mathbf{u}' = \mathbf{v}'$), then

$$\begin{aligned} &\text{either } d_j^- < v_j, v_j' < d_j^+ \text{ then } x_j - c_j = 0;\\ &\text{or } d_j^+ = v_j \ge v_j' \text{ then } x_j \ge c_j;\\ &\text{or } d_j^- = v_j \le v_j' \text{ then } x_j \le c_j. \end{aligned}$$

In all three cases $(c_j - x_j) \cdot (v'_j - v_j) \geq 0$. Summing over all j and using the identity (2.3.3) give:

$$\mathbf{c}^\mathsf{T}(\mathbf{v}' - \mathbf{v}) \geq \mathbf{x}^\mathsf{T}(\mathbf{v}' - \mathbf{v}) = -\mathbf{b}^\mathsf{T}(\mathbf{u}' - \mathbf{u})$$

or

$$-\mathbf{b}^\mathsf{T}\mathbf{u} - \mathbf{c}^\mathsf{T}\mathbf{v} \geq -\mathbf{b}^\mathsf{T}\mathbf{u}' - \mathbf{c}^\mathsf{T}\mathbf{v}'.$$

Hence $\mathbf{u}$ must be maximal.
(Necessity) Application of the following optimal potential algorithm either finds a better solution, i.e., one with a higher cost, or finds an optimal $\mathbf{u}$ and an optimal $\mathbf{x}$ that satisfies the kilter condition. ∎

The following algorithm for solving the optimal potential problem can be considered as a dual counterpart to the optimal flow Algorithm 2.8.1.

Algorithm 2.8.2 (Optimal Potential Algorithm)

Initialization. Find a feasible potential $\mathbf{u}$, and compute its differential $\mathbf{v} = \Delta\mathbf{u}$.

Do: Let the (differential dependent) capacity interval be defined by:

$$C_j = [c_j^-, c_j^+] = \begin{cases} [c_j, c_j] & \text{if } d_j^- < v_j < d_j^+ \\ (-\infty, c_j] & \text{if } d_j^- = v_j < d_j^+ \\ [c_j, \infty) & \text{if } d_j^- < v_j = d_j^+ \\ (-\infty, \infty) & \text{if } d_j^- = v_j = d_j^+ \end{cases}$$

Apply either Algorithm 2.7.1 or Algorithm 2.7.2 to find a feasible flow $\mathbf{x}$ such that $\mathbf{Ex} = \mathbf{b}$ and $x_j \in C_j \quad \forall j$, or a cut $\mathcal{Q} = [\mathcal{S}, \mathcal{N} \setminus \mathcal{S}]$ with $c_\mathcal{Q}^+ < b_\mathcal{S} < \infty$.

If a feasible flow to the above capacity interval is found, return $\mathbf{u}$ as the optimal solution, stop.

Else no feasible flow is found, but a cut $\mathcal{Q}$ with $c_\mathcal{Q}^+ < \infty$ is found instead. Let

$$\eta = \min \begin{cases} d_j^+ - v_j & \text{for arc } j \in \mathcal{Q}^+, \\ v_j - d_j^- & \text{for arc } j \in \mathcal{Q}^-. \end{cases}$$

If $\eta = \infty$, $\mathcal{Q}$ is an unbalanced cut, the solution is infinite, stop.
Else $\eta > 0$ is finite. Change the potential by: $\mathbf{u} \leftarrow \mathbf{u} + \eta \mathbf{e}_{\mathcal{N} \setminus \mathcal{S}}$.
End

End

End ∎

Remark 2.8.3 If a feasible flow is found, kilter condition is satisfied, and by Theorem 2.8.4 the potential must be optimal. If a cut $\mathcal{Q}$ with $c_\mathcal{Q}^+ < b_\mathcal{S} < \infty$ is found, then by definition of the capacity interval, the differential across the cut must be such that $v_j < d_j^+ \ \forall\, j \in \mathcal{Q}^+$, $v_j > d_j^- \ \forall\, j \in \mathcal{Q}^-$, and consequently $\eta > 0$. The new solution has a cost

$$-\mathbf{b}^\top(\mathbf{u} + \eta \mathbf{e}_{\mathcal{N}\setminus\mathcal{S}}) - \mathbf{c}^\top(\mathbf{v} + \eta \mathbf{e}_{\mathcal{Q}}) = -\mathbf{b}^\top \mathbf{u} - \mathbf{c}^\top \mathbf{v} - \eta(\mathbf{b}^\top \mathbf{e}_{\mathcal{N}\setminus\mathcal{S}} + \mathbf{c}^\top \mathbf{e}_{\mathcal{Q}}).$$

Since

$$\begin{aligned} 0 &> \sum_{i\in\mathcal{N}\setminus\mathcal{S}} b_i + \sum_{j\in\mathcal{Q}^+} c_j^+ - \sum_{j\in\mathcal{Q}^-} c_j^- \\ &= \sum_{i\in\mathcal{N}\setminus\mathcal{S}} b_i + \sum_{j\in\mathcal{Q}^+} c_j - \sum_{j\in\mathcal{Q}^-} c_j \\ &= \mathbf{b}^\top \mathbf{e}_{\mathcal{N}\setminus\mathcal{S}} + \mathbf{c}^\top \mathbf{e}_{\mathcal{Q}}, \end{aligned}$$

the new solution has a higher cost than the previous solution. If $\eta = \infty$, the cut is unbalanced, and by Theorem 2.8.3, the optimal solution is unbounded. Otherwise η is finite and if the usual commensurability assumption is satisfied, η will not be ever-diminishing, this process must terminate in a finite solution that gives the optimal potential.

The optimal flow algorithm and optimal potential algorithm are intended to accentuate duality as the driving force, but are not designed to be efficient. In practice, the network simplex (or simplex-on-a-graph) algorithm with appropriate scaling strategies is acknowledged to be the most efficient method for solving linear cost network optimization problems.

Because of duality, there are two versions of the network simplex method: the simplex algorithm for flow, and the simplex algorithm for potential. As the names imply, these are specialization of the simplex algorithm to network flow and potential problems. Like the more general method, the primal algorithm for flow maintains primal feasibility (which corresponds to dual optimality) and works towards primal optimality (which corresponds to dual feasibility). As this book is not intended to be a cookbook for solving problems, the readers may like to refer to specialized texts such as [KH1] or [AMO1] for details of the network simplex algorithm. Some underlying ideas that require some familiarity of linear programming (LP) (see Algorithm 3.1.1 of Chapter 3) are as follows:

• One starts from an initial basic feasible solution, where a basis in this case corresponds to a spanning tree, the basic variables correspond to flow in the tree arcs, and the nonbasic variables correspond to flow in the cotree arcs. The Tucker representation Theorem 2.3.2 asserts that the flows in the tree arcs are uniquely determined by the flows in the cotree arcs. The latter nonbasic variables will take values either on the lower or upper bound of the capacity interval of the cotree arcs.

• In the *pricing step*, the potentials of all nodes are uniquely determined by setting the cost and the differential of each tree arc to be the same. The relative or shadow cost of each cotree arc turns out to be the difference of its cost and differential. If there exists a cotree arc such that its relative cost is negative and its flow is at its lower bound, or if the relative cost is positive and the flow is at its upper bound, then the total cost of flow can be reduced by adjusting the flow of this cotree arc (increase flow if relative cost is negative, decrease flow if the relative cost is positive).

• In the *pivoting step*, the flow change is implemented around a cycle so as to maintain conservation. This cycle includes the cotree arc concerned, and the unique path in the basis tree joining the two end points of the cotree arc. The maximal amount of flow change around this cycle is determined by the smallest slack from the bounds amongst all the arcs in this cycle.

• In the *changing of basis step*, the cotree arc concerned will enter the basis tree, replacing one of the previous tree arc that has its flow hitting one of its bounds. The new spanning tree will thus define a new basis, and the procedure is repeated again from the pricing step until the shadow cost of each cotree arc is such that no reduction in cost is possible, hence the optimal solution.

The simplex algorithm for potential follows in a symmetrical manner.

There is one other well known method for solving the optimal flow or optimal potential problem which warrants special attention. This is more for aesthetic reasons rather than for practical reasons. Duality again plays a key role behind the completely symmetrical nature of the *Out-of-Kilter* algorithm [F3]. Unlike the simplex method, this does not need modification for it to be applicable to either the optimal flow or the optimal potential problem. Furthermore, it does not require the initial solution to be feasible nor to satisfy the optimality condition. The idea is simple: given a flow that satisfies $\nabla \mathbf{x} = \mathbf{b}$ and a potential $\mathbf{u}$, one tries to force each of the flow/differential pair (x_j, v_j) onto the corresponding kilter curve by either changing the flow around some cycle, or changing the potential across some cut. In what follows, we shall present the out-of-kilter algorithm as if we are solving the optimal flow problem, but it does not require much convincing to see that the algorithm can be directly applied to the optimal potential problem.

Algorithm 2.8.3 (The Out-of-Kilter Algorithm)

Initialization. Find any flow that satisfies $\nabla \mathbf{x} = \mathbf{b}$ and any potential $\mathbf{u}$, compute the differential $\mathbf{v} = \Delta \mathbf{u}$.

Do: until all arcs are in kilter, i.e., $(x_j, v_j) \in \Gamma_j \ \forall j \in \mathcal{A}$, stop.

Paint the arcs according to:

$$C_j = [c_j^-, c_j^+] = \begin{cases} \texttt{Green} & \text{if } c_j^- < x_j < c_j^+ \text{ and } \ v_j = d_j; \\ \texttt{Red} & \text{if } c_j^- = x_j \text{ and } v_j < d_j, \text{ or if } c_j^+ = x_j \text{ and } \ v_j > d_j; \\ \texttt{White} & \text{if } c_j^- = x_j \text{ and } v_j = d_j, \text{ or if } x_j < c_j^-, \\ & \text{or if } x_j < c_j^+, \text{ and } v_j > d_j; \\ \texttt{Black} & \text{if } c_j^+ = x_j \text{ and } v_j = d_j, \text{ or if } x_j > c_j^+, \\ & \text{or if } x_j > c_j^-, \text{ and } v_j < d_j. \end{cases}$$

(Essentially, this amounts to painting the interior of the $\mathbb{R}^2$ plane to the left, and the left corner, of the kilter curve as `White`, the interior of the $\mathbb{R}^2$ plane to the right, and the right corner, of the kilter curve as `Black`, the relative interior of both the vertical segments of the kilter curve as `Red`, and the relative interior of the horizontal part of the kilter curve as `Green`.)

Select any arc k that is out-of-kilter (this must be either `Black` or `White`).

If k is `White`, then let $\mathcal{N}^+ \leftarrow \{\omega(k)\}$, $\mathcal{N}^- \leftarrow \{\alpha(k)\}$. Apply the painted network Algorithm 2.2.1 to find a compatible cycle $\mathcal{P} : \omega(k) \rightarrow \alpha(k)$ that includes arc k as part of $\mathcal{P}^+$, or a compatible cut $\mathcal{Q} : \omega(k) \downarrow \alpha(k)$ that includes arc k as part of $\mathcal{Q}^-$.

Else k is `Black`, then let $\mathcal{N}^+ \leftarrow \{\alpha(k)\}$, $\mathcal{N}^- \leftarrow \{\omega(k)\}$. Apply the painted network Algorithm 2.2.1 to find a compatible cycle $\mathcal{P} : \alpha(k) \rightarrow \omega(k)$ that includes arc k as part of $\mathcal{P}^-$, or a compatible cut $\mathcal{Q} : \alpha(k) \downarrow \omega(k)$ that includes arc k as part of $\mathcal{Q}^+$.

End

If (Primal Phase) a compatible cycle $\mathcal{P}$ is found, compute

$$\alpha = \min \begin{cases} c_j^+ - x_j & \text{for arc } j \in \mathcal{P}^+ \text{ and } v_j \geq d_j, \\ c_j^- - x_j & \text{for arc } j \in \mathcal{P}^+ \text{ and } v_j < d_j, \\ c_k^- - x_k & \text{for } \texttt{White} \text{ arc } k \in \mathcal{P}^+ \text{ and } v_k \leq d_k, \\ c_k^+ - x_k & \text{for } \texttt{White} \text{ arc } k \in \mathcal{P}^+ \text{ and } v_k > d_k, \\ x_k - c_k^- & \text{for } \texttt{Black} \text{ arc } k \in \mathcal{P}^- \text{ and } v_k < d_k, \\ x_k - c_k^+ & \text{for } \texttt{Black} \text{ arc } k \in \mathcal{P}^- \text{ and } v_k \geq d_k, \\ x_j - c_j^+ & \text{for arc } j \in \mathcal{P}^- \text{ and } v_j > d_j, \\ x_j - c_j^- & \text{for arc } j \in \mathcal{P}^- \text{ and } v_j \leq d_j. \end{cases}$$

If $\alpha = \infty$, $\mathcal{P}$ is an unbalanced cycle, the solution is unbounded, stop.
Else $\alpha > 0$ is finite. Change the flow by: $\mathbf{x} \leftarrow \mathbf{x} + \alpha \mathbf{e}_{\mathcal{P}}$.
End

Else(Dual Phase) a compatible cut $\mathcal{Q}$ is found, compute

$$\eta = \min \begin{cases} d_j - v_j & \text{for arc } j \in \mathcal{Q}^+ \text{ with } x_j < c_j^+, \\ v_j - d_j & \text{for arc } j \in \mathcal{Q}^- \text{ with } x_j > c_j^-, \\ v_k - d_k & \text{for } \texttt{White} \text{ arc } k \in \mathcal{Q}^- \text{ with } x_k \geq c_k^-, \\ d_k - v_k & \text{for } \texttt{Black} \text{ arc } k \in \mathcal{Q}^+ \text{ with } x_k \leq c_k^+. \end{cases}$$

If $\eta = \infty$, $\mathcal{Q}$ is an unbalanced cut, there is no feasible solution, stop.
Else $\eta > 0$ is finite. Change the potential by: $\mathbf{u} \leftarrow \mathbf{u} + \eta \mathbf{e}_{\mathcal{N} \setminus \mathcal{S}}$.
End

End

End ∎

The usual justification of the out-of-kilter algorithm is based on the concept of a *kilter number.* This is a measure of the *out-of-kilterness* of the flow-differential pair for each arc. If the flow-differential pair for an arc is in-kilter, the corresponding kilter number is zero. Otherwise, the kilter number is strictly positive. There are two possible ways to define this kilter number. If the starting flow is feasible with respect to capacity, then the area of the rectangle subtended by the point (x_j, v_j)

with the kilter curve can be used as one definition of kilter number. If the flow is infeasible with respect to capacity, the horizontal distance from the out-of-kilter point (x_j, v_j) from the kilter curve can also be used as the kilter number. It remains to show that each iteration of the out-of-kilter algorithm strictly reduces the total kilter number, and if the usual commensurability assumption applies, this reduction is a multiple of some positive quantum, and therefore will not be ever-diminishing.

• In the primal phase, the change of flow around a cycle will be such that the flow-differential pair for each arc in the cycle remains on, or moves horizontally closer to the kilter curve, and therefore the kilter number, defined in either way, will be non-increasing for each arc on the cycle, with at least one strictly decreasing. In particular, the arc (or arcs) in the cycle that achieves the minimum in the calculation of η will enter into the kilter curve.

• In the dual phase, the flow-differential pair for each arc on the cut remains on, or moves vertically closer to the kilter curve. In particular, the arc (or arcs) in the cut that achieves the minimum in the calculation of η, will enter into the kilter curve. Thus if the kilter number is measured as the area of rectangle, the kilter number for each arc in the cut will be non-increasing, with at least one strictly decreasing. Otherwise if the kilter number is measured as the horizontal distance, then only the arc (or arcs) in the cut that achieves the minimum η will have its kilter number reduced from some positive value to zero.

• Since the total kilter number decreases strictly as a multiple of a quantum, eventually the algorithm must terminate. If there is some unbalanced cycle or unbalanced cut discovered by the algorithm, then by Theorem 2.8.1 and Theorem 2.8.3, either the optimal solution is unbounded, or the problem is infeasible. Otherwise the finite optimal solution will be obtained after a finite number of iterations. ∎

The linear cost optimal flow problem and the linear cost optimal potential problem are dual to each other under a further assumption. Bearing in mind that adding or subtracting a constant to the objective function of an optimization problem does not change the optimal solution, let c_j, d_j, p_j, and q_j be numbers satisfying the following constraint:

$$p_j + q_j = -c_j d_j. \tag{2.8.5}$$

Definition 2.8.7 (Primal Optimal Flow Problem)

$$\min_{\mathbf{x}} \quad \sum_{j\in\mathcal{A}} d_j x_j + p_j \tag{2.8.6}$$

$$\text{subject to} \quad x_j \geq c_j \;\; \forall j \in \mathcal{A} \tag{2.8.7}$$

$$\text{and} \quad \mathbf{E}\mathbf{x} = \mathbf{b}. \tag{2.8.8}$$

Definition 2.8.8 (Dual Optimal Potential Problem)

$$\max_{\mathbf{u},\mathbf{v}} \quad -\sum_{j\in\mathcal{A}} (c_j v_j + q_j) - \sum_{i\in\mathcal{N}} b_i u_i \tag{2.8.9}$$

$$\text{subject to} \quad v_j \leq d_j \;\; \forall j \in \mathcal{A} \tag{2.8.10}$$

$$\text{and} \quad \mathbf{v} = -\mathbf{E}^{\top}\mathbf{u}. \tag{2.8.11}$$

It is easy to check that the two problems share the same kilter curves (see Figure 2.8.3):

$$\Gamma_j = \{(x_j, v_j) \in \mathbb{R}^2 \mid x_j \geq c_j, v_j \leq d_j, (x_j - c_j)(v_j - d_j) = 0\}$$

and therefore they also share the same optimality condition. Thus solving for the optimal flow to the primal will yield a potential which is the optimal solution to the dual, and symmetrically, solving for the optimal potential to the dual will yield a flow which is the optimal solution to the primal.

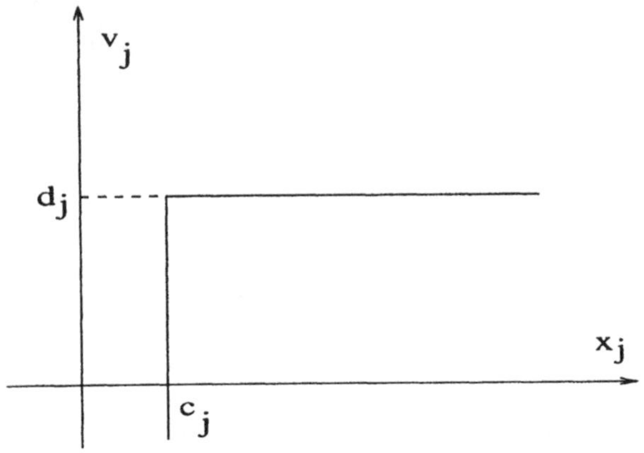

FIGURE 2.8.3 Kilter curve for arc j in optimal flow/optimal potential

The above conclusion can be deduced through the usual weak/strong duality routine.

Theorem 2.8.5 (Weak Duality) If $\mathbf{x}$ is a feasible solution to the primal, and $\mathbf{u}$ is a feasible solution to the dual then the primal cost is always greater or equal to the dual cost.

Proof: Let $\mathbf{x}$ and $\mathbf{u}$ be feasible solutions to the primal and the dual respectively, then from (2.2.3)

$$\mathbf{x}^\top \mathbf{v} = -\mathbf{b}^\top \mathbf{u}. \tag{2.8.12}$$

The difference between the primal cost and the dual cost is then:

$$\begin{aligned}
&\sum_{j \in \mathcal{A}} d_j x_j + p_j - \left[- \sum_{i \in \mathcal{N}} b_i u_i - \sum_{j \in \mathcal{A}} (c_j v_j + q_j) \right], \\
&\quad = \sum_{j \in \mathcal{A}} d_j x_j + c_j v_j - v_j x_j + p_j + q_j, \qquad \text{by (2.8.12)} \\
&\quad = \sum_{j \in \mathcal{A}} d_j x_j + c_j v_j - v_j x_j - c_j d_j, \qquad \text{by (2.8.5)} \\
&\quad = \sum_{j \in \mathcal{A}} (x_j - c_j)(d_j - v_j) \geq 0 \qquad \text{by feasibility.}
\end{aligned} \tag{2.8.13}$$

■

Theorem 2.8.6 (Strong Duality) If both the primal problem and the dual problem have feasible solutions, then the optimal primal cost and the optimal dual cost are finite and share the same value. If only the primal has a feasible solution, it has an unbounded infimum. If only the dual has a feasible solution, it has an unbounded supremum.

Proof: Note that the primal is feasible, and that the primal has a finite minimum as the dual is feasible, by the weak duality. An application of the out-of-kilter algorithm will end up with all arcs satisfying the kilter condition, and hence the inequality in (2.8.13) holds as equality for all arcs. By weak duality, the minimum of the primal and the maximum of the dual must be equal. If the dual is infeasible, it has a supremum of $-\infty$, so has the infimum of the primal by applying the out-of-kilter algorithm. If the primal is infeasible, it has an infimum of ∞, so has the supremum of the dual. ■

Remark 2.8.4 Either by a suitable change in origin, or by setting the parameters $p_j = c_j = q_j = 0$, the primal and dual problems take the form of a standard pair of asymmetric linear programs

$$\begin{aligned} \text{Primal:} \quad & \min \ \mathbf{d}^{\mathsf{T}}\mathbf{x} \quad \text{s.t. } \mathbf{E}\mathbf{x} = \mathbf{b}, \quad \mathbf{x} \geqq \mathbf{0}, \\ \text{Dual:} \quad & \max \ \mathbf{b}^{\mathsf{T}}\mathbf{u} \quad \text{s.t. } -\mathbf{u}^{\mathsf{T}}\mathbf{E} \leqq \mathbf{d}^{\mathsf{T}}. \end{aligned}$$

The kilter condition of $(x_j, v_j) \in \Gamma_j \quad \forall j$ translates into the well-known complementary slackness condition:

$$\mathbf{x} \geqq \mathbf{0}, \quad \mathbf{u}^{\mathsf{T}}\mathbf{E} + \mathbf{d}^{\mathsf{T}} \geqq \mathbf{0}^{\mathsf{T}}, \quad [\mathbf{E}^{\mathsf{T}}\mathbf{u} + \mathbf{d}]^{\mathsf{T}}\mathbf{x} = 0.$$

The results in this section are all special cases of a much more general class of convex optimization problems, which will be taken up in the next section.

2.9 Duality of Convex Optimal Flow and Optimal Potential

We now return to the general optimal flow and optimal potential problems first addressed in Definitions 2.3.4 and 2.3.5. Here we impose the additional assumption that all the underlying functions are convex. This general class of problems include all the problems we have discussed so far as special cases. The resulting duality theory can thus be considered as a generalization of all the duality results discussed in this chapter.

Consider a digraph $\mathcal{G} = (\mathcal{N}, \mathcal{A})$ with a node-arc incidence matrix $\mathbf{E}$. Each arc $j \in \mathcal{A}$ is associated with a capacity interval C_j and a convex cost function $f_j : C_j \to \mathbb{R}$. To extend the domain of f_j to $\mathbb{R}$, we adopt the convention:

$$f_j(\xi) = +\infty \;\forall \xi \notin C_j.$$

In which case we can rewrite C_j as

$$C_j = \{\xi \in \mathbb{R} \mid f_j(\xi) < \infty\}. \tag{2.9.1}$$

Unlike in Section 2.8, here the capacity intervals C_j's need not be closed, although the function f_j's are assumed to be closed. Let the end points of C_j be c_j^- and c_j^+, where c_j^- can be $-\infty$ and c_j^+ can be ∞. In general f_j can be assumed to be continuous on C_j but may not be continuously differentiable. Its left and right derivatives are defined as follows.

Definition 2.9.1 (Left and right derivatives) Let $\xi \in C_j$. The *left and right derivatives* of $f_j : C_j \to \mathbb{R}$ are defined by

$$f'_{j-}(\xi) = \lim_{\zeta \uparrow \xi} \frac{f_j(\xi) - f_j(\zeta)}{\xi - \zeta}, \tag{2.9.2}$$

$$f'_{j+}(\xi) = \lim_{\zeta \downarrow \xi} \frac{f_j(\zeta) - f_j(\xi)}{\zeta - \xi}. \tag{2.9.3}$$

Let

$$\tilde{C}_j = \{\xi \in \mathbb{R} | f'_{j-}(\xi) < +\infty \text{ and } f'_{j+}(\xi) > -\infty\} \subset C_j.$$

Remark 2.9.1 Note that, since f_j is convex, then
(i) For all $\xi \in C_j$,

$$f'_{j-}(\xi) \le f'_{j+}(\xi), \tag{2.9.4}$$

with equality if and only if f_j is differentiable at ξ.
(ii) Both the left and right derivatives are finite in $\tilde{C}_j$ with

$$f'_{j+}(\xi') \le f'_{j-}(\xi'') \quad \text{whenever} \quad \xi' < \xi''. \tag{2.9.5}$$

(iii) f'_{j-} and f'_{j+} are both monotone non-decreasing on C_j.
(iv) f'_{j-} and f'_{j+} form the end points of the subdifferential of f_j, i.e., $\partial f_j(\xi) = [f'_{j-}(\xi), f'_{j+}(\xi)]$.

We extend the domains of f'_{j+} and f'_{j-} from C_j to $\mathbb{R}$ by defining

$$f'_{j-}(\xi) = f'_{j+}(\xi) = +\infty \quad \forall \xi \notin C_j \text{ with } \xi \geq c_j^+, \tag{2.9.6}$$
$$f'_{j-}(\xi) = f'_{j+}(\xi) = -\infty \quad \forall \xi \notin C_j \text{ with } \xi \leq c_j^-. \tag{2.9.7}$$

Under this definition, the inequalities (2.9.4) and (2.9.5) then hold $\forall \xi \in \mathbb{R}$.

Remark 2.9.2 Note that if $c_j^+ \in C_j$, then $f'_{j+}(c_j^+) = +\infty$ but $f'_{j-}(c_j^+)$ can be finite. Similarly if $c_j^- \in C_j$, then $f'_{j-}(c_j^-) = -\infty$ but $f'_{j+}(c_j^-)$ can be finite.

In the previous section, the concept of kilter curve was introduced specifically for the optimality condition of a linear problem. In a more general sense, we can think of a kilter curve as the derivative function of a continuous but non-differentiable function.

Definition 2.9.2 (Kilter curve) The kilter curve Γ_j associated with the convex function f_j is defined to be the set

$$\Gamma_j = \{(\xi, \eta) \in \mathbb{R}^2 \mid f'_{j-}(\xi) \leq \eta \leq f'_{j+}(\xi), \quad \text{or} \quad \eta \in \partial f_j(\xi)\}. \tag{2.9.8}$$

Remark 2.9.3 Note that

$$\begin{aligned} \tilde{C}_j &= \{\xi \in \mathbb{R} \mid \exists \eta \in \mathbb{R} \text{ with } f'_{j-}(\xi) \leq \eta \leq f'_{j+}(\xi)\} \\ &= \{\xi \in \mathbb{R} \mid \exists \eta \in \mathbb{R} \text{ such that } (\xi, \eta) \in \Gamma_j\} \\ &= \text{Projection of } \Gamma_j \text{ onto the horizontal axis.} \end{aligned} \tag{2.9.9}$$

In order to construct a meaningful dual, we associate, in addition to f_j (primal cost) and C_j (capacity interval), with each arc j a *span interval* D_j and a closed convex (dual) cost function $g_j : D_j \to \mathbb{R}$ where g_j is assumed to be continuous but not necessarily differentiable. D_j is not necessarily closed but has end points d_j^- and d_j^+. Let

$$\tilde{D}_j = \{\eta \in \mathbb{R} \mid g'_{j-}(\eta) < +\infty \text{ and } g'_{j+}(\eta) > -\infty\} \subset D_j.$$

To extend the domain of g_j to $\mathbb{R}$, we let $g_j(\eta) = +\infty \ \ \forall \eta \notin D_j$ so that

$$D_j = \{\eta \in \mathbb{R} \mid g_j(\eta) < \infty\}. \tag{2.9.10}$$

We assume that f_j and g_j are conjugate to each other, i.e., $\forall\ j$,

$$g_j(\eta) = \sup_{\xi \in \mathbf{R}} \{\xi\eta - f_j(\xi)\} = -\inf_{\xi \in C_j} \{f_j(\xi) - \xi\eta\}, \tag{2.9.11}$$
$$f_j(\xi) = \sup_{\eta \in \mathbf{R}} \{\xi\eta - g_j(\eta)\} = -\inf_{\eta \in D_j} \{g_j(\eta) - \xi\eta\}. \tag{2.9.12}$$

The duality between f_j and g_j has an obvious geometrical interpretation. If f_j is associated with its kilter curve Γ_j and if g_j is as defined in (2.9.11), then g_j also has a kilter curve which turns out to be the natural inverse of Γ_j, namely

$$\Gamma_j^{-1} = \{(\eta, \xi) \in \mathbb{R}^2 \mid (\xi, \eta) \in \Gamma_j\}. \tag{2.9.13}$$

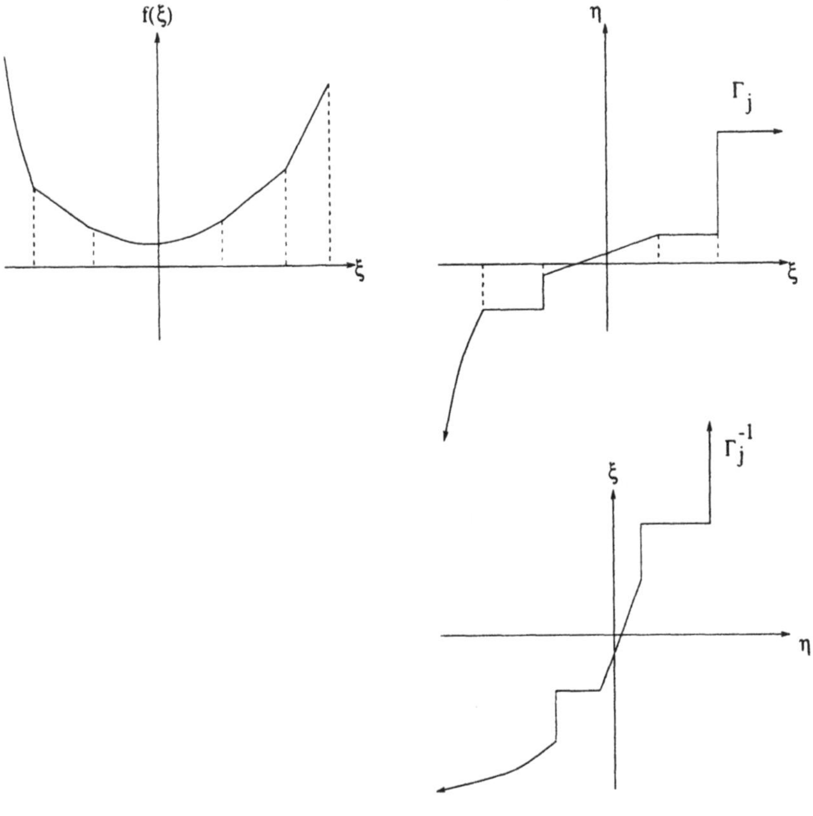

FIGURE 2.9.1 A convex function and its kilter curve

Shortly, Γ_j^{-1} will be shown to be the kilter curve for g_j defined in a completely symmetrical manner:

$$\Gamma_j^{-1} = \{(\eta, \xi) \in \mathbb{R}^2 \mid g'_{j^-}(\eta) \leq \xi \leq g'_{j^+}(\eta)\}. \tag{2.9.14}$$

Thus

$$g'_{j^-}(\eta) \leq \xi \leq g'_{j^+}(\eta) \Leftrightarrow (\eta, \xi) \in \Gamma_j^{-1} \Leftrightarrow (\xi, \eta) \in \Gamma_j \Leftrightarrow f'_{j^-}(\xi) \leq \eta \leq f'_{j^+}(\xi), \tag{2.9.15}$$

and

$$\tilde{D}_j = \{\eta \in \mathbb{R} \mid \exists \xi \in \mathbb{R} \text{ with } (\eta, \xi) \in \Gamma_j^{-1}\} \\ = \text{projection of } \Gamma_j \text{ onto the vertical axis.} \tag{2.9.16}$$

As a result of (2.9.9) and (2.9.16), the end points of C_j and D_j are given by

$$\begin{aligned} d_j^+ &= \lim_{\xi \to +\infty} f'_{j^-}(\xi) = \lim_{\xi \to +\infty} f'_{j^+}(\xi), \\ d_j^- &= \lim_{\xi \to -\infty} f'_{j^-}(\xi) = \lim_{\xi \to -\infty} f'_{j^+}(\xi), \end{aligned} \tag{2.9.17}$$

and

$$\begin{aligned} c_j^+ &= \lim_{\eta \to +\infty} g'_{j^-}(\eta) = \lim_{\eta \to +\infty} g'_{j^+}(\eta), \\ c_j^- &= \lim_{\eta \to -\infty} g'_{j^-}(\eta) = \lim_{\eta \to -\infty} g'_{j^+}(\eta). \end{aligned} \tag{2.9.18}$$

(2.9.17) implies that d_j^+ and d_j^- are the extreme slope values of f_j, and (2.9.18) implies that c_j^+ and c_j^- are the extreme slope values of g_j. Figure 2.9.1 shows an example of a convex function together with its kilter curve, which illustrates that Γ_j and Γ_j^{-1} are mirror images to each other about the line $\xi = \eta$.

The following result is a special case of Fenchel duality discussed in Chapter One that concerns convex functions of a scalar variable. It will be used to establish strong duality as the main result of this section.

Theorem 2.9.1 (Fenchel Duality) Let f_j be a convex function, let g_j be defined by (2.9.11). Then
(i) $f_j(\xi) + g_j(\eta) \geq \xi\eta$,
(ii) g_j is convex,
(iii) $f_j(\xi) + g_j(\eta) = \xi\eta \Leftrightarrow (\xi, \eta) \in \Gamma_j$ (or $(\eta, \xi) \in \Gamma_j^{-1}$).

Proof: (i) Follows directly from (2.9.11).

(ii) For $\lambda \in (0, 1)$, let $\eta = (1 - \lambda)\eta' + \lambda\eta''$. Then $\forall \xi \in \mathbb{R}$, we have

$$\begin{aligned} (1 - \lambda) g_j(\eta') &\geq (1 - \lambda)(\xi\eta' - f_j(\xi)), \\ \lambda g_j(\eta'') &\geq \lambda(\xi\eta'' - f_j(\xi)), \end{aligned}$$

Adding the above, we have

$$(1 - \lambda) g_j(\eta') + \lambda g_j(\eta'') \geq \xi\eta - f_j(\xi) \quad \forall \xi.$$

In particular,

$$(1 - \lambda) g_j(\eta') + \lambda g_j(\eta'') \geq \sup_{\xi}\{\xi\eta - f_j(\xi)\} = g_j(\eta).$$

(iii) Fix η and let $h_j(\xi) = \xi\eta - f_j(\xi)$ then

$$h'_{j+}(\xi) = \eta - f'_{j+}(\xi),$$
$$h'_{j-}(\xi) = \eta - f'_{j-}(\xi).$$

By Theorem 1.2.18, $h(\xi)$ is maximized at ξ if and only if $h'_{j-}(\xi) \geq 0 \geq h'_{j+}(\xi)$. Consequently, ξ maximizes $h_j(\xi)$ if and only if $f'_j(\xi) \leq \eta \leq f'_{j+}(\xi)$, or $(\xi, \eta) \in \Gamma_j$. Similarly, $(\eta, \xi) \in \Gamma_j^{-1}$ if and only if $f_j(\xi) + g_j(\eta) = \xi\eta$. ∎

Example 2.9.1 Let $q > 0$. Consider the following convex function, where the subscript j is suppressed for convenience:

$$f(\xi) = \begin{cases} \frac{1}{2}q\xi^2 + p\xi & \text{if } \xi \in [c^-, c^+] = C, \\ +\infty & \text{otherwise.} \end{cases}$$

We construct a dual cost function and a dual span interval via (2.9.11) and (2.9.17) as follows:

$$g(\eta) = \sup_{\xi \in C}\{\xi\eta - \frac{1}{2}q\xi^2 - p\xi\} = \sup_{\xi \in C}\{h(\xi)\}.$$

The function $h(\xi) = \xi\eta - \frac{1}{2}q\xi^2 - p\xi$ has a minimum at $\hat{\xi}$ where

$$h'(\hat{\xi}) = \eta - q\hat{\xi} - p = 0 \Rightarrow \hat{\xi} = \frac{\eta - p}{q}.$$

If $\hat{\xi} \in [c^-, c^+]$ then

$$\xi^* = \hat{\xi} = \frac{\eta - p}{q},$$

and

$$g(\eta) = h(\xi^*) = \frac{(\eta - p)}{q}\eta - \frac{1}{2}q\left(\frac{(\eta - p)}{q}\right)^2 - p\frac{(\eta - p)}{q}$$
$$= \frac{1}{2q}(\eta - p)^2.$$

If $\hat{\xi} < c^-$, then, $\xi^* = c^-$ and

$$g(\eta) = c^-\eta - \frac{1}{2}q(c^-)^2 - pc^-$$
$$= c^-(\eta - p) - \frac{1}{2}\, q(c^-)^2.$$

If $\hat{\xi} > c^+$, then $\xi^* = c^+$ and

$$g(\eta) = c^+(\eta - p) - \frac{1}{2}q(c^+)^2 \quad \text{(Note } g'(\eta) = c^+\text{)}.$$

Also

$$c^- \leq \hat{\xi} \leq c^+ \Rightarrow c^- \leq \frac{\eta - p}{q} \leq c^+ \Rightarrow c^-q + p \leq \eta \leq c^+q + p,$$
$$\hat{\xi} < c^- \Rightarrow \frac{\eta - p}{q} < c^- \Rightarrow \eta < c^-q + p,$$
$$\hat{\xi} > c^+ \Rightarrow \frac{\eta - p}{q} > c^+ \Rightarrow \eta > c^+q + p,$$

and hence

$$d^+ = \lim_{\xi \to +\infty} f'(\xi) = +\infty,$$
$$d^- = \lim_{\xi \to -\infty} f'(\xi) = -\infty.$$

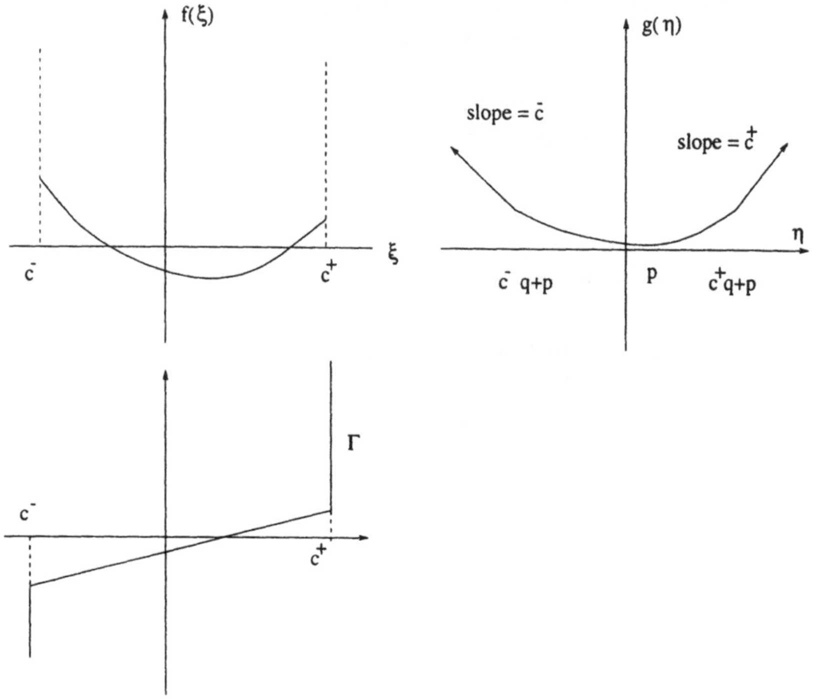

FIGURE 2.9.2 Graphs and kilter curves for Example 2.9.1

In summary, we have

$$g(\eta) = \begin{cases} \frac{1}{2q}(\eta - p)^2 & \text{if } c^- q + p \leq \eta \leq c^+ q + p \text{ (quadratic)} \\ c^-(\eta - p) - \frac{1}{2}qc^- & \text{if } \eta < c^- q + p \text{ (linear)} \\ c^+(\eta - p) - \frac{1}{2}qc^+ & \text{if } \eta > c^+ q + p \text{ (linear)} \end{cases}$$

and $D = (-\infty, +\infty)$.

Having laid the necessary foundation, we now address the duality of convex cost network optimization problems. Two special cases of the optimal flow and potential problems are stated as follows:

Definition 2.9.3 (Convex Optimal Flow Problem P_p)

$$\inf_{\mathbf{x}} \ \Phi(\mathbf{x}) = \sum_{j \in \mathcal{A}} f_j(x_j)$$
$$s.t. \quad \mathbf{E}\mathbf{x} = \mathbf{b} \quad \text{where} \quad b_{\mathcal{N}} = \mathbf{1}^\top \mathbf{b} = 0$$
$$x_j \in C_j \quad \forall\, j \in \mathcal{A}.$$

Definition 2.9.4 (Concave Optimal Potential Problem P_d)

$$\sup_{\mathbf{u}} \ \Psi(\mathbf{u}) = -\mathbf{u}^{\top}\mathbf{b} - \sum_{j} g_j(v_j),$$

$$\text{where} \quad \mathbf{v} = \Delta\mathbf{u} = -\mathbf{E}^{\top}\mathbf{u},$$

$$s.t. \quad v_j \in D_j \quad \forall_j \in \mathcal{A}.$$

Note that Ψ is concave if g_j is convex. It is also assumed that f_j, g_j, C_j, D_j are related in the manner of Fenchel Duality as in (2.9.11), (2.9.12), (2.9.17), and (2.9.18). Let Γ_j and Γ_j^{-1} be the kilter curves for f_j and g_j respectively. The convex cost flow and potential problems are related through the following equilibrium problem.

Definition 2.9.5 The *Network Equilibrium Problem* P_e is given as: Find $\mathbf{x}$ and $\mathbf{u}$ such that $\mathbf{Ex} = \mathbf{b}$ and $(x_j, v_j) \in \Gamma_j \ \ \forall\, j$, or equivalently, $(v_j, x_j) \in \Gamma_j^{-1} \ \ \forall\, j$. Any flow-potential pair $(\mathbf{x}, \mathbf{u})$ that satisfies this condition is said to be in *equilibrium.* We refer to the condition $(x_j, v_j) \in \Gamma_j \ \forall j$, (or $(v_j, x_j) \in \Gamma_j^{-1} \ \forall j$) as the *equilibrium condition*

Remark 2.9.4 It is important to note that one of f_j, g_j and Γ_j determines the other two uniquely.

Theorem 2.9.2 (Convex Network Duality Theorem) If at least one of the problems P_p and P_d has a feasible solution, then

$$\text{inf in Problem } \mathrm{P}_p = \text{sup in Problem } \mathrm{P}_d.$$

Furthermore, Inf in Problem $\mathrm{P}_p = +\infty$ (i.e., there is no feasible solution for Problem P_p) if and only if Problem P_d is unbounded. Similarly, Sup in Problem $\mathrm{P}_d = -\infty$ (i.e., there is no feasible solution for Problem P_d) if and only if Problem P_p is unbounded.

Proof: Since $\mathbf{Ex} = \mathbf{b}$, so $\mathbf{u}^{\top}\mathbf{Ex} = \mathbf{u}^{\top}\mathbf{b}$ or

$$-\mathbf{v}^{\top}\mathbf{x} = \mathbf{u}^{\top}\mathbf{b}. \tag{2.9.19}$$

Since f_j and g_j are conjugate to each other,

$$f_j(x_j) \geq x_j v_j - g_j(v_j) \ \forall j, \tag{2.9.20}$$

hence

$$\begin{aligned}\Phi(x) &= \sum_{j\in\mathcal{A}} f_j(x_j) \geq \mathbf{x}^{\top}\mathbf{v} - \sum_{j\in\mathcal{A}} g_j(v_j)\\ &= -\mathbf{u}^{\top}\mathbf{b} - \sum_{j\in\mathcal{A}} g_j(v_j) = \Psi(\mathbf{u}).\end{aligned}$$

Hence we have the weak duality:

$$\Phi(\mathbf{x}) \geq \Phi(\mathbf{x}^*) \geq \Psi(\mathbf{u}^*) \geq \Psi(\mathbf{u}). \tag{2.9.21}$$

From Theorem 2.9.1(iii), equality holds in (2.9.20) if and only if $(x_j^*, v_j^*) \in \Gamma_j \;\; \forall\, j$. In which case, we have the strong duality

$$\Phi(\mathbf{x}^*) = \Psi(\mathbf{u}^*).$$

∎

As a direct consequence of the duality theorem, we have,

Corollary 2.9.3 (Network Equilibrium Theorem) The flow $\mathbf{x}$ and potential $\mathbf{u}$ satisfy the equilibrium condition if and only if $\mathbf{x}$ solves P_p and $\mathbf{u}$ solves P_d.

The duality between optimal flow and optimal potential is not only aesthetically appealing, it also has practical significance in terms of computational algorithms. The following algorithms for solving the optimal flow and optimal potential problems are generalization of the algorithms discussed in the previous section for linear cost problems.

Algorithm 2.9.1 (Optimal Flow Algorithm for Problem $\mathbf{P}_p$)

Initialization. Find a feasible flow $\mathbf{x}$ using either the feasible flow Algorithm 2.7.1 or the flow rectification Algorithm 2.7.2 .

Do:

Define the flow dependent span interval for arcs j as

$$D_j = [d_j^-, d_j^+] = \partial f_j(x_j) = [f'_{j-}(x_j), f'_{j+}(x_j)] = \{v_j \mid (x_j, v_j) \in \Gamma_j\}.$$

Find a potential $\mathbf{u}$ such that $v_j = [-\mathbf{E}^\top \mathbf{u}]_j$ is feasible with respect to $D_j \;\; \forall\, j$ by either the Feasible Potential Algorithm 2.7.3 or the Tension Rectification Algorithm 2.7.4.

If a feasible potential is found, stop, $(\mathbf{x}, \mathbf{u})$ is optimal.

Else there exists an elementary cycle $\mathcal{P}$ such that $d_{\mathcal{P}}^+ < 0$. Let

$$\rho(t) = \Phi(\mathbf{x} + t\mathbf{e}_{\mathcal{P}}) = \sum_{j \in \mathcal{P}^+} f_j(x_j + t) + \sum_{j \in \mathcal{P}^-} f_j(x_j - t) + \sum_{j \notin \mathcal{P}} f_j(x_j).$$

Then the function ρ has a right derivative:

$$\begin{aligned} \rho'_+(0) &= \sum_{j \in \mathcal{P}^+} f'_{j+}(x_j) - \sum_{j \in \mathcal{P}^-} f'_{j-}(x_j) \\ &= d_{\mathcal{P}}^+ < 0. \end{aligned} \qquad (2.9.22)$$

Consequently $\mathbf{e}_{\mathcal{P}}$ is a descent direction for the current flow $\mathbf{x}$.

Let $\eta = \min\{t > 0 \mid \rho'_+(t) \geq 0\}$, then $\eta > 0$.

If $\eta = \infty$, then $\rho'_+(t) < 0 \;\; \forall t > 0 \Rightarrow \inf = -\infty$. Stop.

Else $\eta < \infty$, $\quad \mathbf{x} \leftarrow \mathbf{x} + \eta \mathbf{e}_{\mathcal{P}}$.
End

End

End ∎

Remark 2.9.5 Note that,

(i) $\mathbf{x} + \eta \mathbf{e}_{\mathcal{P}}$ is feasible, since it satisfies both $\mathbf{Ex} = \mathbf{b}$ and $x_j \in C_j$. The feasibility with respect to capacity is less obvious, but it can be inferred from the fact that $f'_{j+}(c_j^+) = +\infty$ and $f'_{j-}(c_j^-) = -\infty$.
(ii) From (2.9.22), $\Phi(\mathbf{x} + \eta \mathbf{e}_{\mathcal{P}}) < \Phi(\mathbf{x})$ so a strictly reduced cost is guaranteed.
(iii) As it stands, there is no guarantee that the algorithm will terminate in a finite number of iterations, as the cost may be reduced by an ever diminishing amount. As computational issues are beyond the scope of this book, the reader is referred to [R3] for details.

The dual counterpart to the optimal flow algorithm is as follows. Note the striking symmetry with the above algorithm.

Algorithm 2.9.2 (Optimal Potential Algorithm for Problem $\mathbf{P}_d$)

Initialization. Find a feasible potential $\mathbf{u}$ to the optimal potential problem by either the feasible potential Algorithm 2.7.3 or the tension rectification Algorithm 2.7.4.

Do:

Define the potential dependent capacity interval for arc j

$$\begin{aligned} C_j = [c_j^-,\, c_j^+] &= [g'_{j-}(v_j),\; g'_{j+}(v_j)] \\ &= \{x_j \mid (x_j, v_j) \in \Gamma_j\}. \end{aligned}$$

Find a feasible flow $\mathbf{x}$ such that $x_j \in C_j \;\; \forall\, j$ by either the feasible flow Algorithm 2.7.1 or the flow rectification Algorithm 2.7.2.

If a feasible flow $\mathbf{x}$ is found, stop, $(\mathbf{x}, \mathbf{u})$ is optimal.

Else there exists an elementary cut $\mathcal{Q} = [\mathcal{S}, \mathcal{N} \setminus \mathcal{S}]$ with $c_{\mathcal{Q}}^+ < b_{\mathcal{S}}$. Let

$$\begin{aligned} \mu(t) &= \Psi(\mathbf{u} + t\mathbf{e}_{\mathcal{N}\setminus\mathcal{S}}) \\ &= -\mathbf{b}^\top \mathbf{u} - t b_{\mathcal{N}\setminus\mathcal{S}} - \sum_{j \in \mathcal{Q}^+} g_j(v_j + t) - \sum_{j \in \mathcal{Q}^-} g_j(v_j - t) + \sum_{j \notin \mathcal{Q}} g_j(v_j). \end{aligned}$$

Then the function $\mu(t)$ has a right derivative:

$$\begin{aligned}\mu'_+(0) &= -b_{\mathcal{N}\setminus\mathcal{S}} - \left(\sum_{j\in Q^+} g'_{j^+}(v_j) - \sum_{j\in Q^-} g'_{j^-}(v_j)\right)\\ &= b_{\mathcal{S}} - \left(\sum_{j\in Q^+} c_j^+ - \sum_{j\in Q^-} c_j^-\right)\\ &= b_{\mathcal{S}} - c_Q^+ > 0. \end{aligned} \tag{2.9.23}$$

Consequently $\mathbf{e}_{\mathcal{N}\setminus\mathcal{S}}$ is an ascent direction for the current potential $\mathbf{u}$.

Let $\eta = \min\{t > 0 \mid \mu'_+(t) \leq 0\}$, then $\eta > 0$.

If $\eta = \infty$, i.e., $\mu'_+(t) > 0 \;\; \forall t \in \mathbb{R}^+ \;\; \Rightarrow \sup = +\infty$, stop.

Else $\eta < \infty$, $\;\mathbf{u} \leftarrow \mathbf{u} + \eta\mathbf{e}_{\mathcal{N}\setminus\mathcal{S}}$.

End

End

End ∎

Remark 2.9.6 Note that,

(i) $\mathbf{u} + \eta\mathbf{e}_{\mathcal{N}\setminus\mathcal{S}}$ is feasible by virtue of the definition of η.

(ii) From (2.9.23) $\Psi(\mathbf{u} + \eta\mathbf{e}_{\mathcal{N}\setminus\mathcal{S}}) > \Psi(\mathbf{u})$ so a strictly increased cost is guaranteed.

(iii) As in the case of optimal flow, there is no guarantee that the algorithm will terminate in a finite number of iterations.

CHAPTER 3

DUALITY IN LINEAR SYSTEMS

By linear systems, we mean algebraic systems where the underlying functions are linear, and the underlying sets are polyhedral. Such systems are trivially convex, and in addition enjoy many special properties not applicable to nonlinear systems. We begin by summarizing a few key results from the most popular optimization model: *linear programming,* henceforth abbreviated as LP. For readers who have little or no basic knowledge of the subject, we provide a succinct summary of the key theoretical results in Section 3.1. This is followed by the discussion of perhaps the best known duality result in optimization in Section 3.2. The subject of complementary slackness is closely related to LP duality, and this will be studied in Section 3.3. In Section 3.4, theorems of alternatives, otherwise known as transposition theorems, are discussed and shown to be another manifestation of LP duality. In the last two sections, the duality of linear systems is generalized to that of a combinatorial structure in terms of primal and dual supports. This has its origin in matroid theory and leads to a very powerful generalization of network optimization model referred to as monotropic optimization.

3.1 A Crash Course in Linear Programming

As the title implies, Linear Programming (LP) is concerned with optimization problems with a linear cost function and linear constraints. The origin of LP seems to be attributable to the Russian mathematician Kantorovitch [K4] in 1939, although much credit has been attributed to George Dantzig who invented the simplex algorithm for solving LP [D1]. LP enjoys many special properties characterized by polyhedral sets. In spite of the fact that there are apparently superior alternatives for solving LP (see, for example, [FP1]), the simplex method remains well worth studying because of its underlying linear and discrete structure. More importantly in the context of this book, it has a direct bearing to the duality of linear systems.

There are probably as many books written on LP as there are written on a popular subject like Calculus. The reader should be reminded that this is not a specialized book about optimization, and for that matter, even less so for linear programming. As with every other chapter, we only intend to highlight the duality aspects of linear programming. In doing so, we need to repeat some of the standard stuff just so that this book is self-contained. Unlike other books on LP, computational issues and algorithmic development will receive very little attention here. The notation used in this section is pretty much the standard notation of most LP text books, in particular, we follow closely that of [L1]. This unfortunately is somewhat inconsistent with the notation adopted in Chapter Two for network analysis. Readers are urged to exercise caution when switching from the previous chapter to the current one.

We begin by studying the properties of solutions satisfying the following linear constraints: (The upper bound on the variable is included here to allow for broader generality)

$$\mathbf{A}\mathbf{x} = \mathbf{b} \tag{3.1.1}$$

$$\boldsymbol{\ell} \leqq \mathbf{x} \leqq \mathbf{u} \tag{3.1.2}$$

where $\mathbf{x}, \boldsymbol{\ell}, \mathbf{u} \in \mathbb{R}^n, \mathbf{b} \in \mathbb{R}^m, \mathbf{A} \in \mathbb{R}^{m \times n}$ is of full rank. In general $m < n$, otherwise the system (3.1.1) is overdetermined. Since $\mathbf{A}$ is of full rank m, there exist m linearly independent columns of $\mathbf{A}$. We think of these as a set of basis vectors which span $\mathbb{R}^m$. For convenience and without loss of generality, we select the first m columns and denote $\mathbf{B}$ as the nonsingular matrix comprising of these columns. $\mathbf{B}$ is called the *basis matrix* for the system. Let $\mathbf{N}$ be the $m \times (n-m)$ matrix made up of the remaining columns of $\mathbf{A}$, so that $\mathbf{A} = [\mathbf{B} \,\vdots\, \mathbf{N}]$. Partition the vector $\mathbf{x}$ that satisfies (3.1.1) into $\mathbf{x}^\top = [\mathbf{x}_\mathbf{B}^\top \ \mathbf{x}_\mathbf{N}^\top]$, where $\mathbf{x}_\mathbf{B} \in \mathbb{R}^m$ corresponds to the basis $\mathbf{B}$ and $\mathbf{x}_\mathbf{N} \in \mathbb{R}^{n-m}$ corresponds to $\mathbf{N}$. One solution of (3.1.1) can be obtained by assigning arbitrary values to the vector $\mathbf{x}_\mathbf{N}$, then solving the equation

$$\mathbf{B}\mathbf{x}_\mathbf{B} = \mathbf{b} - \mathbf{N}\mathbf{x}_\mathbf{N} \tag{3.1.3}$$

uniquely for the vector $\mathbf{x}_\mathbf{B}$.

Definition 3.1.1 Given the system (3.1.1), let $\mathbf{B}$ be any nonsingular $m \times m$ matrix made up of linearly independent columns of $\mathbf{A}$. Partition the vector that satisfies (3.1.1) into $\mathbf{x}^\top = [\mathbf{x}_\mathbf{B}^\top \ \mathbf{x}_\mathbf{N}^\top]$. Let $[\mathbf{x}_\mathbf{N}]_j = \ell_j$ or u_j and $\mathbf{B}\mathbf{x}_\mathbf{B} = \mathbf{b} - \mathbf{N}\mathbf{x}_\mathbf{N}$. Then $\mathbf{x}$ is called a *basic solution* to (3.1.1) with respect to the basis $\mathbf{B}$. The components of $\mathbf{x}$ associated with $\mathbf{B}$, i.e., $\mathbf{x}_\mathbf{B}$, are called *basic variables* and the components in $\mathbf{x}_\mathbf{N}$ are called *nonbasic variables.* If one or more of the basic variables in a basic solution takes on a value of either the upper or the lower bound, that solution is called a *degenerate basic solution.*

Definition 3.1.2 A solution $\mathbf{x}$ satisfying (3.1.1) and (3.1.2) is said to be *feasible.* Let

$$\mathcal{X} = \{\mathbf{x} \in \mathbb{R}^n \mid \mathbf{A}\mathbf{x} = \mathbf{b}, \boldsymbol{\ell} \leqq \mathbf{x} \leqq \mathbf{u}\}$$

be called the *feasible set*. Clearly $\mathcal{X}$ is a convex polyhedral set. If $u_i < \infty$ and $\ell_i > -\infty\ \forall i$, then $\mathcal{X}$ is bounded and is therefore a polytope. A feasible solution that is also basic is called a *basic feasible solution*. If the solution is degenerate, it is called a *degenerate basic feasible solution*.

The Linear programming problem LP can be stated succinctly as

$$\text{(Problem LP)} \qquad \min_{\mathbf{x}\in\mathcal{X}} \mathbf{c}^\top \mathbf{x},$$

where $\mathbf{x} \in \mathbb{R}^n$ and $\mathbf{c} \in \mathbb{R}^n$ (the cost vector). The key idea behind solving LP problems is that one needs only to search amongst a finite number of basic feasible solutions, which are the extreme points of the convex polyhedral set that defines the feasible set $\mathcal{X}$. This fact is given in the following fundamental theorem of LP, the proof of which has been simplified by assuming that the feasible set $\mathcal{X}$ is bounded. See Remark 3.1.1.

Theorem 3.1.1 (Fundamental Theorem of Linear Programming) Assume that the feasible set $\mathcal{X}$ is a bounded polytope, then there exists a minimum solution of LP at an extreme point of $\mathcal{X}$.

Proof: Since $\mathcal{X}$ is a polytope, Corollary 1.2.14 asserts that $\mathcal{X}$ is the convex hull of the set of all extreme points $\mathcal{E} = \{\mathbf{x}^1, \mathbf{x}^2, \cdots, \mathbf{x}^K\}$. Furthermore it follows from the boundedness of $\mathcal{X}$ that a minimum solution of LP exists. Let $\mathbf{x}^*$ be a minimum solution of LP and let $z^* = \mathbf{c}^\top\mathbf{x}^*$ be the minimum cost. Since $\mathbf{x}^*$ is feasible, $\mathbf{x}^* = \sum_{i=1}^K \lambda_i \mathbf{x}^i = \sum_{i\in\mathcal{I}} \lambda_i \mathbf{x}^i$ where $\mathcal{I} = \{i : \lambda_i > 0\}$, $\lambda_i \geq 0$, and $\sum_{i=1}^K \lambda_i = \sum_{i\in\mathcal{I}} \lambda_i = 1$. Clearly $\mathbf{c}^\top\mathbf{x}^i \geq z^*$. If $\exists k \in \mathcal{I}$ such that $\mathbf{c}^\top\mathbf{x}^k > z^*$, then

$$z^* = \mathbf{c}^\top\mathbf{x}^* = \lambda_k \mathbf{c}^\top\mathbf{x}^k + \sum_{i\in\mathcal{I}\setminus\{k\}} \lambda_i \mathbf{c}^\top\mathbf{x}^i > \sum_{i\in\mathcal{I}} \lambda_i z^* = z^*,$$

a contradiction. This implies that $\mathbf{c}^\top\mathbf{x}^i = z^*\ \ \forall i \in \mathcal{I}$, i.e., every extreme point corresponding to an index in the set $\mathcal{I}$ is a minimum solution. ∎

To facilitate a meaningful algebraic analysis, the geometric concept of an extreme point has to be replaced by an algebraic equivalent.

Theorem 3.1.2 $\mathbf{x}$ is an extreme point of the feasible set $\mathcal{X}$ if and only if $\mathbf{x}$ is a basic feasible solution.

Proof: (Necessity) Let $\mathbf{x}^*$ be an extreme point of $\mathcal{X}$ and let the components of $\mathbf{x}^*$ be, without loss of generality, reordered such that $\ell_i < x_i^* < u_i$, $\ i = 1, 2, \cdots, k$, where $k \leq n$; and $x_i \in \{\ell_i, u_i\}$, $\ i = k+1, \cdots, n$. Reorder the columns of $\mathbf{A}$ accordingly and let $\mathbf{a}^i$ be the ith column of $\mathbf{A}$. If $\mathbf{a}^i, i = 1, 2, \cdots, k$ are not linearly independent, then there exists $\lambda_i \in \mathbb{R}$ and not all zero such that $\sum_{i=1}^k \lambda_i \mathbf{a}^i = \mathbf{0}$. Let

$$\eta = \min\left\{ \min_{1\leq i\leq k} \left\{ \frac{x_i^* - \ell_i}{|\lambda_i|} \mid \lambda_i \neq 0 \right\},\ \ \min_{1\leq i\leq k} \left\{ \frac{u_i - x_i^*}{|\lambda_i|} \mid \lambda_i \neq 0 \right\} \right\}.$$

then clearly $\ell_i \leq x_i^* \pm \eta\lambda_i \leq u_i \quad \forall i$. Let

$$\mathbf{x}^1 = \mathbf{x}^* + \eta\mathbf{y}$$
$$\mathbf{x}^2 = \mathbf{x}^* - \eta\mathbf{y}$$

where $\mathbf{y}^\top = (\lambda_1, \lambda_2, \cdots, \lambda_k, 0, 0, \cdots, 0) \in \mathbb{R}^n$. It is easy to check that $\mathbf{x}^1$ and $\mathbf{x}^2$ are both feasible with respect to the bounds (3.1.2). Since $\mathbf{A}\mathbf{y} = \sum_{i=1}^k \lambda_i \mathbf{a}^i = \mathbf{0}$, they satisfy the constraint (3.1.1) as well. Hence $\mathbf{x}^1, \mathbf{x}^2 \in \mathcal{X}$, but $\mathbf{x}^* = \frac{1}{2}\mathbf{x}^1 + \frac{1}{2}\mathbf{x}^2$, which contradicts that $\mathbf{x}^*$ is an extreme point. Consequently the first k columns of $\mathbf{A}$ must be linearly independent. Since rank $(\mathbf{A}) = m$, therefore $k \leq m$ and hence a basis $\mathbf{B}$ can always be formed from the first m columns of $\mathbf{A}$.

(Sufficiency) Consider a basic feasible solution $\mathbf{x}^\top = (x_1, x_2, \cdots, x_m, \xi_{m+1}, \cdots, \xi_n)$ to (3.1.1), where $\xi_i \in \{\ell_i, u_i\}$, $i = m+1, m+2, \cdots$, then

$$\sum_{i=1}^{m} x_i \mathbf{a}^i = \mathbf{b} - \sum_{i=m+1}^{n} \xi_i \mathbf{a}^i.$$

Suppose there are two *distinct* points in $\mathcal{X}$, $\mathbf{x}^1$ and $\mathbf{x}^2$ say, such that $\mathbf{x} = \lambda\mathbf{x}^1 + (1-\lambda)\mathbf{x}^2$ for $\lambda \in (0,1)$. Since both $\mathbf{x}^1$ and $\mathbf{x}^2$ are feasible, this implies that each of the last $n-m$ components of $\mathbf{x}^1$ and $\mathbf{x}^2$ is identical to the corresponding component of $\mathbf{x}$. To elaborate this further, let $i > m+1$, then

$$[\mathbf{x}]_i = \lambda[\mathbf{x}^1]_i + (1-\lambda)[\mathbf{x}^2]_i.$$

Say if $[\mathbf{x}]_i = \ell_i$ but at least one of $[\mathbf{x}^1]_i$ or $[\mathbf{x}^2]_i$ is strictly greater than ℓ_i, then we have $\ell_i < \ell_i$, a contradiction. A similar argument applies when $[\mathbf{x}]_i = u_i$. Hence,

$$\sum_{i=1}^{m} [\mathbf{x}^1]_i \mathbf{a}^i = \mathbf{b} - \sum_{i=m+1}^{n} \xi_i \mathbf{a}^i, \quad \text{and}$$
$$\sum_{i=1}^{m} [\mathbf{x}^2]_i \mathbf{a}^i = \mathbf{b} - \sum_{i=m+1}^{n} \xi_i \mathbf{a}^i.$$

Subtracting the last two equations from each other, we have

$$\sum_{i=1}^{m} ([\mathbf{x}^1]_i - [\mathbf{x}^2]_i)\mathbf{a}^i = \mathbf{0}.$$

Since the $\mathbf{a}^i$ are linearly independent, the last equation implies that $[\mathbf{x}^1]_i = [\mathbf{x}^2]_i \ \forall i$ and hence $\mathbf{x}^1 = \mathbf{x}^2 = \mathbf{x}$, i.e., $\mathbf{x}$ is extreme. ∎

Although the number of candidate optimal solutions is finite, an exhaustive enumeration of all such solutions can be terribly inefficient if n and m are large, as is often the case. The simplex method exploits the proof to find a much more efficient way of searching through these basic feasible solutions. The simplex method is a special case of a general class of techniques known as the feasible direction search method. It is essentially an efficient way of searching through all the basic feasible solutions. The basic ideas are briefly presented here with some appropriate justification.

We assume, without loss of generality, that $\mathbf{B}$ is the first $m \times m$ non-singular submatrix of $\mathbf{A}$. Partition $\mathbf{A}, \mathbf{x}, \boldsymbol{\ell}, \mathbf{u}$ and $\mathbf{c}$ as

$$\mathbf{A} = [\mathbf{B} \vdots \mathbf{N}], \quad \mathbf{x}^\top = [\mathbf{x}_\mathbf{B}^\top, \mathbf{x}_\mathbf{N}^\top], \quad \boldsymbol{\ell}^\top = [\boldsymbol{\ell}_\mathbf{B}^\top, \boldsymbol{\ell}_\mathbf{N}^\top], \quad \mathbf{u}^\top = [\mathbf{u}_\mathbf{B}^\top, \mathbf{u}_\mathbf{N}^\top], \quad \mathbf{c}^\top = [\mathbf{c}_\mathbf{B}^\top, \mathbf{c}_\mathbf{N}^\top].$$

The full LP problem now reads as

$$\min \ \mathbf{c}_\mathbf{B}^\top \mathbf{x}_\mathbf{B} + \mathbf{c}_\mathbf{N}^\top \mathbf{x}_\mathbf{N}, \tag{3.1.4}$$

$$\text{subject to} \ \mathbf{B}\mathbf{x}_\mathbf{B} + \mathbf{N}\mathbf{x}_\mathbf{N} = \mathbf{b}, \tag{3.1.5}$$

$$\boldsymbol{\ell}_\mathbf{B} \leqq \mathbf{x}_\mathbf{B} \leqq \mathbf{u}_\mathbf{B}, \quad \boldsymbol{\ell}_\mathbf{N} \leqq \mathbf{x}_\mathbf{N} \leqq \mathbf{u}_\mathbf{N}. \tag{3.1.6}$$

Inverting (3.1.5), we have

$$\mathbf{x}_\mathbf{B} = \mathbf{B}^{-1}\mathbf{b} - \mathbf{B}^{-1}\mathbf{N}\mathbf{x}_\mathbf{N}. \tag{3.1.7}$$

Thus if we set the nonbasic variables $\mathbf{x}_\mathbf{N}$ to take values on either the lower or upper bound, then the basic variable $\mathbf{x}_\mathbf{B}$ is uniquely determined. Substituting (3.1.7) into the cost (3.1.4), we have

$$z_0 = \mathbf{c}_\mathbf{B}^\top \mathbf{B}^{-1}\mathbf{b} + (\mathbf{c}_\mathbf{N}^\top - \mathbf{c}_\mathbf{B}^\top \mathbf{B}^{-1}\mathbf{N})\mathbf{x}_\mathbf{N} \tag{3.1.8}$$

which expresses the cost as only dependent on the nonbasic variables $\mathbf{x}_\mathbf{N}$. We define

$$\mathbf{r}^\top = \mathbf{c}_\mathbf{N}^\top - \mathbf{c}_\mathbf{B}^\top \mathbf{B}^{-1}\mathbf{N} \tag{3.1.9}$$

as the *relative cost vector* (also known as *reduced cost or shadow price*). Furthermore, we also refer to the vector

$$\boldsymbol{\lambda}^\top = \mathbf{c}_\mathbf{B}^\top \mathbf{B}^{-1} \tag{3.1.10}$$

as the *dual variable* or *vector of dual variable*, for reasons that will become obvious shortly.

Remember that once the nonbasic variables are set to either the lower or upper bound, the basic variables are uniquely determined from (3.1.7), and together they define a basic solution (which is feasible if $\boldsymbol{\ell} \leqq \mathbf{x}_\mathbf{B} \leqq \mathbf{u}$). The reduced linear LP problem subsequently resides in the lower dimension space of $\mathbb{R}^{n-m}$:

$$\text{(Reduced LP)} \quad \min z_0 = \mathbf{c}_\mathbf{B}^\top \mathbf{B}^{-1}\mathbf{b} + \left(\mathbf{c}_\mathbf{N}^\top - \mathbf{c}_\mathbf{B}^\top \mathbf{B}^{-1}\mathbf{N}\right)\mathbf{x}_\mathbf{N} \tag{3.1.11}$$

$$\text{s.t.} \quad \boldsymbol{\ell}_\mathbf{B} \leqq \mathbf{B}^{-1}\mathbf{b} - \mathbf{B}^{-1}\mathbf{N}\mathbf{x}_\mathbf{N} \leqq \mathbf{u}_\mathbf{B} \tag{3.1.12}$$

$$\text{and} \quad \boldsymbol{\ell}_\mathbf{N} \leqq \mathbf{x}_\mathbf{N} \leqq \mathbf{u}_\mathbf{N}. \tag{3.1.13}$$

Clearly the cost can only be reduced if and only if the corresponding component of the relative cost vector $[x_\mathbf{N}]_k = \ell_k$ and $r_k = [\mathbf{c}_\mathbf{N}^\top - \mathbf{c}_\mathbf{B}^\top \mathbf{B}^{-1}\mathbf{N}]_k$ is strictly negative;

or $[x_{\mathbf{N}}]_k = u_k$ and r_k is strictly positive. The form of this reduced LP immediately implies the following optimality condition.

Theorem 3.1.3 (Optimality condition of LP) Let $\mathbf{x}$ be a basic feasible solution. $\mathbf{x}$ is optimal if and only if $r_j \geq 0\ \forall j$ such that $x_j = \ell_j$ or $r_j \leq 0\ \forall j$ such that $x_j = u_j$.

If the optimality condition is not satisfied, the solution can be improved by increasing or decreasing one of the nonbasic variables $(\mathbf{x_N})_k$ from one of the bounds to some interior point of the interval $[\ell_k, u_k]$. This amounts to changing the nonbasic variable to a basic variable, or *bringing it into the basis.* We call this variable the *entering nonbasic variable.*

In general, changing the value of a nonbasic variable causes all the basic variables to change according to (3.1.7), and at some point, one or more of the basic variables will hit the lower or upper bound (see (3.1.12)). At which time increasing the nonbasic variable further will cause the solution to become infeasible. We call the first basic variable to hit one of the bounds the *blocking variable.* Clearly the blocking basic variable will become nonbasic in the next iteration, and we call this the *leaving basic variable.* If there is more than one basic variable hitting the bound, we make only one of them nonbasic, and the others will remain basic even though they are at the bounds. We call these as *degenerate basic variables.*

To determine the maximal amount of change (η say) in the entering nonbasic variable, x_k say, we note that the feasible direction $\mathbf{d}$ (in the reduced dimension space) that leads to a reduction in cost is $\mathbf{d} = \mathbf{e}^k$ or $\mathbf{d} = -\mathbf{e}^k$ depending on whether $r_k < 0$ or $r_k > 0$. If the blocking variable is also x_k, then it is obvious that the maximal change $\eta = u_k - \ell_k$. If the blocking variable is basic, then the new solution $\mathbf{x} + \eta\mathbf{d}$ must satisfy (3.1.12), i.e.,

$$\boldsymbol{\ell}_{\mathbf{B}} \leqq \mathbf{B}^{-1}\mathbf{b} - \mathbf{B}^{-1}\mathbf{N}(\mathbf{x_N} + \eta\mathbf{d}) \leqq \mathbf{u_B}. \tag{3.1.14}$$

Let $\delta = 1$ if $\mathbf{d} = \mathbf{e}^k$ and $\delta = -1$ if $\mathbf{d} = -\mathbf{e}^k$, and $\mathbf{y} = \mathbf{B}^{-1}[\mathbf{N}]^k$ where $[\mathbf{N}]^k$ is the kth column of $\mathbf{N}$. Then (3.1.14) is equivalent to:

$$\ell_j \leq [\mathbf{x_B}]_j - y_j\delta\eta \leq u_j, \quad j = 1, 2, \cdots, m. \tag{3.1.15}$$

Consequently, to remain feasible, it is necessary that:

$$\text{If } y_j > 0,\ \delta > 0 \quad \text{or} \quad y_j < 0,\ \delta < 0 \quad \text{then} \quad \eta \leq \frac{[\mathbf{x_B}]_j - \ell_j}{|y_j|},$$

$$\text{If } y_j < 0,\ \delta > 0 \quad \text{or} \quad y_j > 0,\ \delta < 0 \quad \text{then} \quad \eta \leq \frac{u_j - [\mathbf{x_B}]_j}{|y_j|}.$$

Once we have determined the entering nonbasic variable and the leaving basic variable, a new basis will be formed by replacing, in the basis matrix $\mathbf{B}$, the column of $\mathbf{A}$ corresponding to the leaving basic variable, by the column of $\mathbf{A}$ corresponding to the entering nonbasic variable. This procedure is called the *pivoting step.* In practice, the pivoting step can be quickly carried out by the usual rank one update trick similar to that of Gaussian elimination. Since this is covered in every book on LP, we shall omit the detail here.

A conceptual outline of the primal simplex algorithm is presented as follows.

Algorithm 3.1.1 (The Primal simplex algorithm)

Initialization. Let $\mathbf{x}^\top = [\mathbf{x}_\mathbf{B}^\top, \mathbf{x}_\mathbf{N}^\top]$ be an initial basic feasible solution, and let $\mathbf{B}$ be the corresponding basis, and $\mathbf{A} = [\mathbf{B} \vdots \mathbf{N}]$.

Do: until $\mathcal{I}_1 \cup \mathcal{I}_2 = \emptyset$, whence the solution is optimal.

(Pricing for the entering nonbasic variable) Let

$$\mathcal{I}_1 = \{k \mid [x_\mathbf{N}]_k = \ell_k \text{ and } r_k = [\mathbf{c}_\mathbf{N}^\top - \mathbf{c}_\mathbf{B}^\top \mathbf{B}^{-1}\mathbf{N}]_k < 0\},$$
$$\mathcal{I}_2 = \{k \mid [x_\mathbf{N}]_k = u_k \text{ and } r_k = [\mathbf{c}_\mathbf{N}^\top - \mathbf{c}_\mathbf{B}^\top \mathbf{B}^{-1}\mathbf{N}]_k > 0\}.$$

Pick $k \in \mathcal{I}_1 \cup \mathcal{I}_2$ and let $\delta = 1$ if $k \in \mathcal{I}_1$, $\delta = -1$ if $k \in \mathcal{I}_2$.

(Ratio test for the leaving basic variable) Let $\mathbf{y} = \mathbf{B}^{-1}[\mathbf{N}]^k$ and

$$\eta = \min \left\{ u_k - \ell_k, \quad \min_{y_j\delta > 0} \left\{ \frac{[\mathbf{x}_\mathbf{B}]_j - \ell_j}{|y_j|}, \infty \right\} \quad \min_{y_j\delta < 0} \left\{ \frac{u_j - [\mathbf{x}_\mathbf{B}]_j}{|y_j|}, \infty \right\} \right\}.$$

Update the solution by:

$$[\mathbf{x}_\mathbf{N}]_k \leftarrow [\mathbf{x}_\mathbf{N}]_k + \eta\delta, \qquad \mathbf{x}_\mathbf{B} \leftarrow \mathbf{x}_\mathbf{B} - \eta\delta\mathbf{y}.$$

(Pivoting for a new basis)

If $\eta \neq u_k - \ell_k$, pick any j such that $y_j\delta > 0$ and $[\mathbf{x}_\mathbf{B}]_j = \ell_j$, or $y_j\delta < 0$ and $[\mathbf{x}_\mathbf{B}]_j = u_j$. Replace the column $\mathbf{a}^j$ in the basis by $\mathbf{a}^k = [\mathbf{N}]^k$.

End

End

Remark 3.1.1 The above formulation of LP is based on the assumption that the feasible set is a bounded polytope. To allow for more generality, the boundedness constraint can be easily removed, in which case it is possible to obtain an unbounded optimal solution with an unbounded cost of $-\infty$. It turns out that this is intimately related to the feasibility of the dual LP problem. This duality aspect is the main focus of the next section.

3.2 Duality in Linear Programming

The subject matter of this section is arguably the best known case of duality. Most Operations Research and some Economics textbooks would go through some length in discussing LP duality. Nevertheless students taking a first course in linear programming often wonder why there is a need to study duality, although many would agree that the symmetric properties of duality are rather nice indeed. The fact is, a firm grasp of duality is essential for understanding the simplex family of methods for solving LP problems.

Just so that the underlying idea does not get obscured by more detail than is necessary, we shall introduce the concept of LP duality via a slightly simpler problem to the one presented in Section 3.2. The duality of the general (upper bounded) LP problem will be studied shortly. Consider the pair of *dual linear programs in symmetric form*:

(Primal LP Problem)

$$\min \quad \mathbf{c}^{\top}\mathbf{x}, \tag{3.2.1}$$

$$\text{s.t. } \mathbf{A}\mathbf{x} \geqq \mathbf{b}, \tag{3.2.2}$$

$$\mathbf{x} \geqq \mathbf{0}, \tag{3.2.3}$$

where $\mathbf{A} \in \mathbb{R}^{m \times n}, \mathbf{c}, \mathbf{x} \in \mathbb{R}^{n}, \mathbf{b} \in \mathbb{R}^{m}$. The corresponding (symmetric) dual is given by:

(Dual LP Problem)

$$\max \quad \boldsymbol{\lambda}^{\top}\mathbf{b}, \tag{3.2.4}$$

$$\text{s.t. } \boldsymbol{\lambda}^{\top}\mathbf{A} \leqq \mathbf{c}^{\top}, \tag{3.2.5}$$

$$\boldsymbol{\lambda} \geqq \mathbf{0}. \tag{3.2.6}$$

where $\boldsymbol{\lambda} \in \mathbb{R}^{m}$. It is important to note that the role of the primal and dual can be reversed. If we call the vector $\mathbf{b}$ in the primal as the *constraint vector*, then one can see that the dual is obtained from the primal by:
(i) interchanging the role of the cost vector $\mathbf{c}$ and the constraint vector $\mathbf{b}$;
(ii) transposing the constraint matrix $\mathbf{A}$;
(iii) reversing the sign of the inequality constraint; and
(iv) changing from minimization to maximization.

Theorem 3.2.1 (Symmetry of LP dual) The dual of the dual problem given by (3.2.4)-(3.2.6) is the primal problem given by (3.2.1)-(3.2.3). In other words, LP duality is symmetric.

Proof: First, we transform the dual into the form of the primal:

(Modified Dual LP Problem)

$$-\left(\min \quad -\mathbf{b}^{\top}\boldsymbol{\lambda}\right),$$

$$\text{s.t. } -\mathbf{A}^{\top}\boldsymbol{\lambda} \geqq -\mathbf{c},$$

$$\boldsymbol{\lambda} \geqq \mathbf{0}.$$

The dual of which is

$$\begin{aligned} &-\left(\max \quad -\mathbf{c}^{\mathsf{T}}\mathbf{y}\right), \\ &\text{s.t. } \mathbf{y}^{\mathsf{T}}(-\mathbf{A}^{\mathsf{T}}) \leqq -\mathbf{b}^{\mathsf{T}}, \\ &\quad \mathbf{y} \geqq \mathbf{0}, \end{aligned}$$

which is equivalent to the primal given by (3.2.1)-(3.2.3). ∎

The LP problem in the form (3.1.1)-(3.1.2) can be transformed into the form in (3.2.1)-(3.2.3). We first note that the lower bound $\boldsymbol{\ell}$ can be assumed to be $\mathbf{0}$ without loss of generality by a simple shift in origin. Then we transform the primal problem

(Primal LP Problem)

$$\min \quad \mathbf{c}^{\mathsf{T}}\mathbf{x}, \tag{3.2.7}$$

$$\text{s.t. } \mathbf{A}\mathbf{x} = \mathbf{b}, \tag{3.2.8}$$

$$\mathbf{0} \leqq \mathbf{x} \leqq \mathbf{u}, \tag{3.2.9}$$

into the asymmetric or non-standard form of (3.2.1)-(3.2.3):

$$\begin{aligned} &\min \quad \mathbf{c}^{\mathsf{T}}\mathbf{x}, \\ &\text{s.t. } \mathbf{A}\mathbf{x} \geqq \mathbf{b}, \\ &\quad -\mathbf{A}\mathbf{x} \geqq -\mathbf{b}, \\ &\quad -\mathbf{x} \geqq -\mathbf{u}, \\ &\quad \mathbf{x} \geqq \mathbf{0}. \end{aligned}$$

We may then construct the symmetric dual to this as

$$\begin{aligned} &\max \quad \boldsymbol{\mu}^{\mathsf{T}}\mathbf{b} - \boldsymbol{\nu}^{\mathsf{T}}\mathbf{b} - \boldsymbol{\rho}^{\mathsf{T}}\mathbf{u}, \\ &\text{s.t. } \boldsymbol{\mu}^{\mathsf{T}}\mathbf{A} - \boldsymbol{\nu}^{\mathsf{T}}\mathbf{A} - \boldsymbol{\rho}^{\mathsf{T}} I \leqq \mathbf{c}^{\mathsf{T}}, \\ &\quad \boldsymbol{\mu} \geqq \mathbf{0}, \ \boldsymbol{\nu} \geqq \mathbf{0}, \ \boldsymbol{\rho} \geqq \mathbf{0}. \end{aligned}$$

Using $\boldsymbol{\lambda} = [\boldsymbol{\mu}^{\mathsf{T}}, \boldsymbol{\nu}^{\mathsf{T}}, \boldsymbol{\rho}^{\mathsf{T}}]^{\mathsf{T}}, \bar{\mathbf{A}} = [\mathbf{A}, \ -\mathbf{A}, \ \mathbf{I}]$, we obtain the dual to (3.2.7)-(3.2.9) in the *asymmetric form*:

(Dual problem)

$$\max \quad \boldsymbol{\lambda}^{\mathsf{T}}\mathbf{b}, \tag{3.2.10}$$

$$\text{s.t. } \boldsymbol{\lambda}^{\mathsf{T}}\bar{\mathbf{A}} \leqq \mathbf{c}^{\mathsf{T}}, \tag{3.2.11}$$

$$\boldsymbol{\lambda} \geqq \mathbf{0}. \tag{3.2.12}$$

Observe that if some of the constraints in the primal problem are changed to equalities, the corresponding components of the variable $\boldsymbol{\lambda}$ in the dual become free variables (i.e., unrestricted by bounds). Also if the components of $\mathbf{x}$ in the primal are free variables, then the corresponding inequalities in $\boldsymbol{\lambda}^{\mathsf{T}}\mathbf{A} \leqq \mathbf{c}^{\mathsf{T}}$ are changed to equalities in the dual. These are consequences of the definition of the symmetric dual and the equivalence of the various forms of LP.

The usual treatment of duality will now be applied to the following asymmetric primal and dual pair of LP problems. A similar result can be obtained for the symmetric form, although the corresponding strong duality is a bit harder to derive. Without loss of generality, the lower bound $\boldsymbol{\ell}$ to $\mathbf{x}$ is assumed to be $\mathbf{0}$ to allow for a slightly simpler notation.

$$\text{(Primal LP)} \qquad \min \quad z_P = \mathbf{c}^\top \mathbf{x} \tag{3.2.13}$$
$$\text{s.t.} \quad \mathbf{A}\mathbf{x} = \mathbf{b} \tag{3.2.14}$$
$$\mathbf{0} \leqq \mathbf{x} \leqq \mathbf{u}. \tag{3.2.15}$$

$$\text{(Dual LP)} \qquad \max \quad z_D = \mathbf{b}^\top \boldsymbol{\lambda} - \mathbf{u}^\top \boldsymbol{\mu} \tag{3.2.16}$$
$$\text{s.t.} \quad \boldsymbol{\lambda}^\top \mathbf{A} - \boldsymbol{\mu}^\top \leqq \mathbf{c}^\top \tag{3.2.17}$$
$$\boldsymbol{\mu} \geqq \mathbf{0}. \tag{3.2.18}$$

Often, we may relax the boundedness constraint and allow the upper bound u_i to take on the value of ∞, in which case it is possible that the optimal solution may become unbounded. In the event that every component of $\mathbf{u}$ is ∞, the primal and dual take on a slightly simplified form:

$$\text{(Primal LP')} \qquad \min \quad z_P = \mathbf{c}^\top \mathbf{x}$$
$$\text{s.t.} \quad \mathbf{A}\mathbf{x} = \mathbf{b}$$
$$\mathbf{x} \geqq \mathbf{0}.$$

$$\text{(Dual LP')} \qquad \max \quad z_D = \mathbf{b}^\top \boldsymbol{\lambda}$$
$$\text{s.t.} \quad \boldsymbol{\lambda}^\top \mathbf{A} \leqq \mathbf{c}^\top.$$

Theorem 3.2.2 (Weak duality) If $\mathbf{x}$ is feasible for the primal (3.2.13)-(3.2.15), $\boldsymbol{\lambda}$ and $\boldsymbol{\mu}$ are feasible for the dual (3.2.16)-(3.2.18), then $\mathbf{c}^\top \mathbf{x} \geq \mathbf{b}^\top \boldsymbol{\lambda} - \mathbf{u}^\top \boldsymbol{\mu}$.

Proof:

$$\begin{aligned} \mathbf{c}^\top \mathbf{x} &\geq (\boldsymbol{\lambda}^\top \mathbf{A} - \boldsymbol{\mu}^\top)\mathbf{x} && \text{by (3.2.15) and (3.2.17)} \\ &= \boldsymbol{\lambda}^\top \mathbf{b} - \boldsymbol{\mu}^\top \mathbf{x} && \text{by (3.2.14)} \\ &\geq \mathbf{b}^\top \boldsymbol{\lambda} - \mathbf{u}^\top \boldsymbol{\mu} && \text{by (3.2.15).} \end{aligned}$$

∎

By convention, if the primal minimization problem is infeasible, the optimal primal cost is ∞; and similarly, if the dual maximization problem is infeasible, the optimal dual objective value is $-\infty$. This leads to the following corollary.

Corollary 3.2.3
(i) The objective value of any feasible dual solution is a lower bound to the cost function value of any feasible primal solution.
(ii) If the primal has an unbounded solution, then the dual is infeasible.
(iii) If the dual has an unbounded solution, then the primal is infeasible.
(iv) If $\mathbf{x}^*$ and $(\boldsymbol{\lambda}^*, \boldsymbol{\mu}^*)$ is feasible for the primal and dual respectively, and if $\mathbf{c}^\top \mathbf{x}^* = \mathbf{b}^\top \boldsymbol{\lambda}^* - \mathbf{u}^\top \boldsymbol{\mu}^*$, then $\mathbf{x}^*$ is optimal for the primal and $(\boldsymbol{\lambda}^*, \boldsymbol{\mu}^*)$ are optimal for the dual.

Theorem 3.2.4 (Strong duality) If any one of the primal or dual problems has a finite optimal solution, so does the other, and both have the same optimal cost.

Proof: By Corollary 3.2.3 (iv), we only need to find a feasible solution for the primal and a feasible solution for the dual such that both yield the same objective values. Let $\mathbf{B}$ be the optimal basis matrix, $\mathbf{c_B}$ be the component of $\mathbf{c}$ corresponding to the optimal basis, $\mathbf{N}$ be the remainder of $\mathbf{A}$ outside the basis, and $\mathbf{c_N}$ be the component of $\mathbf{c}$ corresponding to those outside the basis. Let $\mathbf{x}^*$ be a (finite) feasible solution for the primal, and $(\boldsymbol{\lambda}^*, \boldsymbol{\mu}^*)$ be a (finite) feasible solution to the dual, where

$$(\boldsymbol{\lambda}^*)^\top = \mathbf{c}_\mathbf{B}^\top \mathbf{B}^{-1} \tag{3.2.19}$$

$$\mu_j^* = \begin{cases} (\boldsymbol{\lambda}^*)^\top \mathbf{a}^j - c_j & \text{if } x_j^* = u_j \\ 0 & \text{otherwise} \end{cases} \tag{3.2.20}$$

We need to show that, (i) $(\boldsymbol{\lambda}^*, \boldsymbol{\mu}^*)$ as defined by (3.2.19)-(3.2.20) are feasible for the dual, and (ii) the primal and dual costs are the same. We begin with (ii):

$$\begin{aligned} z_D(\boldsymbol{\lambda}^*, \boldsymbol{\mu}^*) &= \mathbf{c}_\mathbf{B}^\top \mathbf{B}^{-1}\mathbf{b} + \sum_{j:x_j=u_j} \{c_j - (\boldsymbol{\lambda}^*)^\top \mathbf{a}^j\} u_j \\ &= \mathbf{c}_\mathbf{B}^\top \mathbf{B}^{-1}\mathbf{b} + \{\mathbf{c}_\mathbf{N}^\top - \mathbf{c}_\mathbf{B}^\top \mathbf{B}^{-1}\mathbf{N}\} \mathbf{x}_\mathbf{N}^* = z_P(\mathbf{x}^*). \end{aligned}$$

Next, we prove (i), i.e., the feasibility of $(\boldsymbol{\lambda}^*, \boldsymbol{\mu}^*)$. For (3.2.17),

$$(\boldsymbol{\lambda}^*)^\top \mathbf{A} - (\boldsymbol{\mu}^*)^\top = \mathbf{c}_\mathbf{B}^\top \mathbf{B}^{-1}\mathbf{A} - (\boldsymbol{\mu}^*)^\top = \left\{\mathbf{c}_\mathbf{B}^\top \mathbf{B}^{-1}\mathbf{B} \vdots \mathbf{c}_\mathbf{B}^\top \mathbf{B}^{-1}\mathbf{N}\right\} - (\boldsymbol{\mu}^*)^\top.$$

The first m components of $\left\{\mathbf{c}_\mathbf{B}^\top \vdots \mathbf{c}_\mathbf{B}^\top \mathbf{B}^{-1}\mathbf{N}\right\} - \boldsymbol{\mu}^*$ are just $\mathbf{c}_\mathbf{B}^\top$.

The nonbasic component of $\left\{\mathbf{c}_\mathbf{B}^\top \vdots \mathbf{c}_\mathbf{B}^\top \mathbf{B}^{-1}\mathbf{N}\right\} - \boldsymbol{\mu}^*$ corresponding to $x_j^* = u_j$ is

$$(\boldsymbol{\lambda}^*)^\top \mathbf{a}^j - \{(\boldsymbol{\lambda}^*)^\top \mathbf{a}^j - c_j\} = [\mathbf{c_N}]_j.$$

The nonbasic component of $\left\{\mathbf{c}_\mathbf{B}^\top \vdots \mathbf{c}_\mathbf{B}^\top \mathbf{B}^{-1}\mathbf{N}\right\} - \boldsymbol{\mu}^*$ corresponding to $x_j^* = 0$ is

$$(\boldsymbol{\lambda}^*)^\top \mathbf{a}^j - 0 < [\mathbf{c_N}]_j,$$

by virtue of the optimality condition (Theorem (3.1.3)) on $\mathbf{x}^*$. Summarizing the above, we conclude that $(\boldsymbol{\lambda}^*, \boldsymbol{\mu}^*)$ satisfy (3.2.17). For (3.2.18), we note that for $x_j = u_j$,

$$(\boldsymbol{\lambda}^*)^\top \mathbf{a}^j - c_j > 0$$

by virtue of the optimality condition (Theorem (3.1.3)) on $\mathbf{x}^*$. Hence $\boldsymbol{\mu}^*$ satisfies (3.2.18). ∎

The implication of the strong duality theorem is that once the primal is solved, the optimal solution to the dual can be inferred directly from (3.2.19) and (3.2.20).

At any intermediate step of the simplex algorithm, the vector $\boldsymbol{\lambda}^\top = \mathbf{c}_\mathbf{B}^\top \mathbf{B}^{-1}$ is referred to as the *simplex multiplier*. This is not a solution to the dual unless $\mathbf{B}$ is the optimal basis matrix of the primal, and $\mathbf{c}_\mathbf{B}$ is the corresponding cost for the optimal basic variables. The simplex multiplier is used in each step to compute the relative cost coefficients $\mathbf{r}^\top = \mathbf{c}_\mathbf{N}^\top - \boldsymbol{\lambda}^\top \mathbf{N}$ and it has a special economic interpretation. Let $\mathbf{e}^1, \mathbf{e}^2, \cdots, \mathbf{e}^m$ be the m standard unit vectors, i.e., $(\mathbf{e}^i)_j = \delta_{ij}$, the Kronecker delta tensor. Note that

$$\mathbf{e}^j = \mathbf{B}[\mathbf{B}^{-1}]^j \tag{3.2.21}$$

where $[\mathbf{B}^{-1}]^j$ is the jth column of $\mathbf{B}^{-1}$. We say that a vector $\mathbf{v}$ is expressed as a linear combination of the columns of a basis matrix $\mathbf{B}$ when we write $\mathbf{v} = \mathbf{B}\mathbf{y}$ and call $\mathbf{y}$ the *coordinate* of $\mathbf{v}$ with respect to the basis $\mathbf{B}$. If there is a unit cost c_j associated with each basis vector $\mathbf{a}^j$, (the jth column of $\mathbf{B}$), we may ascribe a synthetic cost to the vector $\mathbf{v}$. This is computed as the linear combination of the c_j's associated with the basis, i.e., $\mathbf{c}_\mathbf{B}^\top \mathbf{y}$. Clearly from (3.2.21), $[\mathbf{B}^{-1}]^j$ is the coordinate of $\mathbf{e}^j$ with respect to the basis $\mathbf{B}$, and the corresponding synthetic cost of $\mathbf{e}^j$ is then $\mathbf{c}_\mathbf{B}^\top[\mathbf{B}^{-1}]^j = \lambda_j$. Henceforth we think of the jth component of the simplex multiplier λ_j as the synthetic cost associated with the unit vector $\mathbf{e}^j$.

Now, let's say we want to compute the synthetic cost of a vector $\mathbf{v}$, in particular, those columns of $\mathbf{A}$ excluded from the basis. We first express $\mathbf{v}$ as a linear combination of the standard unit vectors $\mathbf{e}^i$, i.e., $\mathbf{v} = \sum v_i \mathbf{e}^i$, and then compute the cost as $\sum v_i \lambda_i$, since each vector $\mathbf{e}^i$ carries a synthetic cost of λ_i. Thus the simplex multipliers can be used quickly to compute the synthetic cost of any vector expressed in terms of the standard unit vector. Subsequently, the relative cost of the vector can be computed by taking the difference between the true cost and the synthetic cost.

The constraint of the dual $\boldsymbol{\lambda}^\top \mathbf{A} \le \mathbf{c}^\top$ can be interpreted as follows: *optimality of the primal occurs when every vector* $\mathbf{a}^i, i = 1, 2, \cdots, n$ *is cheaper when constructed from the basis than when purchased directly at its own price.* For the vectors in the basis, there is no difference, i.e., zero relative cost; but for the vectors not in the basis, it is always cheaper to synthesize. Note that this also says that *optimality in the primal corresponds to feasibility in the dual.* Conversely, *feasibility in the primal corresponds to optimality in the dual.*

Remark 3.2.1 In view of the dual relationship between primal feasibility and dual optimality, or between dual feasibility and primal optimality, it is not difficult to appreciate that the primal simplex method maintains primal feasibility/dual optimality and works towards dual feasibility/primal optimality. In the linear programming literature, the so-called *dual simplex method* is often presented as an alternative to solving the LP problem. In particular, when the problem at hand has a known basic solution which may not be feasible but does satisfy the optimality condition, then the dual simplex method is particularly advantageous because there is no need to recompute for a basic feasible solution. In a completely symmetrical way, the dual simplex method maintains dual feasibility/primal optimality and works towards primal optimality/dual feasibility. Another way of looking at the dual simplex method is the application of the primal simplex method to the dual problem (see [L1]).

Remark 3.2.2 As an exercise, the reader may want to prove the following assertion: If the primal optimal solution is degenerate, then the dual optimal solution has a non-unique solution. Furthermore, if the primal optimal solution is non-unique, then the dual optimal solution is degenerate.

Sensitivity Analysis

There is another important interpretation to the dual variable $\boldsymbol{\lambda}$, apart from being the synthetic cost of unit vectors. Suppose the optimal basic variable is given by $\mathbf{x_B} = \mathbf{B}^{-1}\mathbf{b} - \mathbf{B}^{-1}\mathbf{N}\mathbf{x_N}$. Assuming nondegeneracy, the vector $\mathbf{b}$ is perturbed by a small amount to $\mathbf{b} + \Delta\mathbf{b}$ such that the optimal basis remains unchanged, and also the nonbasic variables remain unchanged. This will in turn cause a perturbation of $\Delta\mathbf{x_B}$ to the optimal basic variable, where

$$\mathbf{x_B} + \Delta\mathbf{x_B} = \mathbf{B}^{-1}(\mathbf{b} + \Delta\mathbf{b}),$$

while the optimal nonbasic variables remain zero. The corresponding perturbation in the optimal cost is

$$\Delta z = \mathbf{c}_\mathbf{B}^\top \Delta\mathbf{x_B} = \mathbf{c}_\mathbf{B}^\top \mathbf{B}^{-1}\Delta\mathbf{b} = \boldsymbol{\lambda}^\top \Delta\mathbf{b}.$$

Thus λ_j may be interpreted as the sensitivity of the optimal cost when b_j is subjected to a small perturbation, or alternatively, as the *marginal price* of b_j. This ties in also with the meaning of λ_j as the synthetic cost of the unit vector $\mathbf{e}^j$ when it is represented by the basis $\mathbf{B}$, since it directly measures the change in cost due to a small change in b_j.

To wrap up this section, It is instructive to interpret the physical significance of LP duality by a practical example. In particular, the primal and dual LP in (3.2.1)-(3.2.6) together with the weak and strong duality result often have a meaningful economic interpretation.

Example 3.2.1 (The diet problem) This example appears (see [C1]) to be contributed by Stigler in the 1940's, and it has appeared in quite a few linear programming textbooks as a classical example to illustrate LP duality. The following interpretation follows closely that described in [L1]. One can think of the primal problem as one of feeding starving individuals. There are n different foods available in the market and each contains m basic nutrients. Food j costs c_j per unit to buy, and contains $[\mathbf{A}]_{ij}$ units of the ith nutrient. To survive, each individual must receive at least b_i units of the ith nutrient. If we denote the number of units of food j in the diet by x_j, the problem is then to select the $x_i's$ so as to minimize the cost $\mathbf{c}^\top\mathbf{x}$, subject to the nutritional constraint represented by (3.2.2) and the nonnegativity bound (3.2.3).

Next, we interpret the dual as given by (3.2.4)-(3.2.6). Instead of eating food to gain the required nutrients, one can forget about the ecstasy of eating for the moment and think of eating artificially synthesized nutrient pills manufactured by a drug company. The drug company tries to push for the consumption of nutrient pills instead of eating nutritious food. The problem is then to decide on the unit price of the nutrient pills $\lambda_1, \lambda_2, \cdots, \lambda_m$ so as to maximize revenue by the drug company while at the same time remain competitive with real food. To beat food prices, the cost of a unit of food i made from nutrient pills must be less than or equal to c_i, the market price of the food, i.e., $\boldsymbol{\lambda}^\top \mathbf{a}^i \leq c_i$ for all i, or equivalently, $\boldsymbol{\lambda}^\top \mathbf{A} \leqq \mathbf{c}^\top$. Since b_j of the jth nutrient will be bought, the drug company's objective is then to maximize the total revenue $\boldsymbol{\lambda}^\top \mathbf{b}$.

In the context of sensitivity analysis, λ_j represents the maximum price per unit that one would be willing to pay for a small amount of the jth nutrient, since decreasing a unit of nutrient j will reduce the food bill by λ_j dollars. ∎

3.3 Duality and Complementary Slackness

The duality of LP can be looked at in yet another way. The idea of the so-called *complementary slackness* is simple: if a particular inequality in the primal/dual holds as a strict inequality, then there is a corresponding inequality in the dual/primal which holds as an equality, i.e., if one is slack, the other must be tight. This notion also occurs in nonlinear programming where complementary slackness forms an integral part of the Kuhn-Tucker condition, see Chapter Four. For easy reference, the asymmetric primal and dual LP pair (3.2.13)-(3.2.18) of the previous section is restated below.

$$\begin{array}{lll} \text{(Primal LP)} & \min & z_P = \mathbf{c}^\top \mathbf{x} \\ & \text{s.t.} & \mathbf{A}\mathbf{x} = \mathbf{b} \\ & & \mathbf{0} \leqq \mathbf{x} \leqq \mathbf{u}. \end{array} \tag{3.3.1}$$

$$\begin{array}{lll} \text{(Dual LP)} & \max & z_D = \mathbf{b}^\top \boldsymbol{\lambda} - \mathbf{u}^\top \boldsymbol{\mu} \\ & \text{s.t.} & \boldsymbol{\lambda}^\top \mathbf{A} - \boldsymbol{\mu}^\top \leqq \mathbf{c}^\top \\ & & \boldsymbol{\mu} \geqq \mathbf{0}. \end{array} \tag{3.3.2}$$

It is no coincidence that there exists exactly the same number of inequality constraints in the primal and the dual. Another look at the symmetric pair of primal and dual in (3.3.1)-(3.3.2) again confirms this observation. We say that an inequality constraint is *slack* if a strict inequality holds; and *active* if the inequality constraint holds as an equality. An inequality constraint of the primal is usually paired up with a corresponding inequality constraint of the dual. For example, the primal constraint $\mathbf{x} \geqq \mathbf{0}$ corresponds to the dual inequality constraint of $\boldsymbol{\lambda}^\top \mathbf{A} - \boldsymbol{\mu}^\top \leqq \mathbf{c}^\top$; and the primal constraint $\mathbf{x} \leqq \mathbf{u}$ corresponds to the dual inequality constraint of $\boldsymbol{\mu} \geqq \mathbf{0}$. The question of which is paired with which requires some background understanding of the Lagrangian theory, which will be discussed in Chapter Four. Some numerical experiments will confirm that, at optimality, if one inequality constraint in the primal is slack, then the corresponding one in its dual must be active. This is known as *complementary slackness.*

Theorem 3.3.1 (Complementary slackness - asymmetric form) Let $\mathbf{x}$ and $(\boldsymbol{\lambda}, \boldsymbol{\mu})$ be feasible solutions for the (asymmetric) primal (3.3.1) and dual (3.3.2) pair respectively. Then $\mathbf{x}$ is optimal for the primal and $(\boldsymbol{\lambda}, \boldsymbol{\mu})$ is optimal for the dual if and only if

$$(\boldsymbol{\lambda}^\top \mathbf{A} - \boldsymbol{\mu}^\top - \mathbf{c}^\top)\mathbf{x} = 0 \quad \text{and} \quad \boldsymbol{\mu}^\top(\mathbf{u} - \mathbf{x}) = 0.$$

Proof: (Sufficiency) If $(\boldsymbol{\lambda}^\top \mathbf{A} - \boldsymbol{\mu}^\top - \mathbf{c}^\top)\mathbf{x} = 0$ then $\boldsymbol{\lambda}^\top \mathbf{A}\mathbf{x} - \boldsymbol{\mu}^\top \mathbf{x} = \boldsymbol{\lambda}^\top \mathbf{b} - \boldsymbol{\mu}^\top \mathbf{x} = \mathbf{c}^\top \mathbf{x}$. Since $\boldsymbol{\mu}^\top(\mathbf{u} - \mathbf{x}) = 0$, this implies that $\boldsymbol{\lambda}^\top \mathbf{b} - \boldsymbol{\mu}^\top \mathbf{u} = \mathbf{c}^\top \mathbf{x}$, i.e., the primal cost equals to the dual objective value. By Corollary 3.2.3 of the weak duality theorem, $\mathbf{x}$ is optimal for the primal and $(\boldsymbol{\lambda}, \boldsymbol{\mu})$ is optimal for the dual.

(Necessity) Supposed that $\mathbf{x}$ and $(\boldsymbol{\lambda}, \boldsymbol{\mu})$ is optimal respectively for the primal and the dual. Then by the strong duality Theorem 3.2.4, $\boldsymbol{\lambda}^\top \mathbf{b} - \boldsymbol{\mu}^\top \mathbf{u} = \mathbf{c}^\top \mathbf{x}$. This implies that

$$\boldsymbol{\lambda}^\top \mathbf{A}\mathbf{x} - \boldsymbol{\mu}^\top \mathbf{u} - \mathbf{c}^\top \mathbf{x} = 0, \tag{3.3.3}$$

since $\mathbf{A}\mathbf{x} = \mathbf{b}$. Since $\boldsymbol{\mu} \geqq \mathbf{0}$ and $\mathbf{u} \geqq \mathbf{x}$, this also implies that

$$(\boldsymbol{\lambda}^\top \mathbf{A} - \boldsymbol{\mu}^\top - \mathbf{c}^\top)\mathbf{x} \geq 0. \tag{3.3.4}$$

But since the feasibility of $\mathbf{x}$ and $(\boldsymbol{\lambda}, \boldsymbol{\mu})$ requires that

$$\boldsymbol{\lambda}^\top \mathbf{A} - \boldsymbol{\mu}^\top - \mathbf{c}^\top \leqq \mathbf{0}^\top \quad \text{and} \quad \mathbf{x} \geqq \mathbf{0},$$

so

$$(\boldsymbol{\lambda}^\top \mathbf{A} - \boldsymbol{\mu}^\top - \mathbf{c}^\top)\mathbf{x} \leq 0. \tag{3.3.5}$$

Comparing (3.3.4) and (3.3.5), we conclude that

$$(\boldsymbol{\lambda}^\top \mathbf{A} - \boldsymbol{\mu}^\top - \mathbf{c}^\top)\mathbf{x} = 0. \tag{3.3.6}$$

Subtracting (3.3.6) from (3.3.3), we also conclude that $\boldsymbol{\mu}^\top(\mathbf{u} - \mathbf{x}) = 0$. ■

Remark 3.3.1 Let $\mathbf{a}^i$ be the ith column of $\mathbf{A}$, and let $\mathbf{a}_j^\top$ be the jth row of $\mathbf{A}$. Very often the above complementary slackness condition is expressed in the following

equivalent way: $\mathbf{x}$ is optimal for the primal and $(\boldsymbol{\lambda}, \boldsymbol{\mu})$ is optimal for the dual if and only if, for all i:
(i)$x_i > 0 \Rightarrow \boldsymbol{\lambda}^\top \mathbf{a}^i = c_i + \mu_i$,
(ii)$\boldsymbol{\lambda}^\top \mathbf{a}^i - c_i - \mu_i < 0 \Rightarrow x_i = 0$,
(iii)$\mu_i > 0 \Rightarrow u_i = x_i$,
(iv)$x_i < u_i \Rightarrow \mu_i = 0$. ∎

A similar result can be easily established for the LP dual pair in symmetric form. It is stated here without proof.

Theorem 3.3.2 (Complementary slackness - symmetric form) Let $\mathbf{x}$ and $\boldsymbol{\lambda}$ be feasible solutions for the (symmetric) primal LP (3.2.1)-(3.2.3) and dual LP (3.2.4)-(3.2.6) respectively. The $\mathbf{x}$ is optimal for the primal and $\boldsymbol{\lambda}$ is optimal for the dual if and only if,

$$(\boldsymbol{\lambda}^\top \mathbf{A} - \mathbf{c}^\top)\mathbf{x} = 0 \quad \text{and} \quad \boldsymbol{\lambda}^\top(\mathbf{A}\mathbf{x} - \mathbf{b}) = 0.$$

Alternatively, $\mathbf{x}$ is optimal for the primal and $\boldsymbol{\lambda}$ is optimal for the dual if and only if, for all i and j,
(i) $x_i > 0 \Rightarrow \boldsymbol{\lambda}^\top \mathbf{a}^i = c_i$,
(ii)$\boldsymbol{\lambda}^\top \mathbf{a}^i < c_i \Rightarrow x_i = 0$,
(iii)$\lambda_j > 0 \Rightarrow \mathbf{a}_j^\top \mathbf{x} = b_j$,
(iv)$\mathbf{a}_j^\top \mathbf{x} > b_j \Rightarrow \lambda_j = 0$.

In terms of the diet problem of Example 3.2.1 as the primal of a symmetric pair of LP duals, if the optimal diet supplies more than b_j unit of the j nutrient, obviously one would be unwilling to pay anything for small quantities of that nutrient, since its availability would not reduce the cost of the optimal diet. This implies that the marginal price $\lambda_j = 0$ as asserted by part (iv) of Theorem 3.3.2.

Given a feasible solution $(\boldsymbol{\lambda}, \boldsymbol{\mu})$ for the dual (3.3.2), if we can find a feasible solution $\mathbf{x}$ to the primal such that:
(i) $x_i = 0$ if $\boldsymbol{\lambda}^\top \mathbf{a}^i - c_i - \mu_i < 0$, and
(ii) $x_i = u_i$ if $\mu_i > 0$,
then clearly the complementary slackness condition of Theorem 3.3.1 is satisfied, and as such $\mathbf{x}$ is optimal for the primal and $(\boldsymbol{\lambda}, \boldsymbol{\mu})$ is optimal for the dual. To find a feasible solution to the primal problem under these conditions, we set up the *artificial primal problem* :

$$\begin{aligned} \text{(Artificial Primal Problem)} \quad \min \quad & \mathbf{1}^\top \mathbf{y} \\ \text{subject to:} \quad & \mathbf{A}\mathbf{x} + \mathbf{y} = \mathbf{b}, \\ & x_i = 0 \text{ if } \quad \boldsymbol{\lambda}^\top \mathbf{a}^i - c_i - \mu_i < 0, \\ & x_i = u_i \text{ if } \quad \mu_i > 0, \\ & \mathbf{0} \leqq \mathbf{x} \leqq \mathbf{u}, \\ & \mathbf{y} \geqq \mathbf{0}. \end{aligned}$$

Theorem 3.3.3 (Primal-Dual Optimality Theorem) If $(\boldsymbol{\lambda}^*, \boldsymbol{\mu}^*)$ is feasible for the dual, and the optimal solution of the artificial primal problem is $\mathbf{y} = \mathbf{0}$ and $\mathbf{x}^*$, then $\mathbf{x}^*$ is optimal for the primal and $(\boldsymbol{\lambda}^*, \boldsymbol{\mu}^*)$ is optimal for the dual.

Proof: If $\mathbf{y} = \mathbf{0}$ is optimal for the artificial primal, then the corresponding $\mathbf{x}^*$ is feasible for the primal, and the conclusion follows from the complementary slackness condition. ∎

Remark 3.3.2 Of course there is no guarantee that given any feasible $(\boldsymbol{\lambda}, \boldsymbol{\mu})$ for the dual, the artificial primal will always yield $\mathbf{y} = \mathbf{0}$ as the optimal solution. In which case, the dual solution $(\boldsymbol{\lambda}, \boldsymbol{\mu})$ will be adjusted accordingly. This is the basis of the so-called primal-dual algorithm. Although the detail is not provided here, some of the network flow algorithms (e.g., the out-of-kilter algorithm) discussed in the previous chapter would have given the reader a flavor of the underlying mechanism.

3.4 Duality and Theorems of Alternatives

Theorems of Alternatives were first studied by Minkowski, and these are sometimes called *transposition theorems.* These results are not only useful in proving many results in optimization theory, they are also intimately linked to the duality of optimization and variational inequalities [G3]. The Painted Network Theorem discussed in Chapter Two is an example, and the reader should be convinced by now that this theorem is instrumental in proving many of the duality results in Chapter Two. Other network related examples are found in the Feasible Flow Theorem 2.7.1 and the Feasible Potential Theorem 2.7.2. While these results are valid in nonlinear systems too, our interest is only confined to those of linear systems in this chapter. Typically, one has two linear systems, or polyhedral sets, structured in such a way such that exactly one of them is nonempty. The basic tool used in proving these results is usually the Separating Hyperplane Theorem 1.2.8.

Before we go into a formal analysis of theorems of alternatives, more background material in convex analysis is needed.

Definition 3.4.1 (i) $\mathcal{X} \in \mathbb{R}^n$ is said to be a *cone* if $\mathbf{x} \in \mathcal{X} \Rightarrow \lambda\mathbf{x} \in \mathcal{X} \;\; \forall \lambda > 0$.
(ii) If in addition, $\mathcal{X}$ is convex, then $\mathcal{X}$ is said to be a *convex cone.* Alternatively, a cone $\mathcal{X}$ is convex if it is closed under addition, i.e., $\mathbf{x}^1, \mathbf{x}^2 \in \mathcal{X} \Rightarrow \mathbf{x}^1 + \mathbf{x}^2 \in \mathcal{X}$.
(iii) A cone $\mathcal{X} \in \mathbb{R}^n$ is *pointed* if it contains no subspace of $\mathbb{R}^n$ other than the origin.

Remark 3.4.1 Note that:
(i) A convex cone needs not be pointed, e.g., a line through the origin.

(ii) A pointed cone needs not be convex, e.g., two disjoint half rays through the origin.
(iii) A cone $\mathcal{X}$ needs not be closed. But if it is closed and non-empty, then it is necessary that $\mathbf{0} \in \mathcal{X}$.
(iv) Usually the term "cone" means a convex cone by convention.

Definition 3.4.2 Let $\mathcal{X} \in \mathbb{R}^n$ (not necessarily a cone). The *polar cone* (sometimes called a *dual cone*, and in fact more appropriately so) $\mathcal{X}^*$ of $\mathcal{X}$ is defined by $\mathcal{X}^* = \{\mathbf{y} \in \mathbb{R}^n \mid \mathbf{x}^\top \mathbf{y} \leq 0 \ \forall \mathbf{x} \in \mathcal{X}\}$. If $\mathcal{X} = \mathbb{R}^n$, then $\mathcal{X}^* = \{\mathbf{0}\}$ and vice versa. If $\mathcal{X} = \emptyset$ then $\mathcal{X}^*$ is taken to be $\mathbb{R}^n$ by convention. The *bipolar cone* $\mathcal{X}^{**}$ of the set $\mathcal{X}$ is the polar cone of $\mathcal{X}^*$.

Remark 3.4.2 Note that:
(i) A polar cone is a cone.
(ii) If the cone $\mathcal{X}$ is a subspace, $\mathcal{X}^*$ is its orthogonal complement $\{\mathbf{y} \in \mathbb{R}^n \mid \mathbf{x}^\top \mathbf{y} = 0 \ \forall \mathbf{x} \in \mathcal{X}\}$.

Lemma 3.4.1 If $\mathcal{X} \in \mathbb{R}^n$ is nonempty, then $\mathcal{X}^*$ is a closed convex cone, and $\mathcal{X} \subseteq \mathcal{X}^{**}$.

Proof: If $\mathbf{x} \in \mathcal{X}$, then $\mathbf{x}^\top \mathbf{y} \leq 0 \ \forall \mathbf{y} \in \mathcal{X}^*$, which implies that $\mathbf{x} \in \mathcal{X}^{**}$. ∎

Lemma 3.4.2 If $\mathcal{X}^1, \mathcal{X}^2 \in \mathbb{R}^n$ are nonempty, then $\mathcal{X}^1 \subset \mathcal{X}^2 \Rightarrow (\mathcal{X}^2)^* \subset (\mathcal{X}^1)^*$.

Proof: If $\mathbf{y} \in (\mathcal{X}^2)^*$ then $\mathbf{x}^\top \mathbf{y} \leq 0 \ \forall \mathbf{x} \in \mathcal{X}^2$. Since $\mathcal{X}^1 \subset \mathcal{X}^2$, this means that $\mathbf{x}^\top \mathbf{y} \leq 0 \ \forall \mathbf{x} \in \mathcal{X}^1$, or $\mathbf{y} \in (\mathcal{X}^1)^*$. ∎

Theorem 3.4.3 If $\mathcal{X} \in \mathbb{R}^n$ is a nonempty closed convex cone, then $\mathcal{X}^{**} = \mathcal{X}$.

Proof: $\mathcal{X} \subseteq \mathcal{X}^{**}$ by Lemma 3.4.1. We need only to show that $\mathcal{X}^{**} \subseteq \mathcal{X}$. Let $\mathbf{x} \in \mathcal{X}^{**}$. If $\mathbf{x} \notin \mathcal{X}$, then the Separating Hyperplane Theorem 1.2.8 asserts that there exists a hyperplane $\mathcal{H} = \{\mathbf{y} \mid \mathbf{a}^\top \mathbf{y} = b\}$ (with $b \geq 0$ without loss of generality) that separates $\mathbf{x}$ and $\mathcal{X}$, i.e., $\mathbf{a}^\top \mathbf{y} \leq b \ \forall \mathbf{y} \in \mathcal{X}$ and $\mathbf{a}^\top \mathbf{x} > b$. If $\mathbf{a}^\top \mathbf{z} > 0$ for some $\mathbf{z} \in \mathcal{X}$, then, by virtue of $\mathcal{X}$ being a cone, $\mathbf{a}^\top (\lambda \mathbf{z}) > 0$ can be arbitrarily large for $\lambda > 0$ which contradicts that $\mathbf{a}^\top \mathbf{y} \leq b \ \forall \mathbf{y} \in \mathcal{X}$. Hence $\mathbf{a}^\top \mathbf{z} \leq 0 \quad \forall \mathbf{z} \in \mathcal{X}$ and therefore $\mathbf{a} \in \mathcal{X}^*$. Now $\mathbf{x} \in \mathcal{X}^{**} \Rightarrow \mathbf{a}^\top \mathbf{x} \leq 0$ which contradicts $\mathbf{a}^\top \mathbf{x} > 0$. Therefore $\mathbf{x} \in \mathcal{X}$, and thus $\mathcal{X}^{**} \subseteq \mathcal{X}$. ∎

The most famous theorem of alternatives appears to be due to Farkas [F1].

Theorem 3.4.4 (Farkas Lemma) Let $\mathbf{A} \in \mathbb{R}^{m \times n}$ and $\mathbf{b} \in \mathbb{R}^m$. Then there exists $\mathbf{x} \in \mathbb{R}^n, \mathbf{x} \geqq \mathbf{0}$ such that $\mathbf{A}\mathbf{x} = \mathbf{b}$ if and only if $\forall \boldsymbol{\lambda} \in \mathbb{R}^m$, $\boldsymbol{\lambda}^\top \mathbf{A} \geqq \mathbf{0}^\top \Rightarrow \boldsymbol{\lambda}^\top \mathbf{b} \geq 0$. Alternatively, this may be stated in an equivalent form: exactly one of the following systems is non-empty.

$$\text{System I} \qquad \{\mathbf{x} \in \mathbb{R}^n \mid \mathbf{x} \geqq \mathbf{0}, \quad \mathbf{A}\mathbf{x} = \mathbf{b}\}$$
$$\text{System II} \qquad \{\boldsymbol{\lambda} \in \mathbb{R}^m \mid \boldsymbol{\lambda}^\top \mathbf{b} < 0, \ \boldsymbol{\lambda}^\top \mathbf{A} \geqq \mathbf{0}^\top\}$$

Proof: (Necessity) If $\boldsymbol{\lambda} \in \mathbb{R}^m$ and $\boldsymbol{\lambda}^\top \mathbf{A} \geqq \mathbf{0}^\top$ then since $\mathbf{x} \geqq \mathbf{0}$, we have $\boldsymbol{\lambda}^\top \mathbf{A}\mathbf{x} = \boldsymbol{\lambda}^\top \mathbf{b} \geq 0$.

(Sufficiency) If not, let $\mathcal{X} = \{\mathbf{y} \in \mathbb{R}^m \mid \mathbf{y} = \mathbf{A}\mathbf{x} = \sum_{i=1}^n x_i \mathbf{a}^i, \quad x_i \geq 0 \; \forall i.\}$ where $\mathbf{a}^i$ is the ith column of $\mathbf{A}$. Clearly $\mathcal{X}$ is a closed convex cone (in particular, a polyhedral cone). Thus if $\mathbf{b} \notin \mathcal{X}$, i.e, $\mathbf{b} \neq \mathbf{A}\mathbf{x}$ for $\mathbf{x} \geqq \mathbf{0}$, then by Theorem 1.2.8 there exists a separating hyperplane that separates $\mathbf{b}$ from $\mathcal{X}$, i.e., $\exists \boldsymbol{\lambda}$ such that $\boldsymbol{\lambda}^\top \mathbf{b} < 0$ and $\boldsymbol{\lambda}^\top \mathbf{a}^i \geq 0 \; \forall i$ (or, equivalently, $\boldsymbol{\lambda}^\top \mathbf{A} \geqq \mathbf{0}^\top$). ∎

Remark 3.4.3 Note that if $\mathcal{X}$ is as defined in the proof above, then its polar cone is given by $\mathcal{X}^* = \{\mathbf{y} \in \mathbb{R}^m \mid \mathbf{y}^\top \mathbf{A} \leqq \mathbf{0}^\top\}$. Since $\mathcal{X}$ is a closed and convex cone, by Theorem 3.4.3, $\mathcal{X} = \mathcal{X}^{**}$. Thus a given vector $\mathbf{b}$ either belongs to $\mathcal{X} = \mathcal{X}^{**}$ or it does not. System I above says that $\mathbf{b}$ belongs to $\mathcal{X}$, while system II says otherwise.

Corollary 3.4.5 Let $\mathbf{A} \in \mathbb{R}^{m \times n}$ and $\mathbf{b} \in \mathbb{R}^m$. Exactly one of the following systems is non-empty.

$$\begin{array}{ll} \text{System I} & \{\mathbf{x} \in \mathbb{R}^n \mid \mathbf{x} \geqq \mathbf{0}, \quad \mathbf{A}\mathbf{x} \leqq \mathbf{b}\} \\ \text{System II} & \{\boldsymbol{\lambda} \in \mathbb{R}^m \mid \boldsymbol{\lambda}^\top \mathbf{b} < 0, \; \boldsymbol{\lambda}^\top \mathbf{A} \geqq \mathbf{0}^\top, \; \boldsymbol{\lambda} \geqq \mathbf{0}\} \end{array}$$

Proof: System I is equivalent to

$$\left\{ \begin{bmatrix} \mathbf{x} \\ \mathbf{y} \end{bmatrix} \in \mathbb{R}^{m+n} \mid [\mathbf{A}, \; \mathbf{I}] \begin{bmatrix} \mathbf{x} \\ \mathbf{y} \end{bmatrix} = \mathbf{b}, \begin{bmatrix} \mathbf{x} \\ \mathbf{y} \end{bmatrix} \geqq \mathbf{0} \right\}$$

The conclusion follows by replacing the matrix $\mathbf{A}$ in Farkas Lemma by $[\mathbf{A}, \; \mathbf{I}]$. ∎

Corollary 3.4.6 Let $\mathbf{A} \in \mathbb{R}^{m \times n}$, $\mathbf{B} \in \mathbb{R}^{m \times p}$ and $\mathbf{b} \in \mathbb{R}^m$. Exactly one of the following systems is non-empty.

$$\begin{array}{ll} \text{System I} & \left\{ \begin{bmatrix} \mathbf{x} \\ \mathbf{y} \end{bmatrix} \in \mathbb{R}^{n+p} \mid \quad \mathbf{A}\mathbf{x} + \mathbf{B}\mathbf{y} = \mathbf{b}, \; \mathbf{x} \geqq \mathbf{0} \right\} \\ \text{System II} & \{\boldsymbol{\lambda} \in \mathbb{R}^m \mid \boldsymbol{\lambda}^\top \mathbf{b} < 0, \; \boldsymbol{\lambda}^\top \mathbf{B} = \mathbf{0}, \; \boldsymbol{\lambda}^\top \mathbf{A} \geqq \mathbf{0}^\top\} \end{array}$$

Proof: If we replace $\mathbf{y}$ in System I by $\mathbf{u} - \mathbf{v}$ where $\mathbf{u} \geqq \mathbf{0}$, $\mathbf{v} \geqq \mathbf{0}$, then System I is equivalent to

$$\left\{ \begin{bmatrix} \mathbf{x} \\ \mathbf{u} \\ \mathbf{v} \end{bmatrix} \in \mathbb{R}^{m+2p} \mid [\mathbf{A}, \; \mathbf{B}, \; -\mathbf{B}] \begin{bmatrix} \mathbf{x} \\ \mathbf{u} \\ \mathbf{v} \end{bmatrix} = \mathbf{b}, \; \begin{bmatrix} \mathbf{x} \\ \mathbf{u} \\ \mathbf{v} \end{bmatrix} \geqq \mathbf{0} \right\}.$$

The conclusion follows by replacing the matrix $\mathbf{A}$ in Farkas Lemma by $[\mathbf{A}, \; \mathbf{B}, \; -\mathbf{B}]$. ∎

Theorems of alternatives are intimately related to LP duality. For easy reference, we restate the asymmetric primal and dual LP problems as follows:

$$\text{(Primal)} \qquad \min \mathbf{c}^\top \mathbf{x} \quad \text{such that} \quad \mathbf{A}\mathbf{x} = \mathbf{b},\ \mathbf{x} \geqq \mathbf{0}. \tag{3.4.1}$$
$$\text{(Dual)} \qquad \max \mathbf{b}^\top \boldsymbol{\lambda} \quad \text{such that} \quad \boldsymbol{\lambda}^\top \mathbf{A} \leqq \mathbf{c}^\top. \tag{3.4.2}$$

We restate Corollary 3.2.3 (to the weak duality Theorem 3.2.2) as follows:

Corollary 3.4.7 If the primal is feasible and unbounded below, then the dual is infeasible. If the dual is infeasible and the primal is feasible, then the primal is unbounded below. Alternatively, if the primal is feasible, then the primal is unbounded below if and only if the dual is infeasible.

Corollary 3.4.8 If the dual is feasible and unbounded above, then the primal is infeasible. If the primal is infeasible and the dual is feasible, then the dual is unbounded above. Alternatively, if the dual is feasible, then the dual is unbounded above if and only if the primal is infeasible.

A simpler proof of Farkas Lemma (Theorem 3.4.4) can be obtained using these two corollaries for LP weak duality.

Proof: (Farkas Lemma by LP duality) In the primal problem in (3.4.1), let $\mathbf{c} = \mathbf{0}$, so that the primal feasible set is given by System I of Farkas Lemma. Note that $\boldsymbol{\lambda} = \mathbf{0}$ is feasible for the dual problem in (3.4.2). Hence the dual is feasible. By Corollary 3.4.8, the primal is infeasible if and only if the dual is unbounded, if and only if there exists $\boldsymbol{\lambda}$ such that $\mathbf{b}^\top \boldsymbol{\lambda} < 0$ and $\boldsymbol{\lambda}^\top \mathbf{A} \geqq \mathbf{0}^\top$, or there exists $\boldsymbol{\lambda}'$ such that $\mathbf{b}^\top \boldsymbol{\lambda}' > 0$ and $\mathbf{A}^\top \boldsymbol{\lambda}' \leqq \mathbf{0}^\top$, since the solution $t\boldsymbol{\lambda}'$ gives an unbounded dual cost as $t \uparrow \infty$. This means that the dual is unbounded, or system I is infeasible, if and only if the system II is feasible. ∎

Two other well-known theorems of alternatives can also be established in a similar manner by using Corollary 3.4.7 and 3.4.8.

Theorem 3.4.9 (Gale's Transposition Theorem) Let $\mathbf{A} \in \mathbb{R}^{m \times n}$ and $\mathbf{b} \in \mathbb{R}^m$. Exactly one of the following systems is non-empty.

$$\text{System I} \qquad \{\mathbf{x} \in \mathbb{R}^n \mid \mathbf{x} \geqq \mathbf{0},\ \mathbf{c}^\top \mathbf{x} < 0,\ \mathbf{A}\mathbf{x} = \mathbf{0}\}$$
$$\text{System II} \qquad \{\boldsymbol{\lambda} \in \mathbb{R}^m \mid \boldsymbol{\lambda}^\top \mathbf{A} \leqq \mathbf{c}^\top\}$$

Proof: In the primal and dual problems, let $\mathbf{b} = \mathbf{0}$, so that the primal and dual become

$$\text{(Primal)} \qquad \min \mathbf{c}^\top \mathbf{x} \quad \text{such that} \quad \mathbf{A}\mathbf{x} = \mathbf{0},\ \mathbf{x} \geqq \mathbf{0}. \tag{3.4.3}$$
$$\text{(Dual)} \qquad \max \mathbf{c}^\top \boldsymbol{\lambda} \quad \text{such that} \quad \boldsymbol{\lambda}^\top \mathbf{A} \leqq \mathbf{c}^\top. \tag{3.4.4}$$

Since $\mathbf{x} = \mathbf{0}$ is primal feasible, by Corollary 3.4.7, the dual (3.4.4) is infeasible if and only the primal is unbounded below. That is, System II is infeasible if and only if the primal is unbounded below. But the primal is unbounded if and only if there exists $\mathbf{x}$ such that $\mathbf{c}^\top \mathbf{x} < 0$, $\mathbf{A}\mathbf{x} = \mathbf{0}$, and $\mathbf{x} \geqq \mathbf{0}$ since $t\mathbf{x}$ gives an unbounded solution as $t \uparrow \infty$. ∎

Theorem 3.4.10 (Gordan's Transposition Theorem) Let $\mathbf{A} \in \mathbb{R}^{m\times n}$ and $\mathbf{b} \in \mathbb{R}^m$. Exactly one of the following systems is non-empty. (Note that writing $\mathbf{x} \geq \mathbf{0}$ is the same as $\mathbf{x} \geqq \mathbf{0}$ and $\mathbf{x} \neq \mathbf{0}$.)

$$\begin{array}{ll} \text{System I} & \{\mathbf{x} \in \mathbb{R}^n \mid \mathbf{x} \geq \mathbf{0}, \quad \mathbf{Ax} = \mathbf{0}\} \\ \text{System II} & \{\boldsymbol{\lambda} \in \mathbb{R}^m \mid \boldsymbol{\lambda}^\top \mathbf{A} < \mathbf{0}^\top\} \end{array}$$

Proof: In the primal and dual LP, choose $\mathbf{c} = -\mathbf{1}$ (a vector of all ones) and $\mathbf{b} = \mathbf{0}$. Then,

System II is feasible

if and only if the system $\{\boldsymbol{\lambda} \in \mathbb{R}^m \mid \boldsymbol{\lambda}^\top \mathbf{A} \leq -\mathbf{1}^\top\}$ is feasible

if and only if the system$\{\mathbf{x} \mid -\mathbf{1}^\top \mathbf{x} < 0,\ \mathbf{x} \geqq \mathbf{0},\ \mathbf{Ax} = \mathbf{0}\}$ is infeasible by Theorem 3.4.9

if and only if the system$\{\mathbf{x} \mid \mathbf{1}^\top \mathbf{x} > 0,\ \mathbf{x} \geqq \mathbf{0},\ \mathbf{Ax} = \mathbf{0}\}$ is infesible

if and only if the system I is infeasible.

■

We shall end this section with two other well known theorems of alternatives due to Motzkin and Tucker.

Theorem 3.4.11 (Motzkin's Theorem of Alternative) Let $\mathbf{A} \in \mathbb{R}^{m\times n}$, $\mathbf{B} \in \mathbb{R}^{m\times p}$ and $\mathbf{C} \in \mathbb{R}^{m\times q}$. Exactly one of the following systems is non-empty.

$$\text{System I} \quad \left\{ \begin{bmatrix} \mathbf{x} \\ \mathbf{y} \\ \mathbf{z} \end{bmatrix} \in \mathbb{R}^{n+p+q} \mid \quad \mathbf{Ax} + \mathbf{By} + \mathbf{Cz} = \mathbf{0},\ \mathbf{x} \geq \mathbf{0},\ \mathbf{y} \geqq \mathbf{0} \right\}$$

$$\text{System II} \quad \{\boldsymbol{\lambda} \in \mathbb{R}^m \mid \boldsymbol{\lambda}^\top \mathbf{A} > \mathbf{0}^\top,\ \boldsymbol{\lambda}^\top \mathbf{B} \geqq \mathbf{0},\ \boldsymbol{\lambda}^\top \mathbf{C} = \mathbf{0}^\top\}$$

Proof: We first show that if System II is feasible, then System I is infeasible. If System II is feasible, and for $\mathbf{x} \geq \mathbf{0}, \mathbf{y} \geqq \mathbf{0}$ we have

$$\boldsymbol{\lambda}^\top \mathbf{Ax} > 0, \quad \boldsymbol{\lambda}^\top \mathbf{By} \geq 0, \quad \boldsymbol{\lambda}^\top \mathbf{Cz} = 0,$$

or equivalently,

$$\boldsymbol{\lambda}^\top (\mathbf{Ax} + \mathbf{By} + \mathbf{Cz}) > 0$$

contradicting the first equality of System I.

Next, we show that if System II is infeasible, then System I must be feasible. If System II is infeasible, then the system

$$\boldsymbol{\lambda}^\top \mathbf{A} \geqq t\mathbf{1}^\top \text{ for some } t > 0, \quad \boldsymbol{\lambda}^\top \mathbf{B} \geqq \mathbf{0}^\top, \quad \boldsymbol{\lambda}^\top \mathbf{C} = \mathbf{0}$$

is infeasible, or equivalently, the system

$$[\boldsymbol{\lambda}^\top,\ t] \begin{bmatrix} -\mathbf{A} & -\mathbf{B} \\ \mathbf{1}^\top & \mathbf{0}^\top \end{bmatrix} \leqq \mathbf{0}^\top, \qquad [\boldsymbol{\lambda}^\top,\ t] \begin{bmatrix} \mathbf{C} \\ \mathbf{0}^\top \end{bmatrix} = \mathbf{0}^\top, \quad [\boldsymbol{\lambda}^\top,\ t] \begin{bmatrix} 0 \\ 1 \end{bmatrix} > 0,$$

is infeasible, or equivalently, by Corollary 3.4.6, the system

$$\begin{bmatrix} -\mathbf{A} & -\mathbf{B} \\ \mathbf{1}^\top & \mathbf{0}^\top \end{bmatrix} \begin{bmatrix} \mathbf{x} \\ \mathbf{y} \end{bmatrix} + \begin{bmatrix} \mathbf{C} \\ \mathbf{0}^\top \end{bmatrix} \mathbf{v} = \begin{bmatrix} 0 \\ 1 \end{bmatrix}, \quad \mathbf{x} \geqq \mathbf{0},\ \mathbf{y} \geqq \mathbf{0},$$

is feasible, or equivalently, the system

$$-\mathbf{Ax} - \mathbf{By} + \mathbf{Cv} = \mathbf{0}, \quad \mathbf{1}^\top \mathbf{x} = 1, \quad \mathbf{x} \geqq \mathbf{0},\ \mathbf{y} \geqq \mathbf{0},$$

is feasible, or equivalently (after letting $\mathbf{v} = -\mathbf{z}$) the system (System I)

$$\mathbf{Ax} + \mathbf{By} + \mathbf{Cz} = \mathbf{0}, \quad \mathbf{x} \geq \mathbf{0},\ \mathbf{y} \geqq \mathbf{0},$$

is feasible. ∎

Theorem 3.4.12 (Tucker's Theorem of Alternative) Let $\mathbf{A} \in \mathbb{R}^{m\times n}$, $\mathbf{B} \in \mathbb{R}^{m\times p}$ and $\mathbf{C} \in \mathbb{R}^{m\times q}$. Exactly one of the following systems is non-empty.

System I $\quad \left\{ \begin{bmatrix} \mathbf{x} \\ \mathbf{y} \\ \mathbf{z} \end{bmatrix} \in \mathbb{R}^{n+p+q} \mid \mathbf{Ax} + \mathbf{By} + \mathbf{Cz} = \mathbf{0},\ \mathbf{x} > \mathbf{0},\ \mathbf{y} \geqq \mathbf{0} \right\}$

System II $\quad \{\boldsymbol{\lambda} \in \mathbb{R}^m \mid \boldsymbol{\lambda}^\top \mathbf{A} \geq \mathbf{0}^\top,\ \boldsymbol{\lambda}^\top \mathbf{B} \geqq \mathbf{0},\ \boldsymbol{\lambda}^\top \mathbf{C} = \mathbf{0}^\top\}$

Proof: The proof follows closely that of Theorem 3.4.11 and is omitted. ∎

In subsequent chapters, theorems of alternatives will be used over and over again to establish duality results for linear and nonlinear systems.

3.5 Painted Index Theory

This section is concerned with a very powerful generalization of the painted network theory discussed in Chapter Two. While the theoretical foundation is based on the theory of oriented matroids (see, for example, [L4]), it is possible to get away with a simpler version without the need to master all the background material required for a full understanding of matroid theory. The development here follows closely that of Rockafellar [R3], [R7]. The reason why this treatment is separated out from Chapter Two where all the network results are collected is because the current duality theory overlaps significantly with concepts in linear programming presented in the earlier part of this chapter. Naturally, a firm grasp of LP concepts enhances the understanding of the duality of painted index theory.

Under fairly general conditions, almost all the formal properties that have been said about a specialized node-arc incidence matrix $\mathbf{E}$ in Chapter Two also hold true for an arbitrary real matrix. Let $\mathbf{A} \in \mathbb{R}^{m \times n}$ be a real matrix, with $n > m$. We assume that $\mathbf{A}$ is of full rank, i.e., rank($\mathbf{A}$) $= m$. Given real vectors $\mathbf{x} \in \mathbb{R}^n$ and $\mathbf{u} \in \mathbb{R}^m$, let the vectors $\mathbf{y}$ and $\mathbf{v}$ be defined as:

$$\mathbf{y} = \mathbf{A}\mathbf{x} \in \mathbb{R}^m,$$
$$\mathbf{v} = -\mathbf{A}^\top \mathbf{u} \in \mathbb{R}^n.$$

We refer to the variables in $(\mathbf{x}, \mathbf{y})$ as the primal variables, and the variables in $(\mathbf{u}, \mathbf{v})$ as the dual variables. Thus if $\mathbf{A}$ is specialized to a node-arc incidence matrix for a digraph, then the vectors $\mathbf{x}, \mathbf{y}, \mathbf{u}, \mathbf{v}$ represents the flow, divergence, potential, and differential vectors respectively. In the current context, these vectors have no special physical significance and should be considered in their abstract framework.

As in the network case, we have the fundamental identity:

$$\mathbf{y}^\top \mathbf{u} = \mathbf{x}^\top \mathbf{A}^\top \mathbf{u} = -\mathbf{x}^\top \mathbf{v},$$

so that

$$\mathbf{y}^\top \mathbf{u} + \mathbf{x}^\top \mathbf{v} = 0. \tag{3.5.1}$$

Let

$$\mathcal{C} = \{\mathbf{x} \in \mathbb{R}^n \mid \mathbf{A}\mathbf{x} = \mathbf{0}\} \quad \text{and}$$
$$\mathcal{D} = \{\mathbf{v} \in \mathbb{R}^n \mid \mathbf{v} = -\mathbf{A}^\top \mathbf{u} \ \text{ for some } \mathbf{u} \in \mathbb{R}^m\}.$$

be two spaces representing the kernel of the linear mapping represented by $\mathbf{A}$ and the row space of $\mathbf{A}$ respectively. Since given any $\mathbf{x} \in \mathcal{C}$, $\mathbf{A}\mathbf{x} = \mathbf{y} = \mathbf{0}$, so that $\mathbf{x}^\top \mathbf{v} = 0$ by (3.5.1) and therefore $\mathcal{C}$ and $\mathcal{D}$ are orthogonally complementary to each other, i.e.,

$$\mathcal{C} = \{\mathbf{x} \in \mathbb{R}^n \mid \mathbf{v}^\top \mathbf{x} = 0 \ \forall \mathbf{v} \in \mathcal{D}\} = \mathcal{D}^\perp$$
$$\text{and} \quad \mathcal{D} = \{\mathbf{v} \in \mathbb{R}^n \mid \mathbf{v}^\top \mathbf{x} = 0 \ \forall \mathbf{x} \in \mathcal{C}\} = \mathcal{C}^\perp.$$

As in the case of linear programming, we can construct a basis matrix $\mathbf{B}$ from $\mathbf{A}$ by choosing m linearly independent columns of $\mathbf{A}$ to form a non-singular basis matrix $\mathbf{B}$. The remaining $n - m$ columns make up another matrix $\mathbf{N} \in \mathbb{R}^{m \times (n-m)}$. By an appropriate reordering, we may express the matrix $\mathbf{A}$ as $\mathbf{A} = [\mathbf{B},\ \mathbf{N}]$. Correspondingly, we partition the vector $\mathbf{x}^\top = [\mathbf{x}_\mathbf{B}^\top,\ \mathbf{x}_\mathbf{N}^\top]$, into the basic variable $\mathbf{x}_\mathbf{B} \in \mathbb{R}^m$ and the nonbasic variable $\mathbf{x}_\mathbf{N} \in \mathbb{R}^{n-m}$ such that

$$\mathbf{A}\mathbf{x} = [\mathbf{B},\ \mathbf{N}] \begin{bmatrix} \mathbf{x}_\mathbf{B} \\ \mathbf{x}_\mathbf{N} \end{bmatrix} = \mathbf{B}\mathbf{x}_\mathbf{B} + \mathbf{N}\mathbf{x}_\mathbf{N}. \tag{3.5.2}$$

Given any $\mathbf{x} \in \mathcal{C}$, $\mathbf{A}\mathbf{x} = \mathbf{B}\mathbf{x}_\mathbf{B} + \mathbf{N}\mathbf{x}_\mathbf{N} = \mathbf{0}$. Consequently, we may express the basic variable explicitly as a linear function of the nonbasic variable,

$$\mathbf{x}_\mathbf{B} = -\mathbf{B}^{-1}\mathbf{N}\mathbf{x}_\mathbf{N} = \mathbf{T}\mathbf{x}_\mathbf{N} \tag{3.5.3}$$

where $\mathbf{T} = -\mathbf{B}^{-1}\mathbf{N} \in \mathbb{R}^{m\times(n-m)}$ is referred to as the *Tucker Tableau* [T1]. Similarly, let $\mathbf{v} \in \mathcal{D}$ and partition $\mathbf{v}$ in correspondence with $\mathbf{B}$ and $\mathbf{N}$ so that by definition, there exists some $\mathbf{u} \in \mathbb{R}^m$ with

$$\mathbf{v} = \begin{bmatrix} \mathbf{v_B} \\ \mathbf{v_N} \end{bmatrix} = -\mathbf{A}^\top \mathbf{u} = -\begin{bmatrix} \mathbf{B}^\top \\ \mathbf{N}^\top \end{bmatrix} \mathbf{u}. \tag{3.5.4}$$

Thus

$$\mathbf{u} = (-\mathbf{B}^\top)^{-1}\mathbf{v_B},$$

and consequently,

$$\mathbf{v_N} = -\mathbf{N}^\top(-\mathbf{B}^\top)^{-1}\mathbf{v_B} = -\mathbf{T}^\top\mathbf{v_B}. \tag{3.5.5}$$

(3.5.3) and (3.5.5) are referred to as the *Tucker representations* of the linear systems of variables [T1]. Note that the Tucker representation given in (2.3.6)-(2.3.8) is the network specialization of the general case here. Since there exists a maximum of $\binom{n}{m}$ ways of choosing the basis matrix $\mathbf{B}$ (not all of them necessarily give rise to a non-singular $\mathbf{B}$), the number of possible Tucker representations is therefore finite.

When one (or more) column of $\mathbf{B}$ is exchanged with another column (or more) from $\mathbf{N}$, the Tucker tableau $\mathbf{T}$ switches from one form to another. For computational purposes, we restrict the number of changed columns at any time to just one, say $\mathbf{B}$ and $\mathbf{N}$ are changed to $\mathbf{B}'$ and $\mathbf{N}'$ where $\mathbf{B}$ and $\mathbf{B}'$ differs by exactly one column. In which case, the new Tucker tableau $\mathbf{T}' = -(\mathbf{B}')^{-1}\mathbf{N}'$ can be easily computed from the old Tucker tableau $\mathbf{T} = -\mathbf{B}^{-1}\mathbf{N}$ by a simple rank one update procedure commonly known in LP as the *pivoting step.* This is a well-known procedure for readers who are familiar with the simplex algorithm for solving linear programs.

Primal and Dual Support

Let the column index set of $\mathbf{A}$ be $\mathcal{J} = \{1, 2, \cdots, n\}$. The theory of oriented matroid is concerned with combinatorial structures related to subsets of the index sets $\mathcal{J}$ induced by the elements of $\mathcal{C}$ and $\mathcal{D}$. We will not go into a full discussion of matroid theory, but will instead present a slightly more compact version due to Rockafellar [R2], [R8].

Definition 3.5.1 (Primal Support) A subset $\mathcal{P} = \mathcal{P}^+ \cup \mathcal{P}^- \subseteq \mathcal{J}$ is a *primal support* of $\mathcal{C}$ if there exists $\mathbf{x} \in \mathcal{C}$ such that

$$\mathcal{P}^+ = \{j \in \mathcal{J} \mid x_j > 0\},$$
$$\mathcal{P}^- = \{j \in \mathcal{J} \mid x_j < 0\}.$$

Clearly only the trivial vector $\mathbf{x} = \mathbf{0}$ has an empty primal support. A primal support $\mathcal{P}$ is said to be *elementary* if there does not exist a $\mathcal{P}_0 = \mathcal{P}_0^+ \cup \mathcal{P}_0^- \neq \emptyset$ such that $\mathcal{P}_0 \subset \mathcal{P}$ (as a strict subset). A vector in $\mathcal{C}$ is said to be *elementary* if its support is elementary. Given a primal support of $\mathcal{C}$, the associated vector $\mathbf{x} \in \mathcal{C}$ may be non-unique, although it is not hard to show that two vectors corresponding to the same primal support are a scalar multiple of each other. This minor complication can be overcome by identifying a unique associated vector $\mathbf{e}_{\mathcal{P}}$ (generalized incidence vector for the primal support) under the following normalization scheme:

$$\|\mathbf{e}_{\mathcal{P}}\|_1 = |\mathcal{P}|, \quad \text{the cardinality of } \mathcal{P}. \tag{3.5.6}$$

Definition 3.5.2 (Dual Support) A subset $\mathcal{Q} = \mathcal{Q}^+ \cup \mathcal{Q}^- \subseteq \mathcal{J}$ is a *dual support* of $\mathcal{C}$ if there exists $\mathbf{v} \in \mathcal{D}$ such that

$$\begin{aligned} \mathcal{Q}^+ &= \{j \in \mathcal{J} \mid v_j > 0\}, \\ \mathcal{Q}^- &= \{j \in \mathcal{J} \mid v_j < 0\}. \end{aligned}$$

Clearly only the trivial vector $\mathbf{v} = \mathbf{0}$ has an empty dual support. A dual support $\mathcal{Q}$ is said to be *elementary* if there does not exist an $\mathcal{Q}_0 = \mathcal{Q}_0^+ \cup \mathcal{Q}_0^- \neq \emptyset$ such that $\mathcal{Q}_0 \subset \mathcal{Q}$ (as a strict subset). A vector in $\mathcal{D}$ is said to be *elementary* if its support is elementary. Given a dual support of $\mathcal{D}$, the associated vector $\mathbf{v} \in \mathcal{D}$ may be non-unique. This again can be overcome by identifying a unique associated vector $\mathbf{e}_{\mathcal{Q}}$ (generalized incidence vector for the dual support) under the following normalization scheme:

$$\|\mathbf{e}_{\mathcal{Q}}\|_1 = |\mathcal{Q}|, \quad \text{the cardinality of } \mathcal{Q}. \tag{3.5.7}$$

The fact that we are using symbols that are almost identical to that used in Chapter Two is no coincidence. In the event that $\mathbf{A}$ is the node arc incidence matrix of a digraph, then $\mathcal{P}, \mathcal{Q}, \mathbf{e}_{\mathcal{P}}, \mathbf{e}_{\mathcal{Q}}$ reduce to a cycle, a cut, a cycle incidence vector, and a cut incidence vector respectively.

It is possible to show that any $\mathbf{x} \in \mathcal{C}$ with a corresponding support $\mathcal{P}$ can be expressed as a linear combination of generalized incidence vectors (for primal support), each of which corresponds to an elementary primal support. Similarly, any $\mathbf{v} \in \mathcal{D}$ can be expressed as a linear combination of generalized incidence vectors (for dual support) as well. Although the restriction to elementary supports is not necessary for the subsequent analysis, we shall nevertheless make such an assumption for convenience in view of the above.

Definition 3.5.3 (Index painting) A *painting* of the index set $\mathcal{J}$ is a partition of $\mathcal{J}$ into four disjoint subsets, some possibly empty, such that the index in each of the subsets are labelled `Green`, `Red`, `White`, and `Black` respectively.

The following is a generalization of the painted network Theorem 2.2.1, and can be thought of as another theorem of alternative similar to those discussed in the previous sections. This result, however, has a strong combinatorial flavor to it. The qualification of "elementary" as appearing in square parentheses below means that its appearance is optional.

Theorem 3.5.1 (Painted index theorem - combinatorial form) Given any painting of $\mathcal{J}$ and any $\ell \in \mathcal{J}$ that is `Black` or `White`, exactly one of the following is true:

I. There exists an [elementary] primal support $\mathcal{P}$ containing ℓ such that $\forall j \in \mathcal{P}^+$, j is painted `Green` or `White`, and $\forall j \in \mathcal{P}^-$, j is painted `Green` or `Black`.
II. There exists an [elementary] dual support $\mathcal{Q}$ containing ℓ such that $\forall j \in \mathcal{Q}^+$, j is painted `Red` or `Black`, and $\forall j \in \mathcal{Q}^-$, j is painted `Red` or `White`.

(Partial) proof: If there exists an $\mathbf{x} \in \mathcal{C}$ that satisfies the conditions in Alternative I, and there exists an $\mathbf{v} \in \mathcal{D}$ that satisfies the conditions in Alternative II, then x_ℓ and v_ℓ must be of opposite sign and neither is 0; if $j \neq \ell$ is `Green` or `Red` then $x_j v_j = 0$ since either $x_j = 0$ (j is `Red`) or $v_j = 0$ (j is `Green`); if $j \neq \ell$ is `White`, then $x_j \geq 0$ and $v_j \leq 0$; and if $j \neq \ell$ is `Black`, then $x_j \leq 0$ and $v_j \geq 0$. Consequently,

$$\mathbf{x}^\top \mathbf{v} = x_\ell v_\ell + \sum_{\substack{j \texttt{ White}, \\ j \neq \ell}} x_j v_j + \sum_{\substack{j \texttt{ Black}, \\ j \neq \ell}} x_j v_j < 0$$

which contradicts the orthogonality of $\mathbf{x}$ and $\mathbf{v}$. As a result, a constructive procedure that either constructs a primal support or a dual support is sufficient to justify the validity of the theorem. Such a procedure is furnished in the painted index algorithm to be discussed shortly. ∎

A primal support that satisfies the coloring requirement of Alternative I is called a *compatible primal support*. Similarly, a dual support that satisfies the coloring requirement of Alternative II is called a *compatible dual support*. Before we describe the algorithm that finds a compatible primal support or a compatible dual support, it would facilitate understanding to introduce a few more concepts that are direct generalizations of the network case.

Let $\mathbf{B}$ be a basis matrix for $\mathbf{A}$ and let the *basis index set* $\mathcal{F} \subset \mathcal{J}$ comprise of the indices associated with the columns in $\mathbf{B}$. Let $\bar{\mathcal{F}} = \mathcal{J} \setminus \mathcal{F}$, and let $\mathbf{T} = -\mathbf{B}^{-1}\mathbf{N}$ be the corresponding Tucker tableau for $\mathcal{C}$ and $\mathcal{D}$. Given an $\mathbf{x} \in \mathcal{C}$ and its partition into the basic and nonbasic part as in (3.5.2), we can assign a value of 1 to a particular nonbasic variable x_k, $k \in \bar{\mathcal{F}}$, and a value of 0 to every other nonbasic variable x_j, $j \in \bar{\mathcal{F}}$. Consequently the basic variable (depending on k) is given by (3.5.3):

$$\mathbf{x_B} = \mathbf{T}\mathbf{e}^k = \mathbf{t}^k$$

where $\mathbf{t}^k$ is the kth column of $\mathbf{T}$. The resulting vector $\mathbf{x}$ may be normalized accordingly to give an elementary incidence vector, and the corresponding column (primal) support is an elementary one.

On the other hand, given an $\mathbf{v} \in \mathcal{D}$ and its partition into the basic and nonbasic part as in (3.5.4), we can assign a value of 1 to a particular basic (dual) variable v_j, $j \in \mathcal{F}$, and a value of 0 to every other basic v_k, $k \in \mathcal{F}$, and consequently the nonbasic (dual) variable is given by (3.5.5):

$$\mathbf{v_N} = -\mathbf{T}^\top \mathbf{e}^j = -\mathbf{t}_j^\top$$

where $\mathbf{t}_j$ is the jth row of $\mathbf{T}$. The resulting (dual) vector $\mathbf{v}$ may be normalized accordingly to give an elementary incidence vector, and the corresponding row (dual) support is an elementary one.

These observations are formalized as follows.

Definition 3.5.4 (Column support) Given any $k \in \bar{\mathcal{F}}$ (a nonbasic index), there exists an elementary primal support $\mathcal{P}_k$ that contains k in $\mathcal{P}_k^+$ and otherwise contains only elements of $\mathcal{F}$. The corresponding elementary (primal) vector has entries given by $x_k = 1, x_j = 0,\ \forall j \in \bar{\mathcal{F}}, x_i = [\mathbf{T}]_{ik}\ \forall i \in \mathcal{F}$.

Definition 3.5.5 (Row support) Given any $j \in \mathcal{F}$ (a basic index), there exists an elementary dual support $\mathcal{Q}_j$ that contains j in $\mathcal{Q}_j^+$ and otherwise contains only elements of $\bar{\mathcal{F}}$. The corresponding elementary (dual) vector has entries given by $x_j = 1, x_k = 0,\ \forall k \in \mathcal{F}, x_i = -[\mathbf{T}]_{ki}\ \forall i \in \bar{\mathcal{F}}$.

The above statement of the Painted Index Theorem 3.5.1 is known to be in combinatorial form, and is a direct generalization of the Painted Network Theorem 2.2.1. Another equivalent form is expressed purely in terms of the variables $\mathbf{x}$ and $\mathbf{v}$ without references to the primal/dual supports, and is stated below. Viewed in this form, the Painted Index Theorem is just another theorem of alternatives concerning two systems of homogeneous linear equations and linear inequalities. It turns out that this form of the theorem makes it easier to understand the mechanics of the painted index algorithm.

Theorem 3.5.2 (Painted Index Theorem - vector form) Given any painting of $\mathcal{J}$ and any $\ell \in \mathcal{J}$ that is `Black` or `White`, exactly one of the following is true:
I. There exists an [elementary] primal vector $\mathbf{x} \in \mathcal{C}$ such that $x_j \geq 0$ for all `White` j, $x_j \leq 0$ for all `Black` j, and $x_j = 0$ for every `Red` j.
II. There exists an [elementary] dual vector $\mathbf{v} \in \mathcal{D}$ such that $v_j \geq 0$ for all `Black` j, $v_j \leq 0$ for all `White` j, and $v_j = 0$ for every `Green` j.

Proof: We first show the equivalence of Alternative I of Theorem 3.5.1 and Alternative I of Theorem 3.5.2. This is a fairly routine exercise in logic. Alternative I of Theorem 3.5.1 asserts that $x_j > 0 \Rightarrow j$ is `Green` or `White` and $x_j < 0 \Rightarrow j$ is `Green` or `Black`. This is equivalent to saying that if j is `Red` or `Black` $\Rightarrow x_j \leq 0$ and j is `Red` or `White` $\Rightarrow x_j \geq 0$, which in turn is equivalent to saying that if j is `Red` $\Rightarrow x_j = 0$ and j is `White` $\Rightarrow x_j \geq 0$ and j is `Black` $\Rightarrow x_j \leq 0$. A similar argument applies to show the equivalence of Alternative II. ∎

Definition 3.5.6 (Compatible column) Given a Tucker Tableau $\mathbf{T}$, we say that a column $\mathbf{t}^k$ in $\mathbf{T}$ is *compatible* if k is not `Red` and it satisfies the following condition:

- If k is `Green`, then

$$[\mathbf{t}^k]_j \begin{cases} = 0 & \text{if } j \text{ is } \texttt{Red, Black or White} \\ \text{is arbitrary} & \text{if } j \text{ is } \texttt{Green} \end{cases}$$

- If k is `White`, then

$$[\mathbf{t}^k]_j \begin{cases} = 0 & \text{if } j \text{ is } \texttt{Red} \\ \leq 0 & \text{if } j \text{ is } \texttt{Black} \\ \geq 0 & \text{if } j \text{ is } \texttt{White} \\ \text{is arbitrary} & \text{if } j \text{ is } \texttt{Green} \end{cases}$$

• If k is Black, then

$$[\mathbf{t}^k]_j \begin{cases} = 0 & \text{if } j \text{ is Red} \\ \geq 0 & \text{if } j \text{ is Black} \\ \leq 0 & \text{if } j \text{ is White} \\ \text{is arbitrary} & \text{if } j \text{ is Green} \end{cases}$$

Definition 3.5.7 (Compatible row) Given a Tucker tableau $\mathbf{T}$, we say that a row $\mathbf{t}_k$ in $\mathbf{T}$ is *compatible* if k is not Green and it satisfies the following condition:

• If k is Red, then

$$[\mathbf{t}_k]_j \begin{cases} = 0 & \text{if } j \text{ is Green, Black or White} \\ \text{is arbitrary} & \text{if } j \text{ is Red} \end{cases}$$

• If k is Black, then

$$[\mathbf{t}_k]_j \begin{cases} = 0 & \text{if } j \text{ is Green} \\ \geq 0 & \text{if } j \text{ is White} \\ \leq 0 & \text{if } j \text{ is Black} \\ \text{is arbitrary} & \text{if } j \text{ is Red} \end{cases}$$

• If k is White, then

$$[\mathbf{t}_k]_j \begin{cases} = 0 & \text{if } j \text{ is Green} \\ \leq 0 & \text{if } j \text{ is White} \\ \geq 0 & \text{if } j \text{ is Black} \\ \text{is arbitrary} & \text{if } j \text{ is Red} \end{cases}$$

A compatible column corresponds to an elementary primal support $\mathcal{P}$ such that $j \in \mathcal{P}^+ \Rightarrow j$ is Green or White, and $j \in \mathcal{P}^- \Rightarrow j$ is Green or Black. On the other hand, a compatible row corresponds to an elementary dual support $\mathcal{Q}$ such that $j \in \mathcal{Q}^+ \Rightarrow j$ is Red or Black, and $j \in \mathcal{Q}^- \Rightarrow j$ is Red or White.

It helps at this juncture to think of the above arguments in terms of the network problem.

• A basis in this case corresponds to a spanning tree (with due consideration for the rank one deficiency of the node arc incidence matrix).

• An [elementary] primal support corresponds to an [elementary] cycle.

• An [elementary] dual support corresponds to an [elementary] cut.

• $\mathcal{C}$ corresponds to the circulation space and $\mathcal{D}$ corresponds to the differential space.

• A cotree (nonbasic) arc k gives rise to a cycle containing this cotree arc, and with the rest of the (tree) arcs in the cycle forming a unique path between the end points of the cotree arc. The kth column of the Tucker representation gives the orientation (1 for forward, -1 for reverse) of the tree arcs in the cycle.

• A tree (basic) arc gives rise to a cut containing this tree arc, and with the rest of the (cotree) arcs straddling the two components of nodes formed by deleting the tree arc from the tree. The kth row of the Tucker representation gives the orientation (1 for forward, -1 for reverse) of the cotree arcs in the cut.

The idea behind the method for finding a solution to Alternative I or II (of the Painted Index Theorem 3.5.2) is as follows. If we can construct a Tucker Tableau that contains a compatible column that uses ℓ, then a solution to Alternative I is found, otherwise a pivoting procedure can be carried out to try to update to another Tucker tableau that contains a compatible row that uses ℓ, whence a solution to alternative II is found. Repeat this if necessary.

Algorithm 3.5.1 (Painted Index Algorithm)

Initialization: Construct an initial Tucker tableau by choosing an appropriate basis. Identify the given `White` or `Black` index ℓ which could either be used by a row (hence it is basic to begin with) or by a column (hence it is nonbasic to begin with).

Repeat until done

If ℓ is used by a row.

If this row is compatible, terminate the algorithm with a solution to Alternative II.

Else there is an incompatible index k in the ℓ row.

If the k column is compatible, terminate the algorithm with a solution to Alternative I.

Else there is an incompatible index j in the k column. Pivot about the (j,k) entry of the tableau, i.e, replace $\mathbf{a}^j$ by $\mathbf{a}^k$ in the basis and update the tableau.

End

End

Else ℓ is used by a column.

If this column is compatible, terminate the algorithm with a solution to Alternative I.

Else there is an incompatible index j in the ℓ column.

If the j row is compatible, terminate the algorithm with a solution to Alternative II.

Else there is an incompatible index k in the j row. Pivot about the (j,k) entry of the tableau.

End

End

End

End ■

Remaining of the proof of Theorem 3.5.1 We shall now complete the proof of the Painted Index Theorem 3.5.1. It is possible, just as in the case of Simplex algorithm, that the painted index algorithm may go into a cycle and hence fail to terminate. A very simple trick called the Bland's rule will get around this potential difficulty. *The idea is to assign an (arbitrary) priority rank to each of the indices, and whenever there is more than one index that can be chosen during the algorithm, simply choose the one with the highest priority rank.* This rule is also practiced in preventing cycling in the simplex algorithm. Once Bland's rule is adopted, we will show that the algorithm cannot iterate forever. For clarity, we break the somewhat complex argument down in a pointwise manner.

(i) Assuming that the algorithm does not terminate. Since the number of possible Tucker's tableau is finite, i.e., bounded above by $\binom{n}{m}$, there must be at least one cycle of iterations that the algorithm gets itself into.

(ii) After the algorithm has settled in the cycle, each of the indices in $\mathcal{J}$ will either stay as a column index, or stay as a row index, or switch indefinitely between rows and columns. Let $\mathcal{J}_c$ be the set of indices that remain in the columns, $\mathcal{J}_r$ be the set of indices that remain in the rows, and $\mathcal{J}_z$ be the set of indices that switch indefinitely. We will show that $\mathcal{J}_z \neq \emptyset$ leads to a contradiction.

(iii) Since any index in $\mathcal{J}_z$ must be a column index and a row index at different time of the cycle, all indices in $\mathcal{J}_z$ cannot be `Red` or `Green`. Also, $\ell \notin \mathcal{J}_z$ since ℓ is `White` or `Black`.

(iv) Without loss of generality, assume $\ell \in \mathcal{J}_r$, and let q be the lowest rank index in $\mathcal{J}_z$.

(v) At some point of time, assume that q is a column index and is chosen as k (to enter the basis) because the entry $[\mathbf{T}]_{\ell q}$ destroys the compatibility of the ℓ row. This is also the *only* index that can be chosen because otherwise Bland's rule would have been violated. On the other hand, if q is a row index and is chosen as j (to leave the basis) because the entry $[\mathbf{T}]_{qk}$ destroys the compatibility of the k column. Again this is the *only* index that can be chosen because otherwise Bland's rule would have been violated.

(vi) Whatever the original coloring scheme is, it will now be modified such that the color of q is reversed between `White` and `Black`; indices in $\mathcal{J}_c$ are all colored `Red`; and indices in $\mathcal{J}_r$ except that of ℓ are all colored `Green`.

(vii) Under the new coloring scheme, it is easy to check that the ℓ row is compatible in the situation where q is chosen as k. Simultaneously, the k column is compatible in the situation where q is chosen as j. Both these row and column use the index q.

(viii) We have thus constructed a painting scheme where both alternatives to the Painted Index Theorem are true, thus contradicting the argument put forth in the (partial) proof of Theorem 3.5.1 which asserts that both alternatives cannot be true simultaneously.

(ix) The conclusion is that $\mathcal{J}_z$ must be empty and therefore the algorithm must terminate in a finite number of iterations with a solution to either Alternative I or II. ∎

3.6 Duality in Monotropic Optimization

Monotropic optimization is a special case of convex programming and is characterized by three properties: The cost function is convex, separable and the constraints are linear equalities or simple bounds. Such problems includes a huge number of other problems as special cases, for example, linear programming, network programming, and (separable) quadratic programming. A remarkable feature of monotropic optimization is that just about all of the algorithms and duality results in network programming can be generalized to work for monotropic programs. A formal treatment of monotropic optimization requires some extensive and elaborate algorithmic development, as manifested in the constructive proofs of all the theorems in this section. Inclusion of these proofs will involve much more complex notations without shedding more light on duality. For convenience, we shall state without proofs the salient results in monotropic optimization, and refer the reader to [R3] for a detailed analysis.

Although monotropic optimization is, strictly speaking, nonlinear, we have chosen to include it in this chapter mainly because the theoretical foundation is given by the painted index theory. The linear system consideration pertains to the way the linear constraints are being treated.

Let $\mathbf{A} \in \mathbb{R}^{m\times n}$ be a real matrix, $\mathcal{J} = \{1, 2, \cdots, n\}$, $f_j : \mathbb{R} \to \mathbb{R}$, $g_j : \mathbb{R} \to \mathbb{R}$, $j \in \mathcal{J}$ be convex real valued functions, and $C_j = [c_j^-, c_j^+], D_j = [d_j^-, d_j^+]$, $j \in \mathcal{J}$ be real non-empty intervals in $\bar{\mathbb{R}}$, $\mathbf{b} \in \mathbb{R}^m$ be a real vector. In order for duality to work, we shall assume, from the outset, that the functions f_j and g_j, $j \in \mathcal{J}$ are convex conjugate to each other. The primal and dual optimization problems are defined as follows.

Definition 3.6.1 (Primal Monotropic Optimization Problem)

$$\min \quad \Phi(\mathbf{x}) = \sum_{j\in\mathcal{J}} f_j(x_j), \tag{3.6.1}$$

$$\text{s.t.} \quad \mathbf{Ax} = \mathbf{b}, \tag{3.6.2}$$

$$x_j \in C_j \quad \forall j \in \mathcal{J}. \tag{3.6.3}$$

Definition 3.6.2 (Dual Monotropic Optimization Problem)

$$\max \quad \Psi(\mathbf{u}) = -\mathbf{b}^\top \mathbf{u} - \sum_{j\in\mathcal{J}} g_j(v_j), \tag{3.6.4}$$

$$\text{s.t.} \quad \mathbf{u}^\top \mathbf{A} = -\mathbf{v}^\top, \tag{3.6.5}$$

$$v_j \in D_j \quad \forall j \in \mathcal{J}. \tag{3.6.6}$$

Before the issue of optimality is discussed, we should first address the feasibility issue.

Primal and dual feasibility

Let each $j \in \mathcal{J}$ be assigned two non-empty intervals C_j and D_j, the former being the generalization of the capacity interval, and the latter being the generalization of the span interval in the network case.

Definition 3.6.3 (Primal feasibility problem) Find a solution $\mathbf{x} \in \mathbb{R}^n$ that satisfies (3.6.2) and (3.6.3).

Definition 3.6.4 (Dual feasibility problem) Find a solution $\mathbf{u} \in \mathbb{R}^m$ that satisfies (3.6.5) and (3.6.6).

The following results for monotropic optimization are stated in the form of a theorem of alternative, and they include the network feasibility Theorems 2.7.1 and 2.7.2 as special cases. Since the proofs of these results require an extensive discussion of algorithmic development which are beyond the scope of this book, they are omitted from here. Readers are referred to [R3] for details.

Theorem 3.6.1 (Primal Feasibility Theorem) Exactly one of the following is true:
1. There exists a feasible solution to the primal feasibility problem.
2. There exists an [elementary] dual support $\mathcal{Q}$ such that

$$C_j = [c_{\mathcal{Q}}^-, c_{\mathcal{Q}}^+] \subseteq (-\infty, b_{\mathcal{Q}}),$$

where $b_{\mathcal{Q}} = -\mathbf{u}^\top \mathbf{b}$ for any arbitrary $\mathbf{u}$ satisfying $-\mathbf{A}^\top \mathbf{u} = \mathbf{e}_{\mathcal{Q}}$, $\mathbf{e}_{\mathcal{Q}} \in \mathbb{R}^n$ is the (normalized) incidence vector for the dual support, and

$$c_{\mathcal{Q}}^- = \sum_{j \in \mathcal{Q}^+} [\mathbf{e}_{\mathcal{Q}}]_j c_j^- + \sum_{j \in \mathcal{Q}^-} [\mathbf{e}_{\mathcal{Q}}]_j c_j^+,$$
$$c_{\mathcal{Q}}^+ = \sum_{j \in \mathcal{Q}^+} [\mathbf{e}_{\mathcal{Q}}]_j c_j^+ + \sum_{j \in \mathcal{Q}^-} [\mathbf{e}_{\mathcal{Q}}]_j c_j^-.$$

Theorem 3.6.2 (Dual feasibility theorem) Exactly one of the following is true:
1. There exists a feasible solution $\mathbf{u}$ to the dual feasibility problem.
2. There exists an [elementary] primal support $\mathcal{P}$ such that

$$D_j = [d_{\mathcal{P}}^-, d_{\mathcal{P}}^+] \subseteq (-\infty, 0),$$

where $\mathbf{e}_{\mathcal{P}} \in \mathbb{R}^n$ is the (normalized) incidence vector for the primal support, and

$$d_{\mathcal{P}}^- = \sum_{j \in \mathcal{P}^+} [\mathbf{e}_{\mathcal{P}}]_j d_j^- + \sum_{j \in \mathcal{P}^-} [\mathbf{e}_{\mathcal{P}}]_j d_j^+,$$
$$d_{\mathcal{P}}^+ = \sum_{j \in \mathcal{P}^+} [\mathbf{e}_{\mathcal{P}}]_j d_j^+ + \sum_{j \in \mathcal{P}^-} [\mathbf{e}_{\mathcal{P}}]_j d_j^-.$$

Definition 3.6.5 (Monotropic Equilibrium Problem) Let $\Gamma_j,\ j \in \mathcal{J}$ be the kilter curve for f_j. Find $\mathbf{x} \in \mathbb{R}^n$ and $\mathbf{y} \in \mathbb{R}^m$ such that

$$\begin{aligned} \mathbf{Ax} = \mathbf{y} = \mathbf{b}, \\ \text{and} \quad (x_j, y_j) \in \Gamma_j \quad \forall j \in \mathcal{J}. \end{aligned}$$

The following results are generalizations of Theorem 2.9.2 and Corollary 2.9.3. Again the proofs are omitted to avoid a lengthy discussion of algorithmic development.

Theorem 3.6.3 (Weak Duality) $\Phi(\mathbf{x}) \geq \Psi(\mathbf{u})$ for all feasible $\mathbf{x}$ and feasible $\mathbf{u}$.

Theorem 3.6.4 (Strong Duality) If the primal has an unbounded solution, then the dual has no feasible solution. If the dual has an unbounded solution, then the primal has no feasible solution. If both the primal and dual have finite feasible solutions, then

$$\inf \text{ in the primal} = \sup \text{ in the dual}.$$

Theorem 3.6.5 (Monotropic Equilibrium Theorem) The pair of points $\mathbf{x}$ and $\mathbf{u}$ solves the monotropic equilibrium problem if and only if $\mathbf{x}$ solves the primal monotropic optimization problem and $\mathbf{u}$ solves the dual monotropic optimization problem.

CHAPTER 4

DUALITY IN CONVEX NONLINEAR SYSTEMS

It is well known that some nonlinear programming problems can be solved by dual methods (see, for example, [G1], [R9]), although it is not the intention of this book to dwell on methods of solution. This chapter is a natural progression from the previous one in the sense that many of the duality results presented in this chapter include those duality results for linear systems as special cases. However, this generalization is made possible only by the consideration of convex nonlinear systems. Without convexity, the duality result would have been much weaker, at least in the context of Lagrangian duality.

As pointed out in the preface, we shall not pursue convex duality by the generalized approach, but will take a traditional approach instead. There are, in the main, two popular traditional forms of duality for constrained nonlinear optimization. The first is Lagrangian duality, and the second is Wolfe duality. Both will be discussed in some detail. We begin with a duality model for unconstrained problems based on conjugate duality. This can be considered a watered-down version of the generalized Lagrangian duality theory treated in [R2] and [W1]. In Section 4.2, we present some of the better known optimality conditions for constrained optimization so as to pave the way for the traditional Lagrangian duality in the next section. With convexity, we show that all the optimality conditions become necessary and sufficient, and in addition, strong duality holds. In Section 4.4, we specialize the Lagrangian duality to the cases of linear, quadratic and monotropic programming. In Section 4.5, we discuss the other well known form of duality due to Wolfe, and present the first of a series of gap function models.

4.1 Conjugate Duality in Unconstrained Optimization

We shall first state some well known properties of unconstrained optimization. Let $g : \mathbb{R}^n \to \mathbb{R}$ be a differentiable function. The simplest unconstrained optimization problem is stated as follows.

Definition 4.1.1. (Unconstrained optimization problem)

$$\text{(Problem UP)} \qquad \inf_{\mathbf{x}\in\mathbf{R}^n} g(\mathbf{x}).$$

The reader should be familiar with the following results, the proofs of which can be found in any standard textbook in optimization.

Theorem 4.1.1 (Necessary condition for unconstrained minimum) If $\mathbf{x}^*$ is a local minimum (see Definition 1.2.1) for problem UP, then $\nabla g(\mathbf{x}^*) = \mathbf{0}^\top$.

Theorem 4.1.2 (Sufficient condition for unconstrained minimum) Assume that f is twice differentiable. If the gradient $\nabla g(\mathbf{x}^*) = \mathbf{0}^\top$ and the Hessian $\nabla^2 g(\mathbf{x}^*) = \frac{\partial}{\partial \mathbf{x}} \left[\frac{\partial g}{\partial \mathbf{x}}\right]^\top \Big|_{\mathbf{x}^*}$ is positive definite, then $\mathbf{x}^*$ is a local minimum.

Theorem 4.1.3 (Sufficient and necessary optimality condition for convex problems) Assume that g is differentiable and convex (in fact, pseudoconvexity will do). Then $\mathbf{x}^*$ is a global minimum if and only if $\nabla g(\mathbf{x}^*) = \mathbf{0}^\top$.

Remark 4.1.1 Note that Theorem 4.1.3 is the first of a series of equivalent optimality conditions for convex optimization problems.

Traditionally, duality in mathematical programming is meaningful only when constraints are involved. However, it is not too difficult to pretend that all optimization problems are unconstrained with the use of an extended real-valued function. To reduce a constrained problem into an unconstrained one, the following (somewhat frivolous) trick can be used. We assume that the decision variable $\mathbf{x}$ is constrained to lie in some feasible space $\mathcal{X}$ which is a proper subset of $\mathbb{R}^n$. Let $g : \mathcal{X} \to \mathbb{R}$ be a real-valued function. The constrained optimization problem is formulated as

$$\text{(Problem UP}_1\text{)} \qquad \inf_{\mathbf{x}\in\mathcal{X}} g(\mathbf{x}).$$

Now define another function $f : \mathbb{R}^n \to \overline{\mathbb{R}}$ such that

$$f(\mathbf{x}) = \begin{cases} g(\mathbf{x}) & \text{if } \mathbf{x} \in \mathcal{X}, \\ \infty & \text{if } \mathbf{x} \notin \mathcal{X}. \end{cases}$$

Then the constrained problem can simply be reformulated as an unconstrained one:

$$\inf_{\mathbf{x}\in\mathbf{R}^n} f(\mathbf{x}).$$

Some readers will no doubt complain that this doesn't really help to solve the problem. The new function f is neither explicitly defined nor differentiable. Nevertheless from a theoretical point of view, this reformulation is quite acceptable and in fact desirable in establishing some theoretical duality results.

We now present a nifty duality result that is based on conjugate duality. Consider the unconstrained optimization problem (UP$_1$). Furthermore, assume that the function f is differentiable and convex. Under suitable conditions it is possible to construct a closed, proper and convex function $\psi : \mathbb{R}^n \times \mathbb{R}^m \to \overline{\mathbb{R}}$, which is structured in such a way that:

$$\psi(\mathbf{x}, \mathbf{0}) = f(\mathbf{x}).$$

Consider the following perturbation function which is induced from ψ.

Definition 4.1.2 (Perturbation function)

$$p(\mathbf{y}) = \inf_{\mathbf{x} \in \mathbb{R}^n} \psi(\mathbf{x}, \mathbf{y}).$$

Thus $p(\mathbf{0}) = \inf_{\mathbf{x} \in \mathbb{R}^n} \psi(\mathbf{x}, \mathbf{0}) = \inf_{\mathbf{x} \in \mathbb{R}^n} f(\mathbf{x})$ is the least upper bound for the primal problem. Let $\psi^*(\mathbf{u}, \mathbf{v})$ be the Fenchel transform of ψ where $\mathbf{u} \in \mathbb{R}^n, \mathbf{v} \in \mathbb{R}^m$. By Young's inequality (Lemma 1.3.1), we have:

$$\mathbf{u}^\top \mathbf{x} + \mathbf{v}^\top \mathbf{y} \leq \psi(\mathbf{x}, \mathbf{y}) + \psi^*(\mathbf{u}, \mathbf{v}) \quad \forall \mathbf{u}, \mathbf{x} \in \mathbb{R}^n, \ \mathbf{v}, \mathbf{y} \in \mathbb{R}^m. \tag{4.1.1}$$

A pair of primal and dual problems are defined as follows. Note that the primal problem is just the original unconstrained optimization problem UP.

Definition 4.1.3 (Primal and Dual problems)

$$\text{Primal problem} \quad \inf_{\mathbf{x} \in \mathbb{R}^n} \psi(\mathbf{x}, \mathbf{0}),$$

$$\text{Dual problem} \quad \sup_{\mathbf{v} \in \mathbb{R}^m} -\psi^*(\mathbf{0}, \mathbf{v}).$$

Lemma 4.1.4 (Weak Duality) We have

$$\psi(\mathbf{x}, \mathbf{0}) \geq -\psi^*(\mathbf{0}, \mathbf{v}) \quad \forall \mathbf{x} \in \mathbb{R}^n, \mathbf{v} \in \mathbb{R}^m.$$

Proof: Let $\mathbf{y} = \mathbf{0}$ and $\mathbf{u} = \mathbf{0}$ in (4.1.1), we have $0 \leq \psi(\mathbf{x}, \mathbf{0}) + \psi^*(\mathbf{0}, \mathbf{v})$. ∎

A few other results are needed before the strong duality can be established.

Lemma 4.1.5 If ψ is proper and convex, then p is convex.

Proof: ψ being proper implies that there exists $\mathbf{x} \in \mathbb{R}^n, \mathbf{y} \in \mathbb{R}^m, \alpha \in \mathbb{R}$ such that $\psi(\mathbf{x}, \mathbf{y}) \le \alpha$. Subsequently, $p(\mathbf{y}) \le \psi(\mathbf{x}, \mathbf{y}) \le \alpha$ and hence epi(p) is nonempty. For some $(\mathbf{y}, \alpha) \in$ epi(p) and an arbitrary $\epsilon > 0$, there exists $\mathbf{x}' \in \mathbb{R}^n$ such that

$$\psi(\mathbf{x}', \mathbf{y}) < \inf_{\mathbf{x} \in \mathbf{R}^n} \psi(\mathbf{x}, \mathbf{y}) + \epsilon = p(\mathbf{y}) + \epsilon \le \alpha + \epsilon.$$

Thus for any pair of points $(\mathbf{y}^1, \alpha_1)$ and $(\mathbf{y}^2, \alpha_2)$ in epi(p), there exist $\mathbf{x}^1$ and $\mathbf{x}^2$ such that

$$\begin{aligned} \psi(\mathbf{x}^1, \mathbf{y}^1) &< p(\mathbf{y}^1) + \epsilon \le \alpha_1 + \epsilon, \\ \psi(\mathbf{x}^2, \mathbf{y}^2) &< p(\mathbf{y}^2) + \epsilon \le \alpha_2 + \epsilon, \end{aligned}$$

from which we deduce from the convexity of ψ that, for $\lambda \in [0, 1]$,

$$\psi(\lambda\mathbf{x}^1 + (1-\lambda)\mathbf{x}^2, \lambda\mathbf{y}^1 + (1-\lambda)\mathbf{y}^2) \le \lambda\alpha_1 + (1-\lambda)\alpha_2 + \epsilon.$$

Since ϵ is arbitrary, we have

$$\psi(\lambda\mathbf{x}^1 + (1-\lambda)\mathbf{x}^2, \lambda\mathbf{y}^1 + (1-\lambda)\mathbf{y}^2) \le \lambda\alpha_1 + (1-\lambda)\alpha_2.$$

Taking the infimum over all $\lambda\mathbf{x}^1 + (1-\lambda)\mathbf{x}^2$, we conclude that

$$p(\lambda\mathbf{y}^1 + (1-\lambda)\mathbf{y}^2) \le \lambda\alpha_1 + (1-\lambda)\alpha_2,$$

and hence epi(p) is convex. ∎

Lemma 4.1.6 (Fenchel transform of p)

$$p^*(\mathbf{v}) = \psi^*(\mathbf{0}, \mathbf{v}) \quad \forall \mathbf{v} \in \mathbb{R}^m.$$

Proof:

$$\begin{aligned} \psi^*(\mathbf{0}, \mathbf{v}) &= \sup_{\mathbf{x},\mathbf{y}} \{\mathbf{0}^\top\mathbf{x} + \mathbf{v}^\top\mathbf{y} - \psi(\mathbf{x}, \mathbf{y})\} && \text{by definition} \\ &= \sup_{\mathbf{y}} \left\{ \sup_{\mathbf{x}} [\mathbf{v}^\top\mathbf{y} - \psi(\mathbf{x}, \mathbf{y})] \right\} \\ &= \sup_{\mathbf{y}} \left\{ \mathbf{v}^\top\mathbf{y} - \inf_{\mathbf{x}} \psi(\mathbf{x}, \mathbf{y}) \right\} \\ &= \sup_{\mathbf{y}} \{\mathbf{v}^\top\mathbf{y} - p(\mathbf{y})\} = p^*(\mathbf{v}) && \text{by definition.} \end{aligned}$$

∎

Theorem 4.1.7 (Strong Duality) Assuming that $\partial p(\mathbf{0})$ is non-empty, then

$$\inf_{\mathbf{x} \in \mathbf{R}^n} \psi(\mathbf{x}, \mathbf{0}) = \sup_{\mathbf{v} \in \mathbf{R}^m} -\psi^*(\mathbf{0}, \mathbf{v}).$$

Proof: Let $\mathbf{v}^* \in \partial p(\mathbf{0})$, then by Lemma 4.1.5,

$$p(\mathbf{y}) \ge p(\mathbf{0}) + \mathbf{y}^\top\mathbf{v}^* \quad \forall \mathbf{y} \in \mathbb{R}^m,$$

or

$$p(\mathbf{0}) \leq -[\mathbf{y}^\top \mathbf{v}^* - p(\mathbf{y})] \quad \forall \mathbf{y} \in \mathbb{R}^m.$$

It follows that

$$p(\mathbf{0}) \leq \inf_{\mathbf{y}} -[\mathbf{y}^\top \mathbf{v}^* - p(\mathbf{y})] = -\sup_{\mathbf{y}}[\mathbf{y}^\top \mathbf{v}^* - p(\mathbf{y})] = -p^*(\mathbf{v}^*). \tag{4.1.2}$$

From, Lemma 4.1.6, we have $p^*(\mathbf{v}^*) = \psi^*(\mathbf{0}, \mathbf{v}^*)$, so (4.1.2) asserts that

$$\inf_{\mathbf{x}} \psi(\mathbf{x}, \mathbf{0}) = p(\mathbf{0}) \leq -\psi^*(\mathbf{0}, \mathbf{v}^*).$$

The conclusion follows from combining the above inequality with the weak duality Lemma 4.1.4. ∎

Remark 4.1.2 Note that the above conjugate duality result for unconstrained problems is intimately related to the Lagrangian duality for constrained problems to be discussed in Section 4.3, see [MP1],[R2].

4.2 Optimality Conditions for Constrained Optimization

We begin our discussion of constrained optimization with two standard results. No convexity assumption is made just yet, and as such the optimality conditions in this section are only necessary, or only sufficient, but not both.

Definition 4.2.1 Let $f : \mathbb{R}^n \to \mathbb{R}$, $g_i : \mathbb{R}^n \to \mathbb{R}, i = 1, 2, \cdots, m$ and $h_i : \mathbb{R}^n \to \mathbb{R}, i = 1, 2, \cdots, r$ be continuously differentiable functions. The constrained nonlinear programming problem is defined as follows.

$$\text{(Nonlinear Programming Problem NP)} \quad \min_{\mathbf{x} \in \mathbb{R}^n} \ f(\mathbf{x}) \tag{4.2.1}$$

$$\text{subject to} \quad g_i(\mathbf{x}) \leq 0, \quad i = 1, 2, \cdots, m, \tag{4.2.2}$$

$$h_i(\mathbf{x}) = 0, \quad i = 1, 2, \cdots, r. \tag{4.2.3}$$

Let $\mathbf{g}(\mathbf{x}) = [g_1(\mathbf{x}), g_2(\mathbf{x}) \cdots, g_m(\mathbf{x})]^\top$ and $\mathbf{h}(\mathbf{x}) = [h_1(\mathbf{x}), h_2(\mathbf{x}) \cdots, h_r(\mathbf{x})]^\top$.

Ideally, one would like to solve for the global minimum of problem NP, although in the absence of convexity assumptions, one can only hope to settle for a local minimum. The study of constrained optimization has been dominated by Lagrangian theory for as long as the subject existed. Essentially, this amounts to a *scalarization* of the cost and constraint functions into a single scalar function called the *Lagrangian function*, or simply *Lagrangian.* In the case of Lagrangian theory, the scalarization is a *linear* one, i.e., we form *the Lagrangian function by taking a linear combination of the cost and constraint functions.* It may not be obvious at this stage, but this has serious limitations in as far as strong duality is concerned, as will be discussed in the next chapter. But for the time being, it is important to understand NP duality first in the framework of Lagrangian Theory.

Definition 4.2.2 The *Lagrangian* of problem NP is defined as

$$L(\mathbf{x}, \boldsymbol{\lambda}, \boldsymbol{\mu}) = f(\mathbf{x}) + \boldsymbol{\lambda}^{\top}\mathbf{g}(\mathbf{x}) + \boldsymbol{\mu}^{\top}\mathbf{h}(\mathbf{x}) \tag{4.2.4}$$

where $\boldsymbol{\lambda} \in \mathbb{R}^m_+$ and $\boldsymbol{\mu} \in \mathbb{R}^r$ are called the *Lagrange Multipliers* of the problem.

Remark 4.2.1 In some versions of Lagrangian Theory, there is a multiplier λ_0 multiplying the cost f as well. In fact some of the necessary conditions such as the Fritz-John necessary condition (as a precursor to the Kuhn-Tucker necessary condition) is preconditioned on such a general form of Lagrangian. Under a constraint qualification, the multiplier to f in the present context can be normalized to 1, see Theorem 4.3.6.

The Kuhn-Tucker necessary condition for optimality [KT1] is discussed in detail by just about every book on nonlinear programming. Since it does not really contribute much to the understanding of duality, it will only be stated here without proof. Before that, some qualification on the behaviour of the constraints binding the candidate optimal solution is necessary.

Definition 4.2.3 (Regular points) Let the active set $\mathcal{J}$ at $\mathbf{x}^*$ be defined as

$$\mathcal{J}(\mathbf{x}^*) = \{j \mid g_i(\mathbf{x}^*) = 0\}.$$

A *regular point* $\mathbf{x}^*$ of the constraints is such that the gradient vectors of the active constraints, $\nabla g_i(\mathbf{x}^*), \quad i \in \mathcal{J}$ and $\nabla h_j(\mathbf{x}^*), \quad j = 1, 2, \cdots, r$ are all linearly independent.

Theorem 4.2.1 (Kuhn-Tucker necessary condition for optimality) Let $\mathbf{x}^*$ be a regular point of the constraints. If $\mathbf{x}^*$ is also a local minimum of problem NP, then it is necessary that there exists $\boldsymbol{\lambda} \in \mathbb{R}^m_+$ and $\boldsymbol{\mu} \in \mathbb{R}^r$ such that

$$\left.\frac{\partial L}{\partial \mathbf{x}}\right|_{\mathbf{x}^*, \boldsymbol{\lambda}, \boldsymbol{\mu}} = \mathbf{0}^{\top}$$

and

$$\text{(complementary slackness)} \qquad \lambda_i g_i(\mathbf{x}^*) = 0 \quad \forall i = 1, 2, \cdots, m.$$

Proof: See, for example, [L1] or [BS1]. ∎

A point that satisfies the necessary condition in Theorem 4.2.1 is called a *Kuhn-Tucker Point*. This point may not be a minimum point unless additional convexity assumptions are made, see Section 4.4. Therefore the Kuhn-Tucker necessary condition is not very useful for finding the solution, but rather it is used to check if a solution is not optimal. Nonetheless, this condition is the driving force behind many optimization methods for solving constrained optimization problems.

A sufficient condition for optimality is the well-known saddle point condition. While this optimality condition has very little practical application in as far as computing the optimal solution is concerned, it is nevertheless of great theoretical importance because of its direct link with Lagrangian duality. Note that unlike the Kuhn-Tucker condition, there is no assumption made about the differentiability of the cost and constraint functions. This condition is not as popularly known as the Kuhn-Tucker condition, and the proof is included here because of its relevance to duality.

Theorem 4.2.2 (Saddle point sufficient condition) If $\mathbf{x}^* \in \mathbb{R}^n$ and $\boldsymbol{\lambda}^* \in \mathbb{R}^m_+, \boldsymbol{\mu}^* \in \mathbb{R}^r$ are such that

$$L(\mathbf{x}^*, \boldsymbol{\lambda}, \boldsymbol{\mu}) \leq L(\mathbf{x}^*, \boldsymbol{\lambda}^*, \boldsymbol{\mu}^*) \leq L(\mathbf{x}, \boldsymbol{\lambda}^*, \boldsymbol{\mu}^*) \qquad \forall \mathbf{x} \in \mathbb{R}^n, \boldsymbol{\lambda} \in \mathbb{R}^m_+, \boldsymbol{\mu} \in \mathbb{R}^r \quad (4.2.5)$$

then $\mathbf{x}^*$ solves the problem NP. (Note, in addition, that $(\boldsymbol{\lambda}^*, \boldsymbol{\mu}^*)$ is the solution to the Lagrangian dual optimization problem to NP to be defined in the next section.)

Proof: The first inequality implies that:

$$(\boldsymbol{\lambda} - \boldsymbol{\lambda}^*)^\top \mathbf{g}(\mathbf{x}^*) + (\boldsymbol{\mu} - \boldsymbol{\mu}^*)^\top \mathbf{h}(\mathbf{x}^*) \leq 0 \quad \forall \boldsymbol{\lambda} \geqq \mathbf{0}, \ \forall \boldsymbol{\mu}. \qquad (4.2.6)$$

From which we deduce that $g_i(\mathbf{x}^*) \leq 0 \ \forall i$ since otherwise picking $\boldsymbol{\mu} = \boldsymbol{\mu}^*; \lambda_j = \lambda_j^* \ \forall j \neq i$ and $\lambda_i \gg \lambda_i^*$ leads to a contradiction; and $h_j(\mathbf{x}^*) = 0 \ \forall j$ since otherwise picking $\mu_j \gg \mu_j^*$ for $h_j(\mathbf{x}^*) > 0$, or picking $\mu_j \ll \mu_j^*$ for $h_j(\mathbf{x}^*) < 0$ and $\boldsymbol{\lambda} = \boldsymbol{\lambda}^*$, $\mu_k = \mu_k^*, k \neq j$ leads to a contradiction; hence $\mathbf{x}^*$ is feasible. Let $\boldsymbol{\lambda} = \mathbf{0}$, then the first inequality implies that

$$(\boldsymbol{\lambda}^*)^\top \mathbf{g}(\mathbf{x}^*) \geq 0. \qquad (4.2.7)$$

But

$$(\boldsymbol{\lambda}^*)^\top \mathbf{g}(\mathbf{x}^*) \leq 0, \qquad (4.2.8)$$

since $\mathbf{g}(\mathbf{x}^*) \leqq \mathbf{0}$ and $\boldsymbol{\lambda}^* \geqq \mathbf{0}$. (4.2.7) and (4.2.8) together imply that

$$(\boldsymbol{\lambda}^*)^\top \mathbf{g}(\mathbf{x}^*) = 0. \qquad (4.2.9)$$

The second inequality of (4.2.5), (4.2.9) and the feasibility of $\mathbf{x}^*$ imply that

$$f(\mathbf{x}^*) + 0 + 0 \leq f(\mathbf{x}) + (\boldsymbol{\lambda}^*)^\top \mathbf{g}(\mathbf{x}) + (\boldsymbol{\mu}^*)^\top \mathbf{h}(\mathbf{x}) \leq f(\mathbf{x})$$

for all feasible $\mathbf{x}$. Hence $\mathbf{x}^*$ is (globally) optimal for NP. ∎

While the saddle point sufficient condition is aesthetically appealing, it has very little practical use since the condition (4.2.5) is difficult to check. Furthermore, for non-convex problems, this condition is conservative in the sense that the optimal solution sometimes does not satisfy the condition (4.2.5).

To end this section, we note that there are a few other second order (meaning those involving second order derivatives of the cost and constraint functions) optimality conditions. These higher order optimality conditions can be used to construct higher order dual problems, see [M6],[M7].

4.3 Lagrangian Duality

This is probably the second best known duality result in optimization after that of LP duality. For easy reference, we shall restate the primal problem together with its dual as follows.

Definition 4.3.1 (Primal and Dual Nonlinear Program) Let $f : \mathbb{R}^n \to \mathbb{R}$, $\mathbf{g} : \mathbb{R}^n \to \mathbb{R}^m$, $\mathbf{h} : \mathbb{R}^n \to \mathbb{R}^r$ be continuously differentiable functions, $\mathcal{X} \subseteq \mathbb{R}^n$ be a subset of $\mathbb{R}^n$ or $\mathbb{R}^n$ itself. The Lagrangian $L : \mathbb{R}^n \times \mathbb{R}^m \times \mathbb{R}^r \to \mathbb{R}$ is as defined in (4.2.4).

$$\text{(Primal Nonlinear Programming Problem P)} \quad \min_{\mathbf{x}\in\mathcal{X}} \ f(\mathbf{x}) \tag{4.3.1}$$

$$\text{subject to} \quad \mathbf{g}(\mathbf{x}) \leqq \mathbf{0}, \tag{4.3.2}$$

$$\mathbf{h}(\mathbf{x}) = \mathbf{0}. \tag{4.3.3}$$

$$\text{(Dual Nonlinear Programming Problem D)} \quad \max_{\boldsymbol{\lambda}\in\mathbf{R}^m, \boldsymbol{\mu}\in\mathbf{R}^r} \ \phi(\boldsymbol{\lambda}, \ \boldsymbol{\mu}) \tag{4.3.4}$$

$$\text{subject to} \quad \boldsymbol{\lambda} \geqq \mathbf{0}, \tag{4.3.5}$$

$$\text{where} \quad \phi(\boldsymbol{\lambda}, \boldsymbol{\mu}) = \inf_{\mathbf{x}\in\mathcal{X}} L(\mathbf{x}, \boldsymbol{\lambda}, \boldsymbol{\mu}). \tag{4.3.6}$$

The function ϕ as defined in (4.3.6) is called the *dual function.* As in the LP case, it is no coincidence that the number of inequality constraints in the primal and the dual are the same. More specifically, the dual variable $\boldsymbol{\lambda}$ corresponding to the inequality constraint in the primal is restricted in sign, whilst the dual variable $\boldsymbol{\mu}$ corresponding to the equality constraint in the primal is unrestricted in sign.

Theorem 4.3.1 (Concavity of the dual function) Let the dual function $\phi(\boldsymbol{\lambda}, \boldsymbol{\mu})$ be as defined in (4.3.6) and $\mathcal{X}$ be compact, then ϕ is concave.

Proof: By the compactness assumption, ϕ is finite. Let $t \in [0,1]$, then for $(\boldsymbol{\lambda}^1, \boldsymbol{\mu}^1)$ and $(\boldsymbol{\lambda}^2, \boldsymbol{\mu}^2)$,

$$
\begin{aligned}
&\phi(t\boldsymbol{\lambda}^1 + (1-t)\boldsymbol{\lambda}^2, t\boldsymbol{\mu}^1 + (1-t)\boldsymbol{\mu}^2) \\
&= \inf_{\mathbf{x}\in\mathcal{X}} L(\mathbf{x}, t\boldsymbol{\lambda}^1 + (1-t)\boldsymbol{\lambda}^2, t\boldsymbol{\mu}^1 + (1-t)\boldsymbol{\mu}^2) \\
&= \inf_{\mathbf{x}\in\mathcal{X}} tL(\mathbf{x}, \boldsymbol{\lambda}^1, \boldsymbol{\mu}^1) + (1-t)L(\mathbf{x}, \boldsymbol{\lambda}^2, \boldsymbol{\mu}^2) \quad \text{(since } L \text{ is linear in } \boldsymbol{\lambda} \text{ and } \boldsymbol{\mu}) \\
&\geq t \inf_{\mathbf{x}\in\mathcal{X}} L(\mathbf{x}, \boldsymbol{\lambda}^1, \boldsymbol{\mu}^1) + (1-t) \inf_{\mathbf{x}\in\mathcal{X}} L(\mathbf{x}, \boldsymbol{\lambda}^2, \boldsymbol{\mu}^2) \\
&= t\phi(\boldsymbol{\lambda}^1, \boldsymbol{\mu}^1) + (1-t)\phi(\boldsymbol{\lambda}^2, \boldsymbol{\mu}^2),
\end{aligned}
$$

where the inequality arises because the infimum of the sum of two functions is greater than or equal to the sum of the infimum of each of the functions. ∎

Remark 4.3.1 Note that no convexity assumption on the primal is needed at all for the above result.

Lemma 4.3.2 (Weak Duality) If $\mathbf{x}$ is feasible for the primal P, and $(\boldsymbol{\lambda}, \boldsymbol{\mu})$ is feasible for the dual D, then

$$\phi(\boldsymbol{\lambda}, \boldsymbol{\mu}) \leq f(\mathbf{x}).$$

Proof:

$$
\begin{aligned}
\phi(\boldsymbol{\lambda}, \boldsymbol{\mu}) &\leq L(\mathbf{x}, \boldsymbol{\lambda}, \boldsymbol{\mu}) \quad \forall \mathbf{x} \in \mathcal{X} \qquad \text{(by definition)} \\
&= f(\mathbf{x}) + \boldsymbol{\lambda}^{\top}\mathbf{g}(\mathbf{x}) + \boldsymbol{\mu}^{\top}\mathbf{h}(\mathbf{x}) \\
&\leq f(\mathbf{x})
\end{aligned}
$$

since $\mathbf{h}(\mathbf{x}) = \mathbf{0}$, $\mathbf{g}(\mathbf{x}) \leqq \mathbf{0}$, and $\boldsymbol{\lambda} \geqq \mathbf{0}$ by the feasibility of $\mathbf{x}$ and $(\boldsymbol{\lambda}, \boldsymbol{\mu})$. ∎

Corollary 4.3.3 If there exists a feasible $\mathbf{x}^*$ for the primal P and a feasible $(\boldsymbol{\lambda}^*, \boldsymbol{\mu}^*)$ for the dual D such that

$$\phi(\boldsymbol{\lambda}^*, \boldsymbol{\mu}^*) = f(\mathbf{x}^*)$$

then $\mathbf{x}^*$ solves the primal P and $(\boldsymbol{\lambda}^*, \boldsymbol{\mu}^*)$ solves the dual D.

Corollary 4.3.4 If the infimum value of the primal is $-\infty$, then $\phi(\boldsymbol{\lambda}, \boldsymbol{\mu}) = -\infty$ $\forall \boldsymbol{\lambda} \geqq \mathbf{0}$. On the other hand, if the supremum value of the dual is ∞, then the primal is infeasible.

Without any convexity assumption, this is probably as far as Lagrangian duality would go. To derive stronger results for the linear Lagrangian theory, the usual convexity assumption is needed.

Assumption 4.3.1 Assume that $\mathcal{X}$ is a nonempty convex set, $f_0 : \mathbb{R}^n \to \mathbb{R}, \mathbf{g} : \mathbb{R}^n \to \mathbb{R}^m$ are convex functions, and $\mathbf{h} : \mathbb{R}^n \to \mathbb{R}^r$ is an affine function.

Assumption 4.3.2 (Slater constraint qualification) Assume that there exists an $\bar{\mathbf{x}} \in \mathcal{X}$ such that $\mathbf{g}(\bar{\mathbf{x}}) < \mathbf{0}$, $\mathbf{h}(\bar{\mathbf{x}}) = \mathbf{0}$ and $\mathbf{0} \in \text{int } \mathbf{h}(\mathcal{X})$ where $\mathbf{h}(\mathcal{X}) = \{\mathbf{h}(\mathbf{x}) \mid \mathbf{x} \in \mathcal{X}\}$.

The following result is a theorem of alternative for convex nonlinear systems.

Lemma 4.3.5 Let $\mathcal{X}, f_0, \mathbf{g}$, and $\mathbf{h}$ satisfy Assumption 4.3.1. Then exactly one of the following systems is nonempty.

$$\text{System I} \quad \{\mathbf{x} \in \mathcal{X} \mid f_0(\mathbf{x}) < 0, \ \mathbf{g}(\mathbf{x}) \leqq \mathbf{0}, \ \mathbf{h}(\mathbf{x}) = \mathbf{0}\}$$

$$\text{System II} \quad \{(\lambda_0, \boldsymbol{\lambda}, \boldsymbol{\mu}) \mid [\lambda_0, \boldsymbol{\lambda}] \geqq \mathbf{0}, \ \lambda_0 f_0(\mathbf{x}) + \boldsymbol{\lambda}^\top \mathbf{g}(\mathbf{x}) + \boldsymbol{\mu}^\top \mathbf{h}(\mathbf{x}) \geq 0 \quad \forall \mathbf{x} \in \mathcal{X}\}$$

Proof: Consider the set:

$$\Omega = \{(a, \mathbf{b}, \mathbf{c}) \in \mathbb{R}^{1+m+r} \mid f_0(\mathbf{x}) < a, \ \mathbf{g}(\mathbf{x}) \leqq \mathbf{b}, \ \mathbf{h}(\mathbf{x}) = \mathbf{c} \quad \text{for some } \ \mathbf{x} \in \mathcal{X}\}.$$

It is easy to show that Ω is a convex set. If System I is empty, then the point $(0, \mathbf{0}, \mathbf{0}) \notin \Omega$, and hence by the Separating Hyperplane Theorem 1.2.8, there exists a vector $(\lambda_0, \boldsymbol{\lambda}^\top, \boldsymbol{\mu}^\top)^\top \neq \mathbf{0}$ such that

$$\lambda_0 a + \boldsymbol{\lambda}^\top \mathbf{b} + \boldsymbol{\mu}^\top \mathbf{c} \geq 0 \quad \forall (a, \mathbf{b}, \mathbf{c}) \in \overline{\Omega}. \tag{4.3.7}$$

(4.3.7) implies that $[\lambda_0, \boldsymbol{\lambda}] \geqq \mathbf{0}^\top$ since otherwise a contradiction can be obtained by making some elements of $\mathbf{b}$ to be arbitrarily large. For a given $\mathbf{x} \in \mathcal{X}$, let $a = f_0(\mathbf{x})$, $\mathbf{b} = \mathbf{g}(\mathbf{x})$, and $\mathbf{c} = \mathbf{h}(\mathbf{x})$, and clearly

$$(f_0(\mathbf{x}), \ \mathbf{g}(\mathbf{x})^\top, \ \mathbf{h}(\mathbf{x})^\top)^\top \in \overline{\Omega},$$

hence

$$\lambda_0 f_0(\mathbf{x}) + \boldsymbol{\lambda}^\top \mathbf{g}(\mathbf{x}) + \boldsymbol{\mu}^\top \mathbf{h}(\mathbf{x}) \geq 0 \quad \forall \mathbf{x} \in \mathcal{X}, \tag{4.3.8}$$

i.e., System II is nonempty. Conversely, if System II is nonempty, then there exists a vector $(\lambda_0, \boldsymbol{\lambda}^\top, \boldsymbol{\mu}^\top)^\top$ with $[\lambda_0, \boldsymbol{\lambda}] \geqq \mathbf{0}$ such that (4.3.8) holds. In particular, if $\mathbf{x} \in \mathcal{X}$ is feasible, i.e., $\mathbf{g}(\mathbf{x}) \leqq \mathbf{0}$ and $\mathbf{h}(\mathbf{x}) = \mathbf{0}$, then $f_0(\mathbf{x}) \geq 0$, and consequently System I is empty. ∎

Theorem 4.3.6 (Strong Duality) Under Assumption 4.3.1 and Assumption 4.3.2,

$$\inf\{f(\mathbf{x}) \mid \mathbf{x} \in \mathcal{X}, \ \mathbf{g}(\mathbf{x}) \leqq \mathbf{0}, \ \mathbf{h}(\mathbf{x}) = \mathbf{0}\} = \sup\{\phi(\boldsymbol{\lambda}, \boldsymbol{\mu}) \mid \boldsymbol{\lambda} \geqq \mathbf{0}\} \tag{4.3.9}$$

and $(\boldsymbol{\lambda}^*)^\top \mathbf{g}(\mathbf{x}^*) = 0$ where $\mathbf{x}^*$ solves the primal and $(\boldsymbol{\lambda}^*, \boldsymbol{\mu}^*)$ solves the dual.

Proof: Let

$$\alpha = \inf\{f(\mathbf{x}) \mid \mathbf{x} \in \mathcal{X}, \ \mathbf{g}(\mathbf{x}) \leqq \mathbf{0}, \ \mathbf{h}(\mathbf{x}) = \mathbf{0}\},$$

then by definition, the following system is empty:

$$\{\mathbf{x} \in \mathcal{X} \mid f(\mathbf{x}) - \alpha < 0, \ \mathbf{g}(\mathbf{x}) \leqq \mathbf{0}, \ \mathbf{h}(\mathbf{x}) = \mathbf{0}\}.$$

Lemma 4.3.5 asserts that there exists $(\lambda_0, \boldsymbol{\lambda}^\top, \boldsymbol{\mu}^\top) \neq \mathbf{0}^\top, [\lambda_0, \boldsymbol{\lambda}] \geqq \mathbf{0}$ such that

$$\lambda_0(f(\mathbf{x}) - \alpha) + \boldsymbol{\lambda}^\top \mathbf{g}(\mathbf{x}) + \boldsymbol{\mu}^\top \mathbf{h}(\mathbf{x}) \geq 0 \quad \forall \mathbf{x} \in \mathcal{X}. \tag{4.3.10}$$

By Assumption 4.3.2, there exists $\bar{\mathbf{x}} \in \mathcal{X}$ such that $\mathbf{g}(\bar{\mathbf{x}}) < \mathbf{0}$ and $\mathbf{h}(\bar{\mathbf{x}}) = \mathbf{0}$. If $\lambda_0 = 0$ then (4.3.10) implies that $\boldsymbol{\lambda}^\top \mathbf{g}(\bar{\mathbf{x}}) \geq 0$, which is possible only if $\boldsymbol{\lambda} = \mathbf{0}$ since $\mathbf{g}(\bar{\mathbf{x}}) < \mathbf{0}$. Consequently $\boldsymbol{\mu}^\top \mathbf{h}(\mathbf{x}) \geq 0\ \forall \mathbf{x} \in \mathcal{X}$. Since $\mathbf{0} \in \text{int } \mathbf{h}(\mathcal{X})$, we can pick an $\mathbf{x} \in \mathcal{X}$ such that $\mathbf{h}(\mathbf{x}) = -\alpha\boldsymbol{\mu}$ where $\alpha > 0$. Therefore

$$0 \leq \boldsymbol{\mu}^\top \mathbf{h}(\mathbf{x}) = -\alpha\|\boldsymbol{\mu}\|^2 \text{ implying that } \quad \boldsymbol{\mu} = \mathbf{0}.$$

Thus $(\lambda_0, \boldsymbol{\lambda}, \boldsymbol{\mu}) = \mathbf{0}$ if $\lambda_0 = 0$, a contradiction. Hence $\lambda_0 > 0$. Without loss of generality, we may normalize $\lambda_0 = 1$. As a result, we have

$$\phi(\boldsymbol{\lambda}, \boldsymbol{\mu}) = \inf_{\mathbf{x} \in \mathcal{X}} \left\{ f(\mathbf{x}) + \boldsymbol{\lambda}^\top \mathbf{g}(\mathbf{x}) + \boldsymbol{\mu}^\top \mathbf{h}(\mathbf{x}) \right\} \geq \alpha.$$

But by weak duality (Lemma 4.3.2), $\phi(\boldsymbol{\lambda}, \boldsymbol{\mu}) \leq \alpha$, and the conclusion follows.

In (4.3.10) Let $\mathbf{x} = \mathbf{x}^*$ and $(\boldsymbol{\lambda}, \boldsymbol{\mu}) = (\boldsymbol{\lambda}^*, \boldsymbol{\mu}^*)$ to get $(\boldsymbol{\lambda}^*)^\top \mathbf{g}(\mathbf{x}^*) = 0$. ■

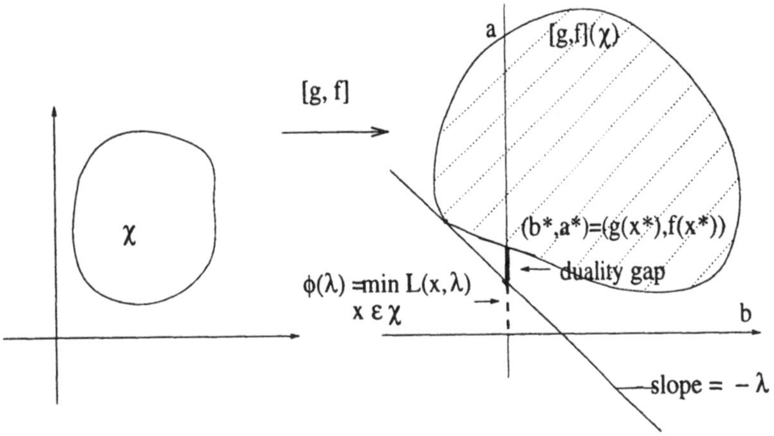

FIGURE 4.3.1 Geometry of Lagrangian duality

Remark 4.3.2 Note that the strong duality result can also be stated in a minimax form (see the minimax theorem of Section 1.6):

$$\inf_{\mathbf{x} \in \mathcal{X}} \sup_{\boldsymbol{\lambda} \geqq \mathbf{0}} L(\mathbf{x}, \boldsymbol{\lambda}, \boldsymbol{\mu}) = \sup_{\boldsymbol{\lambda} \geqq \mathbf{0}} \inf_{\mathbf{x} \in \mathcal{X}} L(\mathbf{x}, \boldsymbol{\lambda}, \boldsymbol{\mu}).$$

Note also that the right hand side of the above is the dual problem D. The left hand side appears to be different from the primal problem P. But upon closer look, $\sup_{\boldsymbol{\lambda} \geqq \mathbf{0}} L(\mathbf{x}, \boldsymbol{\lambda}, \boldsymbol{\mu})$ is only meaningful if $\mathbf{g}(\mathbf{x}) \leqq \mathbf{0}$ and $\mathbf{h}(\mathbf{x}) = \mathbf{0}$. This translates into the constraints of the primal problem P.

Most readers would appreciate the ability to visualize the meaning of weak and strong duality. Since none of us is good at visualizing any geometrical concept in a dimension higher than two, we shall assume that there is only one inequality constraint, so that g is a scalar function. Consider the vector function $[g, f] : \mathcal{X} \to \mathbb{R}^2$ as depicted in Figure 4.3.1. The image for the function $[g, f]$ is

$$\mathcal{W} = \{(b, a) \mid a = f(\mathbf{x}),\ b = g(\mathbf{x}) \text{ for some } \mathbf{x} \in \mathcal{X}\}.$$

Since feasibility requires that $g(\mathbf{x}) \leq 0$, the primal problem NLP is to find the lowest point of $\mathcal{W}$ to the left of the vertical axis in Figure 4.3.1, i.e., the point (b^*, a^*). Note that if we define the perturbation function $p : \mathbb{R} \to \mathbb{R}$ as

$$p(y) = \min\{f(\mathbf{x}) \mid \mathbf{x} \in \mathcal{X},\ g(\mathbf{x}) \leq y\},$$

then p is necessarily monotone and non-increasing since a larger value of y will always yield a lower or equal value for the minimization of f. It is also obvious that the graph of p is just the lower boundary to the set $\mathcal{W}$. Geometrically, a Lagrangian

$$L(\mathbf{x}, \lambda) = f(\mathbf{x}) + \lambda g(\mathbf{x}) = a + \lambda b$$

where $\lambda \geq 0$, is a straight line for a particular value of L. This line $a = L - \lambda b$ has a slope of $-\lambda$ and a vertical intercept of L.

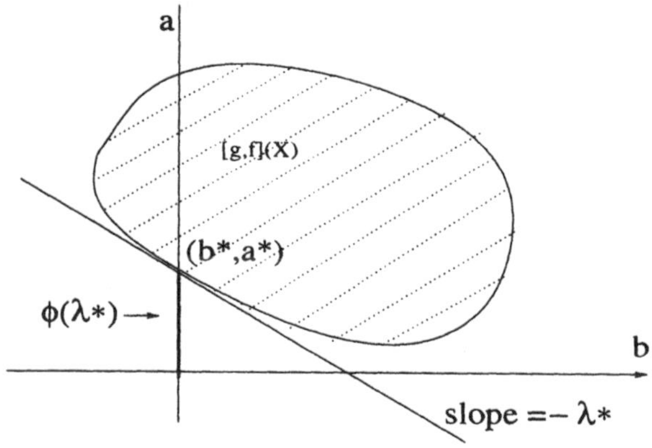

FIGURE 4.3.2 Zero duality gap for convex optimization

For a given value of λ, the function value of the dual function

$$\phi(\lambda) = \inf_{\mathbf{x} \in \mathcal{X}} L(\mathbf{x}, \lambda)$$

is the vertical intercept of any line with slope $-\lambda$ that is still touching some point in $\mathcal{W}$ (the image of some point $\mathbf{x} \in \mathcal{X}$ under the map of $[g, f]$), and clearly this line is determined by the supporting hyperplane to the set $\mathcal{W}$ with slope $-\lambda$. It does not take much convincing that any vertical intercept of a supporting hyperplane is always less than or equal to the value of a of any point $(b, a) = (g(\mathbf{x}), f(\mathbf{x}))$ in $\mathcal{W}$ lying to the left of the vertical axis (and in fact every point in $\mathcal{W}$). *This is weak* *duality*. The (vertical) difference between a^* and the maximum vertical intercept of the supporting hyperplane is then the *duality gap* (see Definition 1.6.3).

If no convexity assumption was made about $\mathcal{X}, f$ and g, very little can be said about the lower boundary of $\mathcal{W}$ apart from that it is monotone non-increasing. In this case, the duality gap may be nonzero, and strong duality may not hold. Note that strong duality may sometimes hold even without the convexity assumption. However, if the convexity Assumption 4.3.1 is made, then by Lemma 4.1.5, the perturbation function p as the lower boundary of $\mathcal{W}$ is convex, and hence the maximum value of the dual function is given by the intercept of the supporting hyperplane that supports $\mathcal{W}$ exactly at the point (b^*, a^*) (see Figure 4.3.2), and this maximum dual function value is the same as the minimum primal cost a^*. This is *strong duality*.

Little would have changed if we regard g as an equality constraint instead. In this case, we are interested in the interception of $\mathcal{W}$ with the vertical axis, rather than the left half portion. Furthermore, the perturbation function (or the lower boundary of $\mathcal{W}$) is not monotone, and the straight line representing the Lagrangian can have a positive or a negative slope. Weak duality clearly still holds. If the convexity Assumption 4.3.1 is made, then the lower boundary of $\mathcal{W}$ is convex, and therefore strong duality must also hold.

Stronger results can be obtained if Assumption 4.3.1 holds. Recall from the previous section that the Kuhn-Tucker condition (Theorem 4.2.1) is only a necessary condition for optimality, whilst the saddle point condition (Theorem 4.2.2) is only a sufficient condition for optimality. If convexity prevails, then the Kuhn-Tucker condition is also sufficient for optimality; and the saddle point condition is also necessary for optimality.

Theorem 4.3.7 (Kuhn-Tucker sufficient condition) Under the convexity Assumption 4.3.1, if there exists $\boldsymbol{\lambda} \in \mathbb{R}_+^m$ and $\boldsymbol{\mu} \in \mathbb{R}^r$ such that

$$\left.\frac{\partial L(\mathbf{x}, \boldsymbol{\lambda}, \boldsymbol{\mu})}{\partial \mathbf{x}}\right|_{\mathbf{x}^*} = \mathbf{0}^\top$$

$$\lambda_i g_i(\mathbf{x}^*) = 0 \quad \forall i = 1, 2, \cdots, m,$$

then $\mathbf{x}^*$ solves (globally) the primal problem P.

Proof: For any $\mathbf{x} \neq \mathbf{x}^*$ that is feasible, we have

$$\begin{aligned}
&f(\mathbf{x})\\
&\geqq f(\mathbf{x}) + \boldsymbol{\lambda}^\top \mathbf{g}(\mathbf{x}) + \boldsymbol{\mu}^\top \mathbf{h}(\mathbf{x}) \quad \text{since } \mathbf{g}(\mathbf{x}) \leqq \mathbf{0},\ \mathbf{h}(\mathbf{x}) = \mathbf{0}, \text{ and } \boldsymbol{\lambda} \geqq \mathbf{0},\\
&\geqq f(\mathbf{x}^*) + \nabla f(\mathbf{x}^*)(\mathbf{x} - \mathbf{x}^*) + \boldsymbol{\lambda}^\top (\mathbf{g}(\mathbf{x}^*) + \nabla \mathbf{g}(\mathbf{x}^*)(\mathbf{x} - \mathbf{x}^*))\\
&\quad + \boldsymbol{\mu}^\top (\mathbf{h}(\mathbf{x}^*) + \nabla \mathbf{h}(\mathbf{x}^*)(\mathbf{x} - \mathbf{x}^*))\\
&= f(\mathbf{x}^*) + \left(\nabla f(\mathbf{x}^*) + \boldsymbol{\lambda}^\top \nabla \mathbf{g}(\mathbf{x}^*) + \boldsymbol{\mu}^\top \nabla \mathbf{h}(\mathbf{x}^*)\right)(\mathbf{x} - \mathbf{x}^*) + \boldsymbol{\lambda}^\top \mathbf{g}(\mathbf{x}^*) + \boldsymbol{\mu}^\top \mathbf{h}(\mathbf{x}^*)\\
&= f(\mathbf{x}^*) \qquad \text{by the assumption of the Theorem.}
\end{aligned}$$

■

Theorem 4.3.8 (Necessity of the saddle point condition) Under Assumption 4.3.1 and Assumption 4.3.2, if $\mathbf{x}^*$ is optimal for the primal P and $(\boldsymbol{\lambda}^*, \boldsymbol{\mu}^*)$ is optimal for the dual D, then

$$L(\mathbf{x}^*, \boldsymbol{\lambda}, \boldsymbol{\mu}) \leq L(\mathbf{x}^*, \boldsymbol{\lambda}^*, \boldsymbol{\mu}^*) \leq L(\mathbf{x}, \boldsymbol{\lambda}^*, \boldsymbol{\mu}^*) \qquad \forall \mathbf{x} \in \mathbb{R}^n, \boldsymbol{\lambda} \in \mathbb{R}^m_+, \boldsymbol{\mu} \in \mathbb{R}^r.$$

Proof: If $\mathbf{x}^*$ is optimal for the primal and $(\boldsymbol{\lambda}^*, \boldsymbol{\mu}^*)$ is optimal for the dual, then by strong duality (Theorem 4.3.6) $f(\mathbf{x}^*) = \phi(\boldsymbol{\lambda}^*, \boldsymbol{\mu}^*)$, and $(\boldsymbol{\lambda}^*)^\top \mathbf{g}(\mathbf{x}^*) = 0$. Furthermore, by the definition of ϕ, and the fact that $\mathbf{h}(\mathbf{x}^*) = 0$, we have,

$$\begin{aligned} L(\mathbf{x}^*, \boldsymbol{\lambda}^*, \boldsymbol{\mu}^*) &= f(\mathbf{x}^*) + (\boldsymbol{\lambda}^*)^\top \mathbf{g}(\mathbf{x}^*) + (\boldsymbol{\mu}^*)^\top \mathbf{h}(\mathbf{x}^*) \\ &= f(\mathbf{x}^*) = \phi(\boldsymbol{\lambda}^*, \boldsymbol{\mu}^*) \\ &\leq L(\mathbf{x}, \boldsymbol{\lambda}^*, \boldsymbol{\mu}^*) \quad \forall \mathbf{x} \in \mathcal{X}, \end{aligned}$$

which is the second inequality of the saddle point condition.

Next, since $\mathbf{g}(\mathbf{x}^*) \leqq \mathbf{0}$ and $\mathbf{h}(\mathbf{x}^*) = \mathbf{0}$, we have, for $\boldsymbol{\lambda} \geqq \mathbf{0}$,

$$\begin{aligned} L(\mathbf{x}^*, \boldsymbol{\lambda}, \boldsymbol{\mu}) &= f(\mathbf{x}^*) + \boldsymbol{\lambda}^\top \mathbf{g}(\mathbf{x}^*) + \boldsymbol{\mu}^\top \mathbf{h}(\mathbf{x}^*) \\ &\leq f(\mathbf{x}^*) \\ &\leq f(\mathbf{x}^*) + (\boldsymbol{\lambda}^*)^\top \mathbf{g}(\mathbf{x}^*) + (\boldsymbol{\mu}^*)^\top \mathbf{h}(\mathbf{x}^*) \\ &= L(\mathbf{x}^*, \boldsymbol{\lambda}^*, \boldsymbol{\mu}^*), \end{aligned}$$

which is the first inequality of the saddle point condition. ∎

Properties of the dual function ϕ

In practice it may be desirable to compute the gradient of the dual function. Although numerical issues are beyond the scope of this book, it is still of some theoretical interest to study some of the properties, especially those concerning differentiability, of the dual function as defined in (4.3.6). First, the notions of subgradient and subdifferential for concave functions are needed.

Definition 4.3.2 Let $\gamma : \mathbb{R}^p \to \overline{\mathbb{R}}$ be a concave function, and let $\boldsymbol{\lambda} \in \text{dom}(\gamma)$. The vector $\mathbf{z} \in \mathbb{R}^p$ is said to be a *subgradient* of γ at $\boldsymbol{\lambda}$ if

$$\mathbf{z}^\top (\boldsymbol{\lambda}' - \boldsymbol{\lambda}) \geq \gamma(\boldsymbol{\lambda}') - \gamma(\boldsymbol{\lambda}) \quad \forall \boldsymbol{\lambda}' \in \mathbb{R}^p.$$

The set

$$\partial\gamma(\boldsymbol{\lambda}) = \{\mathbf{z} \in \mathbb{R}^p \mid \mathbf{z}^\top (\boldsymbol{\lambda}' - \boldsymbol{\lambda}) \geq \gamma(\boldsymbol{\lambda}') - \gamma(\boldsymbol{\lambda}) \ \forall \boldsymbol{\lambda}' \in \mathbb{R}^p\}$$

is called the *subdifferential* of γ at $\boldsymbol{\lambda}$. If $\boldsymbol{\lambda} \notin \text{dom}(\gamma)$ then we let $\partial\gamma(\boldsymbol{\lambda}) = \emptyset$ by default.

To go on, we define the following set-valued function.

Definition 4.3.3 Let

$$\Gamma(\boldsymbol{\lambda}, \boldsymbol{\mu}) = \{\mathbf{y} \mid \mathbf{y} \in \text{argmin}_{\mathbf{x} \in \mathcal{X}} L(\mathbf{x}, \boldsymbol{\lambda}, \boldsymbol{\mu})\}.$$

If the minimum value of $\inf_{\mathbf{x}\in\mathcal{X}} L(\mathbf{x},\boldsymbol{\lambda},\boldsymbol{\mu})$ is not attainable, then $\Gamma(\boldsymbol{\lambda},\boldsymbol{\mu}) = \emptyset$ by convention.

Theorem 4.3.9 Assume that for all $(\boldsymbol{\lambda},\boldsymbol{\mu}) \in \mathbb{R}^m \times \mathbb{R}^r$, $\Gamma(\boldsymbol{\lambda},\boldsymbol{\mu}) \neq \emptyset$. If $\mathbf{x}^* \in \Gamma(\boldsymbol{\lambda}^*,\boldsymbol{\mu}^*)$, then the vector $[(\mathbf{g}(\mathbf{x}^*))^\top,\ (\mathbf{h}(\mathbf{x}^*))^\top]^\top$ is a subgradient of ϕ at $(\boldsymbol{\lambda}^*,\boldsymbol{\mu}^*)$.

Proof: Under the assumption of the theorem, the minimizing $\mathbf{y}$ for the dual function at any point $(\boldsymbol{\lambda},\boldsymbol{\mu})$ is always attainable. Therefore

$$
\begin{aligned}
&\phi(\boldsymbol{\lambda},\boldsymbol{\mu})\\
&= \min_{\mathbf{x}\in\mathcal{X}}\{f(\mathbf{x}) + \boldsymbol{\lambda}^\top\mathbf{g}(\mathbf{x}) + \boldsymbol{\mu}^\top\mathbf{h}(\mathbf{x})\}\\
&\leq f(\mathbf{x}^*) + \boldsymbol{\lambda}^\top\mathbf{g}(\mathbf{x}^*) + \boldsymbol{\mu}^\top\mathbf{h}(\mathbf{x}^*)\\
&= f(\mathbf{x}^*) + (\boldsymbol{\lambda}-\boldsymbol{\lambda}^*)^\top\mathbf{g}(\mathbf{x}^*) + (\boldsymbol{\mu}-\boldsymbol{\mu}^*)^\top\mathbf{h}(\mathbf{x}^*) + (\boldsymbol{\lambda}^*)^\top\mathbf{g}(\mathbf{x}^*) + (\boldsymbol{\mu}^*)^\top\mathbf{h}(\mathbf{x}^*)\\
&= \phi(\boldsymbol{\lambda}^*,\boldsymbol{\mu}^*) + [(\mathbf{g}(\mathbf{x}^*))^\top,(\mathbf{h}(\mathbf{x}^*))^\top]\begin{bmatrix}\boldsymbol{\lambda}-\boldsymbol{\lambda}^*\\ \boldsymbol{\mu}-\boldsymbol{\mu}^*\end{bmatrix}.
\end{aligned}
$$

Thus by Definition 4.3.2, $[(\mathbf{g}(\mathbf{x}^*))^\top,\ (\mathbf{h}(\mathbf{x}^*))^\top]^\top$ is a subgradient of $\phi(\boldsymbol{\lambda}^*,\boldsymbol{\mu}^*)$. ∎

Theorem 4.3.10 Assume that for all $(\boldsymbol{\lambda},\boldsymbol{\mu}) \in \mathbb{R}^m \times \mathbb{R}^r$, and $\Gamma(\boldsymbol{\lambda},\boldsymbol{\mu}) \neq \emptyset$. If $\Gamma(\boldsymbol{\lambda}^*,\boldsymbol{\mu}^*) = \{\mathbf{x}^*\}$ is a singleton, then ϕ is differentiable at $(\boldsymbol{\lambda}^*,\boldsymbol{\mu}^*)$ with gradient $\nabla\phi(\boldsymbol{\lambda}^*,\boldsymbol{\mu}^*) = [(\mathbf{g}(\mathbf{x}^*))^\top,(\mathbf{h}(\mathbf{x}^*))^\top]$.

Proof: For any given $(\boldsymbol{\lambda},\boldsymbol{\mu})$, let $\mathbf{x}^0 \in \Gamma(\boldsymbol{\lambda},\boldsymbol{\mu})$. From the definition of ϕ, we have

$$
\begin{aligned}
\phi(\boldsymbol{\lambda},\boldsymbol{\mu}) &\leq f(\mathbf{x}^*) + \boldsymbol{\lambda}^\top\mathbf{g}(\mathbf{x}^*) + \boldsymbol{\mu}^\top\mathbf{h}(\mathbf{x}^*), &(4.3.11)\\
\phi(\boldsymbol{\lambda}^*,\boldsymbol{\mu}^*) &= f(\mathbf{x}^*) + (\boldsymbol{\lambda}^*)^\top\mathbf{g}(\mathbf{x}^*) + (\boldsymbol{\mu}^*)^\top\mathbf{h}(\mathbf{x}^*), &(4.3.12)\\
\phi(\boldsymbol{\lambda}^*,\boldsymbol{\mu}^*) &\leq f(\mathbf{x}^0) + (\boldsymbol{\lambda}^*)^\top\mathbf{g}(\mathbf{x}^0) + (\boldsymbol{\mu}^*)^\top\mathbf{h}(\mathbf{x}^0), &(4.3.13)\\
\phi(\boldsymbol{\lambda},\boldsymbol{\mu}) &= f(\mathbf{x}^0) + \boldsymbol{\lambda}^\top\mathbf{g}(\mathbf{x}^0) + \boldsymbol{\mu}^\top\mathbf{h}(\mathbf{x}^0). &(4.3.14)
\end{aligned}
$$

Subtracting (4.3.12) from (4.3.11), we have

$$
\phi(\boldsymbol{\lambda},\boldsymbol{\mu}) - \phi(\boldsymbol{\lambda}^*,\boldsymbol{\mu}^*) \leq [(\boldsymbol{\lambda}-\boldsymbol{\lambda}^*)^\top,(\boldsymbol{\mu}-\boldsymbol{\mu}^*)^\top]\begin{bmatrix}\mathbf{g}(\mathbf{x}^*)\\ \mathbf{h}(\mathbf{x}^*)\end{bmatrix}. \tag{4.3.15}
$$

Subtracting (4.3.14) from (4.3.13), we have

$$
\phi(\boldsymbol{\lambda}^*,\boldsymbol{\mu}^*) - \phi(\boldsymbol{\lambda},\boldsymbol{\mu}) \leq [(\boldsymbol{\lambda}^*-\boldsymbol{\lambda})^\top,(\boldsymbol{\mu}^*-\boldsymbol{\mu})^\top]\begin{bmatrix}\mathbf{g}(\mathbf{x}^0)\\ \mathbf{h}(\mathbf{x}^0)\end{bmatrix}. \tag{4.3.16}
$$

Combining (4.3.15) and (4.3.16), we have, with the help of Schwartz inequality,

$$
\begin{aligned}
0 &\geq \phi(\boldsymbol{\lambda},\boldsymbol{\mu}) - \phi(\boldsymbol{\lambda}^*,\boldsymbol{\mu}^*) - [(\boldsymbol{\lambda}-\boldsymbol{\lambda}^*)^\top,(\boldsymbol{\mu}-\boldsymbol{\mu}^*)^\top]\begin{bmatrix}\mathbf{g}(\mathbf{x}^*)\\ \mathbf{h}(\mathbf{x}^*)\end{bmatrix}\\
&\geq [(\boldsymbol{\lambda}-\boldsymbol{\lambda}^*)^\top,(\boldsymbol{\mu}-\boldsymbol{\mu}^*)^\top]\left(\begin{bmatrix}\mathbf{g}(\mathbf{x}^0)\\ \mathbf{h}(\mathbf{x}^0)\end{bmatrix} - \begin{bmatrix}\mathbf{g}(\mathbf{x}^*)\\ \mathbf{h}(\mathbf{x}^*)\end{bmatrix}\right)\\
&\geq -\left\|\begin{bmatrix}\boldsymbol{\lambda}-\boldsymbol{\lambda}^*\\ \boldsymbol{\mu}-\boldsymbol{\mu}^*\end{bmatrix}\right\|\ \left\|\begin{bmatrix}\mathbf{g}(\mathbf{x}^0)\\ \mathbf{h}(\mathbf{x}^0)\end{bmatrix} - \begin{bmatrix}\mathbf{g}(\mathbf{x}^*)\\ \mathbf{h}(\mathbf{x}^*)\end{bmatrix}\right\|.
\end{aligned}
$$

Consequently,

$$0 \geq \frac{\phi(\boldsymbol{\lambda},\boldsymbol{\mu}) - \phi(\boldsymbol{\lambda}^*,\boldsymbol{\mu}^*) - [(\boldsymbol{\lambda}-\boldsymbol{\lambda}^*)^\top, (\boldsymbol{\mu}-\boldsymbol{\mu}^*)^\top] \begin{bmatrix} \mathbf{g}(\mathbf{x}^*) \\ \mathbf{h}(\mathbf{x}^*) \end{bmatrix}}{\| \begin{bmatrix} \boldsymbol{\lambda}-\boldsymbol{\lambda}^* \\ \boldsymbol{\mu}-\boldsymbol{\mu}^* \end{bmatrix} \|}$$

$$\geq -\| \begin{bmatrix} \mathbf{g}(\mathbf{x}^0) \\ \mathbf{h}(\mathbf{x}^0) \end{bmatrix} - \begin{bmatrix} \mathbf{g}(\mathbf{x}^*) \\ \mathbf{h}(\mathbf{x}^*) \end{bmatrix} \|.$$

Let $(\boldsymbol{\lambda},\boldsymbol{\mu}) \to (\boldsymbol{\lambda}^*,\boldsymbol{\mu}^*)$, then by the continuity of $\mathbf{g}$ and $\mathbf{h}$, $\mathbf{g}(\mathbf{x}^0) \to \mathbf{g}(\mathbf{x}^*), \mathbf{h}(\mathbf{x}^0) \to \mathbf{h}(\mathbf{x}^*)$, and hence

$$\lim_{(\boldsymbol{\lambda},\boldsymbol{\mu})\to(\boldsymbol{\lambda}^*,\boldsymbol{\mu}^*)} \frac{\phi(\boldsymbol{\lambda},\boldsymbol{\mu}) - \phi(\boldsymbol{\lambda}^*,\boldsymbol{\mu}^*) - [(\boldsymbol{\lambda}-\boldsymbol{\lambda}^*)^\top, (\boldsymbol{\mu}-\boldsymbol{\mu}^*)^\top] \begin{bmatrix} \mathbf{g}(\mathbf{x}^*) \\ \mathbf{h}(\mathbf{x}^*) \end{bmatrix}}{\| \begin{bmatrix} \boldsymbol{\lambda}-\boldsymbol{\lambda}^* \\ \boldsymbol{\mu}-\boldsymbol{\mu}^* \end{bmatrix} \|} = 0,$$

and the conclusion follows. ∎

4.4 Lagrangian Duality Specialized to LP, QP, and MP

This section is intended to show that many of the previous duality results are special cases of Lagrangian duality. We shall discuss the cases of linear programming, quadratic programming, and monotropic programming in turn.

Linear Programming

Consider the primal and dual pair of LP studied in Chapter Three.

$$\begin{aligned} \text{(Primal LP)} \quad \min \quad & z_P = \mathbf{c}^\top \mathbf{x}, \\ \text{subject to} \quad & \mathbf{A}\mathbf{x} = \mathbf{b}, \\ & \mathbf{0} \leqq \mathbf{x} \leqq \mathbf{u}. \end{aligned} \tag{4.4.1}$$

$$\begin{aligned} \text{(Dual LP)} \quad \max \quad & z_D = \mathbf{b}^\top \boldsymbol{\lambda} - \mathbf{u}^\top \boldsymbol{\mu}, \\ \text{subject to} \quad & \boldsymbol{\lambda}^\top \mathbf{A} - \boldsymbol{\mu}^\top \leqq \mathbf{c}^\top, \\ & \boldsymbol{\mu} \geqq \mathbf{0}. \end{aligned} \tag{4.4.2}$$

Since linear functions are convex (as well as concave), strong Lagrangian duality must hold. The Lagrangian dual to the primal is to maximize the function $\phi(\boldsymbol{\lambda}, \boldsymbol{\mu})$:

$$\begin{aligned}\phi(\boldsymbol{\lambda}, \boldsymbol{\mu}) &= \inf_{\mathbf{x} \geqq \mathbf{0}} \{\mathbf{c}^\top \mathbf{x} + \boldsymbol{\mu}^\top (\mathbf{x} - \mathbf{u}) + \boldsymbol{\lambda}^\top (\mathbf{b} - \mathbf{A}\mathbf{x})\} \\ &= \boldsymbol{\lambda}^\top \mathbf{b} - \boldsymbol{\mu}^\top \mathbf{u} + \inf_{\mathbf{x} \geqq \mathbf{0}} \{(\mathbf{c}^\top + \boldsymbol{\mu}^\top - \boldsymbol{\lambda}^\top \mathbf{A})\mathbf{x}\} \\ &= \begin{cases} \boldsymbol{\lambda}^\top \mathbf{b} - \boldsymbol{\mu}^\top \mathbf{u} & \text{if } \mathbf{c}^\top + \boldsymbol{\mu}^\top - \boldsymbol{\lambda}^\top \mathbf{A} \geqq \mathbf{0}^\top \\ -\infty & \text{otherwise.} \end{cases}\end{aligned}$$

subject to the constraint that $\boldsymbol{\mu} \geqq \mathbf{0}$. This reduces to the dual LP of (4.4.2). It will be just as easy to derive a similar result for the LP dual in symmetric form.

Quadratic Programming

Consider the following quadratic program:

$$\begin{array}{lll} \text{(Primal QP)} & \min & z_P = \dfrac{1}{2}\mathbf{x}^\top \mathbf{Q}\mathbf{x} + \mathbf{p}^\top \mathbf{x} \\ & \text{subject to} & \mathbf{A}^1 \mathbf{x} \leqq \mathbf{b}^1, \\ & & \mathbf{A}^2 \mathbf{x} = \mathbf{b}^2, \end{array}$$

where it is assumed that $\mathbf{Q}$ is symmetric and positive definite. Under this assumption the problem QP is convex, and hence strong duality holds. The Lagrangian of the primal QP is

$$L(\mathbf{x}, \boldsymbol{\lambda}, \boldsymbol{\mu}) = \frac{1}{2}\mathbf{x}^\top \mathbf{Q}\mathbf{x} + \mathbf{p}^\top \mathbf{x} + \boldsymbol{\lambda}^\top (\mathbf{A}^1 \mathbf{x} - \mathbf{b}^1) + \boldsymbol{\mu}^\top (\mathbf{A}^2 \mathbf{x} - \mathbf{b}^2).$$

The dual function is then given by

$$\phi(\boldsymbol{\lambda}, \boldsymbol{\mu}) = \min_{\mathbf{x} \in \mathbf{R}^n} L(\mathbf{x}, \boldsymbol{\lambda}, \boldsymbol{\mu}).$$

Note that L is convex in $\mathbf{x}$ for a given $(\boldsymbol{\lambda}, \boldsymbol{\mu})$, and hence the optimal solution to the above minimization is obtained by taking the first derivative of L with respect to $\mathbf{x}$ and setting it to zero. This yields:

$$\mathbf{x} = -\mathbf{Q}^{-1}(\mathbf{p} + \mathbf{A}^1 \boldsymbol{\lambda} + \mathbf{A}^2 \boldsymbol{\mu}).$$

Substituting this back to L yields an explicit form for the dual function:

$$\phi(\boldsymbol{\lambda}, \boldsymbol{\mu}) = -\frac{1}{2}\mathbf{p}^\top \mathbf{Q}^{-1}\mathbf{p} - \frac{1}{2}(\boldsymbol{\lambda}^\top, \boldsymbol{\mu}^\top)\bar{\mathbf{Q}} \begin{bmatrix} \boldsymbol{\lambda} \\ \boldsymbol{\mu} \end{bmatrix} - (\boldsymbol{\lambda}^\top, \boldsymbol{\mu}^\top) \begin{bmatrix} \mathbf{b}^1 + \mathbf{A}^1 \mathbf{Q}^{-1}\mathbf{p} \\ \mathbf{b}^2 + \mathbf{A}^2 \mathbf{Q}^{-1}\mathbf{p} \end{bmatrix},$$

where

$$\bar{\mathbf{Q}} = \begin{bmatrix} \mathbf{A}^1 \\ \mathbf{A}^2 \end{bmatrix} \mathbf{Q}^{-1}[(\mathbf{A}^1)^\top, (\mathbf{A}^2)^\top].$$

Thus the dual to the primal QP can be succinctly stated as:

$$\text{(Dual QP)} \qquad \max \quad z_D = \phi(\boldsymbol{\lambda}, \boldsymbol{\mu})$$
$$\text{subject to} \qquad \boldsymbol{\lambda} \geqq \mathbf{0}.$$

Note that the dual function is concave in $(\boldsymbol{\lambda}, \boldsymbol{\mu})$ since $\bar{\mathbf{Q}}$ is positive semidefinite, and the only constraint involved is the nonnegativity constraint on $\boldsymbol{\lambda}$. The gradient of $\phi(\boldsymbol{\lambda}, \boldsymbol{\mu})$ is easy to derive analytically. Hence it would appear easier to solve QP in its dual formulation, at least in theory.

Monotropic Optimization

Consider the following primal monotropic program:

$$\text{(Primal monotropic program)} \qquad \inf_{\mathbf{x}\in\mathbb{R}^n} \sum_{j=1}^{n} f_j(x_j)$$
$$\text{subject to} \qquad \mathbf{A}\mathbf{x} = \mathbf{b}$$
$$x_j \in C_j = [c_j^-, c_j^+] \quad \forall j = 1, 2, \cdots, n,$$

where $\mathbf{A} \in \mathbb{R}^{m\times n}, f_j : C_j \to \overline{\mathbb{R}},\ j = 1, \cdots, n$ are convex piecewise differentiable functions. Let $\mathcal{X} = \{\mathbf{x} \in \mathbb{R}^n \mid x_j \in C_j \quad \forall j\}$. The Lagrangian dual function is defined as:

$$\phi(\mathbf{u}) = \inf_{\mathbf{x}\in\mathcal{X}} \sum_{j=1}^{n} f_j(x_j) + \mathbf{u}^\top(\mathbf{A}\mathbf{x} - \mathbf{b}).$$

If we impose the constraint that $\mathbf{v} = -\mathbf{A}^\top\mathbf{u}$, then the cost function for the monotropic dual is given by:

$$\begin{aligned}\phi(\mathbf{u}) &= \inf_{\mathbf{x}\in\mathcal{X}} \sum_{j=1}^{n} [f_j(x_j) - v_j x_j] - \mathbf{u}^\top\mathbf{b} \\ &= -\mathbf{u}^\top\mathbf{b} - \sup_{\mathbf{x}\in\mathcal{X}} \sum_{j=1}^{n} [v_j x_j - f_j(x_j)] \\ &= -\mathbf{u}^\top\mathbf{b} - \sum_{j=1}^{n} g_j(v_j)\end{aligned}$$

where g_j is the conjugate dual of f_j. To determine the bounds $[d_j^-, d_j^+]$ on the dual variables v_j for the dual monotropic program, we consider the maximization of the (concave) function $h_j(\xi) = \xi\eta - f_j(\xi)$.

If $c_j^+ < \infty$, then let $d_j^+ = \infty$. Otherwise if $c_j^+ = \infty$ and

$$d_j^+ = \lim_{\xi\uparrow\infty} f'_{j+}(\xi) = \lim_{\xi\uparrow\infty} f'_{j-}(\xi) < \infty,$$

then let

$$\lim_{\xi\uparrow\infty} h'_{j+}(\xi) = \eta - \lim_{\xi\uparrow\infty} f'_{j+}(\xi) = \eta - d_j^+ > -\infty.$$

Hence if $\eta > d_j^+$ then $\lim_{\xi\uparrow\infty} h'_{j+}(\xi) > 0$, implying that the maximizing $\xi = \infty$. Consequently $g_j(\eta) = \sup h(\xi) = \infty$.

On the other hand, if $c_j^- > -\infty$, then let $d_j^- = -\infty$. Otherwise if $c_j^- = \infty$ and

$$d_j^- = \lim_{\xi\downarrow-\infty} f'_{j-}(\xi) = \lim_{\xi\downarrow-\infty} f'_{j+}(\xi) > -\infty,$$

then

$$\lim_{\xi\downarrow-\infty} h'_{j-}(\xi) = \eta - \lim_{\xi\downarrow-\infty} f'_{j-}(\xi) = \eta - d_j^- < \infty.$$

So if $\eta < d_j^-$ then $\lim_{\xi\downarrow-\infty} h'_{j-}(\xi) < 0$, implying that the maximizing $\xi = -\infty$. Consequently $g_j(\eta) = \sup h(\xi) = \infty$.

Since $g_j(v_j) = \infty$ if $v_j \notin D_j = [d_j^-, d_j^+]$ from the above analysis, we have the following dual monotropic program:

$$\text{(Dual monotropic program)} \qquad \sup_{\mathbf{u}\in\mathbf{R}^m} \; -\mathbf{u}^\top\mathbf{b} - \sum_{j=1}^{n} g_j(v_j)$$

$$\text{subject to} \qquad -\mathbf{A}^\top\mathbf{u} = \mathbf{v}$$

$$\text{and} \qquad v_j \in D_j = [d_j^-, d_j^+] \quad \forall j = 1, 2, \cdots, n.$$

4.5 Wolfe Duality and Gap Functions for Convex Optimization

Another well known form of duality for nonlinear optimization is due to Wolfe [W2]. This is based on the first order Kuhn-Tucker necessary condition. This is a simpler duality result than that of Lagrangian duality, and it has important computational implications.

Consider the primal problem NP as defined in (4.2.1)-(4.2.3), and let the Lagrangian be defined in the usual way (4.2.4). We assume that all functions are differentiable, and that the optimal solution to the problem NP is a regular point with respect to the constraints. The Wolfe dual to the primal problem NP is defined by:

Definition 4.5.1 (Wolfe Dual)

$$\begin{aligned} \text{(Problem WD)} \qquad & \max_{\mathbf{x},\boldsymbol{\lambda},\boldsymbol{\mu}} L(\mathbf{x},\boldsymbol{\lambda},\boldsymbol{\mu}) = f(\mathbf{x}) + \boldsymbol{\lambda}^\top \mathbf{g}(\mathbf{x}) + \boldsymbol{\mu}^\top \mathbf{h}(\mathbf{x}) \\ \text{subject to} \qquad & \nabla f(\mathbf{x}) + \boldsymbol{\lambda}^\top \nabla \mathbf{g}(\mathbf{x}) + \boldsymbol{\mu}^\top \nabla \mathbf{h}(\mathbf{x}) = \mathbf{0}^\top \\ \text{and} \qquad & \boldsymbol{\lambda} \geqq \mathbf{0}. \end{aligned}$$

Compared to Lagrangian duality, Wolfe duality is much easier to establish.

Theorem 4.5.1 (Weak duality) If $\mathbf{x}$ is feasible for the primal NP and $(\mathbf{x},\boldsymbol{\lambda},\boldsymbol{\mu})$ is feasible for the dual WD, then

$$f(\mathbf{x}) \geq L(\mathbf{x},\boldsymbol{\lambda},\boldsymbol{\mu}).$$

Proof: Follows directly from the fact that $\lambda \geqq \mathbf{0}$, $\mathbf{g}(\mathbf{x}) \leqq \mathbf{0}$, and $\mathbf{h}(\mathbf{x}) = \mathbf{0}$ for feasibility. ∎

Theorem 4.5.2 (Strong duality) Assume that $\mathbf{x}^*$ is a regular point with respect to the primal constraints and that Assumption 4.3.1 holds. If $\mathbf{x}^*$ solves the primal NP, then there exists $\boldsymbol{\lambda}^*, \boldsymbol{\mu}^*$ such that $(\mathbf{x}^*,\boldsymbol{\lambda}^*,\boldsymbol{\mu}^*)$ solves the problem WD and

$$\begin{aligned} f(\mathbf{x}^*) &= \min_{\mathbf{x}\in\mathcal{X}} \{f(\mathbf{x}) \mid \mathbf{g}(\mathbf{x}) \leqq \mathbf{0},\ \mathbf{h}(\mathbf{x}) = \mathbf{0}\} \\ &= \max_{\boldsymbol{\lambda} \geqq \mathbf{0}} \left\{ L(\mathbf{x},\boldsymbol{\lambda},\boldsymbol{\mu}) \mid \frac{\partial L}{\partial \mathbf{x}} = \mathbf{0}^\top \right\} \\ &= L(\mathbf{x}^*,\boldsymbol{\lambda}^*,\boldsymbol{\mu}^*). \end{aligned}$$

Proof: If $\mathbf{x}^*$ solves the primal NP and is regular, then the Kuhn-Tucker necessary condition (Theorem 4.2.1) asserts that there exists $(\boldsymbol{\lambda}^*,\boldsymbol{\mu}^*)$ such that

$$\left.\frac{\partial L}{\partial \mathbf{x}}\right|_{(\mathbf{x}^*,\boldsymbol{\lambda}^*,\boldsymbol{\mu}^*)} = \mathbf{0}^\top, \ \boldsymbol{\lambda}^* \geqq \mathbf{0} \ \text{and} \ (\boldsymbol{\lambda}^*)^\top \mathbf{g}(\mathbf{x}^*) = 0,$$

i.e., $(\mathbf{x}^*,\boldsymbol{\lambda}^*,\boldsymbol{\mu}^*)$ is dual feasible. This implies that $f(\mathbf{x}^*) = L(\mathbf{x}^*,\boldsymbol{\lambda}^*,\boldsymbol{\mu}^*)$ directly. By weak duality, $\mathbf{x}^*$ solves the primal NP and $(\mathbf{x}^*,\boldsymbol{\lambda}^*,\boldsymbol{\mu}^*)$ solves the problem WD. ∎

Remark 4.5.1 Note that Wolfe duality is not symmetric, i.e., the dual of the dual is not the primal.

Gap Function for Convex Optimization

In a special case of problem NP where the cost function is convex, and the feasible set is polyhedral and bounded, there is a so-called *gap function* that expresses the duality gap between the primal and the dual as a function of the primal variable only. The term "gap function" for convex optimization was coined by Hearn [H2] who derived many important properties of this function.

Definition 4.5.2 (Hearn's gap function) Let the problem of interest be defined by

$$\text{(Problem CP)} \qquad \min_{\mathbf{x}\in\mathcal{X}} f(\mathbf{x}) \quad \text{where} \quad \mathcal{X} = \{\mathbf{x} \in \mathbb{R}^n \mid \mathbf{A}\mathbf{x} = \mathbf{b},\ \mathbf{x} \geqq \mathbf{0}\} \qquad (4.5.1)$$

and f be convex and differentiable. $\mathcal{X}$ is assumed to be a non-empty and compact set. Let $\mathbf{x}^*$ be the optimal solution to the problem which must exist. Let the gap function be defined by

$$\gamma(\mathbf{x}) = \max_{\mathbf{y}\in\mathcal{X}} \nabla f(\mathbf{x})(\mathbf{x} - \mathbf{y}) = -\min_{\mathbf{y}\in\mathcal{X}} \nabla f(\mathbf{x})(\mathbf{y} - \mathbf{x}). \qquad (4.5.2)$$

The associated Wolfe dual is given by:

$$\begin{aligned} &\max_{\mathbf{x},\boldsymbol{\lambda},\boldsymbol{\mu}} L(\mathbf{x}, \boldsymbol{\lambda}, \boldsymbol{\mu}) = f(\mathbf{x}) - \boldsymbol{\lambda}^\top(\mathbf{A}\mathbf{x} - \mathbf{b}) - \boldsymbol{\mu}^\top \mathbf{x} \\ \text{subject to} \quad & \nabla f(\mathbf{x}) - \boldsymbol{\lambda}^\top \mathbf{A} = \boldsymbol{\mu}^\top, \\ & \boldsymbol{\mu} \geqq \mathbf{0}. \end{aligned}$$

A restricted form of the Wolfe dual can be expressed as:

$$d(\mathbf{x}) = \max_{(\boldsymbol{\lambda},\boldsymbol{\mu})\in\mathcal{D}(\mathbf{x})} L(\mathbf{x}, \boldsymbol{\lambda}, \boldsymbol{\mu}) \qquad (4.5.3)$$

where $\mathcal{D}(\mathbf{x}) = \{(\boldsymbol{\lambda}, \boldsymbol{\mu}) \mid \nabla f(\mathbf{x}) - \boldsymbol{\lambda}^\top \mathbf{A} = \boldsymbol{\mu}^\top,\ \boldsymbol{\mu} \geqq \mathbf{0}\}$. Thus $d(\mathbf{x}^*)$ has the same value as the optimal dual objective function. The meaning of "gap" is now apparent from the following theorem.

Theorem 4.5.3 (Interpretation of gap function)
(i)$\mathbf{x} \in \mathcal{X} \Rightarrow \mathcal{D}(\mathbf{x}) \neq \emptyset$.
(ii) $\gamma(\mathbf{x}) = f(\mathbf{x}) - d(\mathbf{x})$.

Proof: (i) Fix a point $\mathbf{x} \in \mathcal{X}$ and consider the polyhedral set

$$\Lambda = \{\mathbf{z} \mid \mathbf{A}\mathbf{z} = \mathbf{0},\ \mathbf{z} \geqq \mathbf{0},\ \nabla f(\mathbf{x})\mathbf{z} < 0\}.$$

Given some $\mathbf{z}$ such that $\mathbf{A}\mathbf{z} = \mathbf{0}$ and $\mathbf{z} \geqq \mathbf{0}$, then the point $\mathbf{x} + \alpha\mathbf{z}$ satisfies the constraints of problem CP for all $\alpha \geq 0$. This contradicts the assumption that the feasible set $\mathcal{X}$ is compact, unless $\mathbf{z} = \mathbf{0}$. In which case $\nabla f(\mathbf{x})\mathbf{z} = 0$, leading to the conclusion that $\Lambda = \emptyset$. By Gale's Transposition Theorem 3.4.9, $\mathcal{D}(\mathbf{x}) \neq \emptyset$.

(ii) Multiplying the constraint of the Wolfe dual by $\mathbf{x}$, we have

$$\nabla f(\mathbf{x})\mathbf{x} - \boldsymbol{\lambda}^\top \mathbf{A}\mathbf{x} - \boldsymbol{\mu}^\top \mathbf{x} = 0.$$

This allows us to eliminate one of the variables $\boldsymbol{\mu}$ and consequently the restricted Wolfe dual becomes

$$d(\mathbf{x}) = \max_{\boldsymbol{\lambda}} f(\mathbf{x}) + \boldsymbol{\lambda}^\top \mathbf{b} - \nabla f(\mathbf{x})\mathbf{x}, \quad \text{subject to} \quad \nabla f(\mathbf{x}) \geqq \boldsymbol{\lambda}^\top \mathbf{A}. \qquad (4.5.4)$$

The linear programming dual of

$$\max_{\boldsymbol{\lambda}} \boldsymbol{\lambda}^{\top}\mathbf{b}, \quad \text{subject to} \quad \nabla f(\mathbf{x}) \geqq \boldsymbol{\lambda}^{\top}\mathbf{A}$$

is given by

$$\min_{\mathbf{y}} \nabla f(\mathbf{x})\mathbf{y}, \quad \text{subject to} \quad \mathbf{A}\mathbf{y} = \mathbf{b}, \ \mathbf{y} \geqq \mathbf{0},$$

or

$$\min_{\mathbf{y}\in\mathcal{X}} \nabla f(\mathbf{x})\mathbf{y}.$$

We may then rewrite the restricted Wolfe dual as

$$\begin{aligned} d(\mathbf{x}) &= f(\mathbf{x}) - \nabla f(\mathbf{x})\mathbf{x} + \min_{\mathbf{y}\in\mathcal{X}} \nabla f(\mathbf{x})\mathbf{y} \\ &= f(\mathbf{x}) + \min_{\mathbf{y}\in\mathcal{X}} \nabla f(\mathbf{x})(\mathbf{y} - \mathbf{x}) \\ &= f(\mathbf{x}) - \gamma(\mathbf{x}). \end{aligned}$$

∎

Gap functions have several interesting properties, which we shall study with respect to bounds, convexity and differentiability in turn. The first two properties of the following theorem can also be regarded as the defining properties of a gap function.

Theorem 4.5.4 (Bounds on γ)
(i) $\gamma(\mathbf{x}) \geq 0$,
(ii) $\gamma(\mathbf{x}^*) = 0$, where $\mathbf{x}^*$ is the optimal solution to the primal,
(iii) The gap function is minorized by the convex function $f(\mathbf{x}) - f(\mathbf{x}^*)$, i.e.,

$$\gamma(\mathbf{x}) \geq f(\mathbf{x}) - f(\mathbf{x}^*) \ \forall \mathbf{x} \in \mathbb{R}^n.$$

Proof:
(i)

$$\gamma(\mathbf{x}) = \max_{\mathbf{y}} \nabla f(\mathbf{x})(\mathbf{x} - \mathbf{y}) \geq \nabla f(\mathbf{x})(\mathbf{x} - \mathbf{y}) \quad \forall \mathbf{y}$$

and, in particular

$$\gamma(\mathbf{x}) \geq \nabla f(\mathbf{x})(\mathbf{x} - \mathbf{x}) = 0.$$

(ii) Since $\mathbf{x}^*$ is the primal optimal solution, it is necessary that

$$\gamma(\mathbf{x}^*) = \max_{\mathbf{y}\in\mathcal{X}} \nabla f(\mathbf{x}^*)(\mathbf{x}^* - \mathbf{y}) \leq 0,$$

and hence,

$$\nabla f(\mathbf{x}^*)(\mathbf{y} - \mathbf{x}^*) \geq 0 \quad \forall \mathbf{y} \in \mathcal{X}.$$

This together with (i) imply that $\gamma(\mathbf{x}^*) = 0$.

(iii) Since f is convex, we have, for a fixed $\mathbf{x}$,

$$f(\mathbf{y}) \geq f(\mathbf{x}) + \nabla f(\mathbf{x})(\mathbf{y} - \mathbf{x}) \quad \forall \mathbf{y}.$$

Taking the minimum on both sides with respect to $\mathbf{y} \in \mathcal{X}$ gives

$$f(\mathbf{x}^*) = \min_{\mathbf{y}\in\mathcal{X}} f(\mathbf{y}) \geq \min_{\mathbf{y}\in\mathcal{X}}\{f(\mathbf{x}) + \nabla f(\mathbf{x})(\mathbf{y}-\mathbf{x})\} = f(\mathbf{x}) - \gamma(\mathbf{x}).$$

■

Theorem 4.5.5 (Convexity of gap function) Assume that each component of $\nabla f(\mathbf{x})$ is concave in $\mathbf{x}$, and that $\nabla f(\mathbf{x})\mathbf{x}$ is convex. Then $\gamma(\mathbf{x})$ is convex.

Proof: For convenience, define the auxiliary functions

$$\begin{aligned} h(\mathbf{x}) &= \nabla f(\mathbf{x})\mathbf{x} \\ w(\mathbf{z}) &= \max_{\boldsymbol{\lambda}}\{\boldsymbol{\lambda}^\top \mathbf{b} \mid \boldsymbol{\lambda}^\top \mathbf{A} \leq \mathbf{z}^\top\} = \min_{\mathbf{y}\in\mathcal{X}} \mathbf{z}^\top \mathbf{y}, \\ s(\mathbf{x}) &= w(\nabla f(\mathbf{x})). \end{aligned}$$

Then clearly $\gamma(\mathbf{x}) = h(\mathbf{x}) - s(\mathbf{x})$. Note that:
(i) By its first definition, w is a monotone function since, given a larger $\mathbf{z}$, the underlying linear program has a larger feasible set.

(ii) Furthermore w is concave since its second definition asserts that w is the affine majorant of linear functions.

h is assumed convex, so it remains to show that s is concave. Given $\mathbf{x}^1, \mathbf{x}^2 \in \mathbb{R}^n$ and $\lambda \in [0,1]$, we have,

$$\begin{aligned} & s(\lambda\mathbf{x}^1 + (1-\lambda)\mathbf{x}^2) \\ &= w(\nabla f(\lambda\mathbf{x}^1 + (1-\lambda)\mathbf{x}^2) \\ &\geq w(\lambda\nabla f(\mathbf{x}^1) + (1-\lambda)\nabla f(\mathbf{x}^2)) \\ &\qquad\qquad \text{by the monotonicity of } w \text{ and concavity of } \nabla f \\ &\geq \lambda w(\nabla f(\mathbf{x}^1)) + (1-\lambda) w(\nabla f(\mathbf{x}^2)) \quad \text{by the concavity of } w \\ &= \lambda s(\mathbf{x}^1) + (1-\lambda) s(\mathbf{x}^2). \end{aligned}$$

■

Theorem 4.5.6 (Subdifferential of γ) Assume that each component of $\nabla f(\mathbf{x})$ is concave in $\mathbf{x}$, the subdifferential of γ is given by:

$$\partial\gamma(\mathbf{x}) \supseteq \{\nabla f(\mathbf{x}) + (\mathbf{x}-\mathbf{y})^\top \nabla^2 f(\mathbf{x}), \mid \mathbf{y} \in \mathcal{Y}(\mathbf{x})\},$$

where $\mathcal{Y}(\mathbf{x}) = \{\mathbf{y} \in \mathcal{X} \mid \mathbf{y} \in \operatorname{argmin}_{\mathbf{z}\in\mathcal{X}} \nabla f(\mathbf{x})\mathbf{z}\}$, and $\nabla^2 f$ is the Hessian of f.

Proof: Since the components of $\nabla f(\mathbf{x})$ are concave, we have

$$\nabla f(\mathbf{x}) + (\mathbf{z}-\mathbf{x})\nabla^2 f(\mathbf{x}) \geq \nabla f(\mathbf{z}) \quad \forall \mathbf{x}, \mathbf{z} \in \mathcal{X}.$$

Multiplying by $\mathbf{y} \geqq \mathbf{0}$, we have

$$(\mathbf{z}-\mathbf{x})^{\top}\nabla^2 f(\mathbf{x})\mathbf{y} \geq [\nabla f(\mathbf{z}) - \nabla f(\mathbf{x})]\mathbf{y}.$$

Since $\mathbf{y} \in \mathcal{Y}(\mathbf{x})$, $\nabla f(\mathbf{x})\mathbf{y} = w(\nabla f(\mathbf{x}))$ and similarly $\nabla f(\mathbf{z})\mathbf{y} = w(\nabla f(\mathbf{z}))$. Therefore

$$\begin{aligned}(\mathbf{z}-\mathbf{x})^{\top}\nabla^2 f(\mathbf{x})\mathbf{y} &= \mathbf{y}^{\top}\nabla^2 f(\mathbf{x})(\mathbf{z}-\mathbf{x}) \\ &\geq w(\nabla f(\mathbf{z})) - w(\nabla f(\mathbf{x})) \\ &= s(\mathbf{z}) - s(\mathbf{x})\end{aligned}$$

and hence

$$\partial s(\mathbf{x}) \supseteq \{\mathbf{y}^{\top}\nabla^2 f(\mathbf{x}) \mid \mathbf{y} \in \mathcal{Y}(\mathbf{x})\}.$$

Consequently,

$$\begin{aligned}\partial\gamma(\mathbf{x}) &= \partial h(\mathbf{x}) - \partial s(\mathbf{x}) \\ &\supseteq \nabla f(\mathbf{x}) + \mathbf{x}^{\top}\nabla^2 f(\mathbf{x}) - \{\mathbf{y}^{\top}\nabla^2 f(\mathbf{x}), \mid \mathbf{y} \in \mathcal{Y}(\mathbf{x})\} \\ &= \{\nabla f(\mathbf{x}) + (\mathbf{x}-\mathbf{y})^{\top}\nabla^2 f(\mathbf{x}), \mid \mathbf{y} \in \mathcal{Y}(\mathbf{x})\}.\end{aligned}$$

Remark 4.5.2 From the above results, an alternative way of solving the convex optimization problem CP is to solve the unconstrained problem

$$\min_{\mathbf{x}} \gamma(\mathbf{x}).$$

In fact Polyak [P1] has proposed an algorithm for minimizing a non-differentiable function where the optimal cost value (0 in this case) is known *a priori.* see [H2] for detail.

CHAPTER 5

DUALITY IN NONCONVEX SYSTEMS

In Chapter Four, the conventional Lagrangian function used for solving constrained optimization problems is a *linear* combination of the cost and constraint functions. Optimality conditions based on the *linear* Lagrangian theory typically are either necessary, or sufficient, but not both unless the underlying cost and constraint functions are also convex. Recently a more general Lagrangian function is defined in [S4]. This is a unified approach of the conventional Lagrangian duality and the surrogate duality [S2], [S3].

In this chapter, we propose a somewhat different approach to solving nonconvex inequality constrained optimization problems based on a *nonlinear Lagrangian function*. This is a *nonlinear* combination of the cost and constraint functions, which leads to optimality conditions that are both sufficient and necessary, without any convexity assumption. Furthermore, by appropriately defining the dual optimization problem, we show that a zero duality gap will always hold regardless of convexity, contrary to the case of linear Lagrangian duality. This has a clear geometric meaning from the fact that the perturbation function to be defined in (5.1.1) is monotone non-increasing. Thus even though there may not be a supporting hyperplane at the optimal point, there will always be a *supporting cone*. This is related to the concept of nonlinear penalty functions as will be shown in Section 5.7. The supporting cone concept has motivated a series of research (see, [GY1], [GY2], [HY1], [HY2], [YG2], [YL1], [RGY1], [RGY2], [YH1], and [YT1]) on which this chapter is based.

5.1 Examples of Nonzero Duality Gaps

As an introduction to this chapter, we shall provide some intuitive ideas before a formal analysis. The property of zero duality is important in designing algorithms for nonlinear optimization problems. It is well known that there may be a nonzero duality gap between the primal optimization problem and its (linear) Lagrangian dual problem unless the primal problem is convex or satisfies some generalized convexity conditions, see [L6],[JW1],[R2].

Consider the following inequality constrained optimization problem:

$$\text{(Problem P}_\theta) \qquad \inf_{\mathbf{x}\in\mathcal{X}} f_0(\mathbf{x})$$
$$\text{s.t.} \qquad f_i(\mathbf{x}) \le \theta_i, \; i = 1, 2, \cdots, m,$$

where $f_i : \mathbb{R}^n \to \mathbb{R}$, $i = 0, 1, 2, \cdots, m$ are continuous functions, and $\mathcal{X}$ is a subset of $\mathbb{R}^n$. Define the set of all feasible solutions to be

$$\mathcal{X}_0 = \{\mathbf{x} \in \mathcal{X} \mid f_i(\mathbf{x}) \le \theta_i, i = 1, 2, \cdots, m\}.$$

Let $\mathbf{g}(\mathbf{x}) = [f_1(\mathbf{x}), f_2(\mathbf{x}), \cdots, f_m(\mathbf{x})]^\top$ and $\boldsymbol{\theta} = [\theta_1, \cdots, \theta_m]$. The family of perturbed problems associated with problem P_θ is defined by:

$$\text{(Problem P}_\mathbf{y}) \quad \inf_{\mathbf{x}\in\mathcal{X}} f_0(\mathbf{x})$$
$$\text{s.t.} \qquad \mathbf{g}(\mathbf{x}) \leqq \mathbf{y}$$

where the vector $\mathbf{y} = [y_1, y_2, \cdots, y_m]^\top$ is a perturbation to the parameter vector $\boldsymbol{\theta}$ of the original problem P_θ. When $\mathbf{y} = \boldsymbol{\theta}$, the perturbed problem reduces to the original problem P_θ. Let the perturbation function $w : \mathbb{R}^m \to \mathbb{R}$ be defined by

$$w(\mathbf{y}) = \inf\{f_0(\mathbf{x}) \mid \mathbf{g}(\mathbf{x}) \leqq \mathbf{y}, \mathbf{x} \in \mathcal{X}\}. \tag{5.1.1}$$

Using the conventional notion that $\inf \emptyset = +\infty$, then w has an effective domain $\text{dom}(w) = \{\mathbf{y} \mid \exists \mathbf{x} \in \mathcal{X} \text{ s.t. } \mathbf{g}(\mathbf{x}) \leqq \mathbf{y}\}$. Clearly the perturbation function w is a monotone non-increasing function of $\mathbf{y}$. Define the epigraph of $w(\mathbf{y})$ as the set:

$$\text{epi}(w) = \{[\mathbf{y}, \alpha] \mid \mathbf{y} \in \text{dom}(w), \; \alpha \ge w(\mathbf{y})\} \subset \mathbb{R}^{m+1}.$$

At the optimal solution x^*, the conventional Lagrangian (being linear in f_0 and f_i) can be geometrically interpreted as a supporting hyperplane to the set $\text{epi}(w)$, see Figure 5.1.1.

Theorem 5.1.1 If the functions f_i, $i = 0, 1, \cdots, m$ are convex, and $\mathcal{X}$ is convex then the perturbation function $w(\mathbf{y})$ is convex.

Proof: Consider the unconstrained optimization problem:

$$w(\mathbf{y}) = \inf_{\mathbf{x}\in\mathbf{R}^n} \psi(\mathbf{x}, \mathbf{y})$$

where

$$\psi(\mathbf{x}, \mathbf{y}) = \begin{cases} f_0(\mathbf{x}) & \text{if } \mathbf{g}(\mathbf{x}) \leqq \mathbf{y}, \; \mathbf{x} \in \mathcal{X} \\ \infty & \text{otherwise.} \end{cases}$$

By Lemma 4.1.5, we need only to show that $\psi(\mathbf{x}, \mathbf{y})$ is convex in $(\mathbf{x}, \mathbf{y})$.

Consider two points $(\mathbf{x}^1, \mathbf{y}^1)$ and $(\mathbf{x}^2, \mathbf{y}^2)$ for ψ. If for some $i \in \{1, 2, \cdots, m\}$, $f_i(\mathbf{x}^1) > [\mathbf{y}^1]_i$ or $f_i(\mathbf{x}^2) > [\mathbf{y}^2]_i$, then $\psi(\mathbf{x}^1, \mathbf{y}^1) = \infty$ or $\psi(\mathbf{x}^2, \mathbf{y}^2) = \infty$, and the result holds trivially. So we need only to show the case when

$$\mathbf{g}(\mathbf{x}^1) \leqq \mathbf{y}^1 \quad \text{and} \quad \mathbf{g}(\mathbf{x}^2) \leqq \mathbf{y}^2. \tag{5.1.2}$$

Since f_i, $i = 0, 1, \cdots, m$ are all convex and $\mathcal{X}$ is convex, we have, for $\alpha \in (0, 1)$,

$$f_0(\alpha\mathbf{x}^1 + (1-\alpha)\mathbf{x}^2) \le \alpha f_0(\mathbf{x}^1) + (1-\alpha) f_0(\mathbf{x}^2), \tag{5.1.3}$$

$$\mathbf{g}(\alpha\mathbf{x}^1 + (1-\alpha)\mathbf{x}^2) \le \alpha\mathbf{g}(\mathbf{x}^1) + (1-\alpha)\mathbf{g}(\mathbf{x}^2) \le \alpha\mathbf{y}^1 + (1-\alpha)\mathbf{y}^2. \tag{5.1.4}$$

Then,

$$\begin{aligned}\psi(\alpha(\mathbf{x}^1, \mathbf{y}^1) + (1-\alpha)(\mathbf{x}^2, \mathbf{y}^2)) &= \psi(\alpha\mathbf{x}^1 + (1-\alpha)\mathbf{x}^2, \alpha\mathbf{y}^1 + (1-\alpha)\mathbf{y}^2) \\ &= f_0(\alpha\mathbf{x}^1 + (1-\alpha)\mathbf{x}^2) \quad \text{by (5.1.4)} \\ &\le \alpha f_0(\mathbf{x}^1) + (1-\alpha) f_0(\mathbf{x}^2) \quad \text{by (5.1.3)} \\ &= \alpha\psi(\mathbf{x}^1, \mathbf{y}^1) + (1-\alpha)\psi(\mathbf{x}^2, \mathbf{y}^2) \quad \text{by (5.1.2)},\end{aligned}$$

and the proof is complete. ∎

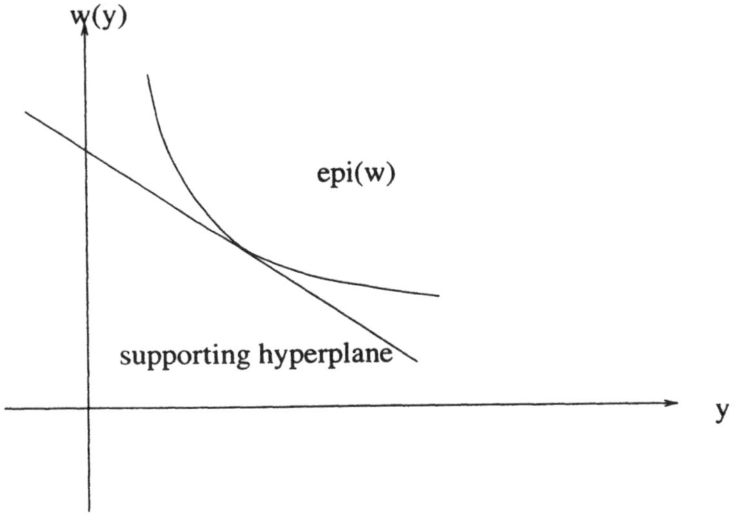

FIGURE 5.1.1 Supporting hyperplane to a convex perturbation function

Consequently, when the functions f_i, $i = 0, 1, \cdots, m$ are convex and $\mathcal{X}$ is convex, the epigraph of w is also convex, and a supporting hyperplane exists at every point of the perturbation function $w(\mathbf{y})$ or the lower boundary of epi(w). When any one of f_i, $i = 0, 1, \cdots, m$ is not convex, then epi(w) may not be convex, and a supporting hyperplane may not exist at some point of the perturbation function, see Figure 5.1.2.

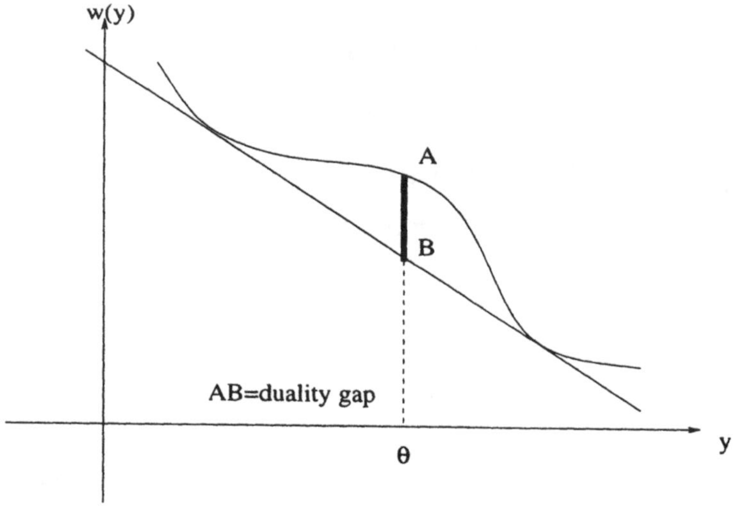

FIGURE 5.1.2 Example of a nonzero duality gap

Example 5.1.1 The following example is a modified version of a problem taken from [L6].

$$\min_{\mathbf{x} \geqq \mathbf{0}} \quad f_0(\mathbf{x}) = 1 - x_1 x_2$$

$$\text{subject to} \quad f_1(\mathbf{x}) = x_1 + 4x_2 \leq 1.$$

It is easy to see geometrically that, at optimality, the inequality constraint is active, and therefore the optimal solution occurs at $\mathbf{x}^* = [\frac{1}{2}, \frac{1}{8}]$ and $f_0(\mathbf{x}^*) = \frac{15}{16}$. The perturbation function can be easily shown to be

$$w(y) = 1 - \frac{y^2}{16},$$

which is clearly nonconvex. Thus if a linear Lagrangian function $L(\mathbf{x}, \lambda) = f_0(\mathbf{x}) + \lambda f_1(\mathbf{x})$ is used, then the dual function

$$\phi_L(\lambda) = \min_{\mathbf{x} \geqq \mathbf{0}} L(\mathbf{x}, \lambda) = \min_{\mathbf{x} \geqq \mathbf{0}} 1 - x_1 x_2 + \lambda(x_1 + 4x_2 - 1)$$

will always have a value of $\phi_L(\lambda) = -\infty$ since L is minimized at $\mathbf{x} = (\infty, \infty)$ for all $\lambda \geq 0$. As a result the duality gap $f_0(\mathbf{x}^*) - \phi_L(\lambda^*) = \infty$.

Furthermore, one can also interpret the (Lagrangian) dual function as the vertical intercept of the supporting hyperplane to the perturbation function with normal $\boldsymbol{\lambda}$ when $\mathbf{y} = \boldsymbol{\theta}$ (see Chapter Four). The duality gap is thus the vertical difference between this vertical intercept and $f_0(\mathbf{x}^*)$. If the perturbation function is nonconvex, then it is easy to see that the duality gap cannot be zero if the supporting hyperplane is not supporting the perturbation function at $\mathbf{y} = \boldsymbol{\theta}$.

Consider the problem P_θ again, but $\mathcal{X}$ is a discrete set. Now the feasible set for Problem $P_\mathbf{y}$ may not change as $\mathbf{y}$ changes in $\mathbb{R}^m$. Thus $w(\mathbf{y})$ is a step function. Thus $w(\mathbf{y})$ is a monotone non-increasing step function of $\mathbf{y}$. The epigraph epi(w) is always nonconvex. In general, a supporting hyperplane does not exist at $\mathbf{y} = \boldsymbol{\theta}$. However, it is easy to see that there is a shifted supporting cone $\mathbf{a} - \mathbb{R}_+^{m+1}$ at any boundary point $\mathbf{a}$ of epi(w). This shifted cone is represented by the weighted Tchebyshev norm.

Example 5.1.2 Consider the example given in page 181 of [BS1]:

$$\begin{aligned} \min \quad & f_0(\mathbf{x}) \\ \text{subject to} \quad & f_1(\mathbf{x}) = 0, \\ & \mathbf{x} = (x_1, x_2) \in \mathcal{X}, \end{aligned}$$

where $\mathcal{X} = \{(0,0), (0,4), (4,4), (4,0), (1,2), (2,1)\}$, $f_0(\mathbf{x}) = -2x_1 + x_2 + 4$, $f_1(\mathbf{x}) = x_1 + x_2 - 3$. It is clear that $\min_{\mathcal{X}_0} f_0(\mathbf{x}) = 1$. If the linear Lagrangian dual is applied, then there is a nonzero duality gap of magnitude 3, see [BS1] since

$$\max_{\lambda} \min_{\mathcal{X}} \{f_0(\mathbf{x}) + \lambda f_1(\mathbf{x})\} = -2.$$

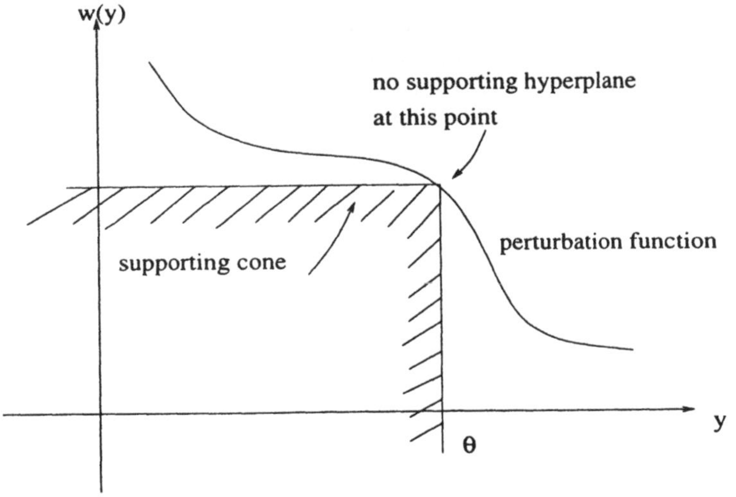

FIGURE 5.1.3 Supporting cone to a nonconvex perturbation function

Our study in this chapter is motivated by the above observation. Specifically, the perturbation function is a non-increasing function, thus *there will always be a supporting cone to the epigraph of the perturbation function,* see Figure 5.1.3. The supporting cone is represented by the weighted Tchebyshev norm. This characterization has also been used to find efficient solutions for multicriteria optimization problems, see [GY3].

5.2 Zero Duality Gap via a Nonlinear Lagrangian Function

Throughout the rest of this chapter, we assume

$$\inf_{\mathbf{x}\in\mathcal{X}} f_0(\mathbf{x}) > 0.$$

We begin by presenting a nonlinear Lagrangian function which relies on the assumption of the existence of an optimal solution. The main reference of this section is [GY1], [GY2]. Consider the problem $\mathrm{P}_{\boldsymbol{\theta}}$. Without loss of generality, we assume in this section that the parameter vector $\boldsymbol{\theta} = [\theta_1, \cdots, \theta_m] \in \text{int } \mathbb{R}_+^m$. The key analytical tool used here is the weighted Tchebyshev norm (strictly speaking, this does not qualify as a norm), which is a scalar-valued function mapping $\mathbb{R}^{m+1}$ to $\mathbb{R}_+$. Given $\mathbf{z} \in \mathbb{R}^{m+1}$ and a weight vector $\mathbf{e} \in \text{int}\mathbb{R}_+^{m+1}$, the weighted Tchebyshev norm of $\mathbf{z}$ is defined as

$$\xi_{\mathbf{e}}(\mathbf{z}) = \max_{0\le i\le m} \{z_i/e_i\}. \tag{5.2.1}$$

It is convenient to normalize the weight vector such that the first component $e_0 = 1$. Define the set

$$\mathcal{E} = \{\mathbf{e} \in \text{int}\mathbb{R}_+^{m+1} \mid e_0 = 1,\ e_i > 0,\ i = 1, 2, \cdots, m\}. \tag{5.2.2}$$

Note that there is no reason why we cannot weigh the vector $\mathbf{z}$ by *multiplying* each component z_i by some weight $t_i > 0$ as in the case of a linear Lagrangian. In fact this approach will be adopted in Section 5.7 where the multiplier takes on a special meaning. In the present case, however, division of z_i by e_i affords a clearer geometrical interpretation as the vector $\mathbf{e}$ turns out to represent the 'direction' of the supporting cone. Note that the weighted Tchebyshev norm is a limiting case ($p = \infty$) of the (weighted) Minkowski metric:

$$\rho(\mathbf{z}; \mathbf{e}) = \left\{\sum_{i=0}^{m} \left[\frac{z_i}{e_i}\right]^p\right\}^{\frac{1}{p}}, \quad \text{when} \mathbf{z} \in \mathbb{R}_+^{m+1}. \tag{5.2.3}$$

There is more than one way of defining the nonlinear Lagrangian function. We begin with a simpler version defined as follows, other versions are discussed in latter sections.

Definition 5.2.1 (Nonlinear Lagrangian Function) The nonlinear Lagrangian function for the (primal) constrained optimization problem P_θ is defined by, given $\mathbf{x} \in \mathcal{X}, \mathbf{e} \in \mathcal{E}$,

$$\mathcal{L}(\mathbf{x}, \mathbf{e}) = \xi_{\mathbf{e}}(\mathbf{f}(\mathbf{x})) = \max_{0\le i\le m} \left\{\frac{f_i(\mathbf{x})}{e_i}\right\}, \tag{5.2.4}$$

where $\mathbf{f}(\mathbf{x}) = [f_0(\mathbf{x}), f_1(\mathbf{x}), f_2(\mathbf{x}), \cdots, f_m(\mathbf{x})]^\top \in \mathbb{R}^{m+1}$.

Let the function $\gamma : \mathcal{E} \to \mathbb{R}$ be defined as follows:

$$\gamma(\mathbf{e}) = \inf_{\mathbf{x}\in\mathcal{X}} \mathcal{L}(\mathbf{x}, \mathbf{e}) = \inf_{\mathbf{x}\in\mathcal{X}} \max_{0\le i\le m} \left\{\frac{f_i(\mathbf{x})}{e_i}\right\}. \tag{5.2.5}$$

Definition 5.2.2 (Nonlinear Lagrangian Dual Function) Assuming that the optimal solution $\mathbf{x}^*$ for problem P_θ exists and $f_0(\mathbf{x}^*) > 0$, then the nonlinear Lagrangian dual function is defined as:

$$\phi(\mathbf{e}) = \begin{cases} \gamma(\mathbf{e}) & \text{if } e_i \geq \dfrac{\theta_i}{f_0(\mathbf{x}^*)} \ \forall i = 1, 2, \cdots, m, \\ -\infty & \text{otherwise.} \end{cases} \tag{5.2.6}$$

Note that there is a peculiar feature of this dual function, namely, given some $\mathbf{e} \in \mathcal{E}$, $\phi(\mathbf{e})$ cannot be computed, since $f_0(\mathbf{x}^*)$ is not known a priori. Fortunately, this does not restrict the usefulness of this duality theory, as we will demonstrate later in the section. In Section 5.5, we present another version of the duality function which is computable. For the rest of this section, the non-computable dual function ϕ as defined in (5.2.6) will be used.

Definition 5.2.3 (The Dual Optimization Problem to Problem P_θ, first version).

$$(\text{Problem } D_\theta) \qquad \sup_{\mathbf{e} \in \mathcal{E}} \phi(\mathbf{e})$$

The following result shows that the maximization of ϕ with respect to $\mathbf{e}$ occurs in a relative interior point of $\mathcal{E}$. As a result, max may be used instead of sup in problem D_θ.

Theorem 5.2.1 The dual function ϕ is maximized at $\mathbf{e} = \mathbf{e}^*$, where $[\mathbf{e}^*]_0 = 1, [\mathbf{e}^*]_i = \dfrac{\theta_i}{f_0(\mathbf{x}^*)}$, $i = 1, 2, \cdots, m$.

Proof: Clearly the dual function cannot be maximized if it takes on the value $-\infty$. We only need to establish that $\gamma(\mathbf{e})$ is a monotone non-increasing function for all $\mathbf{e} \in \mathcal{E}$ such that $e_i \geq \dfrac{\theta_i}{f_0(\mathbf{x}^*)}$ $\forall i = 1, 2, \cdots, m,$. Given $\mathbf{e}^1, \mathbf{e}^2 \in \mathcal{E}$, $\mathbf{e}^1 \leq \mathbf{e}^2$, it is clear that, for a given $\mathbf{x} \in \mathcal{X}$, $\dfrac{f_i(\mathbf{x})}{e_i^1} \geq \dfrac{f_i(\mathbf{x})}{e_i^2}$ $\forall i = 0, 1, \cdots, m$, hence $\max_{0 \leq i \leq m} \left\{ \dfrac{f_i(\mathbf{x})}{e_i^1} \right\} \geq \max_{0 \leq i \leq m} \left\{ \dfrac{f_i(\mathbf{x})}{e_i^2} \right\}$, and consequently $\gamma(\mathbf{e}^1) = \inf_{\mathbf{x} \in \mathcal{X}} \max_{0 \leq i \leq m} \left\{ \dfrac{f_i(\mathbf{x})}{e_i^1} \right\} \geq \gamma(\mathbf{e}^2) = \inf_{\mathbf{x} \in \mathcal{X}} \max_{0 \leq i \leq m} \left\{ \dfrac{f_i(\mathbf{x})}{e_i^2} \right\}$. Thus the dual function must be maximized at the smallest possible $\mathbf{e}$ without taking the value $-\infty$, which is at $\mathbf{e} = \mathbf{e}^*$. ∎

Theorem 5.2.2 (Weak duality) $\phi(\mathbf{e}) \leq f_0(\mathbf{x}) \quad \forall \mathbf{e} \in \mathcal{E}, \ \forall \mathbf{x} \in \mathcal{X}_0$.

Proof: If there exists k such that $e_k < \dfrac{\theta_k}{f_0(\mathbf{x}^*)}$, then $\phi(\mathbf{e}) = -\infty < f_0(\mathbf{x})$ $\forall \mathbf{x}$, and the result holds trivially. Hence we may assume that

$$e_i \geq \frac{\theta_i}{f_0(\mathbf{x}^*)} \ \forall i = 1, 2, \cdots, m. \tag{5.2.7}$$

Then,

$$\begin{aligned}
\phi(\mathbf{e}) = \gamma(\mathbf{e}) &= \min_{\mathbf{x}\in\mathcal{X}} \max_{0\le i\le m} \left\{\frac{f_i(\mathbf{x})}{e_i}\right\} \\
&\le \max\left\{f_0(\mathbf{x}^*), \frac{f_1(\mathbf{x}^*)}{e_1}, \cdots, \frac{f_m(\mathbf{x}^*)}{e_m}\right\} \\
&\le \max\left\{f_0(\mathbf{x}^*), \frac{f_1(\mathbf{x}^*)}{\theta_1} f_0(\mathbf{x}^*), \cdots, \frac{f_m(\mathbf{x}^*)}{\theta_m} f_0(\mathbf{x}^*)\right\} \le f_0(\mathbf{x}^*) \\
&\le f_0(\mathbf{x}).
\end{aligned} \tag{5.2.8}$$

■

Theorem 5.2.3 (Zero duality gap) $f_0(\mathbf{x}^*) = \phi(\mathbf{e}^*)$.

Proof: By definition,

$$\phi(\mathbf{e}^*) = \min_{\mathbf{x}\in\mathcal{X}} \mathcal{L}(\mathbf{x}, \mathbf{e}^*) \tag{5.2.9}$$

Consider the two cases:
Case(i), if $\mathbf{x}$ is feasible, then $\dfrac{f_i(\mathbf{x})}{\theta_i} \le 1 \quad \forall i = 1, 2, \cdots, m$ and $f_0(\mathbf{x}) \ge f_0(\mathbf{x}^*)$, this implies that

$$\mathcal{L}(\mathbf{x}, \mathbf{e}^*) = \max\left\{f_0(\mathbf{x}), \frac{f_1(\mathbf{x})}{\theta_1} f_0(\mathbf{x}^*), \cdots, \frac{f_m(\mathbf{x})}{\theta_m} f_0(\mathbf{x}^*)\right\} = f_0(\mathbf{x}) \ge f_0(\mathbf{x}^*).$$

Case(ii), if $\mathbf{x}$ is infeasible, then there exists k such that $\dfrac{f_k(\mathbf{x})}{\theta_k} > 1$, hence $\mathcal{L}(\mathbf{x}, \mathbf{e}^*) \ge f_0(\mathbf{x}^*)$.
Thus, in both case (i) and case (ii), we have $\mathcal{L}(\mathbf{x}, \mathbf{e}^*) \ge f_0(\mathbf{x}^*)\ \forall \mathbf{x} \in \mathcal{X}$. From (5.2.9), we conclude that

$$\min_{\mathbf{x}\in\mathcal{X}} \mathcal{L}(\mathbf{x}, \mathbf{e}^*) = \phi(\mathbf{e}^*) \ge f_0(\mathbf{x}^*). \tag{5.2.10}$$

Finally, (5.2.10) and the weak duality together imply that $\phi(\mathbf{e}^*) = f_0(\mathbf{x}^*)$. ■

Example 5.2.1 (Example 5.1.1 revisited.)

$$\begin{aligned}
\min_{\mathbf{x} \geqq \mathbf{0}} \quad & f_0(\mathbf{x}) = 1 - x_1 x_2 \\
\text{s.t.} \quad & f_1(\mathbf{x}) = x_1 + 4x_2 \le 1.
\end{aligned}$$

An application of the proposed nonlinear Lagrangian approach yields the following dual function:

$$\phi(e) = \begin{cases} \min_{\mathbf{x}\in\mathcal{X}} \max\left\{1 - x_1 x_2, \dfrac{x_1 + 4x_2}{e}\right\} & \text{if } e \ge e^* = 16/15, \\ -\infty & \text{otherwise.} \end{cases}$$

These dual functions are plotted (in a brute force manner) in Figure 5.2.1. Note that for the dual function,

$$\max_{e>0} \phi(e) = f_0(\mathbf{x}^*) = \frac{15}{16},$$

and therefore the duality gap is zero, even though the perturbation function is nonconvex.

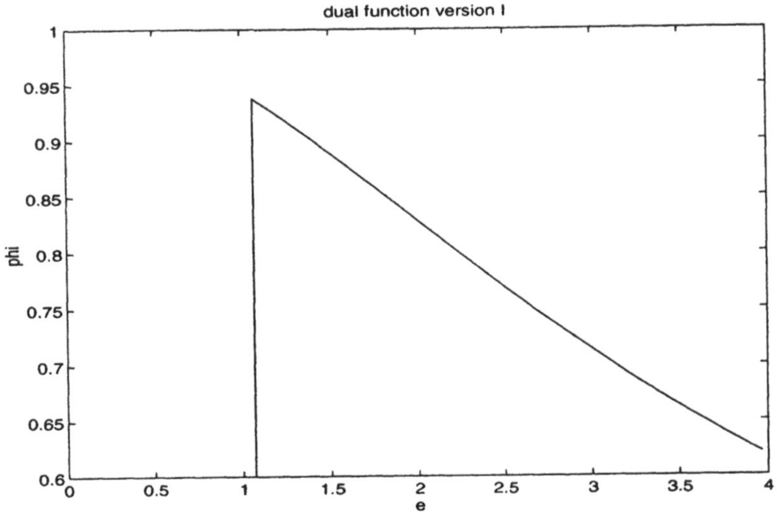

FIGURE 5.2.1 Dual function for Example 5.2.1

5.3 Optimality Conditions via a Monotone Function

We now present an equivalent optimality condition in terms of an unconstrained problem which is both sufficient and necessary. This optimality condition leads directly to a conceptually simple method for solving convex or nonconvex constrained optimization problems. The main references for this section are [GY1],[GY2].

Theorem 5.3.1 (A necessary and sufficient condition for optimality without convexity)
Let $\mathbf{x}^*$ be the (global) optimal solution of problem P_θ (as defined in Section 5.1), and let $\theta_0 = f_0(\mathbf{x}^*) > 0$. Then a solution $\mathbf{x}^0$ solves problem P_θ if and only if $\mathbf{x}^0$ solves the unconstrained problem:

$$(\text{Problem } P_1) \quad \inf_{\mathbf{x}\in\mathcal{X}} \mathcal{L}(\mathbf{x}, \hat{\theta}) = \inf_{\mathbf{x}\in\mathcal{X}} \max_{0\le i\le m} \left\{ \frac{f_i(\mathbf{x})}{\theta_i} \right\},$$

where $\hat{\boldsymbol{\theta}} = [\theta_0, \theta_1, \cdots, \theta_m]$.

Proof: Let us first prove that

$$\inf_{\mathbf{x}\in\mathcal{X}} \max_{0\leq i\leq m} \left\{\frac{f_i(\mathbf{x})}{\theta_i}\right\} = 1.$$

If $\mathbf{x} \in \mathcal{X}$ is feasible, then $f_0(\mathbf{x}) \geq f_0(\mathbf{x}^*)$ and $f_i(\mathbf{x}) \leq \theta_i, i = 1, \cdots, m$. Thus

$$\max_{0\leq i\leq m} \left\{\frac{f_i(\mathbf{x})}{\theta_i}\right\} \geq 1.$$

If $\mathbf{x} \in \mathcal{X}$ is infeasible, then $\exists j \in \{1, 2, \cdots, m\}$ such that $f_j(\mathbf{x}) > \theta_j$, implying that

$$\max_{0\leq i\leq m} \left\{\frac{f_i(\mathbf{x})}{\theta_i}\right\} > 1.$$

Moreover,

$$\max_{0\leq i\leq m} \left\{\frac{f_i(\mathbf{x}^*)}{\theta_i}\right\} = 1.$$

Thus the claim holds.

(Necessity) If $\mathbf{x}^0$ solves problem P_θ, then $\mathbf{x}^0$ must be feasible, and $f_0(\mathbf{x}^0) = f_0(\mathbf{x}^*)$. Hence

$$\max_{0\leq i\leq m} \left\{\frac{f_i(\mathbf{x}^0)}{\theta_i}\right\} = 1,$$

i.e., $\mathbf{x}^0$ solves problem P_1.

(Sufficiency) Assume that $\mathbf{x}^0$ solves problem P_1. Then $\mathbf{x}^0 \in \mathcal{X}$. If $\mathbf{x}^0$ does not solve problem P_θ, then either $\mathbf{x}^0$ is infeasible, or $\mathbf{x}^0$ is feasible and $f_0(\mathbf{x}^0) > f_0(\mathbf{x}^*)$. If $\mathbf{x}^0$ is infeasible, then $\exists j \in \{1, 2, \cdots, m\}$ such that $f_j(\mathbf{x}^0) > \theta_j$, implying that

$$\max_{0\leq i\leq m} \left\{\frac{f_i(\mathbf{x}^0)}{\theta_i}\right\} > 1. \tag{5.3.1}$$

If $\mathbf{x}^0$ is feasible and $f_0(\mathbf{x}^0) > f_0(\mathbf{x}^*)$, then the inequality (5.3.1) still holds. Thus in both cases, $\mathbf{x}^0$ does not solve problem P_1. ∎

The above theorem leads to a conceptually simple method for solving the nonconvex inequality constrained optimization problem P_θ. Consider the following scalar function of a scalar parameter $\theta_0 \in \mathbb{R}_+ \setminus \{0\}$ (Note that θ_i, $i = 1, 2, \cdots .m$ are fixed parameters of the problem P_θ):

$$\psi(\theta_0) = \inf_{\mathbf{x}\in\mathcal{X}} \max_{0\leq i\leq m} \left\{\frac{f_i(\mathbf{x})}{\theta_i}\right\}. \tag{5.3.2}$$

Under appropriate assumptions, the function ψ has some nice properties which are summarized as follows:

Theorem 5.3.2 Assume that $\mathcal{X}_0$ is compact and if $\mathcal{X}$ is unbounded,

$$\lim_{\|\mathbf{x}\|\to\infty, \mathbf{x}\in\mathcal{X}} f_0(\mathbf{x}) = \infty.$$

Let $\theta_0 > 0$. Assume that P_θ admits a solution $\mathbf{x}^*$ with $f_0(\mathbf{x}^*) > 0$ and that for any $\theta_0 > f_0(\mathbf{x}^*)$, there exists $\mathbf{x} \in \mathcal{X}$ such that $f_i(\mathbf{x}) < \theta_i (i = 1, \cdots, m)$, $f_0(\mathbf{x}^*) < f_0(\mathbf{x}) < \theta_0$. Then

(i) $0 < \theta_0 < f_0(\mathbf{x}^*) \Rightarrow \psi(\theta_0) > 1$;
(ii) $\theta_0 > f_0(\mathbf{x}^*) \Rightarrow \psi(\theta_0) < 1$;
(iii) $\theta_0 = f_0(\mathbf{x}^*)$ if and only if $\psi(\theta_0) = 1$;
(iv) $\psi(\theta_0)$ is a monotone non-increasing function of θ_0.

Proof: (i) Assume that $0 < \theta_0 < f_0(\mathbf{x}^*)$. It is clear that

$$\begin{aligned}\psi(\theta_0) &= \inf_{\mathbf{x}\in\mathcal{X}} \max_{0\le i\le m} \left\{\frac{f_i(\mathbf{x})}{\theta_i}\right\} \\ &= \min\{\inf_{\mathbf{x}\in\mathcal{X}_0} \max_{0\le i\le m} \left\{\frac{f_i(\mathbf{x})}{\theta_i}\right\}, \inf_{\mathbf{x}\in\mathcal{X}\setminus\mathcal{X}_0} \max_{0\le i\le m} \left\{\frac{f_i(\mathbf{x})}{\theta_i}\right\}\}.\end{aligned}$$

Thus

$$\inf_{\mathbf{x}\in\mathcal{X}_0} \max_{0\le i\le m} \left\{\frac{f_i(\mathbf{x})}{\theta_i}\right\} \ge \inf_{\mathbf{x}\in\mathcal{X}_0} \max_{0\le i\le m} \left\{\frac{f_0(\mathbf{x}^*)}{\theta_0}, \frac{f_1(\mathbf{x})}{\theta_1}, \cdots, \frac{f_m(\mathbf{x})}{\theta_m}\right\} > 1. \qquad (5.3.3)$$

Now we show that

$$\inf_{x\in\chi\setminus\chi_0} \max_{0\le i\le m} \{\frac{f_i(x)}{\theta_i}\} > 1.$$

Since $\lim_{x\in\chi, \|x\|\to+\infty} f_0(x) = +\infty$, we see that there exists $N > 0$ such that

$$f_0(x) \ge 2\theta_0, \quad \forall x \in A = \{x \in \chi : \|x\| > N\}.$$

Consequently,

$$\inf_{x\in(\chi\setminus\chi_0)\bigcap A} \max_{0\le i\le m} \{\frac{f_i(x)}{\theta_i}\} \ge 2 > 1.$$

So we need only to prove that

$$\inf_{x\in B} \max_{0\le i\le m} \{\frac{f_i(x)}{\theta_i}\} > 1,$$

where $B = \{x \in \chi\setminus\chi_0 : \|x\| \le N\}$. As χ_0 is compact and f_0 is continuous and $0 < \theta_0 < f_0(x^*)$, we deduce that for some $\epsilon > 0$ satisfying

$$\frac{(1-\epsilon)f_0(x^*)}{\theta_0} > 1,$$

there exists $\delta > 0$ such that

$$f_0(x) > (1-\epsilon)f_0(x^*), \quad \forall x \in C = \{x \in B : d(x, \chi_0) < \delta\}.$$

As a result,

$$\inf_{x \in C} \max_{0 \le i \le m} \{\frac{f_i(x)}{\theta_i}\} > 1.$$

Now we need only to show that

$$\inf_{x \in D} \max_{0 \le i \le m} \{\frac{f_i(x)}{\theta_i}\} > 1, \tag{5.3.4}$$

where

$$D = B \backslash C = \{x \in \chi \backslash \chi_0 : d(x, \chi_0) \ge \delta, \|x\| \le N\}.$$

Note that D is compact and that $\max_{0 \le i \le m} \{\frac{f_i(x)}{\theta_i}\}$ is continuous. Thus, the infimum of $\inf_{x \in D} \max_{0 \le i \le m} \{\frac{f_i(x)}{\theta_i}\}$ is attainable. Further notice that

$$\max_{0 \le i \le m} \{\frac{f_i(x)}{\theta_i}\} \ge \max_{1 \le i \le m} \{\frac{f_i(x)}{\theta_i}\} > 1, \quad \forall x \in D.$$

So we obtain

$$\inf_{x \in D} \max_{0 \le i \le m} \{\frac{f_i(x)}{\theta_i}\} = \min_{x \in D} \max_{0 \le i \le m} \{\frac{f_i(x)}{\theta_i}\} > 1. \tag{5.3.5}$$

Thus from (5.3.4) and (5.3.5), $\psi(\theta_0) > 1$.

(ii) By the assumption, if $\theta_0 > f_0(\mathbf{x}^*)$ then there exists $\mathbf{x} \in \mathcal{X}$ such that $f_i(\mathbf{x}) < \theta_i$, $i = 1, 2, \cdots, m$; and $\theta_0 > f_0(\mathbf{x}) > f_0(\mathbf{x}^*)$. Consequently

$$\max_{0 \le i \le m} \left\{ \frac{f_i(\mathbf{x})}{\theta_i} \right\} < 1, \tag{5.3.6}$$

and hence

$$\psi(\theta_0) = \inf_{\mathbf{x} \in \mathcal{X}} \max_{0 \le i \le m} \left\{ \frac{f_i(\mathbf{x})}{\theta_i} \right\} < 1. \tag{5.3.7}$$

(iii) If $\theta_0 = f_0(\mathbf{x}^*)$, it follows from Theorem 5.3.1 that $\psi(\theta_0) = 1$. Assume that $\psi(\theta_0) = 1$. Then it follows from (i) and (ii) that $\theta_0 = f_0(\mathbf{x}^*)$.

(iv) Let $\theta_0^1 \ge \theta_0^2$. Then

$$\frac{f_0(\mathbf{x})}{\theta_0^1} \le \frac{f_0(\mathbf{x})}{\theta_0^2} \quad \forall \mathbf{x} \in \mathcal{X}. \tag{5.3.8}$$

This implies that

$$\max\left\{ \frac{f_0(\mathbf{x})}{\theta_0^1}, \frac{f_i(\mathbf{x})}{\theta_i}, \quad i = 1, 2, \cdots, m \right\} \le \max\left\{ \frac{f_0(\mathbf{x})}{\theta_0^2}, \frac{f_i(\mathbf{x})}{\theta_i}, \quad i = 1, 2, \cdots, m \right\},$$

and hence

$$\psi(\theta_0^1) \le \psi(\theta_0^2). \tag{5.3.9}$$

■

Remark 5.3.1 For the continuity of parametric optimization problems, such as (5.3.2), see [BGKKT1].

Remark 5.3.2 In Theorem 5.3.2, the assumption that for any $\theta_0 > f_0(\mathbf{x}^*)$, there exists $\mathbf{x} \in \mathcal{X}$ such that $f_i(\mathbf{x}) < \theta_i, \theta_0 > f_0(\mathbf{x}) > f_0(\mathbf{x}^*)$ is rather restrictive as it depends on the parameter θ_0. However, the following example shows that this assumption is necessary to assure Theorem 5.3.2 (ii). Let $f_0(x) = x^2 + 1, f_1(x) = (x+1)^3 + 1$, if $x < -1$; $f_1(x) = (x-1)^3 + 1$, if $x > 1$ and $f_1(x) = 1$ otherwise. The assumption does not hold, neither does Theorem 5.3.2 (ii) since $\psi(\theta_0)$ is constant at 1 for all $\theta_0 \in [1, 2]$. Without this assumption, a weaker result is given in Theorem 5.3.3.

Because of these special properties of the function ψ, under the assumptions given in Theorem 5.3.2, solving the constrained optimization problem P_θ is now equivalent to a rather simple problem: find the unique root of a monotone decreasing (scalar) function of a scalar variable:

$$\text{(Problem } P_2) \qquad \text{Find } \theta_0^* = f_0(\mathbf{x}^*) \text{ such that } \psi(\theta_0^*) = 1.$$

At the first glance, this appears to be a trivial problem since it requires little effort to numerically solve for the root of a monotone (strictly monotone at the root by virtue of Theorem 5.3.2 (iii)) scalar function of a scalar variable. However, in practice, this is made non-trivial by the fact that each function call of ψ requires the solution of an unconstrained minimax problem, which is not a trivial problem even if it is unconstrained. Nevertheless there exist several effective methods and software packages that deal with minimax optimization effectively (for example, see [ZT1] or the minimax function in the optimization toolbox of MATLAB, p.18 of [G8]).

A weaker version of Theorem 5.3.2 is as follows.

Theorem 5.3.3 Let $\mathcal{X}_0$ be compact and if $\mathcal{X}$ is bounded,

$$\lim_{\|\mathbf{x}\| \to \infty, \mathbf{x} \in \mathcal{X}} f_0(\mathbf{x}) = \infty.$$

Assume that P_θ admits a solution $\mathbf{x}^*$ with $f_0(\mathbf{x}^*) > 0$.

(i) $0 < \theta_0 < f_0(\mathbf{x}^*) \Rightarrow \psi(\theta_0) > 1$;
(ii) $\theta_0 \geq f_0(\mathbf{x}^*) \Rightarrow \psi(\theta_0) \leq 1$;
(iii) $\theta_0 = f_0(\mathbf{x}^*)$ if and only if θ_0 is the least root of $\psi(\theta_0) = 1$;
(iv) $\psi(\theta_0)$ is a monotone non-increasing function of θ_0.

Proof: The proofs of (i) and (iv) follow from Theorem 5.3.2.
(ii) Let $\mathbf{x} = \mathbf{x}^*$. Then

$$\max_{0 \leq i \leq m} \left\{ \frac{f_i(\mathbf{x}^*)}{\theta_i} \right\} \leq 1,$$

and hence

$$\psi(\theta_0) = \inf_{\mathbf{x} \in \mathcal{X}} \max_{0 \leq i \leq m} \left\{ \frac{f_i(\mathbf{x})}{\theta_i} \right\} \leq 1.$$

(iii) Assume $\theta_0 = f_0(\mathbf{x}^*)$. From Theorem 5.3.1, $\psi(\theta_0) = 1$. It follows from (i) that θ_0 is the least root of $\psi(\theta_0) = 1$. If θ_0 is the least root of $\psi(\theta_0) = 1$, then from (i), $\theta_0 \geq f_0(\mathbf{x}^*)$. It is clear that $\psi(f_0(\mathbf{x}^*)) = 1$. Thus $\theta_0 = f_0(\mathbf{x}^*)$. ∎

Consequently, without the assumptions given in Theorem 5.3.2, solving the constrained optimization problem P_θ is now equivalent to a slightly harder problem: find the least root of a monotone decreasing (scalar) function of a scalar variable:

$$\text{(Problem P}_3\text{)} \qquad \text{Find the least root } \theta_0^* \text{ such that } \psi(\theta_0^*) = 1.$$

5.4 Optimality Conditions via a Monotone Composition Formulation

The analysis in the previous section is based on a minimax formulation. It is clear that the evaluation of the auxiliary function ψ requires the solution of a non-differentiable optimization problem. In this section, the single-parameter approach in Section 5.3 is modified such that the more general subproblem is obtained and that the similar type of necessary and sufficient conditions for global optimum, as well as local optimum, is obtained. The main reference for this section is [YL1].

Consider the following optimization problem P that could be nonconvex:

$$\begin{aligned} &\text{(Problem P)} \qquad \inf_{\mathbf{x}\in\mathcal{X}} \; f_0(\mathbf{x}) \\ &\qquad\qquad\qquad \text{subject to} \quad f_i(\mathbf{x}) \leq 0 \qquad i = 1,\cdots,m, \end{aligned}$$

where $\mathcal{X}$ is a subset of $\mathbb{R}^n$, $f_i : \mathbb{R}^n \to \mathbb{R}$ $(i = 0, 1, \cdots, m)$ are real-valued functions. The feasible set of problem P is

$$\mathcal{X}_0 = \{\mathbf{x} \in \mathcal{X} \mid \; f_i(\mathbf{x}) \leq 0, \quad i = 1, \cdots, m\}.$$

Our study in this section is restricted to the class of optimization problems P that satisfy the following assumptions.

Assumption 5.4.1
(i) f_0 is continuous, and each f_i is continuous on an open set containing $\mathcal{X}$, $i = 1, \cdots, m$;
(ii) $\inf_{\mathbf{x}\in\mathcal{X}} f_0(\mathbf{x}) > 0$;
(iii) $\mathcal{X}$ is connected.

Assumption 5.4.1(ii) states that f_0 is bounded from below on $\mathcal{X}$. The connectedness of $\mathcal{X}$ in Assumption 5.4.1(iii) will be used in the proof of Lemma 5.4.1.

Let $\phi : \mathbb{R}^{1+m} \to \mathbb{R}$ satisfy the following properties, for $\mathbf{y} = (y_0, y_1, \cdots, y_m)^{\mathsf{T}}$:

(A) $\phi(\mathbf{y}) \geq \max_{0 \leq i \leq m} y_i, \quad \forall \mathbf{y} \in \mathbb{R}^{1+m}$;

(B) $\phi(\mathbf{y}) = \max\{0, y_0\}, \quad \forall y_0 \in \mathbb{R}, (y_1, \cdots, y_m) \in \mathbb{R}^m_-$;

(C) ϕ is a nondecreasing function with respect to its first component y_0.

It is clear from Property (B) that $\phi(0, 0, \cdots, 0) = \max\{0, 0\} = 0$.

Example 5.4.1 The following functions satisfy Properties (A), (B) and (C):

$$\phi_\infty(\mathbf{y}) = \max\{0, y_0, y_1, \cdots, y_m\};$$

$$\phi_p(\mathbf{y}) = \left(\sum_{i=0}^{m} [\max\{0, y_i\}]^p\right)^{\frac{1}{p}}, \quad 1 \leq p < \infty;$$

$$\phi_{p1}(\mathbf{y}) = \max\{0, y_0\} + \left(\sum_{i=1}^{m} [\max\{0, q_i(y_i)\}]^p\right)^{\frac{1}{p}}, \quad 1 \leq p < \infty,$$

where $q_i : \mathbb{R} \to \mathbb{R}$ can be any continuous function that satisfies $q_i(z) \leq 0$, if $z < 0$ and $q_i(z) \geq z$, if $z \geq 0$. For example, $q_i(z) = \exp(z) - 1$. ■

It is worth noting that the function defined by

$$\phi(\mathbf{y}) = \max\{y_0, y_1, \cdots, y_m\}$$

does not satisfy Property (B) and that this nonsmooth function is used in Section 5.3 to formulate another type of the auxiliary parametric optimization problems for the problem P.

Let

$$\mathbf{F}(\mathbf{x}, \theta_0) = \left(\frac{f_0(\mathbf{x})}{\theta_0} - 1, f_1(\mathbf{x}), \cdots, f_m(\mathbf{x})\right), \quad \mathbf{x} \in \mathcal{X}, \quad \theta_0 > 0.$$

Definition 5.4.1 Let ϕ be a function that satisfies Properties (A), (B) and (C) and $\theta_0 > 0$. Define an auxiliary function $\Phi : \mathbb{R}^n \times \mathbb{R} \to \mathbb{R}$ as

$$\Phi(\mathbf{x}, \theta_0) = \phi(\mathbf{F}(\mathbf{x}, \theta_0)).$$

An auxiliary problem $P_\phi(\theta_0)$ is defined as follows

$$(\text{Problem } P_\phi(\theta_0)) \quad \inf_{\mathbf{x} \in \mathcal{X}} \Phi(\mathbf{x}, \theta_0).$$

The constraint $\mathbf{x} \in \mathcal{X}$ often represents simple constraints such as lower and upper bounds. The constraint structure in $P_\phi(\theta_0)$ is, in general, much simpler than the constraint structure in P. When $\mathcal{X} = \mathbb{R}^n$, $P_\phi(\theta_0)$ becomes an unconstrained optimization problem.

This form of the auxiliary parametric optimization problem can be considered as a monotone composition formulation since ϕ is a monotone function. $P_\phi(\theta_0)$ can also be rewritten as the following unconstrained optimization problem,

$$\inf_{\mathbf{x}\in\mathbf{R}^n} \phi(\mathbf{F}(\mathbf{x},\theta_0)) + \delta_{\mathcal{X}}(\mathbf{x}),$$

where $\delta_{\mathcal{X}}$ is an indicator function of $\mathcal{X}$: $\delta_{\mathcal{X}}(\mathbf{x}) = 0$, if $\mathbf{x} \in \mathcal{X}$ and $\delta_{\mathcal{X}}(\mathbf{x}) = \infty$, if $\mathbf{x} \notin \mathcal{X}$. Furthermore, if ϕ is a convex function and $\mathcal{X}$ is a convex set, then the problem $P_\phi(\theta_0)$ is a convex composite optimization problem which has been intensively studied in recent years, see [BP1],[Y2]. In particular, if $\phi = \phi_\infty$, problem $P_\phi(\theta_0)$ becomes

$$\inf_{\mathbf{x}\in\mathcal{X}} \max\left\{0, \frac{f_0(\mathbf{x})}{\theta_0} - 1, f_1(\mathbf{x}), \cdots, f_m(\mathbf{x})\right\}.$$

This can be considered as a generalized minimax formulation for solving P. If $\phi = \phi_2$, then the auxiliary problem $P_\phi(\theta_0)$ is given by

$$\inf_{\mathbf{x}\in\mathcal{X}} \{[\max\{0, \frac{f_0(\mathbf{x})}{\theta_0} - 1\}]^2 + \sum_{i=1}^{m}[\max\{0, f_i(\mathbf{x})\}]^2\}^{1/2}.$$

The above problem can be further reduced to the following equivalent form,

$$\inf_{\mathbf{x}\in\mathcal{X}} [\max\{0, \frac{f_0(\mathbf{x})}{\theta_0} - 1\}]^2 + \sum_{i=1}^{m}[\max\{0, f_i(\mathbf{x})\}]^2.$$

Note that the above equivalent formulation of $P_\phi(\theta_0)$ yields a differentiable optimization problem if all the functions f_j's are differentiable.

The following properties of the auxiliary function $\Phi(\mathbf{x},\theta_0)$ will often be used in the sequel.

Lemma 5.4.1 Let $\theta_0 > 0$ and $\mathbf{x} \in \mathcal{X}$. Then $\Phi(\mathbf{x},\theta_0) \geq 0$. Furthermore if $\Phi(\mathbf{x},\theta_0) = 0$, then $\mathbf{x} \in \mathcal{X}_0$.

Proof: Let $\mathbf{x} \in \mathcal{X}$, $\Phi(\mathbf{x},\theta_0) \geq 0$. In fact, if $\mathbf{x}$ is infeasible, then from Property (A), for some $1 \leq i \leq m$,

$$\phi\left(\frac{f_0(\mathbf{x})}{\theta_0} - 1, f_1(\mathbf{x}), \cdots, f_m(\mathbf{x})\right) \geq f_i(\mathbf{x}) > 0. \tag{5.4.1}$$

If $\mathbf{x}$ is feasible, then from Property (B),

$$\phi\left(\frac{f_0(\mathbf{x})}{\theta_0} - 1, f_1(\mathbf{x}), \cdots, f_m(\mathbf{x})\right) = \max\{0, \frac{f_0(\mathbf{x})}{\theta_0} - 1\} \geq 0.$$

Assume that $\Phi(\mathbf{x}, \theta_0) = 0$. If $\mathbf{x}$ is infeasible, then from (5.4.1), $\Phi(\mathbf{x}, \theta_0) > 0$, which is a contradiction. ∎

In the following, a successive solution scheme via a parametric monotone composition formulation is developed for finding the global minimum of optimization problems that could be nonconvex. Specifically, a two-level iterative scheme will be proposed. In the lower level of each iteration, an auxiliary optimization problem with a fixed parameter is solved, while in the upper level the parameter is adjusted such that the least root of the optimal function value of the parametric optimization problem is found.

Definition 5.4.2 Let ϕ be a function that satisfies Properties (A), (B) and (C) and $\theta_0 > 0$. Define $\varphi(\theta_0)$ to be the global optimal value of $P_\phi(\theta_0)$, i.e.,

$$\varphi(\theta_0) = \inf_{\mathbf{x} \in \mathcal{X}} \phi(\mathbf{F}(\mathbf{x}, \theta_0)).$$

The following result provides a necessary and sufficient optimality condition for P.

Theorem 5.4.2 Let $\mathbf{x}^*$ solve the problem P and $\theta_0^* = f_0(\mathbf{x}^*)$. Then $\mathbf{x}^0$ solves P if and only if $\mathbf{x}^0$ solves the problem $P_\phi(\theta_0^*)$.

Proof: From Assumption 5.3.1(ii), $\theta_0^* > 0$. It is clear that $\mathbf{x}^* \in \mathcal{X}_0$ and

$$\Phi(\mathbf{x}^*, \theta_0^*) = \phi\left(\frac{f_0(\mathbf{x}^*)}{\theta_0^*} - 1, f_1(\mathbf{x}^*), \cdots, f_m(\mathbf{x}^*)\right) = \max\{0, \frac{f_0(\mathbf{x}^*)}{\theta_0^*} - 1\} = 0.$$

From Lemma 5.4.1, $\Phi(\mathbf{x}, \theta_0^*) \geq 0, \forall \mathbf{x} \in \mathcal{X}$. Thus the optimal value of problem $P_\phi(\theta_0^*)$ is 0.

If $\mathbf{x}^0$ solves P, then from Property (B), $\Phi(\mathbf{x}^0, \theta_0^*) = 0$, thus $\mathbf{x}^0$ solves $P_\phi(\theta_0^*)$.

If $\mathbf{x}^0$ does not solve P, then there are two cases:

Case 1. $\mathbf{x}^0$ is infeasible, then from Property (A) for some $1 \leq i \leq m$,

$$\phi\left(\frac{f_0(\mathbf{x}^0)}{\theta_0^*} - 1, f_1(\mathbf{x}^0), \cdots, f_m(\mathbf{x}^0)\right) \geq f_i(\mathbf{x}^0) > 0.$$

Case 2. $\mathbf{x}^0$ is feasible, then $f_0(\mathbf{x}^0) > f_0(\mathbf{x}^*) = \theta_0^*$. Thus

$$\phi\left(\frac{f_0(\mathbf{x}^0)}{\theta_0^*} - 1, f_1(\mathbf{x}^0), \cdots, f_m(\mathbf{x}^0)\right) \geq \frac{f_0(\mathbf{x}^0)}{\theta_0^*} - 1 > 0.$$

Then $\mathbf{x}^0$ does not solve $P_\phi(\theta_0^*)$. ∎

Lemma 5.4.3 Assume that $\mathcal{X}_0$ is compact and if $\mathcal{X}$ is unbounded,

$$\lim_{\|\mathbf{x}\| \to \infty, \mathbf{x} \in \mathcal{X}} f_0(\mathbf{x}) = \infty.$$

Let $\mathbf{x}^*$ solve the problem P. If $0 < \theta_0 < f_0(\mathbf{x}^*)$, then $\varphi(\theta_0) > 0$.

Proof: Let $\bar{\mathcal{X}}_0 = \mathcal{X} \setminus \mathcal{X}_0$. Assume that $0 < \theta_0 < f_0(\mathbf{x}^*)$. It is clear that

$$\begin{aligned}\varphi(\theta_0) &= \inf_{\mathbf{x}\in\mathcal{X}} \phi(\mathbf{F}(\mathbf{x},\theta_0)) \\ &= \min\{\inf_{\mathbf{x}\in\mathcal{X}_0} \phi(\mathbf{F}(\mathbf{x},\theta_0)), \inf_{\mathbf{x}\in\bar{\mathcal{X}}_0} \phi(\mathbf{F}(\mathbf{x},\theta_0))\}.\end{aligned}$$

From Property (B), it is easy to see that

$$\inf_{\mathbf{x}\in\mathcal{X}_0} \phi(\mathbf{F}(\mathbf{x},\theta_0)) = \inf_{\mathbf{x}\in\mathcal{X}_0} \max\{0, \frac{f_0(\mathbf{x})}{\theta_0} - 1\} \geq \frac{f_0(\mathbf{x}^*)}{\theta_0} - 1 > 0. \tag{5.4.2}$$

Now we need only to prove that

$$\inf_{x\in\chi_0} \phi(F(x,\theta_0)) > 0.$$

As $\lim_{x\in\chi,\|x\|\to+\infty} f_0(x) = +\infty$, we deduce that there exists $N > 0$ such that

$$f_0(x) \geq 2\theta_0, \quad \forall x \in A = \{x \in \chi : \|x\| > N\}.$$

It follows from Property (A) that

$$\inf_{x\in\chi_0\bigcap A} \phi(F(x,\theta_0)) \geq \inf_{x\in\chi_0\bigcap A} \{\frac{f_0(x)}{\theta_0} - 1\} \geq 1 > 0.$$

Consequently, we need only to show that

$$\inf_{x\in\chi_0\bigcap\bar{A}} \phi(F(x,\theta_0)) > 0,$$

where $\bar{A} = \{x \in \chi : \|x\| \leq N\}$. Since χ_0 is compact and f_0 is continuous and $0 < \theta_0 < f_0(x^*)$, we see that for some $\epsilon > 0$ satisfying

$$\frac{(1-\epsilon)f_0(x^*)}{\theta_0} > 1,$$

there exists $\delta > 0$ such that

$$f_0(x) > (1-\epsilon)f_0(x^*), \quad \forall x \in B = \{x \in \chi_0 \bigcap \bar{A} : d(x,\chi_0) < \delta\}.$$

Thus,

$$\inf_{x\in B} \phi(F(x,\theta_0)) \geq \inf_{x\in B} \{\frac{f_0(x)}{\theta_0}\} > 0. \tag{5.4.3}$$

So we need only to prove that

$$\inf_{x\in C} \phi(F(x,\theta_0)) > 0,$$

where $C = \{x \in \chi_0 \bigcap \bar{A} : d(x,\chi_0) \geq \delta\} = \{x \in \chi_0 : d(x,\chi_0) \geq \delta, \|x\| \leq N\}$.

Note that C is compact and the function $\max\{\frac{f_0(x)}{\theta_0} - 1, f_1(x), \cdots, f_m(x)\}$ is continuous. Hence, the infimum of $\inf_{x\in C} \max\{\frac{f_0(x)}{\theta_0} - 1, f_1(x), \cdots, f_m(x)\}$ is attainable. Moreover,

$$\phi(F(x,\theta_0)) \geq \max\{\frac{f_0(x)}{\theta_0} - 1, f_1(x), \cdots, f_m(x)\} > 0, \quad \forall x \in C.$$

So we have

$$\begin{aligned}\inf_{x\in C} \phi(F(x,\theta_0)) &\geq \inf_{x\in C} \max\{\frac{f_0(x)}{\theta_0} - 1, f_1(x), \cdots, f_m(x)\} \\ &= \min_{x\in C} \max\{\frac{f_0(x)}{\theta_0} - 1, f_1(x), \cdots, f_m(x)\} > 0.\end{aligned} \tag{5.4.4}$$

From (5.4.3) and (5.4.4), the proof is complete. ∎

Theorem 5.4.4 Assume that $\mathcal{X}_0$ is compact and if $\mathcal{X}$ is unbounded,

$$\lim_{\|\mathbf{x}\|\to\infty, \mathbf{x}\in\mathcal{X}} f_0(\mathbf{x}) = \infty.$$

Let $\mathbf{x}^*$ solve the problem P. The following hold:

(i) If $0 < \theta_0 < f_0(\mathbf{x}^*)$, then $\varphi(\theta_0) > 0$.

(ii) If $\theta_0 \geq f_0(\mathbf{x}^*)$, then $\varphi(\theta_0) = 0$.

(iii) $\varphi(\theta_0)$ is a non-increasing and nonnegative function of θ_0.

(iv) $\theta_0 = f_0(\mathbf{x}^*)$ if and only if θ_0 is the solution of the following least root problem

$$\theta_0 = \min\{\theta \mid \varphi(\theta) = 0\}. \tag{5.4.5}$$

Proof: (i) Follows from Lemma 5.4.1.

(ii) Let $\theta_0 \geq f_0(\mathbf{x}^*)$. If $\mathbf{x}$ is feasible, then from Property (B),

$$\phi\left(\frac{f_0(\mathbf{x})}{\theta_0} - 1, f_1(\mathbf{x}), \cdots, f_m(\mathbf{x})\right) = \max\{0, \frac{f_0(\mathbf{x})}{\theta_0} - 1\} \geq 0.$$

If $\mathbf{x}$ is infeasible, then there exists $1 \leq i \leq m$ such that $f_i(\mathbf{x}) > 0$, and so from Property (A),

$$\phi\left(\frac{f_0(\mathbf{x})}{\theta_0} - 1, f_1(\mathbf{x}), \cdots, f_m(\mathbf{x})\right) \geq f_i(\mathbf{x}) > 0.$$

Thus, $\Phi(\mathbf{x}, \theta_0) \geq 0$. Note that

$$\phi\left(\frac{f_0(\mathbf{x}^*)}{\theta_0} - 1, f_1(\mathbf{x}^*), \cdots, f_m(\mathbf{x}^*)\right) = \max\{0, \frac{f_0(\mathbf{x}^*)}{\theta_0} - 1\} = 0.$$

Then $\varphi(\theta_0) = 0$.

(iii) The non-increasing property of $\varphi(\theta_0)$ is assured by Property (C). Results (i) and (ii) together show that $\varphi(\theta_0) \geq 0, \forall \theta_0 > 0$.

(iv) The necessity part follows from Theorem 5.4.1 and the sufficiency part follows from (i) and (ii). ∎

Remark 5.4.1 Although it is intuitively straightforward to find the least root of a nonincreasing function in problem (5.4.5), finding the global optimum of the auxiliary problem $P_\phi(\theta_0)$ is, in general, not an easy task, if the objective function of $P_\phi(\theta_0)$ is not convex.

The proposed method can be considered as a two-level scheme. The lower level is to solve, for a given parameter, an optimization problem with simple constraints or without constraint. The upper level is to check if the parameter θ_0 is the least root of the equation $\varphi(\theta_0) = 0$.

Corollary 5.4.5 Assume $\mathcal{X}_0$ is compact and if $\mathcal{X}$ is unbounded,

$$\lim_{\|\mathbf{x}\| \to \infty, \mathbf{x} \in \mathcal{X}} f_0(\mathbf{x}) = \infty.$$

Let $\mathbf{x}^*$ solve the problem P. Then $\theta_0 = f_0(\mathbf{x}^*)$ if and only if θ_0 is the least root of the equation

$$\varphi_\infty(\theta_0) = 0,$$

where $\varphi_\infty(\theta_0)$ is the optimal value of the auxiliary optimization problem $P_\phi^\infty(\theta_0)$:

$$\inf_{\mathbf{x} \in \mathcal{X}} \max\{0, \frac{f_i(\mathbf{x})}{\theta_0} - 1, f_1(\mathbf{x}), \cdots, f_m(\mathbf{x})\}.$$

Proof: The result follows from Theorem 5.4.4 (iv) by letting $\phi(\mathbf{y}) = \phi_\infty(\mathbf{y})$. ∎

Corollary 5.4.6 Assume $\mathcal{X}_0$ is compact and if $\mathcal{X}$ is unbounded,

$$\lim_{\|\mathbf{x}\| \to \infty, \mathbf{x} \in \mathcal{X}} f_0(\mathbf{x}) = \infty.$$

Let $\mathbf{x}^*$ solve the problem P. Let $0 < p < \infty$. Then $\theta_0 = f_0(\mathbf{x}^*)$ if and only if θ_0 is the least root of the equation

$$\bar{\varphi}_p(\theta_0) = 0, \tag{5.4.6}$$

where $\bar{\varphi}_p(\theta_0)$ is the optimal value of the subproblem $\bar{P}_\phi^p(\theta_0)$:

$$\inf_{\mathbf{x} \in \mathcal{X}} [\max\{0, \frac{f_0(\mathbf{x})}{\theta_0} - 1\}]^p + \sum_{i=0}^{m} [\max\{0, f_i(\mathbf{x})\}]^p. \tag{5.4.7}$$

Proof: From Theorem 5.4.4, $\theta_0 = f_0(\mathbf{x}^*)$ if and only if θ_0 is the least root of the equation

$$\varphi_p(\theta_0) = 0,$$

where $\varphi_p(\theta_0)$ is the optimal value of the auxiliary problem $\mathrm{P}_\phi^p(\theta_0)$:

$$\inf_{\mathbf{x}\in\mathcal{X}} \left([\max\{0, \frac{f_0(\mathbf{x})}{\theta_0} - 1\}]^p + \sum_{i=0}^{m} [\max\{0, f_i(\mathbf{x})\}]^p \right)^{\frac{1}{p}}.$$

This is equivalent to saying that θ_0 is the least root of the equation

$$(\bar{\varphi}_p(\theta_0))^{\frac{1}{p}} = 0,$$

where $\bar{\varphi}_p(\theta_0)$ is the optimal value of the subproblem $\bar{\mathrm{P}}_\phi^p(\theta_0)$. It is clear that the above equation is equivalent to (5.4.6). ∎

Remark 5.4.2 (i) It is much easier to solve $\bar{\mathrm{P}}_\phi^p(\theta_0)$ than $\mathrm{P}_\phi^p(\theta_0)$ since there is no p^{th} root in the objective function of $\bar{\mathrm{P}}_\phi^p(\theta_0)$.
(ii) Assume that $\mathbf{x}^*$ solves P. It is clear that if for $\theta_0 > f_0(\mathbf{x}^*)$, $0 < p_1, p_2 < \infty$, the solution $\mathbf{x}_{p_1}$ of $\bar{\mathrm{P}}_\phi^{p_1}(\theta_0)$ and the solution $\mathbf{x}_{p_2}$ of $\bar{\mathrm{P}}_\phi^{p_2}(\theta_0)$ are feasible for P, then $\bar{\varphi}_{p_1}(\theta_0) = \bar{\varphi}_{p_2}(\theta_0)$. Thus $\varphi_{p_1}(\theta_0) = \varphi_{p_2}(\theta_0)$.

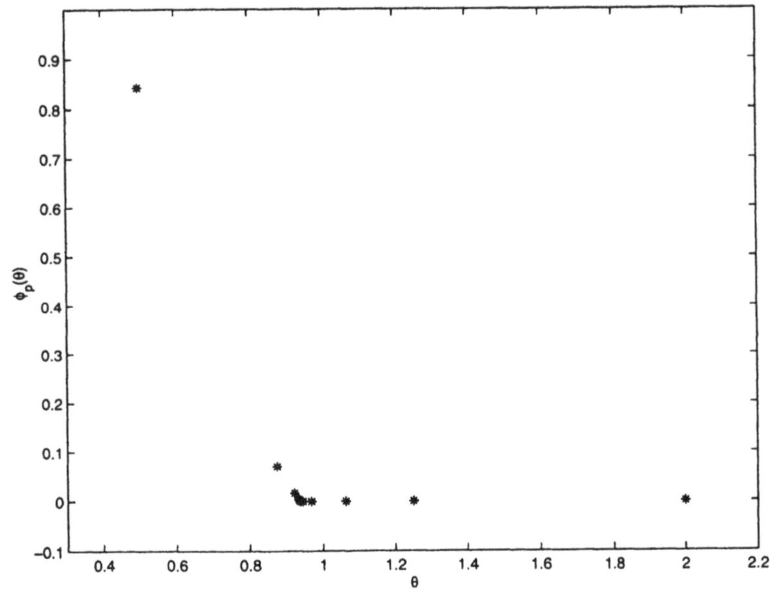

FIGURE 5.4.1 Monotone composition function for Example 5.4.2

Example 5.4.2 (Example 5.1.1 revisited)

$$\begin{aligned} &\min f_0(\mathbf{x}) = 1 - x_1 x_2, \\ &\text{subject to } f_1(\mathbf{x}) = x_1 + 4x_2 - 1 \le 0, \\ &x_1 \ge 0, x_2 \ge 0. \end{aligned}$$

From Theorem 5.4.4, this problem is equivalent to the least root problem

$$\theta_0 = \min\{\theta \mid \varphi_p(\theta) = 0\}$$

where $\varphi_p(\theta_0) = (\bar{\varphi}_p(\theta_0))^{\frac{1}{p}}$ and $\bar{\varphi}_p(\theta_0)$ is the (global) optimal value of the following auxiliary problem $\mathrm{P}^p_\phi(\theta_0)$:

$$\min_{x_1, x_2 \geq 0} [\max\{0, \frac{1 - x_1 x_2}{\theta_0} - 1\}]^p + [\max\{0, x_1 + 4x_2 - 1\}]^p.$$

Problem $\mathrm{P}^p_\phi(\theta_0)$ is solved using MATLAB [ZT1, G8]. In this simple case, the whole function $\varphi_p(\theta_0)$ is plotted in a brute force manner. See Figure 5.4.1. The least root of $\phi_p(\theta) = 0$, θ_0, is found to be 15/16. The optimal solution to $\mathrm{P}_\phi(15/16)$, $\mathbf{x} = (0.5, 0.125)$, is the optimal solution to the original problem. It is verified that all $\varphi_p(\theta_0)$'s are equal for all p. In general, the least root of $\phi_p(\theta) = 0$ can be found by applying the bisection method or the Newton method if the gradient is available.

The direct application of the primal-dual method [L1] would fail in the original setting of this example problem as its perturbation function is nonconvex [L6]. This example problem was solved in [L6] by a pth power Lagrangian method. In the pth power Lagrangian method of [L6], the value of p needs to be chosen sufficiently large in order to convexify the perturbation function. Although a theoretical lower bound can be derived for p [L9], [GY8], how large is large enough for p could be a thorny issue in computational implementation. This nonconvex problem is also solved in Section 5.2 using a nonlinear Lagrangian dual formulation. When adopting the nonlinear Lagrangian dual formulation in Section 5.2, the resulting auxiliary problem is of a minimax type and is generally hard to solve. Note that the successive global optimization method derived in this section works for any $p \geq 1$ when the parametric monotone composition formulation ϕ_p or ϕ'_p is used. In general, p is chosen in such a way that the subproblem $\mathrm{P}_\phi(\theta_0)$ can be easily solved or its global minimum can be found. The resulting auxiliary problem $\mathrm{P}_\phi(\theta_0)$ is also a differentiable optimization problem when all f_j's are differentiable. It is worth noting that the parametric approach developed here and in [GY1] involves a single parameter, although the subproblem in [GY1], like the one in [GY2], is of the minimax type. Thus, the successive global optimization method via the parametric monotone composition formulation developed in this section has some obvious advantages over the global optimization approach proposed in Section 5.2.

To conclude this section, we show that the proposed parametric monotone composition approach is also applicable to searching for strict local minima. This concern is of some practical significance, since in many situations, only a local optimality of $\mathrm{P}_\phi(\theta_0^*)$ can be guaranteed.

Theorem 5.4.7 Consider the problem P. If $\mathbf{x}^0$ is a strict local minimum of P, then $\mathbf{x}^0$ is a strict local minimum of the problem $\mathrm{P}_\phi(\theta_0^*)$ with $\Phi(\mathbf{x}^0, \theta_0^*) = 0$, where $\theta_0^* = f_0(\mathbf{x}^0)$.

Proof: Assume that $\mathbf{x}^0$ is a strict local minimum of P and $\theta_0^* = f_0(\mathbf{x}^0)$. It is clear that

$$\Phi(\mathbf{x}^0, \theta_0^*) = \phi\left(\frac{f_0(\mathbf{x}^0)}{\theta_0^*} - 1, f_1(\mathbf{x}^0), \cdots, f_m(\mathbf{x}^0)\right) = \max\{0, \frac{f_0(\mathbf{x}^0)}{\theta_0^*} - 1\} = 0.$$

There is a neighborhood $\mathcal{N}_1(\mathbf{x}^0)$ of $\mathbf{x}^0$ such that for any $\mathbf{x} \in \mathcal{X}_0 \cap \mathcal{N}_1(\mathbf{x}^0)$ and $\mathbf{x} \neq \mathbf{x}^0$, $f_0(\mathbf{x}) > f_0(\mathbf{x}^0)$. We will show that for any $\mathbf{x} \in \mathcal{X} \cap \mathcal{N}_1(\mathbf{x}^0)$ and $\mathbf{x} \neq \mathbf{x}^0$, $\Phi(\mathbf{x}, \theta_0^*) > 0$.

In fact, if a point $\mathbf{x}$ satisfying $\mathbf{x} \in \mathcal{X} \cap \mathcal{N}_1(\mathbf{x}^0)$ and $\mathbf{x} \neq \mathbf{x}^0$ is infeasible for P, then from Property (A) the following holds for some $1 \leq i \leq m$

$$\phi\left(\frac{f_0(\mathbf{x})}{\theta_0^*} - 1, f_1(\mathbf{x}), \cdots, f_m(\mathbf{x})\right) \geq f_i(\mathbf{x}) > 0,$$

If a point $\mathbf{x}$ satisfying $\mathbf{x} \in \mathcal{X} \cap \mathcal{N}_1(\mathbf{x}^0)$ and $\mathbf{x} \neq \mathbf{x}^0$ is feasible for P, i.e., $\mathbf{x} \in \mathcal{X}_0 \cap \mathcal{N}_1(\mathbf{x}^0)$ and $\mathbf{x} \neq \mathbf{x}^0$, then $f_0(\mathbf{x}) > f_0(\mathbf{x}^0) = \theta_0^*$. Thus from Property (A)

$$\phi\left(\frac{f_0(\mathbf{x})}{\theta_0^*} - 1, f_1(\mathbf{x}), \cdots, f_m(\mathbf{x})\right) \geq \frac{f_0(\mathbf{x})}{\theta_0^*} - 1 > 0.$$

Thus $\mathbf{x}^0$ is a strict local minimum of the problem $\mathrm{P}_\phi(\theta_0^*)$. ∎

Theorem 5.4.8 Consider the problem P. Let $\theta_0^* > 0$. If $\mathbf{x}^0$ is a strict local minimum of the problem $\mathrm{P}_\phi(\theta_0^*)$ and $\Phi(\mathbf{x}^0, \theta_0^*) = 0$, then $\mathbf{x}^0$ is a strict local minimum of P.

Proof: Since $\Phi(\mathbf{x}^0, \theta_0^*) = 0$, from Lemma 5.4.1, $\mathbf{x}^0$ is feasible. It is clear that

$$\Phi(\mathbf{x}^0, \theta_0^*) = \max\{0, \frac{f_0(\mathbf{x})}{\theta_0^*} - 1\} = 0.$$

Thus $\theta_0^* \geq f_0(\mathbf{x}^0)$. Assume that there is a neighborhood $N_2(\mathbf{x}^0)$ of $\mathbf{x}^0$ such that for any $\mathbf{x} \in \mathcal{X} \cap N_2(\mathbf{x}^0)$ and $\mathbf{x} \neq \mathbf{x}^0$,

$$\Phi(\mathbf{x}, \theta_0^*) > \Phi(\mathbf{x}^0, \theta_0^*) = 0. \tag{5.4.8}$$

Note that

$$\Phi(\mathbf{x}, \theta_0^*) = \max\{0, \frac{f_0(\mathbf{x})}{\theta_0^*} - 1\}, \quad \forall \mathbf{x} \in \mathcal{X}_0 \cap N_2(\mathbf{x}^0).$$

By the strict inequality (5.4.8),

$$\Phi(\mathbf{x}, \theta_0^*) = \frac{f_0(\mathbf{x})}{\theta_0^*} - 1 > 0 = \Phi(\mathbf{x}^0, \theta_0^*), \quad \forall \mathbf{x} \in \mathcal{X}_0 \cap N_2(\mathbf{x}^0), \mathbf{x} \neq \mathbf{x}^0.$$

Thus

$$f_0(\mathbf{x}) > \theta_0^* \geq f_0(\mathbf{x}^0), \quad \forall \mathbf{x} \in \mathcal{X}_0 \cap N_2(\mathbf{x}^0), \mathbf{x} \neq \mathbf{x}^0.$$

Then $\mathbf{x}^0$ is a strict local minimum of the problem P. ∎

The condition $\Phi(\mathbf{x}^0, \theta_0^*) = 0$ holds if $\mathbf{x}^0$ is feasible and $\theta_0^* = f_0(\mathbf{x}^0)$. The following result shows that a local minimum of P is actually a global minimum of $\mathrm{P}_\phi(\theta_0^*)$.

Theorem 5.4.9 Consider the problem P. If $\mathbf{x}^0$ is a local minimum of P and $\theta_0^* = f_0(\mathbf{x}^0)$, then $\mathbf{x}^0$ is a global minimum of the problem $P_\phi(\theta_0^*)$ with $\Phi(\mathbf{x}^0, \theta_0^*) = 0$.

Proof: Assume that $\mathbf{x}^0$ is a local minimum of P and $\theta_0^* = f_0(\mathbf{x}^0)$. It is clear that

$$\Phi(\mathbf{x}^0, \theta_0^*) = \phi\left(\frac{f_0(\mathbf{x}^0)}{\theta_0^*} - 1, f_1(\mathbf{x}^0), \cdots, f_m(\mathbf{x}^0)\right) = \max\{0, \frac{f_0(\mathbf{x}^0)}{\theta_0^*} - 1\} = 0.$$

Then from Lemma 5.4.1, $\mathbf{x}^0$ is a global minimum of the problem $P_\phi(\theta_0^*)$ with $\Phi(\mathbf{x}^0, \theta_0^*) = 0$. ∎

The following example shows that a local minimum for $P_\phi(\theta_0^*)$ may not be a local minimum for P if the local minimum of $P_\phi(\theta_0^*)$ is not a strict local minimum.

Example 5.4.3 Consider the optimization problem:

$$\begin{array}{rl} \inf & f_0(x) \\ \text{subject to} & x \in \mathcal{X}, \end{array}$$

where $\mathcal{X} = [0, \infty)$ and $f_0(x) = \cos(x)$, if $0 \le x \le 2\pi$ and $x - 2\pi + 1$, if $2\pi \le x$. Let $\theta_0 = 1$. Then $x^0 = 2\pi$ is a local minimum for $P_\phi(1)$:

$$\begin{array}{rl} \inf & \max\{0, f_0(x) - 1\}, \\ \text{subject to} & x \in \mathcal{X}, \end{array}$$

where $\max\{0, f_0(x) - 1\} = 0$, if $0 \le x \le 2\pi$ and $x - 2\pi$, if $2\pi \le x$. But $x^0 = 2\pi$ is not a local minimum for the original optimization problem.

5.5 Zero Duality Gap via an Alternate Lagrangian Function

We now present an alternate version of nonlinear Lagrangian duality based on a computable dual function. It has the advantage over the previous one in Section 5.2 in that it does not require the assumption of existence of an optimal solution to problem P_θ (as defined in Section 5.1). The main reference for this section is [GY2].

Definition 5.5.1 Let $\theta_0 = 0$. The alternate nonlinear Lagrangian function for problem P_θ (as defined in Section 5.1) is defined by, given $\mathbf{d} \in \mathcal{E}$ (as defined in (5.2.2)) and $\mathbf{x} \in \mathcal{X}$,

$$\mathcal{L}'(\mathbf{x}, \mathbf{d}) = \max_{0 \le i \le m} \left\{ \frac{f_i(\mathbf{x}) - \theta_i}{d_i} \right\}.$$

We now present a computable dual function based on the alternate nonlinear Lagrangian function. This leads to similar weak and strong duality results as in the previous section.

Definition 5.5.2 (An alternate dual function for Problem P_θ) Let $\theta_0 = 0$. The alternate dual function for P_θ is defined as:

$$\phi_1(\mathbf{d}) = \inf_{\mathbf{x}\in\mathcal{X}} \max_{0\leq i\leq m} \left\{ \frac{f_i(\mathbf{x}) - \theta_i}{d_i} \right\},$$

where $\mathbf{d} \in \mathcal{E}$.

Definition 5.5.3 (An alternate dual optimization problem to problem P_θ)

$$(\text{Problem } D_1) \qquad \sup_{\mathbf{d}\in\mathcal{E}} \phi_1(\mathbf{d}).$$

Theorem 5.5.1 (Weak duality) Let $\mathbf{x} \in \mathcal{X}_0$ and $\mathbf{d} \in \mathcal{E}$. Then

$$\phi_1(\mathbf{d}) \leq f_0(\mathbf{x}). \tag{5.5.1}$$

Proof: It follows from the feasibility of $\mathbf{x}$ that

$$f_0(\mathbf{x}) \geq \max_{0\leq i\leq m} \left\{ \frac{f_i(\mathbf{x}) - \theta_i}{d_i} \right\}.$$

By Definition 5.5.2, we have

$$\max_{0\leq i\leq m} \left\{ \frac{f_i(\mathbf{x}) - \theta_i}{d_i} \right\} \geq \phi_1(\mathbf{d}),$$

and the conclusion holds. ∎

Theorem 5.5.2 (Zero duality gap) Assume that $\mathcal{X}_0$ is compact and if $\mathcal{X}$ is unbounded then

$$\lim_{\substack{\mathbf{x}\to\infty \\ \mathbf{x}\in\mathcal{X}}} f_0(\mathbf{x}) = \infty.$$

Then,

$$\inf_{\mathbf{x}\in\mathcal{X}_0} f_0(\mathbf{x}) = \sup_{\mathbf{d}\in\mathcal{E}} \phi_1(\mathbf{d}). \tag{5.5.2}$$

Proof: If (5.5.2) does not hold, then by weak duality (5.5.1), there exists $\epsilon > 0$ such that

$$\inf_{\mathbf{x}\in\mathcal{X}_0} f_0(\mathbf{x}) - \epsilon \geq \phi_1(\mathbf{d}), \quad \forall \mathbf{d} \in \mathcal{E}. \tag{5.5.3}$$

For $\epsilon/5 > 0$, from the definition of infimum, there exists $\hat{\mathbf{x}} \in \mathcal{X}_0$ such that

$$f_0(\hat{\mathbf{x}}) \geq \inf_{\mathbf{x}\in\mathcal{X}_0} f_0(\mathbf{x}) \geq f_0(\hat{\mathbf{x}}) - \frac{\epsilon}{5} > f_0(\hat{\mathbf{x}}) - \epsilon.$$

It is clear that from (5.5.3)

$$f_0(\hat{\mathbf{x}}) - \epsilon \geq \inf_{\mathbf{x}\in\mathcal{X}_0} f_0(\mathbf{x}) - \epsilon \geq \phi_1(\mathbf{d}), \forall \mathbf{d} \in \mathcal{E}. \tag{5.5.4}$$

Then, by the assumption of the Theorem,

$$\begin{aligned}\phi_1(\mathbf{d}) &= \min\left\{\inf_{\mathbf{x}\in\mathcal{X}_0}\max_{0\leq i\leq m}\left\{\frac{f_i(\mathbf{x})-\theta_i}{d_i}\right\}, \inf_{\mathbf{x}\in\mathcal{X}\setminus\mathcal{X}_0}\max_{0\leq i\leq m}\left\{\frac{f_i(\mathbf{x})-\theta_i}{d_i}\right\}\right\} \\ &= \min\left\{\inf_{\mathbf{x}\in\mathcal{X}_0} f_0(\mathbf{x}), \inf_{\mathbf{x}\in\mathcal{X}\setminus\mathcal{X}_0}\max_{0\leq i\leq m}\left\{\frac{f_i(\mathbf{x})-\theta_i}{d_i}\right\}\right\} \\ &\geq \min\left\{f_0(\hat{\mathbf{x}}) - \frac{\epsilon}{5}, \inf_{\mathbf{x}\in\mathcal{X}\setminus\mathcal{X}_0}\max_{0\leq i\leq m}\left\{\frac{f_i(\mathbf{x})-\theta_i}{d_i}\right\}\right\}.\end{aligned} \tag{5.5.5}$$

From (5.5.4) and (5.5.5), we have

$$\phi_1(\mathbf{d}) \geq \inf_{\mathbf{x}\in\mathcal{X}\setminus\mathcal{X}_0}\max_{0\leq i\leq m}\left\{\frac{f_i(\mathbf{x})-\theta_i}{d_i}\right\}, \forall \mathbf{d} \in \mathcal{E}. \tag{5.5.6}$$

Since f_0 is continuous and $\mathcal{X}_0$ is compact, there exists $\delta > 0$ such that if $\mathbf{x} \in \mathcal{X} \setminus \mathcal{X}_0$ and $d(\mathbf{x}, \mathcal{X}_0) \leq \delta$, then

$$-\frac{\epsilon}{2} < f_0(\mathbf{x}) - f_0(\mathbf{x}^2) < \frac{\epsilon}{2}, \text{ for some } \mathbf{x}^2 \in \mathcal{X}_0.$$

Note that $f_0(\mathbf{x}^2) \geq \inf_{\mathbf{x}\in\mathcal{X}_0} f_0(\mathbf{x}) \geq f_0(\hat{\mathbf{x}}) - \epsilon/5$. Then

$$f_0(\hat{\mathbf{x}}) - \frac{4\epsilon}{5} < f_0(\mathbf{x}). \tag{5.5.7}$$

From (5.5.4), (5.5.6) and (5.5.7), we have

$$\phi_1(\mathbf{d}) \geq \inf_{\substack{\mathbf{x}\in\mathcal{X}\setminus\mathcal{X}_0, \\ d(\mathbf{x},\mathcal{X}_0)>\delta}}\max_{0\leq i\leq m}\left\{\frac{f_i(\mathbf{x})-\theta_i}{d_i}\right\}, \forall \mathbf{d} \in \mathcal{E}. \tag{5.5.8}$$

From the assumption, there exists $\Delta > 0$ such that

$$\phi_1(\mathbf{d}) \geq \inf_{\substack{\mathbf{x}\in\mathcal{X}\setminus\mathcal{X}_0, \\ d(\mathbf{x},\mathcal{X}_0)>\delta, \\ \|\mathbf{x}\|\leq\Delta}}\max_{0\leq i\leq m}\left\{\frac{f_i(\mathbf{x})-\theta_i}{d_i}\right\}, \forall \mathbf{d} \in \mathcal{E}.$$

It is clear that there exists $\beta > 0$ such that for any $\mathbf{x} \in \mathcal{X} \setminus \mathcal{X}_0$, $\|\mathbf{x}\| \leq \Delta$ and $d(\mathbf{x}, \mathcal{X}_0) > \delta$, $f_i(\mathbf{x}) - \theta_i > \beta$, for some i. Define

$$\mathcal{B}_i = \{\mathbf{x} \in \mathcal{X} \setminus \mathcal{X}_0 | d(\mathbf{x}, \mathcal{X}_0) > \delta, \|\mathbf{x}\| \leq \Delta, f_i(\mathbf{x}) - \theta_i > \beta\}, \quad i = 1, \cdots, m.$$

Let

$$\hat{d}_i = \begin{cases} \min_{\mathbf{x}\in\mathcal{B}_i} \left\{ \frac{f_i(\mathbf{x})-\theta_i}{f_0(\hat{\mathbf{x}})-\frac{\epsilon}{2}} \right\}, & \text{if } \mathcal{B}_i \neq \emptyset, \\ 1, & \text{if } \mathcal{B}_i = \emptyset. \end{cases}$$

It is clear that $\hat{\mathbf{d}} = (1, \hat{d}_1, \cdots, \hat{d}_m) \in \mathcal{E}$. Moreover,

$$\begin{aligned} \phi_1(\hat{\mathbf{d}}) &\geq \inf_{\substack{\mathbf{x}\in\mathcal{X}\setminus\mathcal{X}_0, \\ d(\mathbf{x},\mathcal{X}_0)>\delta, \\ \|\mathbf{x}\|\leq\Delta}} \max_{0\leq i\leq m} \left\{ \frac{f_i(\mathbf{x}) - \theta_i}{\hat{d}_i} \right\} \\ &\geq \max_{0\leq i\leq m} \frac{f_i(\mathbf{w}) - \theta_i}{\hat{d}_i} - \frac{\epsilon}{4}, \end{aligned}$$

where for some $\mathbf{w} \in \mathcal{X} \setminus \mathcal{X}_0, \|\mathbf{x}\| \leq \Delta, d(\mathbf{w}, \mathcal{X}_0) > \delta$. Now there exists $k_{\mathbf{w}}$ such that $\mathbf{w} \in \mathcal{B}_{k_{\mathbf{w}}}$ and

$$\max_{0\leq i\leq m} \frac{f_i(\mathbf{w}) - \theta_i}{\hat{d}_i} \geq \frac{f_{k_{\mathbf{w}}}(\mathbf{w}) - \theta_{k_{\mathbf{w}}}}{\hat{d}_{k_{\mathbf{w}}}}.$$

Thus

$$\phi_1(\hat{\mathbf{d}}) \geq \frac{f_{k_{\mathbf{w}}}(\mathbf{w}) - \theta_{k_{\mathbf{w}}}}{\hat{d}_{k_{\mathbf{w}}}} - \frac{\epsilon}{4},$$

and hence

$$\phi_1(\hat{\mathbf{d}}) \geq (f_{k_{\mathbf{w}}}(\mathbf{w}) - \theta_{k_{\mathbf{w}}}) \left(\min_{\mathbf{v}\in B_{k_{\mathbf{w}}}} \frac{f_{k_{\mathbf{w}}}(\mathbf{v}) - \theta_{k_{\mathbf{w}}}}{f_0(\hat{\mathbf{x}}) - \frac{\epsilon}{2}} \right)^{-1} - \frac{\epsilon}{4}.$$

Since $\mathbf{w} \in \mathcal{B}_{k_{\mathbf{w}}}$, we have

$$\phi_1(\hat{\mathbf{d}}) \geq f_0(\hat{\mathbf{x}}) - \frac{3\epsilon}{4}.$$

This contradicts (5.5.4) and the proof is complete. ∎

Corollary 5.5.3 If there exists $\mathbf{x}^0 \in \mathcal{X}_0$ and $\mathbf{d}^* \in \mathcal{E}$ such that

$$f_0(\mathbf{x}^0) = \phi_1(\mathbf{d}^*),$$

then $\mathbf{x}^0$ is an optimal solution for P_θ and $\mathbf{d}^*$ is an optimal solution for D_1.

Proof: This follows from Theorem 5.5.2. ∎

5.6 Zero Duality Gap for a Discrete Optimization Problem

Our focus on duality has been confined to the context of continuous optimization so far. Interestingly, the nonlinear Lagrangian approach is also applicable to a certain class of discrete optimization problems. Unlike the continuous case, the duality for discrete optimization problems is not as well developed, but there are some useful results in the literature, see [NW1] for example. In this section, we consider a variant of the problem P_0 (i.e., set $\boldsymbol{\theta} = \mathbf{0}$ in $P_{\boldsymbol{\theta}}$) where the variable space $\mathcal{X}$ is discrete. The main reference of this section is [YG2].

Consider the following discrete inequality constrained optimization problem P:

$$\text{(Problem P)} \qquad \inf_{\mathbf{x}\in\mathcal{X}} \; f_0(\mathbf{x})$$
$$\text{subject to} \quad f_i(\mathbf{x}) \leq 0, \quad i = 1, \cdots, m,$$

where $\mathcal{X} \subset \mathbb{R}^n$ is a discrete (finite or infinite) set, $f_i : \mathcal{X} \to \mathbb{R}$, $i = 0, 1, 2, \cdots, m$ are real-valued functions. Since $\mathcal{X}$ may have an infinite (but countable) number of elements, 'inf' is used instead of 'min'.

Let the set of feasible solutions of problem P be

$$\mathcal{X}_0 = \{\mathbf{x} \in \mathbb{R}^n \mid \mathbf{x} \in \mathcal{X}, f_i(\mathbf{x}) \leq 0, \quad i = 1, 2, \cdots, m\}$$

and

$$\mathcal{E} = \{\mathbf{e} = (1, e_1, \cdots, e_m)^\top \mid e_i > 0, i = 1, \cdots, m\}.$$

The set $\mathcal{E}$ will be the feasible set for the dual optimization problem to be constructed in Definition 5.6.3.

Definition 5.6.1 The nonlinear Lagrangian function of problem P is defined as

$$N(\mathbf{u}, \mathbf{e}) = \max_{0\leq i\leq m} \left\{ \frac{f_i(\mathbf{u})}{e_i} \right\},$$

where $\mathbf{e} \in \mathcal{E}, \mathbf{u} \in \mathcal{X}$.

Note that the function $N(\mathbf{u}, \mathbf{e})$ can be interpreted as a weighted Tchebyshev metric of the vector $(f_0(\mathbf{u}), f_1(\mathbf{u}), \cdots, f_m(\mathbf{u}))^\top$.

Definition 5.6.2 The nonlinear Lagrangian dual function for problem P is defined as follows

$$\phi(\mathbf{e}) = \inf_{\mathbf{u}\in\mathcal{X}} N(\mathbf{u}, \mathbf{e}),$$

where $\mathbf{e} \in \mathcal{E}$.

Theorem 5.6.1 The dual function $\phi(\mathbf{e})$ is nonincreasing, i.e., for $\mathbf{e}^1, \mathbf{e}^2 \in \mathcal{E}$, $\mathbf{e}^1 - \mathbf{e}^2 \in \mathbb{R}^{m+1}_+$, $\phi(\mathbf{e}^1) \leq \phi(\mathbf{e}^2)$.

Proof: Follows directly from the definition of ϕ. ∎

It is clear that computing the nonlinear dual function $\phi(\mathbf{e})$ is equivalent to solving a minimax discrete optimization problem or a convex composite discrete optimization problem. This, in general, is not an easy task. Thus the difficulty in the following dual optimization problem is to evaluate the dual function $\phi(\mathbf{e})$. However, the constraint has a simple structure.

Definition 5.6.3 The dual optimization problem D of problem P is defined as

$$\begin{array}{lrl} \text{(Problem D)} & \sup & \phi(\mathbf{e}) \\ & \text{subject to} & \mathbf{e} \in \mathcal{E}. \end{array}$$

Remark 5.6.1 Observe that while the primal problem P is a discrete optimization problem, the dual optimization problem D is continuous in nature. This peculiar feature is not without precedent, one other example is found in the max flow min cut duality of Section 2.4.

Theorem 5.6.2 (Weak duality) Assume that $f_0(\mathbf{x}) \geq 0, \forall \mathbf{x} \in \mathcal{X}_0$. Let $\mathbf{x} \in \mathcal{X}_0$ and $\mathbf{e} \in \mathcal{E}$. Then

$$f_0(\mathbf{x}) \geq \phi(\mathbf{e}). \tag{5.6.1}$$

Proof: It follows from the assumption and $\mathbf{x} \in \mathcal{X}_0$ that

$$f_0(\mathbf{x}) \geq \max_{0 \leq i \leq m} \left\{ \frac{f_i(\mathbf{x})}{e_i} \right\} = N(\mathbf{x}, \mathbf{e}).$$

Thus

$$f_0(\mathbf{x}) \geq N(\mathbf{x}, \mathbf{e}) \geq \inf_{\mathbf{u} \in \mathcal{X}} N(\mathbf{u}, \mathbf{e}) = \phi(\mathbf{e}).$$

Thus, (5.6.1) holds. ∎

Note that an equality constraint can be trivially converted into two inequality constraints. Thus our results are applicable to optimization problems with both inequality and equality constraints, see Example 5.6.1.

We now establish a property of zero duality gap for the present discrete optimization problem and their nonlinear Lagrangian dual problem. We consider both the cases where the discrete set $\mathcal{X}$ is a set of finite number of elements (Theorem 5.6.3), or a set of infinite number of elements (Theorem 5.6.4).

Theorem 5.6.3 (Zero duality gap) Assume that $\mathcal{X}$ is a discrete set with a finite number of elements and $\mathcal{X}_0 \neq \emptyset$. If $f_0(\mathbf{x}) \geq 0, \forall \mathbf{x} \in \mathcal{X}_0$, then

$$\min_{\mathbf{x} \in \mathcal{X}_0} f_0(\mathbf{x}) = \sup_{\mathbf{e} \in \mathcal{E}} \phi(\mathbf{e}). \tag{5.6.2}$$

Note that here 'min' is used since $\mathcal{X}$ is a discrete and finite set of elements.

Proof: If the equality does not hold, then by weak duality, there exists $\epsilon > 0$ such that

$$\min_{\mathbf{x} \in \mathcal{X}_0} f_0(\mathbf{x}) - \epsilon \geq \phi(\mathbf{e}), \quad \forall \mathbf{e} \in \mathcal{E}. \tag{5.6.3}$$

For $\epsilon/4 > 0$, there exists $\hat{\mathbf{x}} \in \mathcal{X}_0$ such that

$$f_0(\hat{\mathbf{x}}) \geq \min_{\mathbf{x}\in\mathcal{X}_0} f_0(\mathbf{x}) \geq f_0(\hat{\mathbf{x}}) - \frac{\epsilon}{4}. \tag{5.6.4}$$

It is clear from (5.6.2) and (5.6.3) that

$$f_0(\hat{\mathbf{x}}) - \epsilon \geq \min_{\mathbf{x}\in\mathcal{X}_0} f_0(\mathbf{x}) - \epsilon \geq \phi(\mathbf{e}), \forall \mathbf{e} \in \mathcal{E}. \tag{5.6.5}$$

Then, by the assumption and (5.6.4),

$$\begin{aligned}
\phi(\mathbf{e}) &= \min\left\{\min_{\mathbf{x}\in\mathcal{X}_0}\max_{0\leq i\leq m}\left\{\frac{f_i(\mathbf{x})}{e_i}\right\}, \min_{\mathbf{x}\in\mathcal{X}\setminus\mathcal{X}_0}\max_{0\leq i\leq m}\left\{\frac{f_i(\mathbf{x})}{e_i}\right\}\right\} \\
&= \min\left\{\min_{\mathbf{x}\in\mathcal{X}_0} f_0(\mathbf{x}), \min_{\mathbf{x}\in\mathcal{X}\setminus\mathcal{X}_0}\max_{0\leq i\leq m}\left\{\frac{f_i(\mathbf{x})}{e_i}\right\}\right\} \qquad (5.6.6)\\
&\geq \min\left\{f_0(\hat{\mathbf{x}}) - \frac{\epsilon}{4}, \min_{\mathbf{x}\in\mathcal{X}\setminus\mathcal{X}_0}\max_{0\leq i\leq m}\left\{\frac{f_i(\mathbf{x})}{e_i}\right\}\right\}.
\end{aligned}$$

From (5.6.5) and (5.6.6), we have

$$\phi(e) \geq \min_{\mathbf{x}\in\mathcal{X}\setminus\mathcal{X}_0}\max_{0\leq i\leq m}\left\{\frac{f_i(\mathbf{x})}{e_i}\right\}, \forall \mathbf{e} \in \mathcal{E}. \tag{5.6.7}$$

Let $i \in \{1, \cdots, m\}$ be fixed and $\gamma > 0$ be given. One of the the following two cases may happen.

Case I. There exists $\mathbf{x} \in \mathcal{X} \setminus \mathcal{X}_0$ such that $f_i(\mathbf{x}) > 0$. Since $\mathcal{X} \setminus \mathcal{X}_0$ has only a finite number of elements, there exists $\tau_i > 0$ such that $f_i(\mathbf{x}) > \tau_i$, for any $\mathbf{x} \in \mathcal{X} \setminus \mathcal{X}_0$ and $f_i(\mathbf{x})$ being positive. In this case, let e_i be such that

$$\frac{\tau_i}{\gamma} \geq e_i.$$

Then

$$\frac{f_i(\mathbf{x})}{e_i} \geq \gamma, \quad \forall \mathbf{x} \in \mathcal{X} \setminus \mathcal{X}_0 \text{ and } f_i(\mathbf{x}) \text{ being positive.}$$

Case II. For all $\mathbf{x} \in \mathcal{X} \setminus \mathcal{X}_0$, $f_i(\mathbf{x}) \leq 0$. Let $\mathbf{x} \in \mathcal{X} \setminus \mathcal{X}_0$. There always exists k such that $f_k(\mathbf{x}) > 0$, i.e., Case I holds. Thus

$$\max_{0\leq i\leq m}\left\{\frac{f_i(\mathbf{x})}{e_i}\right\} \geq \gamma, \quad \forall \mathbf{x} \in \mathcal{X} \setminus \mathcal{X}_0.$$

So

$$\min_{\mathbf{x}\in\mathcal{X}\setminus\mathcal{X}_0}\max_{0\leq i\leq m}\left\{\frac{f_i(\mathbf{x})}{e_i}\right\} \geq \gamma.$$

Note that γ remains arbitrary, so

$$\min_{\mathbf{x}\in\mathcal{X}\setminus\mathcal{X}_0}\max_{0\le i\le m}\left\{\frac{f_i(\mathbf{x})}{e_i}\right\}\to+\infty,\quad \text{as } \mathbf{e}\downarrow(1,\mathbf{0})^\top \text{ in } \mathcal{E},$$

which is a contradiction to (5.6.7). ∎

Next we consider the case when $\mathcal{X}$ is a discrete set but has an infinite number of elements. The proof of the next result is similar to that of Theorem 4.2 in [GY2] but with some modification to allow for the case when $\mathcal{X}$ is discrete but has an infinite number of elements. In [GY2], the case where $\mathcal{X}$ is a connected set was considered.

Theorem 5.6.4 (Zero duality gap) Assume that $\mathcal{X}$ is a discrete set, but has an infinite number of elements. If $\mathcal{X}$ is unbounded,

$$\lim_{\substack{\mathbf{x}\to\infty\\ \mathbf{x}\in\mathcal{X}}} f_0(\mathbf{x})=\infty.$$

Let f_0 be continuous and f_i, $i=1,\cdots,m$ be continuous functions defined on an open set containing $\mathcal{X}$. Then

$$\inf_{\mathbf{x}\in\mathcal{X}_0} f_0(\mathbf{x})=\sup_{\mathbf{e}\in\mathcal{E}}\phi(\mathbf{e}). \tag{5.6.8}$$

Proof: If (5.6.8) does not hold, then by weak duality (5.6.1) as in the first part of the proof for Theorem 5.6.3, there exist $\epsilon>0$ and $\hat{\mathbf{x}}\in\mathcal{X}_0$ such that

$$f_0(\hat{\mathbf{x}})\ge\inf_{\mathbf{x}\in\mathcal{X}_0} f_0(\mathbf{x})\ge f_0(\hat{\mathbf{x}})-\frac{\epsilon}{4}. \tag{5.6.9}$$

$$f_0(\hat{\mathbf{x}})-\epsilon\ge\inf_{\mathbf{x}\in\mathcal{X}_0} f_0(\mathbf{x})-\epsilon\ge\phi(\mathbf{e}),\forall\mathbf{e}\in\mathcal{E}, \tag{5.6.10}$$

and

$$\phi(\mathbf{e})\ge\inf_{\mathbf{x}\in\mathcal{X}\setminus\mathcal{X}_0}\max_{0\le i\le m}\left\{\frac{f_i(\mathbf{x})}{e_i}\right\},\forall\mathbf{e}\in\mathcal{E}. \tag{5.6.11}$$

It follows from the continuity of f_0 and the boundedness of $\mathcal{X}_0$ that there exists $\delta>0$ such that if $\mathbf{x}\in\mathcal{X}\setminus\mathcal{X}_0$ and $d(\mathbf{x},\mathcal{X}_0)\le\delta$, then

$$-\frac{\epsilon}{2}<f_0(\mathbf{x})-f_0(\mathbf{x}^2)<\frac{\epsilon}{2},\text{ for some }\mathbf{x}^2\in\mathcal{X}_0.$$

Note that $f_0(\mathbf{x}^2)\ge\inf_{\mathbf{x}\in\mathcal{X}_0}f_0(\mathbf{x})\ge f_0(\hat{\mathbf{x}})-\epsilon/4$. Then

$$f_0(\hat{\mathbf{x}})-\frac{3\epsilon}{4}<f_0(\mathbf{x}). \tag{5.6.12}$$

From (5.6.9)-(5.6.11), we have

$$\phi(\mathbf{e})\ge\inf_{\substack{\mathbf{x}\in\mathcal{X}\setminus\mathcal{X}_0,\\ d(\mathbf{x},\mathcal{X}_0)>\delta}}\max_{0\le i\le m}\left\{\frac{f_i(\mathbf{x})}{e_i}\right\},\forall\mathbf{e}\in\mathcal{E}. \tag{5.6.13}$$

By the assumption, there exists $\Delta > 0$ such that

$$\phi(\mathbf{e}) \geq \inf_{\substack{\mathbf{x}\in\mathcal{X}\setminus\mathcal{X}_0,\\ d(\mathbf{x},\mathcal{X}_0)>\delta,\\ \|\mathbf{x}\|\leq\Delta}} \max_{0\leq i\leq m} \left\{\frac{f_i(\mathbf{x})}{e_i}\right\}, \forall \mathbf{e}\in\mathcal{E}.$$

It is clear that there exists $\beta > 0$ such that for any $\mathbf{x} \in \mathcal{X} \setminus \mathcal{X}_0, \|\mathbf{x}\| \leq \Delta$ satisfying $d(\mathbf{x}, \mathcal{X}_0) > \delta$, $f_i(\mathbf{x}) > \beta$, for some i. Define

$$\mathcal{B}_i = \{\mathbf{x} \in \mathcal{X} \setminus \mathcal{X}_0, \|\mathbf{x}\| \leq \Delta : d(\mathbf{x}, \mathcal{X}_0) > \delta, f_i(\mathbf{x}) > \beta\}.$$

Let

$$\hat{e}_i = \begin{cases} \min_{\mathbf{x}\in\mathcal{B}_i}\left\{\frac{f_i(\mathbf{x})}{f_0(\hat{\mathbf{x}})-\frac{\epsilon}{2}}\right\}, & \text{if } \mathcal{B}_i \neq \emptyset,\\ 1, & \text{if } \mathcal{B}_i = \emptyset.\end{cases}$$

It is clear that $\hat{\mathbf{e}} = (1, \hat{\mathbf{e}}_1, \cdots, \hat{\mathbf{e}}_m) \in \mathcal{E}$. Moreover,

$$\begin{aligned}\phi(\hat{\mathbf{e}}) &\geq \inf_{\substack{\mathbf{x}\in\mathcal{X}\setminus\mathcal{X}_0,\\ d(\mathbf{x},\mathcal{X}_0)>\delta,\\ \|\mathbf{x}\|\leq\Delta}} \max_{0\leq i\leq m} \left\{\frac{f_i(\mathbf{x})}{\hat{\mathbf{e}}_i}\right\}\\ &\geq \max_{0\leq i\leq m} \frac{f_i(\mathbf{w})}{\hat{e}_i} - \frac{\epsilon}{3},\end{aligned}$$

where for some $\mathbf{w} \in \mathcal{X} \setminus \mathcal{X}_0, \|\mathbf{x}\| \leq \Delta, d(\mathbf{w}, \mathcal{X}_0) > \delta$ from the definition of 'inf'. Now there exists $k_{\mathbf{w}}$ such that $\mathbf{w} \in \mathcal{B}_{k_{\mathbf{w}}}$ and

$$\max_{0\leq i\leq m} \frac{f_i(\mathbf{w})}{\hat{\mathbf{e}}_i} \geq \frac{f_{k_{\mathbf{w}}}(\mathbf{w})}{\hat{\mathbf{e}}_{k_w}}.$$

So

$$\phi(\hat{\mathbf{e}}) \geq \frac{f_{k_{\mathbf{w}}}(\mathbf{w})}{\hat{e}_{k_{\mathbf{w}}}} - \frac{\epsilon}{3},$$

and hence

$$\phi(\hat{\mathbf{e}}) \geq f_{k_{\mathbf{w}}}(\mathbf{w}) \left(\min_{\mathbf{v}\in\mathcal{B}_{k_{\mathbf{w}}}} \frac{f_{k_{\mathbf{w}}}(\mathbf{v})}{f_0(\hat{\mathbf{x}}) - \frac{\epsilon}{2}}\right)^{-1} - \frac{\epsilon}{3}.$$

Since $\mathbf{w} \in \mathcal{B}_{k_{\mathbf{w}}}$, we have

$$\phi(\hat{\mathbf{e}}) \geq f_0(\hat{\mathbf{x}}) - \frac{5\epsilon}{6} > f_0(\hat{\mathbf{x}}) - \epsilon.$$

This contradicts (5.6.10) and the proof is complete. ∎

Corollary 5.6.5 (Zero duality gap) Assume that $\mathcal{X}$ is a discrete and bounded set, but has an infinite number of elements. Let f_i, $i = 0, 1, \cdots, m$ be continuous functions defined on an open set containing $\mathcal{X}$. Then

$$\inf_{\mathbf{x}\in\mathcal{X}_0} f_0(\mathbf{x}) = \sup_{\mathbf{e}\in\mathcal{E}} \phi(\mathbf{e}).$$

Proof: Clearly f_0 is uniformly continuous on an open set containing $\mathcal{X}$. The result follows from Theorem 5.6.4. ∎

Example 5.6.1 Consider Example 5.1.2 again. We apply the nonlinear Lagrangian dual formulation to the problem, and calculate the dual cost. To do this, we rewrite the optimization problem as

$$\begin{aligned} \min \quad & f_0(\mathbf{x}) \\ \text{subject to} \quad & f_1(\mathbf{x}) \le 0, \\ & f_2(\mathbf{x}) \le 0, \\ & \mathbf{x} = (x_1, x_2) \in \mathcal{X}, \end{aligned}$$

where $\mathcal{X} = \{(0,0), (0,4), (4,4), (4,0), (1,2), (2,1)\}$, $f_0(\mathbf{x}) = -2x_1 + x_2 + 4$, $f_1(\mathbf{x}) = x_1 + x_2 - 3$, $f_2(\mathbf{x}) = -x_1 - x_2 + 3$.

Then the dual function is

$$\phi(\mathbf{e}) = \min_{\mathbf{u}\in\mathcal{X}} \max_{0\le i\le 2} \left\{ \frac{f_i(\mathbf{u})}{e_i} \right\},$$

where $e_0 = 1, e_i > 0,\ i = 1, 2$. It is not difficult to verify that

$$\min_{\mathbf{x}\in\mathcal{X}_0} f_0(\mathbf{x}) = 1$$

and

$$\max_{\mathbf{e}\in\mathcal{E}} \phi(\mathbf{e}) = 1,$$

where $e_0 = 1$. Hence there exists a zero duality gap when the nonlinear Lagrangian dual formulation is applied.

5.7 Zero Duality Gap via an Extended Lagrangian Function

In this section, we relate the concept of a nonlinear Lagrangian to that of penalty functions. To do so, the Lagrangian function has to be modified by a suitable composition with an increasing positively homogeneous function. The main references of this section are [RGY1], [RGY2].

Consider the following inequality constrained optimization problem:

$$\text{(Problem P)} \qquad \inf f_0(\mathbf{x}) \quad \text{subject to } \mathbf{x} \in \mathcal{X},\ f_i(\mathbf{x}) \le 0,\ i \in I,$$

where $\mathcal{X} \subseteq \mathbb{R}^n$, $I = \{1, \ldots, m\}$, and $f_i : \mathcal{X} \to \mathbb{R}$, $i \in \{0\} \cup I$ are continuous functions.

Let $\mathcal{Y} \subseteq \mathbb{R}^{1+m}$. A function $p : \mathcal{Y} \to \bar{\mathbb{R}}$ is said to be *monotone* if $\mathbf{y}^1 - \mathbf{y}^2 \in \mathbb{R}^{1+m}_+$ implies $p(\mathbf{y}^1) \ge p(\mathbf{y}^2)$. In this section we shall consider continuous monotone functions p that are defined either on the space $\mathbb{R}^{1+m}$ or on the cone $\mathbb{R}^{1+m}_+$ and which enjoy the following properties:

Property (A) There exist positive numbers $a_o, a_i, \cdots a_m$ with $a_o = 1$ such that for $\bar{\mathbf{y}} = (y_0, y_1, \cdots, y_m)^\top$ belonging to the domain of p with $y_0 \in \mathbb{R}_+$ we have:

$$p(\bar{\mathbf{y}}) \ge \max_{1 \le i \le m} a_i y_i. \tag{5.7.1}$$

Property (B) For $y_0 \in \mathbb{R}_+$,

$$p(y_0, 0, \cdots, 0) = y_0. \tag{5.7.2}$$

We shall now present some examples. First we consider functions defined on the entire space $\mathbb{R}^{1+m}$.

Example 5.7.1 The function

$$p(y_0, \mathbf{y}) = \max\,(y_0, a_1 y_1, \cdots, a_m y_m)$$

with positive $a_1, \cdots, a_m$ is a monotone function with Properties (A) and (B). Let $\psi(y_0, \mathbf{y})$ be an arbitrary monotone function defined on $\mathbb{R}^{1+m}$ such that $\psi(y_0, 0) = y_0$. Then the function

$$p(y_0, \mathbf{y}) = \max\,(\psi(y_0, y), a_1 y_1, \cdots, a_m y_m)$$

is monotone and possesses Properties (A) and (B). In particular the function

$$p(y_0, \mathbf{y}) = \max\,(y_0 + c_1 y_1 + \cdots + c_m y_m, a_1 y_1, \cdots, a_m y_m)$$

with nonnegative c_i and positive a_i enjoys these properties.

We now consider an example of monotone functions with Properties (A) and (B) defined on the cone $\mathbb{R}^{1+m}_+$.

Example 5.7.2 Let $\bar{\mathbf{y}} = (y_0, \mathbf{y}) \in \mathbb{R}^{1+m}_+$ with $\mathbf{y} = (y_1, \cdots, y_m)^\top$. The following monotone functions have Properties (A) and (B):

$$p_k(\bar{\mathbf{y}}) = \left(\sum_{i=o}^{m} y_i^k \right)^{\frac{1}{k}}, \quad 0 < k < +\infty, \tag{5.7.3}$$

$$p_\infty(\bar{\mathbf{y}}) = \max_{0 \le i \le m} y_i, \tag{5.7.4}$$

$$p(\bar{\mathbf{y}}) = p_\infty(\bar{\mathbf{y}}) + \prod_{i=0}^{m} y_i. \tag{5.7.5}$$

It is easy to see that any convex combination of monotone functions with Properties (A) and (B) also enjoys these properties.

Remark 5.7.1 Note that the function p_k as defined in (5.7.3) is not monotone on $\mathbb{R}^{1+m}$ (although it is monotone on the cone $\mathbb{R}_+^{1+m}$). For this reason, we will consider the modified Lagrangian function using the monotone functions defined on $\mathbb{R}^{1+m}$ and the modified penalty function using the monotone function defined on $\mathbb{R}_+^{1+m}$ respectively.

Define the set of all feasible solutions for Problem P to be

$$\mathcal{X}_0 = \{\mathbf{x} \in \mathcal{X} : f_i(\mathbf{x}) \le 0, i \in I\}.$$

Let us consider the vector-valued function $\mathbf{F} : \mathcal{X} \times \mathbb{R}^{1+m} \to \mathbb{R}^{1+m}$ defined as follows, for $\mathbf{x} \in \mathcal{X}$, $\mathbf{t} \in \mathbb{R}^{1+m}$:

$$\mathbf{F}(\mathbf{x}, \mathbf{t}) = (t_0 f_0(\mathbf{x}), t_1 f_1(\mathbf{x}), \cdots, t_m f_m(\mathbf{x})).$$

Let $p : \mathbb{R}^{1+m} \to \bar{\mathbb{R}}$ be a monotone function with Properties (A) and (B). We now define a modified Lagrangian function, dual function and dual problem, corresponding to p for the problem P.

Definition 5.7.1 (Modified Lagrangian function for P) The modified Lagrangian function corresponding to p for the problem P is defined by,

$$\mathcal{L}(\mathbf{x}, d) = p(\mathbf{F}(\mathbf{x}, d)), \qquad \mathbf{x} \in \mathcal{X}, \quad \mathbf{t} \in \mathcal{T} = \{\mathbf{t} \in \mathbb{R}^{1+m} \mid t_0 = 1, t_i \ge 0, i \in I\}.$$

Definition 5.7.2 (The dual function for P) The dual function for P corresponding to p is defined by,

$$q(\mathbf{t}) = \inf_{\mathbf{x} \in \mathcal{X}} \mathcal{L}(\mathbf{x}, \mathbf{t}), \quad \mathbf{t} \in \mathcal{T}.$$

Definition 5.7.3 (The dual problem to P) The following problem:

$$\text{(Problem D)} \qquad \sup q(\mathbf{t}) \quad \text{subject to } \mathbf{t} \in \mathcal{T}$$

is called the dual problem to the problem P corresponding to the function p.

Note that the constraints of the above dual problem have a simple structure.

Remark 5.7.2 Note that $\mathcal{L}(\mathbf{x}, \mathbf{t})$ is an *increasing positively homogeneous (IPH) function*. The theory of IPH functions will be employed in our development. This is the main reason why multipliers t_i is used here, as opposed to the use of divisors e_i in

Section 5.2. In Section 5.2, we are more concerned with the geometric interpretation of the supporting cone. Here we are more concerned with the properties of IPH functions.

Remark 5.7.3 Let $p : \mathbb{R}^{1+m} \to \bar{\mathbb{R}}$ be a monotone function with Properties (A) and (B). If p is concave on $\mathbb{R}^{1+m}$ then the dual function q is also concave. Thus the dual problem involves the maximization of a concave function over a convex set.

Let M_P be the optimal value of the primal problem P and M_D be the optimal value of the dual problem D, that is

$$M_P = \inf\{f_o(\mathbf{x}) : \mathbf{x} \in \mathcal{X}_0\}, \quad M_D = \sup\{q(\mathbf{t}) : \mathbf{t} \in \mathcal{T}\}.$$

Properties (A) and (B) of a monotone function p allow us to establish the following result.

Lemma 5.7.1 Let p be a monotone function with Properties (A) and (B). Then $p(\mathbf{F}(\mathbf{x}, \mathbf{t})) = f_0(\mathbf{x})$ for all $\mathbf{x} \in \mathcal{X}_0$ and $\mathbf{t} \in \mathcal{T}$.

Proof: Since $\mathbf{x} \in \mathcal{X}_0$ it follows that $f_i(\mathbf{x}) \le 0$ $(i \in I)$. Applying the monotonicity of p and Property (B) we conclude that

$$p(f_0(\mathbf{x}), t_1 f_1(\mathbf{x}), \cdots, t_m f_m(\mathbf{x})) \le p(f_0(\mathbf{x}), 0, \cdots, 0) = f_0(\mathbf{x}).$$

Also Property (A) shows that for some numbers $a_i > 0$, $(i \in I)$

$$\begin{aligned} p(\mathbf{F}(\mathbf{x}, \mathbf{t})) &= p(f_0(\mathbf{x}), t_1 f_1(\mathbf{x}), \cdots, t_m f_m(\mathbf{x})) \\ &\ge \max\{f_0(\mathbf{x}), a_1 t_1 f_1(\mathbf{x}), \cdots, a_m t_m f_m(\mathbf{x})\} \\ &= f_0(\mathbf{x}). \end{aligned}$$

Thus the desired result follows. ∎

We now establish the weak duality result and the zero duality gap result for the optimization problem P and its modified Lagrangian dual problem D.

Lemma 5.7.2 (Weak duality) $M_P \ge M_D$.

Proof: Let $\mathbf{x} \in \mathcal{X}_0$ and $\mathbf{t} \in \mathbb{R}^{1+m}, t_0 = 1$. It follows from Lemma 5.7.1 that $p(\mathbf{F}(\mathbf{x}, d)) = f_0(\mathbf{x})$. Therefore $f_0(\mathbf{x}) \ge \inf_{\mathbf{x}' \in \mathcal{X}} p(\mathbf{F}(\mathbf{x}', \mathbf{t})) = q(\mathbf{t})$. Thus

$$M_D = \sup_{\mathbf{t} \in \mathcal{T}} q(\mathbf{t}) \le \inf_{\mathbf{x} \in \mathcal{X}_0} f_0(\mathbf{x}) = M_P.$$

∎

Theorem 5.7.3 (Zero duality gap) Assume that $\mathcal{X}_0$ is a compact set. Furthermore if $\mathcal{X}$ is unbounded, then we assume also that the function f_0 is such that:

$$\lim_{\substack{\mathbf{x} \to \infty, \\ \mathbf{x} \in \mathcal{X}}} f_0(\mathbf{x}) = \infty.$$

Then $M_P = M_D$.

Proof: It follows from Lemma 5.7.2 that $M_P \geq M_D$. Assume $M_P > M_D$. Then there exist $\epsilon > 0$ and $\tilde{\mathbf{x}} \in \mathcal{X}_0$ such that

$$M_P - 3\epsilon > q(\mathbf{t}) \quad \text{for all } \mathbf{t} \in \mathcal{T} \tag{5.7.6}$$

and

$$f_0(\tilde{\mathbf{x}}) \geq M_P > f_0(\tilde{\mathbf{x}}) - 2\epsilon. \tag{5.7.7}$$

Assume that ϵ is sufficiently small so that $f_0(\tilde{\mathbf{x}}) - 2\epsilon > 0$. It follows from (5.7.6) and (5.7.7) that

$$f_0(\tilde{\mathbf{x}}) - 3\epsilon \geq M_P - 3\epsilon > q(\mathbf{t}) \text{ for all } \mathbf{t} \in \mathcal{T}. \tag{5.7.8}$$

Applying Lemma 5.7.1 and (5.7.7) we deduce that

$$\begin{aligned} q(\mathbf{t}) &= \min\{\inf_{\mathbf{x}\in\mathcal{X}_0} p(\mathbf{F}(\mathbf{x},\mathbf{t})), \inf_{\mathbf{x}\in\overline{\mathcal{X}\setminus\mathcal{X}_0}} p(\mathbf{F}(\mathbf{x},\mathbf{t}))\} \\ &= \min\{M_P, \inf_{\mathbf{x}\in\overline{\mathcal{X}\setminus\mathcal{X}_0}} p(\mathbf{F}(\mathbf{x},\mathbf{t}))\} \\ &\geq \min\{f_0(\tilde{\mathbf{x}}) - 2\epsilon, \inf_{\mathbf{x}\in\overline{\mathcal{X}\setminus\mathcal{X}_0}} p(\mathbf{F}(\mathbf{x},\mathbf{t}))\} \\ &\geq \min\{f_0(\tilde{\mathbf{x}}) - 3\epsilon, \inf_{\mathbf{x}\in\overline{\mathcal{X}\setminus\mathcal{X}_0}} p(\mathbf{F}(\mathbf{x},\mathbf{t}))\}, \end{aligned}$$

where $\overline{\mathcal{X}\setminus\mathcal{X}_0}$ is the closure of the set $\mathcal{X}\setminus\mathcal{X}_0$. Combining this inequality with (5.7.8), we obtain

$$q(\mathbf{t}) \geq \inf_{\mathbf{x}\in\overline{\mathcal{X}\setminus\mathcal{X}_0}} p(\mathbf{F}(\mathbf{x},\mathbf{t})) \quad \text{for all } \mathbf{t} \in \mathcal{T}. \tag{5.7.9}$$

Since f_0 is continuous and $\mathcal{X}_0$ is compact it follows that we can find $\delta > 0$ such that for each $\mathbf{x} \in \overline{\mathcal{X}\setminus\mathcal{X}_0}$ with the property $\rho(\mathbf{x}, \mathcal{X}_0) = \min\{\|\mathbf{x} - \mathbf{x}^0\| : \mathbf{x}^0 \in \mathcal{X}_0\} \leq \delta$, the inequality

$$f_0(\mathbf{x}') < f_0(\mathbf{x}) + \epsilon$$

holds for some $\mathbf{x}' \in \mathcal{X}_0$. It follows from (5.7.7) that $f_0(\mathbf{x}') \geq M_P \geq f_0(\tilde{\mathbf{x}}) - 2\epsilon$. So

$$f_0(\tilde{\mathbf{x}}) - 3\epsilon < f_0(\mathbf{x}) \text{ for all } \mathbf{x} \in \mathbb{R}^{1+m} \setminus \mathcal{X}_0,\ \rho(\mathbf{x}, \mathcal{X}_0) \leq \delta. \tag{5.7.10}$$

Let $\mathcal{X}^\delta = \{\mathbf{x} \in \overline{\mathcal{X}\setminus\mathcal{X}_0} : \rho(\mathbf{x}, \mathcal{X}_0) \geq \delta\}$ and

$$M_1(\mathbf{t}) = \inf_{\mathbf{x}\in\overline{\mathcal{X}\setminus\mathcal{X}_0},\ \rho(\mathbf{x},\mathcal{X}_0)\leq\delta} p(\mathbf{F}(\mathbf{x},\mathbf{t})),$$

$$M_2(\mathbf{t}) = \inf_{\mathbf{x}\in\overline{\mathcal{X}\setminus\mathcal{X}_0},\ \rho(\mathbf{x},\mathcal{X}_0)>\delta} p(\mathbf{F}(\mathbf{x},\mathbf{t})) = \inf_{\mathbf{x}\in\mathcal{X}^\delta} p(\mathbf{F}(\mathbf{x},\mathbf{t})).$$

It follows from (5.7.7) that $M_1(\mathbf{t}) \geq f_0(\tilde{\mathbf{x}}) - 3\epsilon$. So, by applying (5.7.9) we have for all $\mathbf{t} \in \mathcal{T}$:

$$q(\mathbf{t}) = \min\{M_1(\mathbf{t}), M_2(\mathbf{t})\} \geq \min\{f_0(\tilde{\mathbf{x}}) - 3\epsilon, M_2(\mathbf{t})\}. \tag{5.7.11}$$

Combining this inequality with (5.7.8), we have,

$$q(\mathbf{t}) \geq M_2(\mathbf{t}) = \inf_{\mathbf{x}\in\mathcal{X}^\delta} p(\mathbf{F}(\mathbf{x},\mathbf{t})) \text{ for all } \mathbf{t} \in \mathcal{T}.$$

It follows from Property (A) for p that there exist numbers $a_i > 0$, $(i = 0, 1\cdots, m)$ with $a_0 = 1$ such that for $\mathbf{t} \in \mathcal{T}$, we have:

$$p(\mathbf{F}(\mathbf{x},\mathbf{t})) = p(f_0(\mathbf{x}), t_1 f_1(\mathbf{x}), \cdots, t_m f_m(\mathbf{x})) \geq \max_{0\leq i\leq m} a_i t_i f_i(\mathbf{x}) \geq f_0(\mathbf{x})$$

Since $\lim_{\substack{\mathbf{x}\in\mathcal{X},\\ \mathbf{x}\to\infty}} f_0(\mathbf{x}) = +\infty$, it follows that there exists a number $\Delta > 0$ such that:

$$q(\mathbf{t}) \geq \inf_{\mathbf{x}\in\mathcal{X}^\delta} p(\mathbf{F}(\mathbf{x},\mathbf{t})) = \inf_{\mathbf{x}\in\mathcal{X}^\delta, \|\mathbf{x}\|\leq\Delta} p(\mathbf{F}(\mathbf{x},\mathbf{t})) \text{ for all } \mathbf{t} \in \mathcal{T}. \tag{5.7.12}$$

Let $g(\mathbf{x}) = \max_{1\leq i\leq m} f_i(\mathbf{x})$. It follows directly from the definition of the set $\mathcal{X}_0$ that $\mathcal{X}_0 = \{\mathbf{x} \in \mathcal{X} : g(\mathbf{x}) \leq 0\}$. Since the set $\{\mathbf{x} \in \mathcal{X}^\delta : \|\mathbf{x}\| \leq \Delta\}$ is compact and the function g is continuous we conclude that:

$$\min\{g(\mathbf{x}) : \ \mathbf{x} \in \mathcal{X}^\delta, \|\mathbf{x}\| \leq \Delta\} = 2\beta > 0.$$

Thus for any $\mathbf{x} \in \mathcal{X}^\delta$, $\|\mathbf{x}\| \leq \Delta$ there exist some i such that $\mathbf{x} \in \mathcal{B}_i$ where

$$\mathcal{B}_i = \{\mathbf{x} \in \mathcal{X}^\delta, \|\mathbf{x}\| \leq \Delta : f_i(\mathbf{x}) > \beta\}.$$

Let $\tilde{t}_0 = 1$ and for $i \in I$:

$$\tilde{t}_i = \begin{cases} \sigma_i, & \text{if } \mathcal{B}_i \neq \emptyset, \\ 1, & \text{if } \mathcal{B}_i = \emptyset \end{cases}$$

where

$$\sigma_i = \max_{\mathbf{x}\in\mathcal{B}_i} \frac{f_0(\tilde{\mathbf{x}}) - 2\epsilon}{a_i f_i(\mathbf{x})}.$$

Recall that ϵ have been chosen sufficiently small so that $f(\tilde{\mathbf{x}}) - 2\epsilon > 0$. So $\tilde{\mathbf{t}} = (\tilde{t}_0, \tilde{t}_1 \cdots, \tilde{t}_m) \in \mathbb{R}^{1+m}$. Applying (5.7.12) and (5.7.1) we can find a vector $\mathbf{u} \in \mathcal{X}^\delta$, $\|\mathbf{u}\| \leq \Delta$ such that

$$q(\tilde{\mathbf{t}}) = \inf_{\mathbf{x}\in\mathcal{X}^\delta, \ \|\mathbf{x}\|\leq\Delta} p(\mathbf{F}(\mathbf{x},\tilde{\mathbf{t}})) \geq p(\mathbf{F}(\mathbf{u},\tilde{\mathbf{t}})) - \epsilon.$$

Let i_u be the index that achieves the max in (5.7.1) for a given $\mathbf{u}$. There exists an index i_u such that $\mathbf{u} \in \mathcal{B}_{i_u}$. It follows from (5.7.1) that

$$p(\mathbf{F}(\mathbf{u},\tilde{\mathbf{t}})) \geq a_{i_u} \tilde{t}_{i_u} f_{i_u}(\mathbf{u}).$$

Thus

$$q(\tilde{\mathbf{t}}) \geq a_{i_u} \tilde{t}_{i_u} f_{i_u}(\mathbf{u}) - \epsilon = a_{i_u} f_{i_u}(\mathbf{u}) \max_{\mathbf{x}\in\mathcal{B}_{i_u}} \frac{f_0(\tilde{\mathbf{x}}) - 2\epsilon}{a_{i_u} f_{i_u}(\mathbf{x})} - \epsilon \geq f_0(\tilde{\mathbf{x}}) - 3\epsilon.$$

This contradicts (5.7.8) and the proof is complete. ∎

Remark 5.7.4 Recall that due to Assumption 5.7.1, the function f_0 is positive on the set $\mathcal{X}$. This assumption plays a crucial role in establishing the zero duality gap property.

Next we consider a type of modified penalty functions. Consider the problem P. The set $\mathcal{X}_0$ of feasible solutions for P can be represented in the following form:

$$\mathcal{X}_0 = \{\mathbf{x} \in \mathcal{X} : f_i^+(\mathbf{x}) = 0, \, i \in I\},$$

where $f_i^+(\mathbf{x}) = \max\{f_i(\mathbf{x}), 0\}$. Thus the problem P is equivalent to the following problem P^+:

$$(\text{Problem } \mathrm{P}^+) \qquad \inf f_0(\mathbf{x}) \qquad \text{subject to } \mathbf{x} \in \mathcal{X}, \; f_i^+(\mathbf{x}) = 0, \quad i \in I. \qquad (5.7.13)$$

Due to Assumption 5.7.1 we have $f_0^+(\mathbf{x}) = f_0(\mathbf{x})$. Since $f_i^+(\mathbf{x}) \geq 0$ for all $i = 0, 1, \cdots, m,$ dual functions and dual problems can be constructed by using the monotone functions p defined only on the cone $\mathbb{R}_+^{1+m}$. Let

$$\mathbf{F}^+(\mathbf{x}, \mathbf{t}) = (t_0 f_0(\mathbf{x}), t_1 f_1^+(\mathbf{x}), \cdots, t_m f_m^+(\mathbf{x})), \qquad \mathbf{x} \in \mathcal{X}, \; \mathbf{t} \in \mathcal{T}.$$

In this section, let p be defined on $\mathbb{R}_+^{1+m}$ with Properties (A) and (B). The dual function $\mathcal{L}^+$ for the problem P^+, corresponding to the monotone function p, has the following form:

$$\mathcal{L}^+(\mathbf{x}, \mathbf{t}) = p(\mathbf{F}^+(\mathbf{x}, \mathbf{t})), \qquad \mathbf{x} \in \mathcal{X}, \; \mathbf{t} \in \mathcal{T}. \qquad (5.7.14)$$

In particular if $p(\mathbf{y}) = \sum_{i=0}^m y_i$ then

$$\mathcal{L}^+(\mathbf{x}, \mathbf{t}) = t_0 f_0(\mathbf{x}) + \sum_{i=1}^m t_i \max\{f_i(\mathbf{x}), 0\}. \qquad (5.7.15)$$

We can consider the function $\mathcal{L}^+$ defined by (5.7.14) as a penalty function for problem P. Thus the function $\mathcal{L}^+$ for the problem P^+ corresponding to the monotone function p can be considered as a *modified penalty function* for the problem P, corresponding to p.

The dual function q^+ corresponding to p can be represented as follows:

$$q^+(\mathbf{t}) = \inf_{\mathbf{x} \in \mathcal{X}} \mathcal{L}^+(\mathbf{x}, \mathbf{t}) \equiv \inf_{\mathbf{x} \in \mathcal{X}} p(t_0 f_0(\mathbf{x}), t_1 f_1^+(\mathbf{x}), \cdots, t_m f_m^+(\mathbf{x})). \qquad (5.7.16)$$

The dual problem D^+ for the problem P^+ has the following form:

$$(\text{Problem } \mathrm{D}^+) \qquad \sup q^+(\mathbf{t}) \qquad \text{subject to } \mathbf{t} \in \mathcal{T}$$

where q^+ is the dual function defined by (5.7.16). It is easy to check that Lemma 5.7.1, Lemma 5.7.2 (Weak duality) and Theorem 5.7.3 (Zero duality gap) hold for dual problem D^+ corresponding to the monotone function p with Properties (A) and (B) defined on $\mathbb{R}^{1+m}$.

We shall consider two examples to illustrate the concept.

Example 5.7.3 Consider the monotone function p_k defined on $\mathbb{R}_+^{1+m}$ for $0 < k < +\infty$:

$$p_k(\mathbf{y}) = \left(\sum_{i=0}^{m} y_i^k\right)^{\frac{1}{k}}, \qquad \mathbf{y} = (y_0, y_1, \cdots, y_m) \in \mathbb{R}_+^{1+m}. \tag{5.7.17}$$

The modified penalty function for the problem P can be constructed as follows:

$$\mathcal{L}_k^+(\mathbf{x}, \mathbf{t}) = \left(t_0 f_0^k(\mathbf{x}) + \sum_{i=1}^{m} t_i (f_i^+(\mathbf{x}))^k\right)^{\frac{1}{k}}, \quad \mathbf{x} \in \mathcal{X},\ \mathbf{t} \in \mathcal{T}$$

The dual function for Problem P^+ is now defined as:

$$q_k^+(\mathbf{t}) = \inf_{\mathbf{x}\in\mathcal{X}} \left(f_0^k(\mathbf{x}) + \sum_{i=1}^{m} t_i (f_i^+(\mathbf{x}))^k\right)^{\frac{1}{k}}.$$

The dual problem D_k^+ to problem P has the following form:

$$(\text{Problem } D_k^+) \qquad \sup q_k^+(\mathbf{t}) \text{ subject to } \mathbf{t} \in \mathcal{T}.$$

If the conditions of Theorem 5.7.3 hold, then

$$\inf_{\mathbf{x}\in\mathcal{X}_0} f_0(\mathbf{x}) = \sup_{\mathbf{t}\in\mathcal{T}} \inf_{\mathbf{x}\in\mathcal{X}} \left(f_0^k(\mathbf{x}) + \sum_{i=1}^{m} t_i (f_i^+(\mathbf{x}))^k\right)^{\frac{1}{k}}.$$

If $0 < k \leq 1$ then the function p_k is concave so the function q_k is also concave.

Example 5.7.4 If we let

$$p_\infty(\mathbf{y}) = \max_{0\leq i\leq m} y_i, \qquad \mathbf{y} = (y_0, y_1, \cdots, y_m)^\top \in \mathbb{R}^{1+m}, \tag{5.7.18}$$

then we obtain the following modified Lagrangian function:

$$\mathcal{L}_\infty(\mathbf{x}, \mathbf{t}) = \max_{0\leq i\leq m} t_i f_i(\mathbf{x}). \tag{5.7.19}$$

If we let

$$p_{+\infty}(\mathbf{y}) = \max_{0\leq i\leq m} y_i, \qquad \mathbf{y} = (y_0, y_1, \cdots, y_m)^\top \in \mathbb{R}_+^{1+m}, \tag{5.7.20}$$

then we obtain the following modified penalty function:

$$\mathcal{L}_{+\infty}(\mathbf{x}, \mathbf{t}) = \max_{0 \le i \le m} t_i f_i^+(\mathbf{x}). \tag{5.7.21}$$

If the conditions of Theorem 5.7.3 hold then the zero duality gap property will hold for both the modified Lagrangian function and the modified penalty function. The first of these can be represented as follows:

$$\inf_{\mathbf{x} \in \mathcal{X}_0} f_0(\mathbf{x}) = \sup_{\mathbf{t} \in \mathbf{R}^m} \inf_{\mathbf{x} \in \mathcal{X}} \max\{f_0(\mathbf{x}), \max_{1 \le i \le m} t_i f_i(\mathbf{x})\}. \tag{5.7.22}$$

The zero duality gap property for the modified penalty function has the following form:

$$\inf_{\mathbf{x} \in \mathcal{X}_0} f_0(\mathbf{x}) = \sup_{\mathbf{t} \in \mathbf{R}^m} \inf_{\mathbf{x} \in \mathcal{X}} \max\{f_0(\mathbf{x}), \max_{1 \le i \le m} t_i f_i^+(\mathbf{x})\}. \tag{5.7.23}$$

Since $\max_{1 \le i \le m} t_i f_i^+(\mathbf{x}) = \max(0, \max_{1 \le i \le m} t_i f_i(\mathbf{x}))$ and $f_0(\mathbf{x}) > 0$, (5.7.23) is reduced to (5.7.22). Thus a zero duality gap for the modified penalty functions coincides with a zero duality gap for the modified Lagrangian function. The equality (5.7.22) has been proven in Theorem 5.5.2.

Let $g(\mathbf{x}) = \max_{i \in I} f_i(\mathbf{x})$. The set $\mathcal{X}_0$ of all feasible solutions of the problem P can be represented by using only one constraint function g, namely $\mathcal{X}_0 = \{\mathbf{x} \in \mathcal{X} : g(\mathbf{x}) \le 0\}$. The dual function $\bar{q}$ in this case has the following form:

$$\bar{q}(t_0, t_1) = \inf_{\mathbf{x} \in \mathcal{X}} \max\{t_0 f_0(\mathbf{x}), t_1 g(\mathbf{x})\}, \qquad t_0, t_1 \ge 0.$$

We now compare this function with the dual function q defined for the Problem P using the given constraint functions $f_1, \cdots, f_m$. We have for both the modified Lagrangian function and the modified penalty functions:

$$q(t_0, t_1, \cdots, t_m) = \inf_{\mathbf{x} \in \mathcal{X}} \max\{t_0 f_0(\mathbf{x}), t_1 f_1(\mathbf{x}), \cdots, t_m f_m(\mathbf{x})\}, \quad t_0, t_1, \cdots t_m \ge 0.$$

Clearly

$$\bar{q}(t_0, t) = q(t_0, t, \cdots, t).$$

Since q is a monotone function, it follows that:

$$\sup_{t > 0} \bar{q}(1, t) = \sup_{t_1, \cdots, t_m > 0} q(1, t_1, \cdots, t_m).$$

Thus the value $\bar{M}_D$ of the dual problem with respect to one constraint g coincides with the value M_D of the dual problem with respect to the given constraints $f_1, \cdots, f_m$.

CHAPTER 6

DUALITY IN VARIATIONAL INEQUALITIES

The study of variational inequalities was initiated by reformulating variational problems with unilateral constraints for partial differential operators as a variational inequality problem. There are several books in the literature that are dedicated to variational inequalities. In [KS1], the classical theory of variational inequalities and its applications in partial differential equations are given. Variational inequalities are subsequently found to be a very useful tool in optimization [HP1], [G5], as well as in modelling traffic equilibria [D2],[CM1]. In [N1], various solution methods for solving variational inequalities and their applications in network economics have been investigated in detail. In [CH1], a basic theory of variational inequalities and its applications in mechanics are given. However, there are very few books, if any at all, that are dedicated to the study of variational inequalities in the context of optimization, even though the studies of variational inequalities and vector variational inequalities have attracted much attention recently in the optimization community.

A recent survey paper by [HP1] gives an extensive summary of current research and open questions of variational inequalities in optimization. Issues such as the existence and uniqueness of solutions and solution methods for variational inequalities and vector variational inequalities have been the most active areas of study, see [HP1]. In this chapter, our concern is mainly in line with the main theme of this book, namely the study of duality and gap functions for variational inequalities. Many of the results presented herein have only appeared in the literature recently.

6.1 Duality in Variational Inequalities

In this section we discuss an early duality result in variational inequalities due to Mosco's [M1]. To be consistent with the rest of this book, the results herein are presented in finite dimensional spaces although Mosco's original result in [M1] was given in locally convex Hausdorff topological spaces. We shall recall from Chapter One the definition of a variational inequality.

Definition 6.1.1 (Variational inequality problem) Given a closed convex set $\mathcal{K} \subset \mathbb{R}^n$, and a vector-valued function $\mathbf{F} : \mathbb{R}^n \to \mathbb{R}^n$, the *variational inequality problem* is defined as follows.

$$\text{(Problem VI)} \qquad \text{Find a point } \mathbf{x}^0 \in \mathcal{K} \text{ such that}$$
$$\mathbf{F}(\mathbf{x}^0)^\top (\mathbf{x} - \mathbf{x}^0) \geq 0, \quad \forall \mathbf{x} \in \mathcal{K}. \tag{6.1.1}$$

The variational inequality (6.1.1) can be related to a fixed point problem arising from a particular optimization problem. For a given $\mathbf{x}^0$ consider the following optimization problem

$$\min_{\mathbf{x} \in \mathcal{K}} \mathbf{F}(\mathbf{x}^0)^\top \mathbf{x}.$$

Note that since $\mathbf{x}^0$ is fixed, the above problem has a linear cost function. Thus if $\mathcal{K}$ is a polyhedral set, the problem is a linear program. If the minimum solution of this optimization problem turns out to be also $\mathbf{x}^0$, then $\mathbf{x}^0$ solves the variational inequality (6.1.1) by definition. View in this way, the solution to the variational inequality (6.1.1) is also the fixed point of the following map:

$$\mathbf{x} \mapsto \operatorname*{argmin}_{\mathbf{y} \in \mathcal{K}} \mathbf{F}(\mathbf{x})^\top \mathbf{y}.$$

As it turns out, the duality of variational inequalities is more appropriately discussed in a more general formulation than that given in Definition 6.1.1. *The extended variational inequality problem* is defined as follows, and this includes the variational inequality problem presented in Definition 6.1.1 as a special case.

Definition 6.1.2 (Extended variational inequality problem) Given a vector-valued function $\mathbf{F} : \mathbb{R}^n \to \mathbb{R}^n$, and an extended real-valued lower semi-continuous proper convex function $f : \mathbb{R}^n \to \overline{\mathbb{R}}$, the *extended variational inequality problem* is defined as follows.

$$\text{(Problem EVI)} \qquad \text{Find a point } \mathbf{x}^0 \in \mathbb{R}^n \text{ such that}$$
$$\mathbf{F}(\mathbf{x}^0)^\top (\mathbf{x} - \mathbf{x}^0) \geq f(\mathbf{x}^0) - f(\mathbf{x}), \quad \forall \mathbf{x} \in \mathbb{R}^n. \tag{6.1.2}$$

The solution to the extended variational inequality (6.1.2) is also the fixed point of the following map:

$$\mathbf{x} \mapsto \operatorname*{argmin}_{\mathbf{y} \in \mathcal{K}} \mathbf{F}(\mathbf{x})^\top \mathbf{y} + f(\mathbf{y}).$$

The problem EVI was first studied in the context of partial differential equations by Stampacchia [S1]. Note that the problem EVI reduces to the problem VI if we let f be the indicator function $\delta_{\mathcal{K}}(\mathbf{x})$ for the convex ground set $\mathcal{K}$, i.e., $\delta_{\mathcal{K}}(\mathbf{x}) = 0$ if $\mathbf{x} \in \mathcal{K}$ and $\delta_{\mathcal{K}}(\mathbf{x}) = +\infty$ if $\mathbf{x} \notin \mathcal{K}$. In general, the effective domain dom(f) of the convex function f implicitly defines the convex ground set $\mathcal{K}$ in the context of problem VI, since if $\mathbf{x} \notin \mathcal{K}$, i.e., $f(\mathbf{x}) = +\infty$, then the inequality (6.1.2) always holds, whilst if $\mathbf{x}^0 \notin \mathcal{K}$, then the inequality (6.1.2) never hold.

Definition 6.1.3 (Adjoint function) Let $\mathbf{F} : \mathbb{R}^n \to \mathbb{R}^n$ be a one-to-one (injective) function. The *adjoint function* $\mathbf{F}^\dagger : \mathbb{R}^n \to \mathbb{R}^n$ of $\mathbf{F}$ is defined by

$$\mathbf{F}^\dagger : \mathbf{u} \mapsto -\mathbf{F}^{-1}(-\mathbf{u}), \quad \mathbf{u} \in -\text{Range}(\mathbf{F}). \tag{6.1.3}$$

The dual to the extended variational inequality problem EVI is now defined in terms of the adjoint function of $\mathbf{F}$ and the Fenchel transform of f as follows.

Definition 6.1.4 Given $\mathbf{F}^\dagger : -\text{Range}(\mathbf{F}) \to \mathbb{R}^n$ as defined by (6.1.3), and $f^* : \mathbb{R}^n \to \overline{\mathbb{R}}$ as the Fenchel transform of f, the *dual extended variational inequality problem* is defined as follows.

(Problem DEVI) Find a point $\mathbf{u}^0 \in -\text{Range}(\mathbf{F})$ such that

$$\mathbf{F}^\dagger(\mathbf{u}^0)^\top(\mathbf{u} - \mathbf{u}^0) \geq f^*(\mathbf{u}^0) - f^*(\mathbf{u}), \quad \forall \mathbf{u} \in -\text{Range}(\mathbf{F}). \tag{6.1.4}$$

The following duality result was first obtained by Mosco [M1] for a locally convex Hausdorff topological space. The original proof requires the use of subdifferential and other results in abstract notions. The following proof in [CGY1] is a simpler one which requires only the definitions laid down so far.

Theorem 6.1.1 (Duality of extended variational inequality)

(i) $\mathbf{x}^0$ solves EVI if and only if $\mathbf{u}^0 = -\mathbf{F}(\mathbf{x}^0)$ solves DEVI.

(ii) The primal and dual variational inequalities (6.1.2) and (6.1.4) hold if and only if $\mathbf{u}^0 = -\mathbf{F}(\mathbf{x}^0)$ (or, equivalently, $\mathbf{x}^0 = -\mathbf{F}^\dagger(\mathbf{u}^0)$), and

$$f(\mathbf{x}^0) + f^*(\mathbf{u}^0) - (\mathbf{x}^0)^\top \mathbf{u}^0 = 0. \tag{6.1.5}$$

Proof:
(i) Let $\mathbf{x}^0$ solves EVI and let $\mathbf{u}^0 = \mathbf{F}(\mathbf{x}^0)$. Then

$$\mathbf{F}(\mathbf{x}^0)^\top(\mathbf{x} - \mathbf{x}^0) \geq f(\mathbf{x}^0) - f(\mathbf{x}), \quad \forall \mathbf{x} \in \mathbb{R}^n,$$

or

$$-\mathbf{F}(\mathbf{x}^0)^\top \mathbf{x}^0 - f(\mathbf{x}^0) \geq -\mathbf{F}(\mathbf{x}^0)^\top \mathbf{x} - f(\mathbf{x}), \quad \forall \mathbf{x} \in \mathbb{R}^n,$$

implying that

$$-\mathbf{F}(\mathbf{x}^0)^\top \mathbf{x}^0 - f(\mathbf{x}^0) = \max_{\mathbf{x} \in \mathbb{R}^n} \{-\mathbf{F}(\mathbf{x}^0)^\top \mathbf{x} - f(\mathbf{x})\} = f^*(-\mathbf{F}(\mathbf{x}^0)) = f^*(\mathbf{u}^0). \tag{6.1.6}$$

If $\mathbf{u}^0 = -\mathbf{F}(\mathbf{x}^0)$ does not solve the problem DEVI, then there exists an $\mathbf{u} \in \mathbb{R}^n$ such that

$$\mathbf{F}^\dagger(\mathbf{u}^0)^\top(\mathbf{u} - \mathbf{u}^0) < f^*(\mathbf{u}^0) - f^*(\mathbf{u}).$$

By (6.1.6), we have

$$(\mathbf{u}^0)^\top \mathbf{x}^0 < -f(\mathbf{x}^0) - f^*(\mathbf{u}),$$

or

$$f(\mathbf{x}^0) + f^*(\mathbf{u}) - (\mathbf{x}^0)^\top \mathbf{u} < 0,$$

contradicting the Young's inequality (Lemma 1.3.1). Thus $\mathbf{u}^0 = -\mathbf{F}(\mathbf{x}^0)$ solves the problem DEVI.

To prove that $\mathbf{u}^0$ solves the problem DEVI implies $\mathbf{x}^0$ solves the problem EVI, we proceed as before with the role of $f, \mathbf{F}, \mathbf{x}, \mathbf{x}^0$ replaced by $f^*, \mathbf{F}^\dagger, \mathbf{u}, \mathbf{u}^0$ respectively, and by noting that the biconjugate of f is itself, since f is a closed or lower-semi-continuous function by Corollary 1.3.5.

(ii) The proof follows directly from (6.1.6). ∎

As the problem VI is a special case of the problem EVI, we may now establish the duality of the problem VI as a Corollary of Theorem 6.1.1. First, we shall relate the indicator function to its Fenchel transform.

Definition 6.1.5 (Support function of a set) The *support function* $\sigma_{\mathcal{K}} : \mathbb{R}^n \to \overline{\mathbb{R}}$ of a set $\mathcal{K} \subseteq \mathbb{R}^n$ is defined by

$$\sigma_{\mathcal{K}}(\mathbf{u}) = \sup\{\mathbf{u}^\top \mathbf{x} \mid \mathbf{x} \in \mathcal{K}\}, \quad \forall \ \mathbf{u} \in \mathbb{R}^n.$$

Definition 6.1.6 (Dual variational inequality) Given $\mathbf{F}^\dagger : -\text{Range}(\mathbf{F}) \to \mathbb{R}^n$ in Definition 6.1.3, and $\sigma_{\mathcal{K}} : \mathbb{R}^n \to \overline{\mathbb{R}}$ as the support function of $\mathcal{K}$ as in Definition 6.1.5, the *dual variational inequality* problem is defined as follows.

(Problem DVI) Find a point $\mathbf{u}^0 \in -\text{Range}(\mathbf{F})$ such that

$$\mathbf{F}^\dagger(\mathbf{u}^0)^\top(\mathbf{u} - \mathbf{u}^0) \geq \sigma_{\mathcal{K}}(\mathbf{u}^0) - \sigma_{\mathcal{K}}(\mathbf{u}), \quad \forall \mathbf{u} \in \mathbb{R}^n. \tag{6.1.7}$$

Corollary 6.1.2 (Duality of Variational Inequality)

(i) $\mathbf{x}^0$ solves the problem VI if and only if $\mathbf{u}^0 = -\mathbf{F}(\mathbf{x}^0)$ solves the problem DVI.

(ii) The variational inequalities (6.1.1) and (6.1.7) hold if and only if $\mathbf{u}^0 = -\mathbf{F}(\mathbf{x}^0)$ (or, equivalently, $\mathbf{x}^0 = -\mathbf{F}^\dagger(\mathbf{u}^0)$), and

$$\sigma_{\mathcal{K}}(\mathbf{u}^0) = {\mathbf{x}^0}^\top \mathbf{u}^0. \tag{6.1.8}$$

Proof: Follows directly from Theorem 6.1.1, together with the facts that the problem EVI reduces to the problem VI when f is the indicator function of $\mathcal{K}$, and that $\sigma_{\mathcal{K}}$ is the Fenchel transform of the indicator function, i.e., $\sigma_{\mathcal{K}} = \delta^*_{\mathcal{K}}$. (6.1.8) follows directly from (6.1.5). ∎

As an application of the duality of EVI, we present a result which relates the critical point of a function to a particular variational inequality. This result is due to Ekeland [E2].

Consider the extended real-valued function:

$$\alpha(\mathbf{x}) = \frac{1}{2}\mathbf{x}^\top \mathbf{A}\mathbf{x} + f(\mathbf{x}),$$

where $\mathbf{A} \in \mathbb{R}^{n\times n}$ is symmetric but not necessarily positive semidefinite, and $f : \mathbb{R}^n \to \overline{\mathbb{R}}$ is a lower semicontinuous convex function. A point $\mathbf{x}^0$ is called a *critical point* of the function α if

$$0 \in \mathbf{A}\mathbf{x}^0 + \partial f(\mathbf{x}^0).$$

Define the function $\beta : \mathbb{R}^n \to \overline{\mathbb{R}}$ by

$$\beta(\mathbf{x}) = \frac{1}{2}\mathbf{x}^\top \mathbf{A}\mathbf{x} + f^*(-\mathbf{A}\mathbf{x}).$$

Theorem 6.1.3 If $\mathbf{x}^0$ is a critical point of the function α, then it is also a critical point of β. If $\mathbf{y}^0$ is a critical point of the function β, and if the interior of $\mathbf{A}(\mathbb{R}^n) + \mathrm{dom}(f^*)$ contains the origin in $\mathbb{R}^n$, then there exists some $\mathbf{w}^0 \in \ker(\mathbf{A}) = \{\mathbf{x} \mid \mathbf{A}\mathbf{x} = \mathbf{0}\}$ such that $\mathbf{x}^0 = \mathbf{y}^0 - \mathbf{w}^0$ is a critical point of the function α. Moreover,

$$\beta(\mathbf{y}^0) = -\alpha(\mathbf{x}^0).$$

Proof: If $\mathbf{x}^0$ is a critical point of α, then

$$\mathbf{0} \in \mathbf{A}\mathbf{x}^0 + \partial f(\mathbf{x}^0)$$

or

$$-\mathbf{A}\mathbf{x}^0 \in \partial f(\mathbf{x}^0).$$

By Corollary 1.3.7, we have

$$\mathbf{x}^0 \in \partial f^*(-\mathbf{A}\mathbf{x}^0). \tag{6.1.9}$$

Premultiplying (6.1.9) by $\mathbf{A}$, and noting that $-\mathbf{A}\partial f^*(-\mathbf{A}\mathbf{x}^0) \subseteq \partial g(\mathbf{x}^0)$ where $g(\mathbf{x}) = f^*(-\mathbf{A}\mathbf{x})$, we have $\mathbf{0} \in \mathbf{A}\mathbf{x}^0 + \partial g(\mathbf{x}^0)$, and hence $\mathbf{x}^0$ is a critical point of β.

Conversely, if $\mathbf{y}^0$ is a critical point of β and if the condition $\mathbf{0} \in \mathbf{A}(\mathbb{R}^n) + \mathrm{dom}(f^*)$ holds, then

$$\mathbf{0} \in \mathbf{A}\mathbf{y}^0 - \mathbf{A}\partial f^*(-\mathbf{A}\mathbf{y}^0)$$

and hence there exists some $\mathbf{x}^0 \in \partial f^*(-\mathbf{A}\mathbf{y}^0)$ such that

$$\mathbf{A}\mathbf{y}^0 = \mathbf{A}\mathbf{x}^0. \tag{6.1.10}$$

Let $\mathbf{w}^0 = \mathbf{y}^0 - \mathbf{x}^0 \in \ker(\mathbf{A})$ and apply Corollary 1.3.7 again to obtain

$$-\mathbf{A}\mathbf{y}^0 \in \partial f(\mathbf{x}^0). \tag{6.1.11}$$

By (6.1.10), (6.1.11) reduces to

$$\mathbf{0} \in \mathbf{A}\mathbf{x}^0 + \partial f(\mathbf{x}^0)$$

or $\mathbf{x}^0$ is a critical point of α.

Lastly, by (6.1.11) and Theorem 1.3.6, we have

$$f^*(-\mathbf{A}\mathbf{y}^0) + f(\mathbf{x}^0) + (\mathbf{y}^0)^\top \mathbf{A}\mathbf{x}^0 = 0,$$

which is the same as

$$f^*(-\mathbf{A}\mathbf{y}^0) + \frac{1}{2}(\mathbf{y}^0)^\top \mathbf{A}\mathbf{y}^0 + f(\mathbf{x}^0) + \frac{1}{2}(\mathbf{x}^0)^\top \mathbf{A}\mathbf{x}^0 = 0,$$

or $\alpha(\mathbf{x}^0) + \beta(\mathbf{y}^0) = 0$. ∎

Under the stronger condition that the matrix $\mathbf{A}$ is non-singular, (and hence $\ker(\mathbf{A}) = \{\mathbf{0}\}$), this result can be related to a pair of primal and dual EVI.

Theorem 6.1.4 $\mathbf{x}^0$ is a critical point of the function α if and only if $\mathbf{x}^0$ is a solution of the following EVI with $\mathbf{F}(\mathbf{x}) = \mathbf{A}\mathbf{x}$:

$$(\mathbf{A}\mathbf{x}^0)^\top(\mathbf{x} - \mathbf{x}^0) \geq f(\mathbf{x}^0) - f(\mathbf{x}) \quad \forall \mathbf{x} \in \mathbb{R}^n. \tag{6.1.12}$$

Proof: By definition, $\mathbf{x}^0$ is a critical point of α if and only if

$\mathbf{0} \in \mathbf{A}\mathbf{x}^0 + \partial f(\mathbf{x}^0)$

if and only if, $-\mathbf{A}\mathbf{x}^0 \in \partial f(\mathbf{x}^0)$

if and only if, by Theorem 1.3.6, $f(\mathbf{x}^0) + f^*(-\mathbf{A}\mathbf{x}^0) + (\mathbf{x}^0)^\top \mathbf{A}\mathbf{x}^0 = 0$,

if and only if, $f(\mathbf{x}^0) - (\mathbf{x}^0)^\top \mathbf{A}\mathbf{x} - f(\mathbf{x}) + (\mathbf{x}^0)^\top \mathbf{A}\mathbf{x}^0 \leq 0, \quad \forall \mathbf{x} \in \mathbb{R}^n$,

by the definition of f^*

if and only if $\mathbf{x}^0$ solves the EVI: $(\mathbf{x}^0)^\top \mathbf{A}(\mathbf{x} - \mathbf{x}^0) \geq f(\mathbf{x}^0) - f(\mathbf{x}), \quad \forall \mathbf{x} \in \mathbb{R}^n$.

∎

Theorem 6.1.5 Assuming that $\mathbf{A}$ is non-singular. Then $\mathbf{x}^0$ is a critical point of the function β if and only if $\mathbf{u}^0 = -\mathbf{A}\mathbf{x}^0$ is a solution of the following dual extended variational inequality to (6.1.12):

$$(\mathbf{A}^{-1}\mathbf{u}^0)^\top(\mathbf{u} - \mathbf{u}^0) \geq f^*(\mathbf{u}^0) - f^*(\mathbf{u}) \quad \forall \mathbf{u} \in \mathbb{R}^n. \tag{6.1.13}$$

Proof: Let the primal and dual variables $\mathbf{u}$ and $\mathbf{x}$ be related by $\mathbf{u} = \mathbf{F}(\mathbf{x}) = -\mathbf{A}\mathbf{x}$. Then

$$\beta(\mathbf{x}) = \beta(-\mathbf{A}^{-1}\mathbf{u}) = \gamma(\mathbf{u}) = \frac{1}{2}\mathbf{u}^\top \mathbf{A}^{-1}\mathbf{u} + f^*(\mathbf{u}).$$

$\mathbf{x}^0 = -\mathbf{A}^{-1}\mathbf{u}^0$ is a critical point of β if and only if $\mathbf{u}^0$ is a critical point of γ,

$$\begin{aligned}
&\text{if and only if} \quad \mathbf{0} \in \mathbf{A}^{-1}\mathbf{u}^0 + \partial f^*(\mathbf{u}^0),\\
&\text{if and only if} \quad -\mathbf{A}^{-1}\mathbf{u}^0 \in \partial f^*(\mathbf{u}^0) \qquad \text{by Corollary 1.3.7},\\
&\text{if and only if} \quad -f^*(\mathbf{u}^0) - (\mathbf{u}^0)^\top \mathbf{A}^{-1}\mathbf{u}^0 \geq -f^*(\mathbf{u}) + \mathbf{u}^\top \mathbf{A}^{-1}\mathbf{u}^0 \quad \forall \mathbf{u} \in \mathbb{R}^n\\
&\qquad\qquad\qquad\qquad\qquad\qquad \text{by definition of } \partial f^*,\\
&\text{if and only if} \quad (\mathbf{u}^0)^\top \mathbf{A}^{-1}(\mathbf{u} - \mathbf{u}^0) \geq f^*(\mathbf{u}^0) - f^*(\mathbf{u}) \quad \forall \mathbf{u} \in \mathbb{R}^n,
\end{aligned}$$

i.e., $\mathbf{u}^0$ solves the DEVI (6.1.13). ∎

6.2 Gap Functions for Variational Inequalities

The concept of a gap function has already been introduced in Chapter Four for a convex optimization problem. The meaning of "gap" is interpreted as the difference between the cost function and the restricted Wolfe dual. The concept of gap functions is also applicable to variational inequalities problems and the first such gap function was introduced by Auslender [A1] for a general class of variational inequality problems involving convex functions. This concept is later extended to quasi-variational inequalities in [G4]. In this section, we shall formalize the general concept of a gap function for variational inequalities and illustrate the concept by a number of examples. The main reference for this section is [CGY2].

Let $\mathbf{F} : \mathbb{R}^n \to \mathbb{R}^n$ and $\mathcal{K} \subseteq \mathbb{R}^n$ be a closed convex set. For ease of reference, we restate the variational inequality problem of Definition 6.1.1 as follows.

$$\begin{aligned}
&\text{(Problem VI)} \quad \text{Find} \quad \mathbf{x} \in \mathcal{K}\\
&\quad \text{such that} \quad \mathbf{F}(\mathbf{x})^\top (\mathbf{y} - \mathbf{x}) \geq 0 \quad \forall \mathbf{y} \in \mathcal{K}.
\end{aligned} \tag{6.2.1}$$

Definition 6.2.1 (Defining properties of gap functions) A function $\gamma : \mathbb{R}^n \to \mathbb{R}$ is said to be a *gap function* for the VI (6.2.1) if it satisfies the following properties.

(i)$\gamma(\mathbf{x}) \geq \mathbf{0},\ \forall \mathbf{x} \in \mathcal{K}$,

(ii)$\gamma(\mathbf{x}^0) = 0$ if and only if $\mathbf{x}^0$ solves the problem VI.

As a result of the defining properties of a gap function, solving the problem VI is equivalent to solving the following unconstrained optimization problem:

(Equivalent optimization problem) $\min_{\mathbf{x}\in\mathbf{R}^n} \gamma(\mathbf{x})$.

Auslender [A1] defined one such gap function for the problem VI as follows:

Definition 6.2.2 (Auslender's gap function)

$$\gamma_A(\mathbf{x}) = \max_{\mathbf{y}\in\mathcal{K}} \mathbf{F}(\mathbf{x})^\top(\mathbf{x}-\mathbf{y}). \tag{6.2.2}$$

For consistency, let the effective domain of $\mathbf{F}$ be $\mathrm{dom}(\mathbf{F}) = \mathcal{K}$.

Theorem 6.2.1 The function γ_A is a gap function.

Proof: (i) Let $\mathbf{y} = \mathbf{x}$, then $\gamma(\mathbf{x}) \geq \mathbf{F}(\mathbf{x})^\top(\mathbf{x}-\mathbf{x}) = 0$.

(ii) If $\gamma_A(\mathbf{x}^0) = 0$, then

$$\max_{\mathbf{y}\in\mathcal{K}} \mathbf{F}(\mathbf{x}^0)^\top(\mathbf{x}^0-\mathbf{y}) = 0$$

implying that the VI (6.2.1) holds. Conversely, if the VI (6.2.1) holds, then

$$\min_{\mathbf{y}\in\mathcal{K}} \mathbf{F}(\mathbf{x}^0)^\top(\mathbf{y}-\mathbf{x}^0) \geq 0,$$

or

$$\max_{\mathbf{y}\in\mathcal{K}} \mathbf{F}(\mathbf{x}^0)^\top(\mathbf{x}^0-\mathbf{y}) \leq 0,$$

and equality holds when $\mathbf{y} = \mathbf{x}^0$. So $\gamma_A(\mathbf{x}^0) = 0$. ∎

In general the Auslander's gap function is not differentiable. The property of non-differentiability of this gap function poses a major difficulty for minimizing the gap function as a method for solving the problem VI. Fukushima [F4] subsequently proposed a method of minimizing a differentiable gap function for solving VI.

Given the function $\mathbf{F}: \mathbb{R}^n \to \mathbb{R}^n$ and the closed convex set $\mathcal{K}$, let the function $\mathbf{H}: \mathbb{R}^n \to \mathbb{R}^n$ be defined by

$$\mathbf{H}(\mathbf{x}) = \mathrm{Proj}_{\mathcal{K}}(\mathbf{x}-\mathbf{F}(\mathbf{x})) = \operatorname*{argmin}_{\mathbf{y}\in\mathcal{K}} \|\mathbf{y}-(\mathbf{x}-\mathbf{F}(\mathbf{x}))\|. \tag{6.2.3}$$

Note that the minimum is a global one by virtue of the convexity of $\mathcal{K}$.

Lemma 6.2.2 $\mathbf{x}$ solves the VI (6.2.1) if and only if $\mathbf{x}$ is a fixed point of the mapping $\mathbf{H}$, i.e., $\mathbf{H}(\mathbf{x}) = \mathbf{x}$.

Proof: $\mathbf{H}(\mathbf{x}) = \mathbf{x}$ if and only if, by the inequality (1.2.1) of the Separating Hyperplane Theorem 1.2.8,

$$(\mathbf{x}-(\mathbf{x}-\mathbf{F}(\mathbf{x}))^\top(\mathbf{y}-\mathbf{x}) \geq 0 \quad \forall \mathbf{y}\in\mathcal{K},$$

or

$$\mathbf{F}(\mathbf{x})^\top(\mathbf{y}-\mathbf{x}) \geq 0 \quad \forall \mathbf{y}\in\mathcal{K}.$$

Hence the conclusion follows. ∎

Definition 6.2.3 (Fukushima's gap function)

$$\gamma_F(\mathbf{x}) = -\mathbf{F}(\mathbf{x})^\top(\mathbf{H}(\mathbf{x}) - \mathbf{x}) - \frac{1}{2}(\mathbf{H}(\mathbf{x}) - \mathbf{x})^\top(\mathbf{H}(\mathbf{x}) - \mathbf{x}). \tag{6.2.4}$$

Theorem 6.2.3 The function γ_F is a gap function.

Proof: By completing square, the Fukushima's gap function can be rewritten as

$$\gamma_F(\mathbf{x}) = \frac{1}{2}\left\{\|\mathbf{F}(\mathbf{x})\|^2 - \|\mathbf{H}(\mathbf{x}) - (\mathbf{x} - \mathbf{F}(\mathbf{x})\|^2\right\}. \tag{6.2.5}$$

Since $\|\mathbf{F}(\mathbf{x})\|$ is the distance between the point $\mathbf{x} - \mathbf{F}(\mathbf{x})$ (which may not be in $\mathcal{K}$) and $\mathbf{x} \in \mathcal{K}$, and $\|\mathbf{H}(\mathbf{x}) - (\mathbf{x} - \mathbf{F}(\mathbf{x})\|$ is the distance between the point $\mathbf{x} - \mathbf{F}(\mathbf{x})$ and its projection $\mathbf{H}(\mathbf{x})$ in $\mathcal{K}$, it is obvious that the first distance is always greater or equal to the second, hence $\gamma_F(\mathbf{x}) \geq 0 \quad \forall \mathbf{x}$. Furthermore, the two distances are equal if and only if $\mathbf{x} = \mathbf{H}(\mathbf{x})$, whence $\gamma_F(\mathbf{x}) = 0$. The conclusion follows directly from Lemma 6.2.2. ∎

Corollary 6.2.4 $\mathbf{x}$ solves the VI (6.2.1) if and only if it solves the optimization problem:

$$\min_{\mathbf{x} \in \mathbf{R}^n} \gamma_F(\mathbf{x}).$$

Unlike the Auslender's gap function, the Fukushima's gap function enjoys some nice properties which make it a tractable alternative to solving the VI.

Theorem 6.2.5 If $\mathbf{F}$ is continuous, then γ_F is also continuous. Furthermore, if $\mathbf{F}$ is differentiable, then γ_F is also differentiable with

$$\nabla\gamma_F(\mathbf{x}) = \mathbf{F}(\mathbf{x})^\top - (\mathbf{H}(\mathbf{x}) - \mathbf{x})^\top(\nabla\mathbf{F}(\mathbf{x}) - I). \tag{6.2.6}$$

Proof: The continuity part is trivial by the definition of γ_F and the continuity of the projection function $\mathbf{H}$. For the differentiability part, let

$$h(\mathbf{x}, \mathbf{y}) = \mathbf{F}(\mathbf{x})^\top(\mathbf{y} - \mathbf{x}) + \frac{1}{2}\|\mathbf{y} - \mathbf{x})\|^2.$$

It is easy to see from (6.2.4) that the Fukushima's gap function can be rewritten as

$$\gamma_F(\mathbf{x}) = -\min\{h(\mathbf{x}, \mathbf{y}) \mid \mathbf{y} \in \mathcal{K}\},$$

where the minimum is uniquely attained at $\mathbf{y} = \mathbf{H}(\mathbf{x})$. By Theorem 1.7 of Chapter 4 of [A1], γ_F is differentiable. Lastly, taking the derivative of h with respect to the first argument, we have

$$\frac{\partial h}{\partial \mathbf{x}} = (\mathbf{y} - \mathbf{x})^\top\nabla\mathbf{F}(\mathbf{x}) - \mathbf{F}(\mathbf{x})^\top + (\mathbf{x} - \mathbf{y})^\top.$$

The conclusion follows from

$$\nabla \gamma_F(\mathbf{x}) = \left[\frac{\partial h}{\partial \mathbf{x}}\right]\Bigg|_{(\mathbf{x},\mathbf{H}(\mathbf{x}))}.$$

The differentiability of γ_F and its explicit gradient formulae make the alternative optimization method to solving the problem VI attractive. However, since the gap function γ_F is in general nonconvex, minimization of the gap function may run into the risk of getting stuck in some local minimum that does not give the minimum cost of 0. Under a further convexity assumption on the function $\mathbf{F}$, this difficulty can be alleviated.

Theorem 6.2.6 Assume that $\mathbf{F}$ is differentiable and the Jacobian $\nabla \mathbf{F}(\mathbf{x})$ is positive definite for all $\mathbf{x} \in \mathcal{K}$. If $\mathbf{x}$ is a stationary point for the equivalent optimization problem in Corollary 6.2.4, i.e.,

$$\nabla \gamma_F(\mathbf{x})(\mathbf{y} - \mathbf{x}) \geq 0 \quad \forall \mathbf{y} \in \mathcal{K}, \tag{6.2.7}$$

then $\mathbf{x}$ is a global minimum to the optimization problem as well as a solution to the VI (6.2.1).

Proof: Substituting the gradient $\nabla \gamma_F$ of (6.2.6) into (6.2.7) and letting $\mathbf{y} = \mathbf{H}(\mathbf{x})$, we have

$$(\mathbf{F}(\mathbf{x}) + \mathbf{H}(\mathbf{x}) - \mathbf{x})^\top (\mathbf{H}(\mathbf{x}) - \mathbf{x}) \geq (\mathbf{H}(\mathbf{x}) - \mathbf{x})^\top \nabla \mathbf{F}(\mathbf{x})(\mathbf{H}(\mathbf{x}) - \mathbf{x}). \tag{6.2.8}$$

From the definition of $\mathbf{H}$, the inequality (1.2.1) of the Separating Hyperplane Theorem 1.2.8, and the fact that $\mathbf{x} \in \mathcal{K}$, we have

$$(\mathbf{x} - \mathbf{F}(\mathbf{x}) - \mathbf{H}(\mathbf{x}))^\top (\mathbf{x} - \mathbf{H}(\mathbf{x})) \leq 0, \tag{6.2.9}$$

or,

$$(\mathbf{F}(\mathbf{x}) + \mathbf{H}(\mathbf{x}) - \mathbf{x})^\top (\mathbf{H}(\mathbf{x}) - \mathbf{x}) \leq 0. \tag{6.2.10}$$

(6.2.8) and (6.2.10) together imply that

$$(\mathbf{H}(\mathbf{x}) - \mathbf{x})^\top \nabla \mathbf{F}(\mathbf{x})(\mathbf{H}(\mathbf{x}) - \mathbf{x}) \leq 0. \tag{6.2.11}$$

Since $\nabla \mathbf{F}(\mathbf{x})$ is positive definite, (6.2.11) can only hold as an equality when $\mathbf{x} = \mathbf{H}(\mathbf{x})$, and by Corollary 6.2.4 and Lemma 6.2.2, $\mathbf{x}$ solves the VI (6.2.1) and the equivalent optimizaton problem (6.2.5) globally. ∎

Hitherto, we have not included the constraints that defined the ground set $\mathcal{K}$ explicitly into the analysis of gap functions. In [G5], Giannessi proposed a gap function which explicitly incorporates the constraints that define the ground set $\mathcal{K}$. Let $\mathcal{K} = \{\mathbf{y} \in \mathbb{R}^n \mid g_i(\mathbf{y}) \leq 0, \quad i = 1, 2, \cdots, m\}$ where $g_i : \mathbb{R}^n \to \mathbb{R}$ are convex functions, and we write $\mathbf{g}(\mathbf{x}) = (g_1(\mathbf{x}), g_2(\mathbf{x}), \cdots, g_m(\mathbf{x}))^\top$.

Definition 6.2.4 (Giannessi's gap function)

$$\gamma_G(\mathbf{x}) = \inf_{\boldsymbol{\lambda} \in \mathbf{R}_+^m} \sup_{\mathbf{y} \in \mathbf{R}^n} \left\{ \mathbf{F}(\mathbf{x})^\top (\mathbf{x} - \mathbf{y}) + \boldsymbol{\lambda}^\top \mathbf{g}(\mathbf{y}) \right\}.$$

Under the Slater constraint qualification, i.e., $\text{int}(\mathcal{K}) \neq \emptyset$, γ_G can be shown [G5] to satisfy the defining properties of gap functions. Here we present some analysis for a special case where the constraints are linear, and hence the ground set is polyhedral.

Theorem 6.2.7 If $\mathbf{g}(\mathbf{x}) = \mathbf{A}\mathbf{x} - \mathbf{b}$ for some $\mathbf{A} \in \mathbb{R}^{m \times n}$ and $\mathbf{b} \in \mathbb{R}^m$, then the function γ_G reduces to the Auslender's gap function γ_A.

Proof:

$$\begin{aligned}
\gamma_G(\mathbf{x}) &= \inf_{\boldsymbol{\lambda} \in \mathbf{R}_+^m} \sup_{\mathbf{y} \in \mathbf{R}^n} \left\{ \mathbf{F}(\mathbf{x})^\top (\mathbf{x} - \mathbf{y}) + \boldsymbol{\lambda}^\top (\mathbf{A}\mathbf{y} - \mathbf{b}) \right\} \\
&= \mathbf{F}(\mathbf{x})^\top \mathbf{x} + \inf_{\boldsymbol{\lambda} \in \mathbf{R}_+^m} \left\{ -\boldsymbol{\lambda}^\top \mathbf{b} + \sup_{\mathbf{y} \in \mathbf{R}^n} \left[(\boldsymbol{\lambda}^\top \mathbf{A} - \mathbf{F}(\mathbf{x})^\top) \mathbf{y} \right] \right\} \\
&= \mathbf{F}(\mathbf{x})^\top \mathbf{x} + \inf_{\substack{\boldsymbol{\lambda} \in \mathbf{R}_+^m \\ \boldsymbol{\lambda}^\top \mathbf{A} = \mathbf{F}(\mathbf{x})^\top}} \left\{ -\boldsymbol{\lambda}^\top \mathbf{b} \right\}
\end{aligned}$$

since

$$\sup_{\mathbf{y} \in \mathbf{R}^n} \left[(\boldsymbol{\lambda}^\top \mathbf{A} - \mathbf{F}(\mathbf{x})^\top) \mathbf{y} \right] = \infty \quad \text{unless} \quad \boldsymbol{\lambda}^\top \mathbf{A} = \mathbf{F}(\mathbf{x})^\top.$$

Hence

$$\begin{aligned}
\gamma_G(\mathbf{x}) &= \mathbf{F}(\mathbf{x})^\top \mathbf{x} - \sup_{\substack{\boldsymbol{\lambda} \in \mathbf{R}_+^m, \\ \boldsymbol{\lambda}^\top \mathbf{A} = \mathbf{F}(\mathbf{x})^\top}} \left\{ \boldsymbol{\lambda}^\top \mathbf{b} \right\} \\
&= \mathbf{F}(\mathbf{x})^\top \mathbf{x} - \inf_{\mathbf{A}\mathbf{y} \leqq \mathbf{b}} \mathbf{F}(\mathbf{x})^\top \mathbf{y} \quad \text{by LP duality} \qquad (6.2.12) \\
&= \mathbf{F}(\mathbf{x})^\top \mathbf{x} + \sup_{\mathbf{y} \in \mathcal{K}} -\mathbf{F}(\mathbf{x})^\top \mathbf{y} \\
&= \sup_{\mathbf{y} \in \mathcal{K}} \mathbf{F}(\mathbf{x})^\top (\mathbf{x} - \mathbf{y}) = \gamma_A(\mathbf{x}).
\end{aligned}$$

∎

Under further assumptions, it is possible to say a bit more about the convexity of the Auslender's gap function. From (6.2.12), we have

$$\gamma_A(\mathbf{x}) = \mathbf{F}(\mathbf{x})^\top \mathbf{x} - \psi(\mathbf{F}(\mathbf{x})) \qquad (6.2.13)$$

where $\psi(\mathbf{z}) = \inf_{\mathbf{y} \in \mathcal{K}} \mathbf{z}^\top \mathbf{y}$. Now ψ is a concave functions since for $\mathbf{z}^1, \mathbf{z}^2 \in \mathbb{R}^n$, $\lambda \in (0, 1)$, we have

$$\begin{aligned}
\psi(\lambda \mathbf{z}^1 + (1 - \lambda)\mathbf{z}^2) &= \inf_{\mathbf{y} \in \mathcal{K}} (\lambda \mathbf{z}^1 + (1 - \lambda)\mathbf{z}^2)^\top \mathbf{y} \\
&\geq \lambda \inf_{\mathbf{y} \in \mathcal{K}} (\mathbf{z}^1)^\top \mathbf{y} + (1 - \lambda) \inf_{\mathbf{y} \in \mathcal{K}} (\mathbf{z}^2)^\top \mathbf{y} \\
&= \lambda \psi(\mathbf{z}^1) + (1 - \lambda)\psi(\mathbf{z}^2).
\end{aligned}$$

Theorem 6.2.8 If $\mathbf{F}(\mathbf{x}) = \mathbf{A}\mathbf{x} + \mathbf{b}$ where $\mathbf{A}$ is positive semidefinite, then the gap function $\gamma_G(\mathbf{x})$ is convex.

Proof: The first term in (6.2.13) is quadratic in $\mathbf{A}$ and is convex. The second term is the composition of a concave function with an affine function and is therefore concave. ∎

Remark 6.2.1 Note that if $\mathbf{F}^\top = \nabla \mathbf{f}$ for some convex differentiable function $\mathbf{f}$, then the gap function for the variational inequality (6.2.1) as defined by (6.2.2) reduces to the gap function (4.5.2) for the optimization problem

$$\min_{\mathbf{x}\in\mathcal{K}} \mathbf{f}(\mathbf{x}).$$

6.3 Primal and Dual Gap Functions for Extended Variational Inequalities

In this section, we study the relationships between gap functions and the duality of extended variational inequalities. We seek to provide a meaningful interpretation to the corresponding extended gap function, and show that properties (i) and (ii) of gap functions given in Definition 6.2.1 can essentially be interpreted as weak duality and strong duality respectively. Various properties of the gap function are derived subsequently. The main reference for this section is [CGY2].

Auslender's definition of gap function is extended for the problem EVI (Definition 6.1.2) in the following manner.

Definition 6.3.1 (Extended gap function)

$$\gamma(\mathbf{x}) = \max_{\mathbf{y}\in\mathbf{R}^n} \{ \mathbf{F}(\mathbf{x})^\top (\mathbf{x} - \mathbf{y}) + f(\mathbf{x}) - f(\mathbf{y}) \} \tag{6.3.1}$$

This form of definition readily leads to an alternative form involving the Fenchel transform of f:

$$\gamma(\mathbf{x}) = f(\mathbf{x}) + f^*(-\mathbf{F}(\mathbf{x})) + \mathbf{x}^\top \mathbf{F}(\mathbf{x}) \tag{6.3.2}$$

As with the Auslender's gap function, this extended gap function satisfies the following defining properties:

Theorem 6.3.1 The function γ is a gap function for the problem EVI, i.e.,

(i) $\gamma(\mathbf{x}) \geq 0\ \forall \mathbf{x} \in \mathbb{R}^n$, and

(ii) $\gamma(\mathbf{x}^0) = 0$ if and only if $\mathbf{x}^0$ solves the problem EVI.

Proof: (i) Follows directly from Young's inequality (Lemma 1.3.1) and this can be interpreted as a form of weak duality.

(ii) Follows from Theorem 6.1.1(ii). This can be interpreted as a form of strong duality. The meaning of "gap" is now apparent. ∎

The above gap function is defined for the problem EVI and can be considered a primal gap function. We may further define the dual gap function for the dual problem DEVI as follows:

Definition 6.3.2 (Dual gap function)

$$\psi(\mathbf{u}) = \max_{\mathbf{v} \in \mathbf{R}^n} \{\ \mathbf{F}^{\dagger}(\mathbf{u})^{\top}(\mathbf{u} - \mathbf{v}) + f^*(\mathbf{u}) - f^*(\mathbf{v})\ \},\ \ \mathbf{u} \in -\text{Range}(\mathbf{F}). \tag{6.3.3}$$

It is easy to show that ψ satisfies the following defining properties of a gap function:

Theorem 6.3.2 The function ψ is a gap function for the problem DEVI, i.e.,

(i)$\psi(\mathbf{u}) \geq 0\ \forall \mathbf{u} \in -\text{Range}(\mathbf{F})$, and

(ii)$\psi(\mathbf{u}^0) = 0$ if and only if $\mathbf{u}^0$ solves the problem DEVI.

The primal gap function and dual gap function are related to each other as follows:

Theorem 6.3.3 Let $\mathbf{x} \in \mathbb{R}^n$ and $\mathbf{u} \in -\text{Range}(\mathbf{F})$. Then

(i) $\gamma(\mathbf{x}) = 0$ if and only if $\psi(-\mathbf{F}(\mathbf{x})) = 0$, if and only if $f(\mathbf{x}) + f^*(-\mathbf{F}(\mathbf{x})) - \mathbf{x}^{\top}(-\mathbf{F}(\mathbf{x})) = 0$, and

(ii)$\psi(\mathbf{u}) = 0$ if and only if $\gamma(-\mathbf{F}^{\dagger}(\mathbf{u})) = 0$ if and only if $f(-\mathbf{F}^{\dagger}(\mathbf{u})) + f^*(\mathbf{u}) - \mathbf{u}^{\top}(-\mathbf{F}^{\dagger}(\mathbf{u})) = 0$.

Proof: (i) $\gamma(\mathbf{x}) = 0$ if and only if $\mathbf{x}$ solves EVI, if and only if $\mathbf{u} = -\mathbf{F}(\mathbf{x})$ solves DEVI, if and only if $\psi(-\mathbf{F}(\mathbf{x})) = 0$, if and only if

$$\max_{\mathbf{v} \in \mathbf{R}^n} \{(-\mathbf{F}^{-1}(\mathbf{F}(\mathbf{x}))^{\top}(-\mathbf{F}(\mathbf{x}) - \mathbf{v}) - f^*(\mathbf{v})\ \} + f^*(-\mathbf{F}(\mathbf{x})) = 0,$$

i.e.,

$$\max_{\mathbf{v} \in \mathbf{R}^n} \{\mathbf{x}^{\top}\mathbf{v} - f^*(\mathbf{v})\} + f^*(-\mathbf{F}(\mathbf{x})) - \mathbf{x}^{\top}(-\mathbf{F}(\mathbf{x})) = 0,$$

if and only if

$$f(\mathbf{x}) + f^*(-\mathbf{F}(\mathbf{x})) - \mathbf{x}^{\top}(-\mathbf{F}(\mathbf{x})) = 0.$$

The proof for (ii) is similar. ∎

Gap functions furnish a natural optimization method for solving the problem EVI, since the solution of EVI is also the global minimum of the gap function, and furthermore this solution yields the known optimum value of zero. To be able to solve this optimization problem effectively, we need to understand a bit more about the convexity and smoothness of gap functions.

Definition 6.3.3 A (vector-valued) function $\mathbf{F} : \mathbb{R}^n \to \mathbb{R}^n$ is *monotone* if

$$\forall \mathbf{x}, \mathbf{y} \in \mathbb{R}^n, \quad (\mathbf{F}(\mathbf{x}) - \mathbf{F}(\mathbf{y}))^\top (\mathbf{x} - \mathbf{y}) \geq 0,$$

and *strictly monotone* if

$$\forall \mathbf{x} \neq \mathbf{y} \in \mathbb{R}^n, \quad (\mathbf{F}(\mathbf{x}) - \mathbf{F}(\mathbf{y}))^\top (\mathbf{x} - \mathbf{y}) > 0.$$

Theorem 6.3.4 If $\mathbf{F}$ is affine and monotone, then the gap function γ is convex.

Proof: It suffices to show that each term in (6.3.2) is convex. f is by definition convex. If $\mathbf{F}(\mathbf{x})$ is affine and monotone, then $\mathbf{F}(\mathbf{x}) = \mathbf{A}\mathbf{x} + \mathbf{b}$ where $\mathbf{A}$ is positive semidefinite. Clearly $\mathbf{x}^\top \mathbf{F}(\mathbf{x}) = \mathbf{x}^\top \mathbf{A}\mathbf{x} + \mathbf{x}^\top \mathbf{b}$ is convex. It remains to show that $f^*(-\mathbf{F}(\mathbf{x}))$ is convex. Now f^* is the Fenchel conjugate of a convex function and hence is convex by Theorem 1.3.2. If $\mathbf{F}$ is affine, so is $-\mathbf{F}$. It is well-known that the composition of a convex function with an affine function $f^* \circ (-\mathbf{F})$ is also convex. ∎

Remark 6.3.1 To have γ strictly convex, it suffices to require either f to be strictly convex, or $\mathbf{F}$ to be strictly monotone.

Remark 6.3.2 Note that in the context of convex optimization, Theorem 4.5.5 requires that $\mathbf{x}^\top \mathbf{F}(\mathbf{x})$ be convex, $\mathcal{K}$ be polyhedral, and $\mathbf{F}$ be concave in order for the gap function to be convex. The last concavity requirement stems from the fact that, without f, the function f^* is monotone, hence if $\mathbf{F}$ is concave, $f^* \circ (-\mathbf{F})$ is convex. Clearly the concavity of $\mathbf{F}$ is a weaker condition than the linearity of $\mathbf{F}$ as required by Theorem 6.3.4. Unfortunately, with the presence of a convex function f, the conjugate function f^* is, in general, no longer monotone, hence requiring $\mathbf{F}$ to be concave would not have made a difference.

Another sufficient condition requiring only the concavity of $\mathbf{F}$ is as follows:

Theorem 6.3.5 If $\mathbf{F}(\mathbf{x})^\top \mathbf{x}$ is a convex function of $\mathbf{x}$, $\mathbf{F}(\mathbf{x})$ is a concave function and

$$f(\mathbf{x}) = \begin{cases} \mathbf{c}^\top \mathbf{x} + d, & \text{if } \mathbf{x} \in \mathcal{K}, \\ +\infty, & \text{otherwise,} \end{cases}$$

where $\mathcal{K} = \{\mathbf{x} : A\mathbf{x} = \mathbf{b}, \mathbf{x} \geqq 0\}$ is a polyhedral set, then the gap function γ is convex.

Proof: By (6.3.2) and the assumptions of the theorem, it suffices to prove that f^* is monotone since the composition of a monotone function and a convex function is convex. By the definition of Fenchel transform and the duality of linear programming (see Section 3.3),

$$\begin{aligned} f^*(\mathbf{y}) &= \max_{\mathbf{x} \in \mathbf{R}^n} \{\mathbf{x}^\top \mathbf{y} - f(\mathbf{x})\} \\ &= \max_{\mathbf{x} \in \mathcal{K}} \{\mathbf{x}^\top \mathbf{y} - \mathbf{c}^\top \mathbf{x} - d\} \\ &= -d + \min_{\lambda^\top A \geqq \mathbf{y}^\top - \mathbf{c}^\top} \{\lambda^\top \mathbf{b}\}. \end{aligned}$$

If $\mathbf{y}^1 \geqq \mathbf{y}^2$, then the feasible set of the above linear program corresponding to $\mathbf{y}^1$ is included in the feasible set corresponding to $\mathbf{y}^2$. Consequently,

$$f^*(\mathbf{y}^1) \geq f^*(\mathbf{y}^2),$$

and hence f^* is monotone. ∎

Before investigating the differentiability of the gap function γ, let us recall some differentiability properties of a proper convex function [R1].

Definition 6.3.4 A proper convex function $f : \mathbb{R}^n \to \overline{\mathbb{R}}$ is said to be *essentially differentiable* if the following three conditions hold:
(i) $\text{int}(\text{dom}(f))$ is not empty;
(ii) f is differentiable on $\text{int}(\text{dom}(f))$;
(iii) $\lim_{i\to+\infty} \|\nabla f(\mathbf{x}^i)\| = +\infty$ whenever $\{\mathbf{x}^i\}$ is a sequence in $\text{int}(\text{dom}(f))$ converging to a boundary point of $\text{int}(\text{dom}(f))$.

A proper convex function f is said to be essentially strictly convex if f is strictly convex on every convex subset of $\{\mathbf{x} \mid \partial f(\mathbf{x}) \neq \emptyset\} = \text{dom}(\partial f)$.

It follows (see [R1]) that *a closed proper convex function is essentially strictly convex if and only if its conjugate is essentially differentiable.*

Theorem 6.3.6 If $\mathbf{F}$ is differentiable on $\mathbb{R}^n$ and f is essentially differentiable and essentially strictly convex (hence ∇f is monotone and injective), then the gap function γ is differentiable on $\text{int}(\text{dom}(f)) \cap \mathbf{F}^{-1}(\text{int}(-\text{dom}(f^*)))$. Furthermore the gradient of γ can be explicitly computed to be

$$\nabla\gamma(\mathbf{x}) = \nabla f(\mathbf{x}) - [(\nabla f)^{-1}(-\mathbf{F}(\mathbf{x}))]^\top \nabla\mathbf{F}(\mathbf{x}) + \mathbf{F}(\mathbf{x})^\top + \mathbf{x}^\top \nabla\mathbf{F}(\mathbf{x}). \qquad (6.3.4)$$

Proof: Let $\mathbf{x} \in \text{int}(\text{dom}(f)) \cap \mathbf{F}^{-1}(\text{int}(-\text{dom}(f^*)))$. Then $\mathbf{F}$ is differentiable at $\mathbf{x}$ and f is differentiable at $\mathbf{x}$, thus the gradient of the first and last term of γ in (6.3.2) is obvious. We now turn to the differentiability of the Fenchel conjugate f^*. From Corollary 1.3.7, we have $\mathbf{u} \in \partial f(\mathbf{x})$ if and only if $\mathbf{x} \in \partial f^*(\mathbf{u})$. Since f is essentially strictly convex and $-\mathbf{F}(\mathbf{x}) \in \text{int}(\text{dom}(f^*))$, f^* is differentiable at $-\mathbf{F}(\mathbf{x})$. Then $\partial f(\mathbf{x})$ is a singleton, and $\partial f(\mathbf{x}) = \{\nabla f(\mathbf{x})\}$. Since f is strictly convex, then ∇f is strictly monotone and hence f is injective, thus the derivative of $f^*(\mathbf{u})$ is $\nabla f^*(\mathbf{u}) = \mathbf{x}$, where $\nabla f(\mathbf{x}) = \mathbf{u}$ or $\mathbf{x} = (\nabla f)^{-1}(\mathbf{u}) = \nabla f^*(\mathbf{u})$. Thus the gradient of the second term in the gap function of (6.3.2) can be obtained by chain rule to be

$$\nabla f^*(-\mathbf{F}(\mathbf{x})) = -[(\nabla f)^{-1}(-\mathbf{F}(\mathbf{x}))]^\top \nabla\mathbf{F}(\mathbf{x}).$$

∎

Since the above gradient formula is explicit, we may use it to find the minimum of the gap function. Hence the problem EVI can be solved using standard gradient descent methods such as Newton, Quasi-Newton, or conjugate gradient methods. Unlike the case of the Fukushima gap function where the gradient formulae (6.2.6) requires the solution of another optimization problem, the above gradient formula does not require the solution of any optimization problem.

Example 6.3.1 Consider the following asymmetric EVI problem in $\mathbb{R}^3$. Let $\mathcal{K}$ be a polyhedral set

$$\mathcal{K} = \{\mathbf{x} \in \mathbb{R}^3 \mid A\mathbf{x} \leqq \mathbf{b}, \ \ \mathbf{0} \leqq \mathbf{x} \leqq \mathbf{u}\}$$

where $A = \begin{pmatrix} 1 & 2 & 1 \\ -1 & 3 & 4 \end{pmatrix}$, $\mathbf{b} = \begin{pmatrix} 7 \\ 5 \end{pmatrix}$, $\mathbf{u} = \begin{pmatrix} 2 \\ 3 \\ 4 \end{pmatrix}$. Let $\mathbf{F}(\mathbf{x}) = \begin{pmatrix} 5 & 1 & 1 \\ 7 & 10 & -1 \\ -5 & 5 & 4 \end{pmatrix} \mathbf{x} +$ $\begin{pmatrix} -10 \\ 15 \\ -20 \end{pmatrix}$ and $f(\mathbf{x}) = 0.1||\mathbf{x}||^2$. Note that in this case, $\mathbf{F}$ is linear and monotone, by Theorem 6.3.5 the gap function is convex. Using the unconstrained optimization function *fminu* from MATLAB, and with an arbitrary initial starting point, the solution converges to $x^* = (0.3636, \ 0, \ 0.9060)^\top$ almost instantly after a mere few iterations on a workstation.

6.4 Gap Functions and Dual Fenchel Optimization

If we extend the extended gap function as defined in Definition 6.3.1 even further, it is possible to produce more interesting results for another variant of the gap function. We relate these to a pair of Fenchel optimization problems.

Let $\Omega : \mathbb{R}^n \times \mathbb{R}^n \to \mathbb{R}$ be a function satisfying the following assumptions:

Assumption 6.4.1

(i) $\Omega(\mathbf{x}, \mathbf{y}) \geq 0 \ \ \forall \ (\mathbf{x}, \mathbf{y}) \in \mathbb{R}^n \times \mathbb{R}^n$ and $\Omega(\mathbf{x}, \mathbf{x}) = 0 \ \forall \mathbf{x} \in \mathbb{R}^n$;

(ii) For all $\mathbf{x} \in \mathbb{R}^n$, $\Omega(\mathbf{x}, \mathbf{y})$ is convex in the second argument;

(iii) $\mathbf{0} = \nabla_2\Omega(\mathbf{x}, \mathbf{x}) \ \forall \mathbf{x} \in \mathbb{R}^n$, where $\nabla_2\Omega(\mathbf{x}, \mathbf{x})$ is the gradient of $\Omega(\mathbf{x}, \mathbf{y})$ with respect to the second argument, and evaluated at $\mathbf{y} = \mathbf{x}$.

Consider the extended variational inequality problem EVI (6.1.2) together with another variant of the gap function defined as follows. (Note that the inclusion of the Ω term was also discussed in [ZM1] in the context of the less general variational inequality problem VI, albeit with some subtle difference.)

$$\gamma(\mathbf{x}) = \max_{\mathbf{y}\in\mathbf{R}^n} \{ \mathbf{F}(\mathbf{x})^\top(\mathbf{x}-\mathbf{y}) + f(\mathbf{x}) - f(\mathbf{y}) - \Omega(\mathbf{x},\mathbf{y}) \} \tag{6.4.1}$$

We shall establish that this is also a gap function in the following Theorem.

Theorem 6.4.1 Assume that $f(\mathbf{x})$ is differentiable. Then γ is a gap function of the problem EVI, i.e., $\gamma(\mathbf{x})$ satisfies the Definition 6.2.1.

Proof: Let

$$\Phi(\mathbf{x},\mathbf{y}) = \mathbf{F}(\mathbf{x})^\top(\mathbf{y}-\mathbf{x}) + f(\mathbf{y}) - f(\mathbf{x}) + \Omega(\mathbf{x},\mathbf{y}).$$

Then,

$$\gamma(\mathbf{x}) = \max_{\mathbf{y}\in\mathbf{R}^n}[-\Phi(\mathbf{x},\mathbf{y})] = -\min_{\mathbf{y}\in\mathbf{R}^n}\Phi(\mathbf{x},\mathbf{y}). \tag{6.4.2}$$

Since $\Phi(\mathbf{x},\mathbf{x}) = 0 \ \forall\mathbf{x}\in\mathbb{R}^n$, then $\min_{\mathbf{y}\in\mathbf{R}^n}\Phi(\mathbf{x},\mathbf{y}) \le 0 \ \forall\mathbf{x}\in\mathbb{R}^n$, and hence $\gamma(\mathbf{x}) \ge 0 \ \forall\mathbf{x}\in\mathbb{R}^n$.

Next, suppose that $\mathbf{x}$ solves the problem EVI. Then by Assumption 6.4.1 (i), we have

$$\mathbf{F}(\mathbf{x})^\top(\mathbf{y}-\mathbf{x}) + f(\mathbf{y}) - f(\mathbf{x}) + \Omega(\mathbf{x},\mathbf{y}) \ge \mathbf{F}(\mathbf{x})^\top(\mathbf{y}-\mathbf{x}) + f(\mathbf{y}) - f(\mathbf{x}) \ge 0 \ \forall\mathbf{y}\in\mathbb{R}^n,$$

i.e., $\min_{\mathbf{y}\in\mathbf{R}^n}\Phi(\mathbf{x},\mathbf{y}) \ge 0$, or $\gamma(\mathbf{x}) = -\min_{\mathbf{y}\in\mathbf{R}^n}\Phi(\mathbf{x},\mathbf{y}) \le 0$. Thus $\gamma(\mathbf{x}) = 0$ since $\gamma(\mathbf{y}) \ge 0 \ \forall\mathbf{y}\in\mathbb{R}^n$.

Conversely, suppose $\gamma(\mathbf{x}) = 0$, i.e., $\max_{\mathbf{y}\in\mathbf{R}^n}[-\Phi(\mathbf{x},\mathbf{y})] = 0$, which implies that $-\Phi(\mathbf{x},\mathbf{y}) \le 0 \ \forall\mathbf{y}\in\mathbb{R}^n$, or $\Phi(\mathbf{x},\mathbf{y}) \ge 0 \ \forall\mathbf{y}\in\mathbb{R}^n$. Observe that $\Phi(\mathbf{x},\mathbf{x}) = 0$, which implies that $\mathbf{x}$ is a solution of the following optimization problem:

$$\min_{\mathbf{y}\in\mathbf{R}^n}\ \Phi(\mathbf{x},\mathbf{y}).$$

Since by Assumption 6.4.1 (ii) and the assumption that f is convex, it is clear that $\Phi(\mathbf{x},\mathbf{y})$ is convex in $\mathbf{y}$. Consequently, the solution $\mathbf{x}$ of the convex optimization problem also solves the following variational inequality:

$$\text{Find } \mathbf{x}\in\mathbb{R}^n \text{ such that} \quad \nabla\Phi_2(\mathbf{x},\mathbf{x})^\top(\mathbf{y}-\mathbf{x}) \ge 0 \ \forall\mathbf{y}\in\mathbb{R}^n,$$

where $\nabla\Phi_2(\mathbf{x},\mathbf{x})$ is the gradient of $\Phi(\mathbf{x},\mathbf{y})$ with respect to the second argument, and evaluated at $\mathbf{y}=\mathbf{x}$.

Hence we have

$$\mathbf{F}(\mathbf{x})^\top(\mathbf{y}-\mathbf{x}) + \nabla f(\mathbf{x})^\top(\mathbf{y}-\mathbf{x}) + \nabla_2\Omega(\mathbf{x},\mathbf{x})^\top(\mathbf{y}-\mathbf{x}) \ge 0 \ \forall\mathbf{y}\in\mathbb{R}^n.$$

Since f is proper convex, $f(\mathbf{y}) - f(\mathbf{x}) \ge \nabla f(\mathbf{x})^\top(\mathbf{y}-\mathbf{x}) \ \forall\mathbf{y}\in\mathbb{R}^n$. Thus,

$$\mathbf{F}(\mathbf{x})^\top(\mathbf{y}-\mathbf{x}) + f(\mathbf{y}) - f(\mathbf{x}) + \nabla_2\Omega(\mathbf{x},\mathbf{x})^\top(\mathbf{y}-\mathbf{x}) \ge 0 \ \forall\mathbf{y}\in\mathbb{R}^n.$$

Finally, since $\mathbf{0} = \nabla_2\Omega(\mathbf{x}, \mathbf{x})$ by Assumption 6.4.1(iii), we have:

$$\mathbf{F}(\mathbf{x})^\top(\mathbf{y}-\mathbf{x}) + f(\mathbf{y}) - f(\mathbf{x}) \geq 0 \ \forall \mathbf{y} \in \mathbb{R}^n,$$

i.e., $\mathbf{x}$ solves the problem EVI. ∎

The gap function as defined in (6.4.1) has some nice smoothness properties similar to the smooth optimization problem proposed in [F4]. Furthermore, it represents an explicit primal Fenchel optimization problem, and the gap function of the corresponding dual variational inequality problem represents an explicit dual Fenchel optimization problem, where the primal and dual Fenchel optimization problems of Theorem 1.3.9 are restated as follows.

$$\text{(Primal Fenchel optimization problem } \mathrm{P}_p) \quad \min_{\mathbf{y}\in\mathbf{R}^n} [\alpha(\mathbf{y}) - \beta(\mathbf{y})\,],$$

where $\alpha : \mathbb{R}^n \to \mathbb{R}$ is a convex function, and $\beta : \mathbb{R}^n \to \mathbb{R}$ is a concave function.

$$\text{(Dual Fenchel optimization problem } \mathrm{P}_d) \quad \max_{\mathbf{v}\in\mathbf{R}^n} [\,\beta^*(\mathbf{v}) - \alpha^*(\mathbf{v})\,],$$

where $\alpha^* : \mathbb{R}^n \to \mathbb{R}$ and $\beta^* : \mathbb{R}^n \to \mathbb{R}$ are the Fenchel transforms of α and β respectively.

We specialize the gap function as defined in (6.4.1) to the following form:

$$\gamma(\mathbf{x}) = \max_{\mathbf{y}\in\mathbf{R}^n} \{\mathbf{F}(\mathbf{x})^\top(\mathbf{x}-\mathbf{y}) + f(\mathbf{x}) - f(\mathbf{y}) - \frac{1}{2}(\mathbf{y}-\mathbf{x})^\top\mathbf{Q}(\mathbf{y}-\mathbf{x})\,\} \qquad (6.4.3)$$

where $\mathbf{Q} \in \mathbb{R}^n \times \mathbb{R}^n$ is positive definite, and hence the quadratic term in (6.4.3) satisfies Assumption 6.4.1. This gap function is identified with the problem EVI as defined in (6.1.2). Similarly, the following can be shown to be a gap function for the dual extended variational inequality defined in (6.1.4):

$$\psi(\mathbf{u}) = \max_{\mathbf{v}\in\mathbf{R}^n} \{\,\mathbf{F}^\dagger(\mathbf{u})^\top(\mathbf{u}-\mathbf{v}) + f^*(\mathbf{u}) - f^*(\mathbf{v}) - \frac{1}{2}(\mathbf{v}-\mathbf{u})^\top\mathbf{Q}^{-1}(\mathbf{v}-\mathbf{u})\,\}. \quad (6.4.4)$$

Let the primal gap function γ and the dual gap function ψ be identified, respectively, with the following pair of optimization problems:

$$\text{(Problem P}_1) \quad \min_{\mathbf{y}\in\mathbf{R}^n} \{\mathbf{F}(\mathbf{x})^\top(\mathbf{y}-\mathbf{x}) + f(\mathbf{y}) - f(\mathbf{x}) + \frac{1}{2}(\mathbf{y}-\mathbf{x})^\top\mathbf{Q}(\mathbf{y}-\mathbf{x})\,\} \quad (6.4.5)$$

and

$$\text{(Problem P}_2) \quad \max_{\mathbf{v}\in\mathbf{R}^n} \{\,-[\mathbf{F}^\dagger(\mathbf{u})^\top(\mathbf{v}-\mathbf{u}) + f^*(\mathbf{v}) - f^*(\mathbf{u}) + \frac{1}{2}(\mathbf{v}-\mathbf{u})^\top\mathbf{Q}^{-1}(\mathbf{v}-\mathbf{u})\,]\,\}. \quad (6.4.6)$$

Theorem 6.4.2 If $\mathbf{x}$ solves EVI and $\mathbf{u}$ solves DEVI, then P_1 is the dual Fenchel optimization problem of P_2.

Proof: Let the functions α and β be defined by:

$$\alpha(\mathbf{y}) = \mathbf{F}(\mathbf{x})^\top(\mathbf{y}-\mathbf{x}) + \frac{1}{2}(\mathbf{y}-\mathbf{x})^\top \mathbf{Q}(\mathbf{y}-\mathbf{x}), \tag{6.4.7}$$

and

$$\beta(\mathbf{y}) = f(\mathbf{x}) - f(\mathbf{y}). \tag{6.4.8}$$

Clearly α is convex and β is concave (in $\mathbf{y}$). Thus the cost function of P_1 is given by $\alpha(\mathbf{y}) - \beta(\mathbf{y})$. We only need to show that $\max[\beta^*(\mathbf{v}) - \alpha^*(\mathbf{v})]$ is equivalent to P_2.

$$\begin{aligned}\alpha^*(\mathbf{v}) &= \max_{\mathbf{y}\in\mathbf{R}^n}\{\mathbf{v}^\top\mathbf{y} - \mathbf{F}(\mathbf{x})^\top(\mathbf{y}-\mathbf{x}) - \frac{1}{2}(\mathbf{y}-\mathbf{x})^\top Q(\mathbf{y}-\mathbf{x})\ \}\\ &= \max_{\mathbf{y}\in\mathbf{R}^n}\{(\mathbf{v}-\mathbf{F}(\mathbf{x}))^\top\mathbf{y} + \mathbf{F}(\mathbf{x})^\top\mathbf{x} - \frac{1}{2}(\mathbf{y}-\mathbf{x})^\top Q(\mathbf{y}-\mathbf{x})\ \}.\end{aligned}$$

Using the fact that the Fenchel dual of $\frac{1}{2}\mathbf{x}^\top Q\mathbf{x}$ is $\frac{1}{2}\mathbf{u}^\top Q^{-1}\mathbf{u}$, we let $\mathbf{u} = \mathbf{y} - \mathbf{x}$ to get:

$$\begin{aligned}\alpha^*(\mathbf{v}) &= \max_{\mathbf{u}\in\mathbf{R}^n}\{(\mathbf{v}-\mathbf{F}(\mathbf{x})^\top\mathbf{u} - \frac{1}{2}\mathbf{u}^\top Q\mathbf{u}\ \} + (\mathbf{v}-\mathbf{F}(\mathbf{x}))^\top\mathbf{x} + \mathbf{F}(\mathbf{x})^\top\mathbf{x}\\ &= \frac{1}{2}(\mathbf{v}-\mathbf{F}(\mathbf{x}))^\top Q^{-1}(\mathbf{v}-\mathbf{F}(\mathbf{x})) + (\mathbf{v}-\mathbf{F}(\mathbf{x}))^\top\mathbf{x} + \mathbf{F}(\mathbf{x})^\top\mathbf{x}.\end{aligned}$$

Remembering that if $\mathbf{x}$ solves EVI and $\mathbf{u}$ solves DEVI, then $\mathbf{u} = -\mathbf{F}(\mathbf{x})$, $\mathbf{x} = -\mathbf{F}^\dagger(\mathbf{u})$, and $\mathbf{x}^\top\mathbf{u} = f(\mathbf{x}) + f^*(\mathbf{u})$, consequently,

$$\alpha^*(\mathbf{v}) = \frac{1}{2}(\mathbf{v}+\mathbf{u})^\top Q^{-1}(\mathbf{v}+\mathbf{u}) - (\mathbf{v}+\mathbf{u})^\top\mathbf{F}^\dagger(\mathbf{u}) - f(\mathbf{x}) - f^*(\mathbf{u}),$$

or

$$\alpha^*(-\mathbf{v}) = \frac{1}{2}(\mathbf{v}-\mathbf{u})^\top Q^{-1}(\mathbf{v}-\mathbf{u}) + (\mathbf{v}-\mathbf{u})^\top\mathbf{F}^\dagger(\mathbf{u}) - f(\mathbf{x}) - f^*(\mathbf{u}).$$

Next,

$$\begin{aligned}\beta^*(-\mathbf{v}) &= \min_{\mathbf{y}\in\mathbf{R}^n}\{-\mathbf{v}^\top\mathbf{y} - f(\mathbf{x}) + f(\mathbf{y})\ \}\\ &= -\max_{\mathbf{y}\in\mathbf{R}^n}\{\mathbf{v}^\top\mathbf{y} - f(\mathbf{y})\ \} - f(\mathbf{x})\\ &= -f^*(\mathbf{v}) - f(\mathbf{x}).\end{aligned}$$

Adding $-\alpha^*(-\mathbf{v})$ and $\beta^*(-\mathbf{v})$ to get

$$\beta^*(-\mathbf{v}) - \alpha^*(-\mathbf{v}) = -[\ \mathbf{F}^\dagger(\mathbf{u})^\top(\mathbf{v}-\mathbf{u}) + f^*(\mathbf{v}) - f^*(\mathbf{u}) + \frac{1}{2}(\mathbf{v}-\mathbf{u})^\top Q^{-1}(\mathbf{v}-\mathbf{u})\ ].$$

Finally, since we are maximizing over the whole of $\mathbb{R}^n$,

$$\max_{\mathbf{v}\in\mathbf{R}^n}\{\beta^*(-\mathbf{v}) - \alpha^*(-\mathbf{v})\} \quad \text{is the same as} \quad \max_{\mathbf{v}\in\mathbf{R}^n}\{\beta^*(\mathbf{v}) - \alpha^*(\mathbf{v})\}.$$

This completes the proof. ∎

Under further assumptions, it is also possible to establish a convex lower bound to the gap function. The following result is an extension of Theorem 3.3 of [ZM1] to the problem EVI.

Definition 6.4.1 Let $f : \mathbb{R}^n \to \overline{\mathbb{R}}$ be a scalar-valued lower-semi-continuous proper convex function. $\mathbf{F}$ is said to be *strongly pseudo-monotone* with respect to f and with modulus μ, if there exists a positive constant μ such that

$$\mathbf{F}(\mathbf{x})^\top(\mathbf{y}-\mathbf{x}) \geq f(\mathbf{x}) - f(\mathbf{y})$$

implies that

$$\mathbf{F}(\mathbf{x})^\top(\mathbf{y}-\mathbf{x}) \geq \mu\|\mathbf{x}-\mathbf{y}\|^2 + f(\mathbf{x}) - f(\mathbf{y}).$$

Theorem 6.4.3 In addition to Assumption 6.4.1, we further assume that:

(i)Ω is continuously differentiable with respect to the second argument;

(ii)$\nabla_2\Omega(\mathbf{x},\mathbf{x}) = 0 \ \ \forall \mathbf{x}$;

(iii) the gradient of Ω with respect to the second argument, $\nabla_2\Omega$, is Lipschitz continuous in the second argument with constant L_Ω;

(iv)$\mathbf{F}$ is strongly pseudo-monotone with respect to f and with modulus μ.

Let $\mathbf{x}^0$ be a solution to the problem EVI. If (i)-(iv) hold, then there exists a positive constant α such that

$$\gamma(\mathbf{x}) \geq \alpha\|\mathbf{x}-\mathbf{x}^0\|^2, \ \forall x \in \mathbb{R}^n.$$

Proof. Since $\mathbf{x}^0$ solves the problem EVI, we have

$$\mathbf{F}(\mathbf{x}^0)^\top(\mathbf{y}-\mathbf{x}^0) \geq f(\mathbf{x}^0) - f(\mathbf{y}) \ \ \forall \mathbf{y}. \tag{6.4.9}$$

By Assumption (iv), this implies that

$$\mathbf{F}(\mathbf{x}^0)^\top(\mathbf{y}-\mathbf{x}^0) \geq \mu\|\mathbf{x}^0-\mathbf{y}\|^2 + f(\mathbf{x}^0) - f(\mathbf{y}) \ \ \forall \mathbf{y}. \tag{6.4.10}$$

Let

$$\mathbf{x}^t = \mathbf{x} + t(\mathbf{x}^0 - \mathbf{x}), \quad t \in (0,1). \tag{6.4.11}$$

Then, by the convexity of f, we have

$$f(\mathbf{x}^t) \leq t f(\mathbf{x}^0) + (1-t) f(\mathbf{x}). \tag{6.4.12}$$

By the convexity of Ω with respect to the second argument (Assumption 6.4.1(ii)), we have,

$$\begin{aligned}\Omega(\mathbf{x},\mathbf{x}^t) - \Omega(\mathbf{x},\mathbf{x}) &\leq \nabla_2\Omega(\mathbf{x},\mathbf{x}^t)^\top(\mathbf{x}^t-\mathbf{x}) \\ &= (\nabla_2\Omega(\mathbf{x},\mathbf{x}^t) - \nabla_2\Omega(\mathbf{x},\mathbf{x}))^\top(\mathbf{x}^t-\mathbf{x}) \quad \text{(Assumption (ii))} \\ &\leq L_\Omega\|\mathbf{x}^t-\mathbf{x}\|^2. \quad \text{(Assumption (iii))}\end{aligned} \tag{6.4.13}$$

By definition of the gap function in (6.4.1), we have, $\forall t \in (0,1)$,

$$\begin{aligned}
\gamma(\mathbf{x}) &\geq \mathbf{F}(\mathbf{x})^\top(\mathbf{x}-\mathbf{x}^t) + f(\mathbf{x}) - f(\mathbf{x}^t) - \Omega(\mathbf{x},\mathbf{x}^t) \\
&\geq \mathbf{F}(\mathbf{x})^\top(\mathbf{x}-\mathbf{x}^t) + t[f(\mathbf{x}) - f(\mathbf{x}^0)] - (\Omega(\mathbf{x},\mathbf{x}^t) - \Omega(\mathbf{x},\mathbf{x})) \\
&\qquad\qquad\qquad \text{(from Assumption (ii) and (6.4.12))} \\
&\geq t[\mathbf{F}(\mathbf{x})^\top(\mathbf{x}-\mathbf{x}^0) + f(\mathbf{x}) - f(\mathbf{x}^0)] - L_\Omega\|\mathbf{x}^t - \mathbf{x}\|^2 \\
&\qquad\qquad\qquad \text{(from (6.4.11) and (6.4.13))} \\
&\geq t\mu\|\mathbf{x}-\mathbf{x}^0\|^2 - t^2 L_\Omega\|\mathbf{x}-\mathbf{x}^0\|^2 \quad \text{(from (6.4.9) and Assumption (iv))} \\
&= (t\mu - t^2 L_\Omega)\|\mathbf{x}-\mathbf{x}^0\|^2.
\end{aligned}$$

Since the unconstrained maximum of $(t\mu - t^2 L_\Omega)$ occurs at $\mu/(2L_\Omega)$, we choose

$$t = \min\{1, \frac{\mu}{2L_\Omega}\},$$

to obtain

$$\gamma(\mathbf{x}) \geq \alpha\|\mathbf{x}-\mathbf{x}^0\|^2,$$

where

$$0 < \alpha = \begin{cases} \mu - L_\Omega, & \text{if } \mu \geq 2L_\Omega \\ \frac{\mu^2}{4L_\Omega}, & \text{otherwise.} \end{cases}$$

■

CHAPTER 7

ELEMENTS OF MULTICRITERIA OPTIMIZATION

The optimization and variational inequality problems we have discussed hitherto are invariably concerned with a single cost or objective function $f(\mathbf{x})$ of the decision vector $\mathbf{x}$. There is much theoretical interests, as well as motivations by practical considerations, to consider the case of an optimization problem where the cost function $\mathbf{f}(\mathbf{x}) = (f_1(\mathbf{x}), f_2(\mathbf{x}), \cdots, f_p(\mathbf{x}))^\top, \quad p > 1$ is a vector-valued function. Thus a *vector optimization, or multicriteria optimization, or multiobjective optimization problem*, is concerned with the optimization of a vector-valued function, possibly subject to one or more constraints.

The notion of optimization essentially deals with the comparison of the magnitude of similar objects. Most of us should feel quite comfortable when there is only one thing to compare. For example, is this apple better (hence more optimal) than that one? In the real life, however, one is often faced with the problem of comparing apples with oranges, or objects that are measured in different ways or scales, and most of us would feel quite uneasy about that. As it turns out, the notion of optimality can be generalized to vector-valued functions in a meaningful and rigorous way, provided that we adopt a systematic way of comparing or ordering vectors. The subject of multicriteria optimization is not new of course and can be dated back to 1896 when Pareto introduced the notion of a Pareto-optimal solution. Unlike scalar optimization where the optimal solution is usually a point, the solution to a multicriteria optimization problem is an infinite set, and hence it is usually much harder to solve a multicriteria optimization problem than a scalar one.

From a theoretical point of view, it turns out that many of the theoretical properties of scalar optimization can be generalized to the case of multicriteria optimization, albeit in a non-trivial way. In this regard, duality is probably the best example of these generalizable properties. Chapter Eight and Nine are thus devoted to the discussion of these duality results.

Some prerequisite material required for Chapter Eight and Nine will be discussed in some detail in this chapter first. Section 7.1 is a generalization of Section 1.2, where the notion of convexity is extended to the case of vector and set-valued functions. Section 7.2 is a generalization of Section 1.3 where the concept of conjugate duality is extended to the case of vector and functions. In Section 7.3, we lay the foundation of multicriteria optimization, and present some fundamental results pertaining to multicriteria optimization in Section 7.4. Related references for this chapter are found in [CP1],[CGY4],[G7],[G9],[GW1], [GY4],[GY5],[GY6],[J1],[L2], [SNT1], [Y4].

7.1 Elements of Multicriteria Convex Analysis

Since we will be dealing with vector-valued cost functions, it is necessary to compare or order vectors in a meaningful way. The following ordering relationships for vectors will be used for the next three chapters. These are a formalization of the ordering relationships $<, \leq, \leqq, >, \geq, \geqq$ previously introduced in Chapter Two.

Definition 7.1.1 (Ordering of vectors) Let C be a closed and convex cone in $\mathbb{R}^p$. Given $\boldsymbol{\xi}, \boldsymbol{\eta} \in \mathbb{R}^p$, the vector ordering relationships $\leq_C, \leq_{C\setminus\{0\}}, \not\leq_{C\setminus\{0\}}$, are defined as:

$$\boldsymbol{\xi} \leq_C \boldsymbol{\eta} \Longleftrightarrow \boldsymbol{\eta} - \boldsymbol{\xi} \in C;$$
$$\boldsymbol{\xi} \leq_{C\setminus\{0\}} \boldsymbol{\eta} \Longleftrightarrow \boldsymbol{\eta} - \boldsymbol{\xi} \in C \setminus \{\mathbf{0}\};$$
$$\boldsymbol{\xi} \not\leq_{C\setminus\{0\}} \boldsymbol{\eta} \Longleftrightarrow \boldsymbol{\eta} - \boldsymbol{\xi} \notin C \setminus \{\mathbf{0}\};$$

If int$C \neq \emptyset$, the following orderings $\leq_{\text{int } C}, \not\leq_{\text{int } C}$, are also defined

$$\boldsymbol{\xi} \leq_{\text{int } C} \boldsymbol{\eta} \Longleftrightarrow \boldsymbol{\eta} - \boldsymbol{\xi} \in \text{int}C;$$
$$\boldsymbol{\xi} \not\leq_{\text{int } C} \boldsymbol{\eta} \Longleftrightarrow \boldsymbol{\eta} - \boldsymbol{\xi} \notin \text{int}C;$$

The vector ordering relationships $\geq_C, \geq_{C\setminus\{0\}}, \not\geq_{C\setminus\{0\}}, \geq_{\text{int } C} \not\geq_{\text{int } C}$, are defined similarly.

Some basic relations for working with the above ordering relationships are given as follows. The proofs are straightforward.

Lemma 7.1.1 Let C be a closed and convex cone in $\mathbb{R}^p$. Then

$$\boldsymbol{\xi} \not\leq_{C\setminus\{0\}} \boldsymbol{\eta} \quad \text{if and only if} \quad \boldsymbol{\xi} + \boldsymbol{\zeta} \not\leq_{C\setminus\{0\}} \boldsymbol{\eta} + \boldsymbol{\zeta}, \forall \boldsymbol{\xi}, \boldsymbol{\eta}, \boldsymbol{\zeta} \in \mathbb{R}^p;$$

$$\boldsymbol{\xi} \not\leq_{C\setminus\{0\}} \boldsymbol{\eta} \quad \text{if and only if} \quad \lambda\boldsymbol{\xi} \not\leq_{C\setminus\{0\}} \lambda\boldsymbol{\eta}, \forall \boldsymbol{\xi}, \boldsymbol{\eta} \in \mathbb{R}^p, \lambda > 0;$$

$$\boldsymbol{\xi} \not\leq_{\text{int } C} \boldsymbol{\eta} \quad \text{if and only if} \quad \boldsymbol{\xi} + \boldsymbol{\zeta} \not\leq_{\text{int } C} \boldsymbol{\eta} + \boldsymbol{\zeta}, \forall \boldsymbol{\xi}, \boldsymbol{\eta}, \boldsymbol{\zeta} \in \mathbb{R}^p;$$

$$\boldsymbol{\xi} \not\leq_{\text{int } C} \boldsymbol{\eta} \quad \text{if and only if} \quad \lambda\boldsymbol{\xi} \not\leq_{\text{int } C} \lambda\boldsymbol{\eta}, \forall \boldsymbol{\xi}, \boldsymbol{\eta} \in \mathbb{R}^p, \lambda \geq 0.$$

Lemma 7.1.2 Let $\mathcal{C}$ be a closed and convex cone in $\mathbb{R}^p$. We have

(i) If $\mathbf{y}^i \leq_{\mathcal{C}\setminus\{\mathbf{0}\}} \mathbf{z}, \quad i = 1, 2, \cdots$ and $\lim_{i\to\infty} \mathbf{y}^i = \mathbf{y}$, then $\mathbf{y} \leq_{\mathcal{C}\setminus\{\mathbf{0}\}} \mathbf{z}$.

(ii) The following hold:

$$\boldsymbol{\xi} \leq_{\mathcal{C}} \boldsymbol{\eta} \leq_{\mathcal{C}\setminus\{\mathbf{0}\}} \boldsymbol{\zeta} \;\Rightarrow\; \boldsymbol{\xi} \leq_{\mathcal{C}\setminus\{\mathbf{0}\}} \boldsymbol{\zeta};$$

$$\boldsymbol{\xi} \leq_{\mathcal{C}\setminus\{\mathbf{0}\}} \boldsymbol{\eta} \leq_{\mathcal{C}} \boldsymbol{\zeta} \;\Rightarrow\; \boldsymbol{\xi} \leq_{\mathcal{C}\setminus\{\mathbf{0}\}} \boldsymbol{\zeta};$$

$$\boldsymbol{\xi} \not\leq_{\text{int}\,\mathcal{C}} \boldsymbol{\eta} \geq_{\text{int}\,\mathcal{C}} \boldsymbol{\zeta} \;\Rightarrow\; \boldsymbol{\zeta} \not\geq_{\text{int}\,\mathcal{C}} \boldsymbol{\xi};$$

$$\boldsymbol{\xi} \not\leq_{\text{int}\,\mathcal{C}} \boldsymbol{\eta} \geq_{\mathcal{C}\setminus\{\mathbf{0}\}} \boldsymbol{\zeta} \;\Rightarrow\; \boldsymbol{\zeta} \not\geq_{\mathcal{C}\setminus\{\mathbf{0}\}} \boldsymbol{\xi};$$

$$\boldsymbol{\xi} \not\leq_{\mathcal{C}\setminus\{\mathbf{0}\}} \boldsymbol{\eta} \geq_{\mathcal{C}} \boldsymbol{\zeta} \;\Rightarrow\; \boldsymbol{\zeta} \not\geq_{\mathcal{C}\setminus\{\mathbf{0}\}} \boldsymbol{\xi};$$

$$\boldsymbol{\xi} \not\geq_{\text{int}\,\mathcal{C}} \boldsymbol{\eta} \leq_{\text{int}\,\mathcal{C}} \boldsymbol{\zeta} \;\Rightarrow\; \boldsymbol{\zeta} \not\leq_{\text{int}\,\mathcal{C}} \boldsymbol{\xi};$$

$$\boldsymbol{\xi} \not\leq_{\text{int}\,\mathcal{C}} \boldsymbol{\eta} \leq_{\mathcal{C}\setminus\{\mathbf{0}\}} \boldsymbol{\zeta} \;\Rightarrow\; \boldsymbol{\zeta} \not\leq_{\mathcal{C}\setminus\{\mathbf{0}\}} \boldsymbol{\xi};$$

$$\boldsymbol{\xi} \not\geq_{\mathcal{C}\setminus\{\mathbf{0}\}} \boldsymbol{\eta} \leq_{\mathcal{C}} \boldsymbol{\zeta} \;\Rightarrow\; \boldsymbol{\zeta} \not\leq_{\mathcal{C}\setminus\{\mathbf{0}\}} \boldsymbol{\xi}.$$

Definition 7.1.2 (Ordering of sets) Let $\mathcal{C}$ be a closed and convex cone in $\mathbb{R}^p$ with $\text{int}\mathcal{C} \neq \emptyset$. Given $\mathcal{A}, \mathcal{B} \subseteq \mathbb{R}^p$, the set ordering relationships $\leqq, \leq, <, \not\leq, \not<$, are defined as:

$$\mathcal{A} \leq_{\mathcal{C}} \mathcal{B} \Longleftrightarrow \boldsymbol{\xi} \leq_{\mathcal{C}} \boldsymbol{\eta},\ \forall \boldsymbol{\xi} \in \mathcal{A},\ \boldsymbol{\eta} \in \mathcal{B};$$
$$\mathcal{A} \leq_{\mathcal{C}\setminus\{\mathbf{0}\}} \mathcal{B} \Longleftrightarrow \boldsymbol{\xi} \leq_{\mathcal{C}\setminus\{\mathbf{0}\}} \boldsymbol{\eta},\ \forall \boldsymbol{\xi} \in \mathcal{A},\ \boldsymbol{\eta} \in \mathcal{B};$$
$$\mathcal{A} \leq_{\text{int}\,\mathcal{C}} \mathcal{B} \Longleftrightarrow \boldsymbol{\xi} \leq_{\text{int}\,\mathcal{C}} \boldsymbol{\eta},\ \forall \boldsymbol{\xi} \in \mathcal{A},\ \boldsymbol{\eta} \in \mathcal{B};$$
$$\mathcal{A} \not\leq_{\mathcal{C}\setminus\{\mathbf{0}\}} \mathcal{B} \Longleftrightarrow \boldsymbol{\xi} \not\leq_{\mathcal{C}\setminus\{\mathbf{0}\}} \boldsymbol{\eta},\ \forall \boldsymbol{\xi} \in \mathcal{A},\ \boldsymbol{\eta} \in \mathcal{B};$$
$$\mathcal{A} \not\leq_{\text{int}\,\mathcal{C}} \mathcal{B} \Longleftrightarrow \boldsymbol{\xi} \not\leq_{\text{int}\,\mathcal{C}} \boldsymbol{\eta},\ \forall \boldsymbol{\xi} \in \mathcal{A},\ \boldsymbol{\eta} \in \mathcal{B}.$$

The set ordering relationships $\geq_{\mathcal{C}}, \geq_{\mathcal{C}\setminus\{\mathbf{0}\}}, \geq_{\text{int}\,\mathcal{C}}, \not\geq_{\mathcal{C}\setminus\{\mathbf{0}\}}, \not\geq_{\text{int}\,\mathcal{C}}$, are defined similarly.

When we have two vectors $\mathbf{y},\ \mathbf{z} \in \mathbb{R}^p$ with $\mathbf{z} \leq_{\text{int}\,\mathcal{C}} \mathbf{y}$, we say that $\mathbf{y}$ is *strongly dominated* below by $\mathbf{z}$; and if $\mathbf{z} \leq_{\mathcal{C}\setminus\{\mathbf{0}\}} \mathbf{y}$, we say that $\mathbf{y}$ is *dominated* below by $\mathbf{z}$. Thus an *efficient or minimal* point is one that is not dominated below by any other point in $\mathcal{Y}$, and a *weakly efficient or weakly minimal point* is one that is not strongly dominated below by any other point in $\mathcal{Y}$.

Definition 7.1.3 Let $\mathcal{C}$ be a closed and convex cone in $\mathbb{R}^p$ which induces all the ordering relationships as defined in Definition 7.1.1 and 7.1.2. Given a set $\mathcal{Y} \subset \mathbb{R}^p$.
(i) A point $\mathbf{y}^*$ in $\mathcal{Y}$ is said to be a *minimal point* or *efficient point* if there exists no $\mathbf{y} \in \mathcal{Y}$ such that $\mathbf{y} \leq_{\mathcal{C}\setminus\{\mathbf{0}\}} \mathbf{y}^*$. The set of all minimal points is called the *efficient frontier* of the set $\mathcal{Y}$:

$$\min_{\mathcal{C}\setminus\{\mathbf{0}\}}\mathcal{Y} = \{\mathbf{y}^* \mid \nexists \mathbf{y} \in \mathcal{Y} \text{ such that } \mathbf{y} \leq_{\mathcal{C}\setminus\{\mathbf{0}\}} \mathbf{y}^*\}.$$

Similarly, a point $\mathbf{y}^*$ in $\mathcal{Y}$ is said to be a *maximal point* if there exists no $\mathbf{y} \in \mathcal{Y}$ such that $\mathbf{y} \geq_{C\setminus\{\mathbf{0}\}} \mathbf{y}^*$. The set of all maximal points is denoted by

$$\max_{C\setminus\{\mathbf{0}\}}\mathcal{Y} = \{\mathbf{y}^* \mid \nexists \mathbf{y} \in \mathcal{Y} \text{ such that } \mathbf{y} \geq_{C\setminus\{\mathbf{0}\}} \mathbf{y}^*\}.$$

(ii) Assume further that int$C \neq \emptyset$. A point $\mathbf{y}^*$ in $\mathcal{Y}$ is said to be a *weakly minimal point* or *weakly efficient point* if there exists no $\mathbf{y} \in \mathcal{Y}$ such that $\mathbf{y} \leq_{\text{int } C} \mathbf{y}^*$. The set of all weakly minimal points is called the *weakly minimal or weakly efficient frontier* of the set $\mathcal{Y}$:

$$\min_{\text{int } C}\mathcal{Y} = \{\mathbf{y}^* \mid \nexists \mathbf{y} \in \mathcal{Y} \text{ such that } \mathbf{y} \leq_{\text{int } C} \mathbf{y}^*\}.$$

Similarly, a point $\mathbf{y}^*$ in $\mathcal{Y}$ is said to be a *weakly maximal point* if there exists no $\mathbf{y} \in \mathcal{Y}$ such that $\mathbf{y} \geq_{\text{int } C} \mathbf{y}^*$. The set of all maximal points is denoted by

$$\max_{\text{int } C}\mathcal{Y} = \{\mathbf{y}^* \mid \nexists \mathbf{y} \in \mathcal{Y} \text{ such that } \mathbf{y} \geq_{\text{int } C} \mathbf{y}^*\}.$$

Definition 7.1.4 Given a set $\mathcal{Y} \subset \mathbb{R}^p$. A point $\mathbf{y}^* \in \mathcal{Y}$ is said to be a *properly minimal point or properly efficient point* of $\mathcal{Y}$ if it is minimal and there exists a finite scalar $M > 0$ such that for each i,

$$\frac{y_i^* - y_i}{y_j - y_j^*} \leq M,$$

for some j such that $y_j > y_j^*$ whenever $\mathbf{y} \in \mathcal{Y}$ and $y_i < y_i^*$.

Definition 7.1.5 Let C be a closed and convex cone in $\mathbb{R}^p$ and $\mathcal{X} \subset \mathbb{R}^p$.
(i) $\mathcal{X}$ is said to be C-bounded if there exists $\mathbf{b} \in -C$ such that $\mathcal{X} \subseteq \mathbf{b} + C$.
(ii) The set $\mathcal{X}$ is said to be strongly C-bounded if there exists $\mathbf{b} \in -C$ such that $\mathcal{X} \subset \mathbf{b} + \text{int } C$.
(iii) The set $\mathcal{X}$ is said to be strongly C-closed if $\mathcal{X} + \text{cl}(C)$ is closed.

Remark 7.1.1 Note that the set of all properly minimal points is a subset of $\min_{C\setminus\{\mathbf{0}\}}\mathcal{Y}$, and $\min_{C\setminus\{\mathbf{0}\}}\mathcal{Y} \subseteq \min_{\text{int } C}(\mathcal{Y})$. Note also that if C has a non-empty interior, then C-boundedness implies strong C-boundedness. This is certainly the case if $C = \mathbb{R}^p_+$.

Theorem 7.1.3 Let $C = \mathbb{R}^p$. If $\mathcal{Y}$ is $\mathbb{R}^p_+$-bounded and closed (resp., $\mathbb{R}^p_-$-bounded and closed), then $\min_{C\setminus\{\mathbf{0}\}}\mathcal{Y} \neq \emptyset$ (resp., $\max_{C\setminus\{\mathbf{0}\}} \mathcal{Y} \neq \emptyset$).

Proof: Refer to [SNT1]. ∎

Lemma 7.1.4 Let $\mathcal{Y}^1, \mathcal{Y}^2 \subseteq \mathbb{R}^p$.

$$\min_{C\setminus\{\mathbf{0}\}}(\mathcal{Y}^1 + \mathcal{Y}^2) \subseteq \min_{C\setminus\{\mathbf{0}\}}(\mathcal{Y}^1) + \min_{C\setminus\{\mathbf{0}\}}(\mathcal{Y}^2).$$

Proof: Let $\mathbf{z} \in \min_{C\setminus\{\mathbf{0}\}}(\mathcal{Y}^1 + \mathcal{Y}^2)$, then $\mathbf{z} = \mathbf{y}^1 + \mathbf{y}^2$ for some $\mathbf{y}^1 \in \mathcal{Y}^1$ and $\mathbf{y}^2 \in \mathcal{Y}^2$. Suppose $\mathbf{y}^1 \notin \min_{C\setminus\{\mathbf{0}\}}(\mathcal{Y}^1)$, then there exists $\hat{\mathbf{y}} \in \mathcal{Y}^1$ such that $\hat{\mathbf{y}} \leq \mathbf{y}^1$.

The vector $\hat{\mathbf{y}} + \mathbf{y}^2 \in \mathcal{Y}^1 + \mathcal{Y}^2$ is such that $\hat{\mathbf{y}} + \mathbf{y}^2 \leq \mathbf{y}^1 + \mathbf{y}^2$, contradicting that $\mathbf{y}^1 + \mathbf{y}^2 \in \min_{C\setminus\{\mathbf{0}\}}(\mathcal{Y}^1 + \mathcal{Y}^2)$. Hence $\mathbf{y}^1 \in \min_{C\setminus\{\mathbf{0}\}}(\mathcal{Y}^1)$. Similarly we can prove that $\mathbf{y}^2 \in \min_{C\setminus\{\mathbf{0}\}}(\mathcal{Y}^2)$, and therefore $\mathbf{y}^1 + \mathbf{y}^2 \in \min_{C\setminus\{\mathbf{0}\}}(\mathcal{Y}^1) + \min_{C\setminus\{\mathbf{0}\}}(\mathcal{Y}^2)$. ∎

Definition 7.1.6 (Convexity of vector-valued functions) Let $\mathcal{X}$ be a convex subset in $\mathbb{R}^n$ and C a closed and convex cone in $\mathbb{R}^p$. A vector-valued function $\mathbf{f} : \mathcal{X} \to \mathbb{R}^p$ is said to be *C-convex* (quite often we simply call it convex, without the prefix C) on $\mathcal{X}$ if for $\mathbf{x}^1, \mathbf{x}^2 \in \mathcal{X}$, $t \in (0,1)$,

$$\mathbf{f}(t\mathbf{x}^1 + (1-t)\mathbf{x}^2) \leq_C t\mathbf{f}(\mathbf{x}^1) + (1-t)\mathbf{f}(\mathbf{x}^2).$$

$\mathbf{f}$ is said to be *strictly convex* on $\mathcal{X}$ if for $\mathbf{x}^1, \mathbf{x}^2 \in \mathcal{X}$, $t \in (0,1)$,

$$\mathbf{f}(t\mathbf{x}^1 + (1-t)\mathbf{x}^2) \leq_{\text{int } C} t\mathbf{f}(\mathbf{x}^1) + (1-t)\mathbf{f}(\mathbf{x}^2).$$

When $C = \mathbb{R}^p_+$, a vector-valued function $\mathbf{f} : \mathbb{R}^n \longrightarrow \mathbb{R}^p$ is said to be $\mathbb{R}^p_+$*-convex* (or just convex) if each component f_i is convex, and strictly $\mathbb{R}^p_+$*-convex* if each component f_i is strictly convex.

Definition 7.1.7 Let $\mathcal{X}$ be a convex subset in $\mathbb{R}^n$ and $\mathbf{f} : \mathcal{X} \to \mathbb{R}^p$ be a vector-valued function.

(i) The *directional derivative* of the function $\mathbf{f}$ at $\mathbf{x} \in \mathcal{X}$ in the direction $\mathbf{d} \in \mathbb{R}^n$ is defined, if it exists, by:

$$\mathbf{f}'(\mathbf{x}; \mathbf{d}) = \lim_{t \downarrow 0} \frac{\mathbf{f}(\mathbf{x} + t\mathbf{d}) - \mathbf{f}(\mathbf{x})}{t}.$$

(ii) $\mathbf{f}$ is said to be *Gateaux differentiable* at $\mathbf{x}$ if there exists a $p \times n$ matrix $\nabla\mathbf{f}(\mathbf{x})$ such that for any $\mathbf{d} \in \mathbb{R}^n$, $\mathbf{f}'(\mathbf{x}; \mathbf{d})$ exist and $\mathbf{f}'(\mathbf{x}; \mathbf{d}) = \nabla\mathbf{f}(\mathbf{x})\mathbf{d}$.

(iii) If $\mathbf{f}$ is Gateaux differentiable at every $\mathbf{x}$ of $\mathcal{X}$, then $\mathbf{f}$ is said to *Gateaux differentiable on $\mathcal{X}$*.

Lemma 7.1.5 Let $\mathcal{X}$ be a convex subset in $\mathbb{R}^n$ and $\mathbf{f} : \mathcal{X} \to \mathbb{R}^p$ be a vector-valued function. Assume that $\mathbf{f}$ is Gateaux differentiable on $\mathcal{X}$. Then f is convex on $\mathcal{X}$ if and only if for every $\mathbf{x}, \mathbf{y} \in \mathcal{X}$,

$$\mathbf{f}(\mathbf{x}^2) \geq_C \mathbf{f}(\mathbf{x}^1) + \nabla\mathbf{f}(\mathbf{x}^1)(\mathbf{x}^2 - \mathbf{x}^1). \tag{7.1.1}$$

Proof: By definition, if $\mathbf{f}$ is convex, then for all $t \in (0,1)$,

$$\mathbf{f}(\mathbf{x}^1 + t(\mathbf{x}^2 - \mathbf{x}^1)) \leq_C t\mathbf{f}(\mathbf{x}^2) + (1-t)\mathbf{f}(\mathbf{x}^1)$$

$$\frac{\mathbf{f}(\mathbf{x}^1 + t(\mathbf{x}^2 - \mathbf{x}^1)) - \mathbf{f}(\mathbf{x}^1)}{t} \leq_C \mathbf{f}(\mathbf{x}^2) - \mathbf{f}(\mathbf{x}^1).$$

(7.1.1) holds by taking the limit as $t \downarrow 0$ and Lemma 7.1.2.

Conversely, if (7.1.1) holds, then for $\mathbf{x}^1, \mathbf{x}^2 \in \mathcal{X}$, $t \in [0,1]$,

$$\mathbf{f}(\mathbf{x}^1) \geq_C \mathbf{f}(t\mathbf{x}^1 + (1-t)\mathbf{x}^2) + \nabla \mathbf{f}(t\mathbf{x}^1 + (1-t)\mathbf{x}^2)(\mathbf{x}^1 - (t\mathbf{x}^1 + (1-t)\mathbf{x}^2)) \quad (7.1.2)$$
$$\mathbf{f}(\mathbf{x}^2) \geq_C \mathbf{f}(t\mathbf{x}^1 + (1-t)\mathbf{x}^2) + \nabla \mathbf{f}(t\mathbf{x}^1 + (1-t)\mathbf{x}^2)(\mathbf{x}^2 - (t\mathbf{x}^1 + (1-t)\mathbf{x}^2)) \quad (7.1.3)$$

The conclusion follows from taking the sum of $t \times (7.1.2) + (1-t) \times (7.1.3)$.

■

Definition 7.1.8 Let $\mathcal{X}$ be a subset in $\mathbb{R}^n$ and $\mathbf{f} : \mathcal{X} \to \mathcal{Y}$ be a Gateaux differentiable vector-valued function. $\mathbf{f}$ is said to be *C-invex on $\mathcal{X}$* if for every $\mathbf{x}^1, \mathbf{x}^2 \in \mathcal{X}$, there exists a function $\eta(\mathbf{x}^2, \mathbf{x}^1) \in \mathcal{X}$ such that $\mathbf{x}^1 + t\eta(\mathbf{x}^2, \mathbf{x}^1) \in \mathcal{X}, \forall t \in (0,1)$, and

$$\mathbf{f}(\mathbf{x}^2) \geq_C \mathbf{f}(\mathbf{x}^1) + \nabla \mathbf{f}(\mathbf{x}^1)\eta(\mathbf{x}^2, \mathbf{x}^1).$$

The notion of convexity is well-defined for scalar and vector-valued functions using the ordering relationships in Definition 7.1.1. However, such a definition by ordering cannot be extended in a straightforward way to set-valued mappings, and inclusion type relationships must be used. It appears that the notion of convex set-valued mapping is slightly more complicated and more care is needed to deal with it. The notion of type I convexity (and concavity) set-valued mappings as defined below first appeared in [CGY1] in the context of a Hausdorff space. Let the power set of $\mathbb{R}^p$ be denoted by $2^{\mathbb{R}^p}$.

Definition 7.1.9 Let $\mathcal{X}$ be a closed and convex subset of $\mathbb{R}^n$ and C be a closed convex cone in $\mathbb{R}^n$. Let $\mathbf{x}^1, \mathbf{x}^2 \in \mathcal{X}$ and let $t \in (0,1)$. A set-valued function $\mathbf{f} : \mathcal{X} \longrightarrow 2^{\mathbb{R}^p}$ is said to be:
(i) *Type I C-convex* if and only if , $\mathbf{f}(t\mathbf{x}^1 + (1-t)\mathbf{x}^2) \subseteq t\mathbf{f}(\mathbf{x}^1) + (1-t)\mathbf{f}(\mathbf{x}^2) - C$;
(ii) *Type II C-convex* if and only if , $t\mathbf{f}(\mathbf{x}^1) + (1-t)\mathbf{f}(\mathbf{x}^2) \subseteq \mathbf{f}(t\mathbf{x}^1 + (1-t)\mathbf{x}^2) + C$;
(iii) *Type I C-concave* if and only if , $t\mathbf{f}(\mathbf{x}^1) + (1-t)\mathbf{f}(\mathbf{x}^2) \subseteq \mathbf{f}(t\mathbf{x} + (1-t)\mathbf{x}^2) - C$;
(iv) *Type II C-concave* if and only if , $\mathbf{f}(t\mathbf{x}^1 + (1-t)\mathbf{x}^2) \subseteq t\mathbf{f}(\mathbf{x}^1) + (1-t)\mathbf{f}(\mathbf{x}^2) + C$.

Quite often the function is simply referred to as convex or concave without the prefix C.

Remark 7.1.2 Type II convexity and concavity have been used previously in the literature, see, for example, [SNT1] and [L2], where it was acknowledged that if $\mathbf{f}$ is type II $\mathbb{R}^p_+$-convex, then $-\mathbf{f}$ is not necessarily type II $\mathbb{R}^p_+$-concave. However, from the fact that given two sets $\mathcal{X}^1, \mathcal{X}^2$, $\mathcal{X}^1 \subseteq \mathcal{X}^2 + C$ if and only if $-\mathcal{X}^1 \subseteq -\mathcal{X}^2 - C$, it is not difficult to see that

$\mathbf{f}$ is type I $\mathbb{R}^p_+$-convex if and only if $-\mathbf{f}$ is type II $\mathbb{R}^p_+$-concave;

and similarly,

$\mathbf{f}$ is type II $\mathbb{R}^p_+$-convex if and only if $-\mathbf{f}$ is type I $\mathbb{R}^p_+$-concave.

Remark 7.1.3 If $\mathbf{f}$ is a vector-valued function, then both type I and type II convexity (concavity) are equivalent to that given in Definition 7.1.7.

Definition 7.1.10 Let $\mathcal{X}$ be a closed and convex subset of $\mathbb{R}^n$ and $\mathcal{C}$ be a closed convex cone in $\mathbb{R}^n$. The *epigraph* of a set-valued function $\mathbf{f} : \mathcal{X} \longrightarrow 2^{\mathbb{R}^p}$ is defined as:

$$\text{epi}(\mathbf{f}) = \{(\mathbf{x}, \mathbf{y}) \in \mathbb{R}^n \times \mathbb{R}^p \mid \mathbf{x} \in \mathcal{X},\ \mathbf{y} \in \mathbf{f}(\mathbf{x}) + \mathcal{C}\}.$$

Theorem 7.1.6 Let $\mathcal{X}$ be a closed and convex subset of $\mathbb{R}^n$, $\mathcal{C}$ be a closed convex cone in $\mathbb{R}^n$, and $\mathbf{f} : \mathcal{X} \longrightarrow 2^{\mathbb{R}^p}$ be a set-valued function on $\mathbb{R}^p$. Then $\mathbf{f}$ is type II $\mathcal{C}$-convex if and only if epi($\mathbf{f}$) is convex.

Proof: See Proposition 2.2.3 of [SNT1]. ∎

Definition 7.1.11 (Monotonicity of vector-valued functions) A function $\mathbf{f} : \mathbb{R}^n \rightarrow \mathbb{R}^n$ is said to be *monotone* if

$$(\mathbf{f}(\mathbf{x}^1) - \mathbf{f}(\mathbf{x}^2))^\top (\mathbf{x}^1 - \mathbf{x}^2) \geq 0 \quad \forall \mathbf{x}^1, \mathbf{x}^2 \in \mathbb{R}^n.$$

Definition 7.1.12 Let $\mathcal{X}$ be a subset of $\mathbb{R}^n$. A matrix-valued function $\mathbf{F} : \mathcal{X} \rightarrow \mathbb{R}^{n \times m}$ is said to be *affine* if

$$\mathbf{F}(\alpha \mathbf{x}^1 + \beta \mathbf{x}^2) = \alpha \mathbf{F}(\mathbf{x}^1) + \beta \mathbf{F}(\mathbf{x}^2), \quad \forall \mathbf{x}^1, \mathbf{x}^2 \in \mathcal{X}, \quad \alpha, \beta \in \mathbb{R},\ \alpha + \beta = 1.$$

7.2 Vector Conjugate Duality

The concept of conjugate duality as discussed in Section 1.3 can be readily generalized to vector-valued functions, and we call the generalization as *vector conjugate duality,* where there exists more than one possible version. Like in the scalar case, vector conjugate duality plays a very important role in the duality of multicriteria optimization and vector variational inequalities. While the results here are presented for finite dimensional vector spaces, corresponding abstract versions for infinite dimensional spaces exist as well, see [L2]. For our application, we present two versions of conjugate duals, one for the matrix case and another for the vector case. Let $\mathcal{C}$ be a closed and convex cone of $\mathbb{R}^p$.

Definition 7.2.1 (Type I Conjugate or Fenchel transform of a vector-valued function, matrix version) Let $\mathbf{f} : \mathbb{R}^n \rightarrow \bar{\mathbb{R}}^p$ be a vector-valued function.

(i) The *Type I conjugate or Type I Fenchel Transform, (or simply Fenchel transform)* of $\mathbf{f}$ is defined by a set-valued function: $\mathbf{f}^* : \mathbb{R}^{p\times n} \to 2^{\mathbb{R}^p}$:

$$\mathbf{f}^*(\mathbf{U}) = \max_{C\setminus\{0\}} \{\mathbf{Ux} - \mathbf{f}(\mathbf{x}) \mid \mathbf{x} \in \mathbb{R}^n\}, \quad \mathbf{U} \in \mathbb{R}^{p\times n}.$$

(ii) The *Type I weak conjugate or Type I weak Fenchel Transform* of $\mathbf{f}$ is defined by a set-valued function: $\mathbf{f}_w^* : \mathbb{R}^{p\times n} \to 2^{\mathbb{R}^p}$

$$\mathbf{f}_w^*(\mathbf{U}) = \max_{\text{int } C} \{\mathbf{Ux} - \mathbf{f}(\mathbf{x}) \mid \mathbf{x} \in \mathbb{R}^n\}, \quad \mathbf{U} \in \mathbb{R}^{p\times n}.$$

(iii) The Type I Fenchel transform $\mathbf{f}^{**}$ of $\mathbf{f}^*$ is called the *Type I bi-Fenchel transform or Type I biconjugate* of $\mathbf{f}$:

$$\mathbf{f}^{**}(\mathbf{x}) = \max_{C\setminus\{0\}} \cup_{\mathbf{U}\in\mathbb{R}^{p\times n}} \{\mathbf{Ux} - \mathbf{f}^*(\mathbf{U})\ \}, \quad \mathbf{x} \in \mathbb{R}^n.$$

(iv) The Type I weak Fenchel transform $\mathbf{f}^{**}$ of $\mathbf{f}_w^*$ is called the *Type I weak bi-Fenchel transform or Type I weak biconjugate* of $\mathbf{f}$:

$$\mathbf{f}_w^{**}(\mathbf{x}) = \max_{\text{int } C} \cup_{\mathbf{U}\in\mathbb{R}^{p\times n}} \{\mathbf{Ux} - \mathbf{f}_w^*(\mathbf{U})\ \}, \quad \mathbf{x} \in \mathbb{R}^n.$$

Lemma 7.2.1 (Young's inequality for set-valued Fenchel transform, matrix version)

$$\mathbf{f}(\mathbf{x}) + \mathbf{f}^*(\mathbf{U}) - \mathbf{Ux} \not\leq_{C\setminus\{0\}} \{\mathbf{0}\} \qquad \forall \mathbf{x} \in \mathbb{R}^n \text{ and } \mathbf{U} \in \mathbb{R}^{p\times n}.$$

$$\mathbf{f}(\mathbf{x}) + \mathbf{f}_w^*(\mathbf{U}) - \mathbf{Ux} \not\leq_{\text{int } C} \{\mathbf{0}\} \qquad \forall \mathbf{x} \in \mathbb{R}^n \text{ and } \mathbf{U} \in \mathbb{R}^{p\times n}.$$

Proof: Follows directly from the definitions. ∎

Definition 7.2.2 (Type I subgradient of vector-valued function, matrix version) Let $\mathcal{X}$ be a convex subset of $\mathbb{R}^n$. Let $\mathbf{f} : \mathcal{X} \to \mathbb{R}^p$ be a convex function. Let $\mathbf{U} \in \mathbb{R}^{p\times n}$ be a matrix.

(i) $\mathbf{U}$ is said to be a *Type I weak subgradient* of $\mathbf{f}$ at $\mathbf{x}^0 \in \mathcal{X}$ if

$$\mathbf{f}(\mathbf{x}) - \mathbf{f}(\mathbf{x}^0) - \mathbf{U}(\mathbf{x} - \mathbf{x}^0) \not\leq_{\text{int } C} \mathbf{0}, \quad \forall \mathbf{x} \in \mathcal{X}.$$

(ii) $\mathbf{U}$ is said to be a *Type I subgradient* of $\mathbf{f}$ at $\mathbf{x}^0 \in \mathcal{X}$ if

$$\mathbf{f}(\mathbf{x}) - \mathbf{f}(\mathbf{x}^0) - \mathbf{U}(\mathbf{x} - \mathbf{x}^0) \not\leq_{C\setminus\{0\}} \mathbf{0}, \quad \forall \mathbf{x} \in \mathcal{X}.$$

(iii) $\mathbf{U}$ is said to be a *Type I strong subgradient* of $\mathbf{f}$ at $\mathbf{x}^0 \in \mathcal{X}$ if

$$\mathbf{f}(\mathbf{x}) - \mathbf{f}(\mathbf{x}^0) - \mathbf{U}(\mathbf{x} - \mathbf{x}^0) \geq_C \mathbf{0}, \quad \forall \mathbf{x} \in \mathcal{X}.$$

(iv) The set of all Type I weak subgradient of $\mathbf{f}$ at $\mathbf{x}^0$ is denoted by $\partial_w \mathbf{f}(\mathbf{x}^0)$, the set of all Type I subgradient of $\mathbf{f}$ at $\mathbf{x}^0$ is denoted by $\partial \mathbf{f}(\mathbf{x}^0)$ and the set of all Type I strong subgradient of $\mathbf{f}$ at $\mathbf{x}^0$ is denoted by $\partial_s \mathbf{f}(\mathbf{x}^0)$. Furthermore, if the specification of Type I or Type II is absent, by default it is regarded as Type I.

Lemma 7.2.2

$$\mathbf{U} \in \partial \mathbf{f}(\mathbf{x}) \text{ if and only if } \mathbf{Ux} - \mathbf{f}(\mathbf{x}) \in \mathbf{f}^*(\mathbf{U}),$$

$$\mathbf{U} \in \partial_w \mathbf{f}(\mathbf{x}) \text{ if and only if } \mathbf{Ux} - \mathbf{f}(\mathbf{x}) \in \mathbf{f}^*_w(\mathbf{U}).$$

Proof: We shall do it only for the case of type I subgradient, the proof for type I weak-subgradient follows in a similar manner. By definition, $\mathbf{U}$ is a type I subgradient of $\mathbf{f}$ at $\mathbf{x}$ if and only if

$$\begin{aligned} & \mathbf{f}(\mathbf{y}) - \mathbf{f}(\mathbf{x}) - \mathbf{U}(\mathbf{y} - \mathbf{x}) \not\leq_{C\setminus\{\mathbf{0}\}} \mathbf{0} \quad \forall \mathbf{y} \in \mathcal{X}, \\ \text{if and only if} \quad & \mathbf{Ux} - \mathbf{f}(\mathbf{x}) \not\leq_{C\setminus\{\mathbf{0}\}} \mathbf{Uy} - \mathbf{f}(\mathbf{y}) \quad \forall \mathbf{y} \in \mathcal{X}, \\ \text{if and only if} \quad & \mathbf{Ux} - \mathbf{f}(\mathbf{x}) \in \mathbf{f}^*(\mathbf{U}). \end{aligned}$$

∎

Definition 7.2.3 (Type II Fenchel transform of a vector-valued function, vector version) Let $\mathbf{f} : \mathbb{R}^n \to \bar{\mathbb{R}}^p$ be a vector-valued function. Denote the vector $\mathbf{1} = (1, 1, \cdots, 1)^\top \in \mathbb{R}^p$.

(i) The *Type II Fenchel Transform* of $\mathbf{f}$ is defined by a set-valued function: $\mathbf{f}^*_{\mathbf{1}} : \mathbb{R}^p \to 2^{\mathbb{R}^p}$ as follows:

$$\mathbf{f}^*_{\mathbf{1}}(\mathbf{y}) = \max\nolimits_{C\setminus\{\mathbf{0}\}}\{[\mathbf{y}^\top \mathbf{x}]\mathbf{1} - \mathbf{f}(\mathbf{x}) \mid \mathbf{x} \in \mathbb{R}^n\}, \quad \mathbf{y} \in \mathbb{R}^n$$

(ii) The *weak Type II Fenchel Transform* of $\mathbf{f}$ is defined by a set-valued function: $\mathbf{f}^*_{w\mathbf{1}} : \mathbb{R}^p \to 2^{\mathbb{R}^p}$ as follows:

$$\mathbf{f}^*_{w\mathbf{1}}(\mathbf{y}) = \max\nolimits_{\text{int } C}\{[\mathbf{y}^\top \mathbf{x}]\mathbf{1} - \mathbf{f}(\mathbf{x}) \mid \mathbf{x} \in \mathbb{R}^n\}, \quad \mathbf{y} \in \mathbb{R}^n$$

(iii) The Type II Fenchel Transform $\mathbf{f}^{**}_{\mathbf{1}}$ of $\mathbf{f}^*_{\mathbf{1}}$ is called the *Type II bi-Fenchel Transform or Type II biconjugate* of $\mathbf{f}$ and the weak Type II Fenchel Transform $\mathbf{f}^{**}_{w\mathbf{1}}$ of $\mathbf{f}^*_{w\mathbf{1}}$ is called the *Type II weak bi-Fenchel Transform or Type II biconjugate* of $\mathbf{f}$.

Lemma 7.2.3 (Young's inequality for set-valued Fenchel transform, vector version)

$$\mathbf{f}(\mathbf{x}) + \mathbf{f}^*_{w\mathbf{1}}(\mathbf{y}) - [\mathbf{y}^\top \mathbf{x}]\mathbf{1} \not\leq_{\text{int } C} \{\mathbf{0}\} \qquad \forall \mathbf{x}, \mathbf{y} \in \mathbb{R}^n.$$

$$\mathbf{f}(\mathbf{x}) + \mathbf{f}^*_{\mathbf{1}}(\mathbf{y}) - [\mathbf{y}^\top \mathbf{x}]\mathbf{1} \not\leq_{C\setminus\{\mathbf{0}\}} \{\mathbf{0}\} \qquad \forall \mathbf{x}, \mathbf{y} \in \mathbb{R}^n.$$

Proof: Follows directly from the definitions. ∎

Definition 7.2.4 (Type II subgradient of vector-valued functions, vector version) Let $\mathcal{X}$ be a convex subset of $\mathbb{R}^n$, $\mathbf{f} : \mathcal{X} \to \mathbb{R}^p$ be a convex function. and $\mathbf{y} \in \mathbb{R}^p$ be a vector.

(i) $\mathbf{y}$ is said to be a *Type-II weak subgradient* of $\mathbf{f}$ at $\mathbf{x}^0 \in \mathcal{X}$ if

$$\mathbf{f}(\mathbf{x}) - \mathbf{f}(\mathbf{x}^0) - [\mathbf{y}^\top(\mathbf{x} - \mathbf{x}^0)]\mathbf{1} \not\leq_{\text{int } C} \mathbf{0}, \quad \forall \mathbf{x} \in \mathcal{X}.$$

(ii) $\mathbf{y}$ is said to be a *Type-II subgradient* of $\mathbf{f}$ at $\mathbf{x}^0 \in \mathcal{X}$ if

$$\mathbf{f}(\mathbf{x}) - \mathbf{f}(\mathbf{x}^0) - [\mathbf{y}^\top(\mathbf{x} - \mathbf{x}^0)]\mathbf{1} \not\leq_{C\setminus\{\mathbf{0}\}} \mathbf{0}, \quad \forall \mathbf{x} \in \mathcal{X}.$$

(iii) $\mathbf{y}$ is said to be a *Type-II strong subgradient* of $\mathbf{f}$ at $\mathbf{x}^0 \in \mathcal{X}$ if

$$\mathbf{f}(\mathbf{x}) - \mathbf{f}(x_0) - [\mathbf{y}^\top(\mathbf{x} - \mathbf{x}^0)]\mathbf{1} \geq_C \mathbf{0}, \quad \forall \mathbf{x} \in \mathcal{X}.$$

(iv) The set of all type-II weak subgradient of $\mathbf{f}$ at $\mathbf{x}^0$ is denoted by $\partial_{w1}\mathbf{f}(\mathbf{x}^0)$, the set of all type-II subgradient of $\mathbf{f}$ at $\mathbf{x}^0$ is denoted by $\partial_1\mathbf{f}(\mathbf{x}^0)$ and the set of all type-II strong subgradient of $\mathbf{f}$ at $\mathbf{x}^0$ is denoted by $\partial_{s1}\mathbf{f}(\mathbf{x}^0)$.

Lemma 7.2.4

$$\mathbf{y} \in \partial_1\mathbf{f}(\mathbf{x}) \text{ if and only if } [\mathbf{y}^\top\mathbf{x}]\mathbf{1} - \mathbf{f}(\mathbf{x}) \in \mathbf{f}_1^*(\mathbf{y}).$$
$$\mathbf{y} \in \partial_{w1}\mathbf{f}(\mathbf{x}) \text{ if and only if } [\mathbf{y}^\top\mathbf{x}]\mathbf{1} - \mathbf{f}(\mathbf{x}) \in \mathbf{f}_{w1}^*(\mathbf{y}).$$

Proof: We shall do it only for the case of type-II subgradient, the proof for type-II weak-subgradient follows in a similar manner. By definition, $\mathbf{y}$ is a type-II subgradient of $\mathbf{f}$ at $\mathbf{x}$ if and only if

$$\mathbf{f}(\mathbf{z}) - \mathbf{f}(\mathbf{x}) - [\mathbf{y}^\top(\mathbf{z} - \mathbf{x})]\mathbf{1} \not\leq_{C\setminus\{\mathbf{0}\}} \mathbf{0} \quad \forall \mathbf{z} \in \mathcal{X}$$
$$\text{if and only if } [\mathbf{y}^\top\mathbf{x}]\mathbf{1} - \mathbf{f}(\mathbf{x}) \not\leq_{C\setminus\{\mathbf{0}\}} [\mathbf{y}^\top\mathbf{z}]\mathbf{1} - \mathbf{f}(\mathbf{z}) \quad \forall \mathbf{z} \in \mathcal{X}$$
$$\text{if and only if } [\mathbf{y}^\top\mathbf{x}]\mathbf{1} - \mathbf{f}(\mathbf{x}) \in \mathbf{f}_1^*(\mathbf{y}).$$

∎

It is well-known that the Fenchel transform of a scalar-valued function is convex in the usual definition (Theorem 1.3.2). However, a similar result (see [CGY1]) is not so well known in the case of vector-valued functions.

Theorem 7.2.5 Let $\mathbf{f} : \mathbb{R}^n \to \bar{\mathbb{R}}^p$ be convex. The Fenchel transform of $\mathbf{f}$ is type I convex.

Proof:

$$\begin{aligned}
&\mathbf{f}^*(t\mathbf{U}^1 + (1-t)\mathbf{U}^2) \\
&\quad = \max\nolimits_{C\setminus\{\mathbf{0}\}} \{(t\mathbf{U}^1 + (1-t)\mathbf{U}^2)\mathbf{x} - \mathbf{f}(\mathbf{x}) \mid \mathbf{x} \in \mathbb{R}^n\} \\
&\quad \subseteq \{(t\mathbf{U}^1 + (1-t)\mathbf{U}^2)\mathbf{x} - \mathbf{f}(\mathbf{x}) \mid \mathbf{x} \in \mathbb{R}^n\} \\
&\quad = \{t(\mathbf{U}^1\mathbf{x} - \mathbf{f}(\mathbf{x})) + (1-t)(\mathbf{U}^2\mathbf{x} - \mathbf{f}(\mathbf{x})) \mid \mathbf{x} \in \mathbb{R}^n\} \\
&\quad \subseteq t\{\mathbf{U}^1\mathbf{x} - \mathbf{f}(\mathbf{x}) \mid \mathbf{x} \in \mathbb{R}^n\} + (1-t)\{\mathbf{U}^2\mathbf{x} - \mathbf{f}(\mathbf{x}) \mid \mathbf{x} \in \mathbb{R}^n\} \\
&\quad \subseteq t \max\nolimits_{C\setminus\{\mathbf{0}\}} \{\mathbf{U}^1\mathbf{x} - \mathbf{f}(\mathbf{x}) \mid \mathbf{x} \in \mathbb{R}^n\} - C \\
&\qquad\qquad + (1-t) \max\nolimits_{C\setminus\{\mathbf{0}\}} \{\mathbf{U}^2\mathbf{x} - \mathbf{f}(\mathbf{x}) \mid \mathbf{x} \in \mathbb{R}^n\} - C \\
&\quad = t\mathbf{f}^*(\mathbf{U}^1) + (1-t)\mathbf{f}^*(\mathbf{U}^2) - C.
\end{aligned}$$

Thus $\mathbf{f}^*$ is type I convex. ∎

7.3 Scalarization of Multicriteria Optimization

We first formalize the notion of optimality of a multicriteria optimization problem VO. Essentially, we consider two sets: a feasible set $\mathcal{X} \subseteq \mathbb{R}^n$ and its image under a vector mapping $\mathbf{f}$, $\mathcal{Y} = \mathbf{f}(\mathcal{X}) \subseteq \mathbb{R}^p$. In this section, let $\mathcal{C} = \mathbb{R}^p_+$.

Definition 7.3.1 (Multicriteria Optimization Problem VO)

Let $\mathbf{f} = (f_1, f_2, \cdots, f_p)^\top : \mathbb{R}^n \to \mathbb{R}^p$ be a vector-valued function, and $\mathcal{X} \subseteq \mathbb{R}^n$ be the feasible set.

$$\text{(Problem VO)} \qquad \min_{\mathbf{R}^p_+\setminus\{0\}}\{\mathbf{f}(\mathbf{x}) \mid \mathbf{x} \in \mathcal{X}\} = \min_{\mathbf{R}^p_+\setminus\{0\}}\mathbf{f}(\mathcal{X}).$$

Ideally, one would like to find a solution $\mathbf{x}^*$ to the problem VO such that $\mathbf{f}(\mathbf{x}^*) \leq_C \mathbf{f}(\mathbf{x})$ either for all $\mathbf{x}$ in a local neighborhood of $\mathbf{x}^*$, or for all $\mathbf{x} \in \mathcal{X}$, i.e., $f_i(\mathbf{x}^*) \leq f_i(\mathbf{x}) \quad i = 1, 2, \cdots, p$. Such a point is called an *utopia point*. Unfortunately, utopia points rarely exist, as most multicriteria optimization problems invariably have conflicting cost functions. We have to settle for a much weaker notion of optimality here. We will discuss this in a more general framework of the minimal elements of a set in a space of dimension higher than two.

Definition 7.3.2 Consider the function $\mathbf{f} : \mathbb{R}^n \to \mathbb{R}^p$ and the set $\mathcal{X} \subseteq \mathbb{R}^n$ for the problem VO.

(i) A point $\mathbf{x}^* \in \mathcal{X}$ is said to be a *minimal solution (or efficient solution, or Pareto minimal solution, or non-dominated solution)* of VO if there exists no $\mathbf{x} \in \mathcal{X}$ such that $\mathbf{f}(\mathbf{x}) \leq_{\mathbf{R}^p_+\setminus\{0\}} \mathbf{f}(\mathbf{x}^*)$. The set of minimal solutions is denoted by $\text{argmin}_{\mathbf{R}^p_+\setminus\{0\}}(\mathbf{f}, \mathcal{X})$. $\mathbf{f}(\mathbf{x}^*)$ is called a *minimal point* or *efficient point* if $\mathbf{x}^*$ is a minimal solution. The set $\min_{\mathbf{R}^p_+\setminus\{0\}}\mathbf{f}(\mathcal{X}) = \{\mathbf{f}(\mathbf{x}^*) \mid \mathbf{x}^* \in \text{argmin}_{\mathbf{R}^p_+\setminus\{0\}}(\mathbf{f}, \mathcal{X})\}$ is called the *minimal frontier* or *efficient frontier*.

(ii) A point $\mathbf{x}^* \in \mathcal{X}$ is said to be a *weakly minimal solution (or weakly efficient solution, or weakly Pareto minimal solution, or non-strongly-dominated solution)* of VO if there exists no $\mathbf{x} \in \mathcal{X}$ such that $\mathbf{f}(\mathbf{x}) \leq_{\text{int } \mathbf{R}^p_+} \mathbf{f}(\mathbf{x}^*)$. The set of weakly minimal solutions is denoted by $\text{argmin}_{\text{int } \mathbf{R}^p_+}(\mathbf{f}, \mathcal{X})$. $\mathbf{f}(\mathbf{x}^*)$ is called a *weakly minimal point* if $\mathbf{x}^*$ is a weakly minimal solution. The set $\min_{\text{int } \mathbf{R}^p_+}\mathbf{f}(\mathcal{X}) = \{\mathbf{f}(\mathbf{x}^*) \mid \mathbf{x}^* \in \text{argmin}_{\text{int } \mathbf{R}^p_+}(\mathbf{f}, \mathcal{X})\}$ is called the *weakly minimal frontier*.

(iii) A point $\mathbf{x}^* \in \mathcal{X}$ is said to be a *properly minimal solution (in the sense of Geoffrion)* of VO if it is minimal and if there exists a finite scalar $M > 0$ such that for each i,

$$\frac{f_i(\mathbf{x}^*) - f_i(\mathbf{x})}{f_j(\mathbf{x}) - f_j(\mathbf{x}^*)} \leq M,$$

for some j such that $f_j(\mathbf{x}) > f_j(\mathbf{x}^*)$ whenever $f_i(\mathbf{x}^*) < f_i(\mathbf{x})$. $\mathbf{f}(\mathbf{x}^*)$ is called a *properly minimal point* if $\mathbf{x}^*$ is a properly minimal solution. The set of all properly minimal points is called the *properly minimal frontier*.

Remark 7.3.1 Note that there are several other versions of proper minimality, see [SNT1]. The above definition due to Geoffrion [G5] appears to be most popular.

Definition 7.3.3 In view of Definition 7.3.2 (ii), we define the following weaker version of problem VO.

$$\text{(Problem WVO)} \qquad \min_{\text{int } \mathbf{R}^p_+}\{\mathbf{f}(\mathbf{x}) \mid \mathbf{x} \in \mathcal{X}\} = \min_{\text{int } \mathbf{R}^p_+} \mathbf{f}(\mathcal{X}).$$

We say that we

- *solve problem VO* if we find the set of all minimal solutions $\operatorname{argmin}_{\mathbf{R}^p_+\setminus\{\mathbf{0}\}}(\mathbf{f},\mathcal{X})$ and the minimal frontier $\min_{\mathbf{R}^p_+\setminus\{\mathbf{0}\}}\mathbf{f}(\mathcal{X})$;

- *solve problem WVO* if we find the set of all weakly minimal solutions $\operatorname{argmin}_{\text{int } \mathbf{R}^p_+}(\mathbf{f},\mathcal{X})$ and the weakly minimal frontier $\min_{\text{int } \mathbf{R}^p_+}\mathbf{f}(\mathcal{X})$.

The solution of multicriteria optimization problems is a set. This is in general a much more difficult problem than the scalar case. Only a few special cases of problem VO have been solved fully. Two such cases are found in the linear case [Y4] and the quadratic case [GY4]. Apart from these there are few satisfactory results in finding the complete set of minimal solutions and minimal frontier numerically for general nonlinear problems. For the rest of this section, we shall present some important theoretical results for solving the multicriteria optimization problem.

In practice, multicriteria optimization problems are often reduced to scalar optimization problems by composing the vector cost function with a so-called *utility or value function.* There are two fundamental requirements of all utility functions used for scalarizing multicriteria optimization problems:

- Requirement 1. They should cover all minimal solutions for any multicriteria optimization problem, i.e., all minimal solutions can be computed by solving some scalarized optimization problem.

- Requirement 2. Solutions to the scalarized optimization problem should also be minimal solutions to the multicriteria optimization problem.

Definition 7.3.4 Let $\mathbb{R}^p_+$ be a closed and convex cone in $\mathbb{R}^p$, and $\mathbf{y}^1, \mathbf{y}^2 \in \mathbb{R}^p$. A function $\psi : \mathbb{R}^p \to \mathbb{R}$ is said to be:
(i) *monotone* if $\mathbf{y}^1 \geq_{\mathbf{R}^p_+} \mathbf{y}^2 \Rightarrow \psi(\mathbf{y}^1) \geq \psi(\mathbf{y}^2)$,
(ii) *strictly monotone* if $\mathbf{y}^1 \geq_{\text{int } \mathbf{R}^p_+} \mathbf{y}^2 \Rightarrow \psi(\mathbf{y}^1) > \psi(\mathbf{y}^2)$, and
(iii) *strongly monotone* if $\mathbf{y}^1 \geq_{\mathbf{R}^p_+\setminus\{\mathbf{0}\}} \mathbf{y}^2 \Rightarrow \psi(\mathbf{y}^1) > \psi(\mathbf{y}^2)$.

The following function proves to be very useful both in the scalarization of multicriteria optimization problems as well as in the scalarization of vector variational inequalities to be discussed in Chapter Nine. The original version due to Gerstewitz (see [GW1]) is published in German. Its first appearance in English is found in [L2]. This result is further studied and applied in [CGY4] and [GY3].

Definition 7.3.5 Given fixed $\mathbf{e} \in \text{ int } \mathbb{R}^p_+$ and $\mathbf{a} \in \mathbb{R}^p$, the Gerstewitz's function $\xi_{\mathbf{ea}} : \mathbb{R}^p \to \mathbb{R}$ is defined by:

$$\xi_{\mathbf{ea}}(\mathbf{y}) = \min\left\{t \in \mathbb{R} \mid \mathbf{y} \in \mathbf{a} + t\mathbf{e} - \mathbb{R}^p_+\right\} \text{ for } \mathbf{y} \in \mathbb{R}^p$$
$$= \max_{1\le i\le p}\left\{\frac{y_i - a_i}{e_i}\right\}.$$

Definition 7.3.6 A function $\phi : \mathbb{R}^p \to \mathbb{R}$ is a *utility function* if it is strongly monotone, and a *weak utility function* is it is only strictly monotone.

The corresponding scalar optimization problem is then defined by, given the utility function $\phi : \mathbb{R}^p \to \mathbb{R}$,

$$\text{(Problem } \phi\text{-Opt)} \qquad \min \phi \circ \mathbf{f}(\mathbf{x}) \quad \text{subject to } \mathbf{x} \in C.$$

The most popular scalarization method appears to be the linear scalarization method, where the utility function is linear, resulting in the scalar optimization problem:

$$\text{(Problem } \boldsymbol{\lambda}\text{-Opt)} \qquad \min \boldsymbol{\lambda}^\top \mathbf{f}(\mathbf{x}), \quad \text{subject to } \mathbf{x} \in C$$

where $\boldsymbol{\lambda} \in \mathbb{R}^p_+$ (which happens to be the dual cone of itself).

Despite its overwhelming popularity, the linear scalarization method is really not very useful when dealing with non-convex problems since it does not satisfy Requirement 1, i.e, it does not guarantee to compute all the minimal solutions.

Another powerful yet not so popular utility function is the Gerstewitz's function (or alternatively, weighted Tchebyshev norm) as defined in Definition 7.3.5. This gives rise to a nonlinear scalarization. The corresponding scalar optimization problem is given by:

$$\text{(Problem } \xi\text{-Opt)} \qquad \min \xi_{\mathbf{ea}}(\mathbf{f}(\mathbf{x})) \quad \text{subject to } \mathbf{x} \in C,$$

for some $\mathbf{e} \in \text{int } \mathbb{R}^p_+$, $\mathbf{a} \in \mathbb{R}^p$. Note that $\xi_{\mathbf{ea}}$ is only strictly monotone, hence it qualifies only as a weak utility function. We shall present some results from [GY3] which justify that this is, in some sense, the best for analyzing weak multicriteria optimization problems.

Both the above utility functions are special cases of the Minkowski metric $d_{\rho,\boldsymbol{\lambda}}(\cdot, \mathbf{a}) : \mathcal{Y} \subset \mathbb{R}^p \to \mathbb{R}_+$,

$$d_{\rho,\boldsymbol{\lambda}}(\mathbf{y}, \mathbf{a}) = \begin{cases} \{\sum_{i=1}^p [\lambda_i(y_i - a_i)]^\rho\}^{\frac{1}{\rho}}, & \text{if } 1 \le \rho < \infty; \\ \max_{1\le i\le p} \ \lambda_i(y_i - a_i), & \text{if } \rho = \infty, \end{cases}$$

where $\mathbf{y} \in \mathcal{Y}$, $\mathcal{Y}$ is a $\mathbb{R}^p_+$-bounded set, $\mathbf{a} \in \mathbb{R}^p, \boldsymbol{\lambda} \in \mathbb{R}^p_+$. This very large class of utility functions was previously proposed by [CP1] in the context of goal programming. It is not difficult to see that, when $\rho = 1, \mathbf{a} = \mathbf{0}$, (in fact $\mathbf{a}$ can be anything!) scalarization by $d_{\rho,\boldsymbol{\lambda}}(\cdot, \mathbf{a}) : \mathcal{Y} \subset \mathbb{R}^p \to \mathbb{R}_+$ reduces to the linear scalarization.

When $\rho = \infty, \boldsymbol{\lambda} = (\frac{1}{e_1}, \cdots, \frac{1}{e_p})$ then scalarization by $d_{\rho,\boldsymbol{\lambda}}(\cdot, \mathbf{a}) : \mathcal{Y} \subset \mathbb{R}^p \to \mathbb{R}_+$ reduces to the $\xi_{\mathbf{ea}}$ scalarization. It is easy to check that the utility function $d_{\rho,\boldsymbol{\lambda}}(\cdot, \mathbf{a}) : \mathcal{Y} \subset \mathbb{R}^p \to \mathbb{R}_+$ is strongly monotone for all $1 < \rho < \infty$, or for $\rho = 1, \boldsymbol{\lambda} \in \text{int } \mathbb{R}^p_+$. For $\rho = \infty$, it is only strictly monotone.

Theorem 7.3.1 (Sufficient condition for minimal solution)
(i) For any utility function ϕ, a solution of ϕ-Opt is also a minimal solution of VO.
(ii) For any weak utility function ϕ, a solution of ϕ-Opt is also a weakly minimal solution of VO.

Proof: (i) For notational convenience, let $\mathcal{Y} = \mathbf{f}(\mathcal{X}) = \cup_{\mathbf{x}\in\mathcal{X}}\mathbf{f}(\mathbf{x})$ and hence

$$\min_{\mathbb{R}^p_+\setminus\{\mathbf{0}\}}(\mathcal{Y}) = \min_{\mathbb{R}^p_+\setminus\{\mathbf{0}\}}(\mathbf{f}(\mathcal{X})).$$

If a point $\mathbf{y} = \mathbf{f}(\mathbf{x}^1) \notin \min_{\mathbb{R}^p_+\setminus\{\mathbf{0}\}}(\mathcal{Y})$, then there exists an $\mathbf{z} = \mathbf{f}(\mathbf{x}^2)$ such that $\mathbf{z} \leq_{\mathbb{R}^p_+\setminus\{\mathbf{0}\}} \mathbf{y}$. By the strong monotonicity property of utility functions, this implies that $\phi(\mathbf{z}) < \phi(\mathbf{y})$, hence, $\phi(\mathbf{y}) > \min \mathbf{f}(\mathcal{X})$. The proof for (ii) is similar, using only strict monotonicity. ∎

The converse to Theorem 7.3.1 is not so obvious, and in fact is "not true" in general. In order for this to be true, the most common trick is to assume that the set $\mathbf{f}(\mathcal{X})$ is cone-convex, i.e., the set $\mathbf{f}(\mathcal{X}) + \mathbb{R}^p_+$ is a convex set in $\mathbb{R}^p$. Such a condition holds when one is dealing with convex multicriteria optimization, i.e., $\mathbf{f}$ is convex and $\mathcal{X}$ is convex.

Theorem 7.3.2 (Necessary condition for minimal solution in convex problems) Assume that $\mathbf{f}(\mathcal{X}) + \mathbb{R}^p_+$ is convex, then
(i) There exists some utility function ϕ such that a properly minimal solution of VO is also a solution of ϕ-Opt.
(ii) There exists some weak utility function ϕ such that a weakly minimal solution of VO is also a solution of ϕ-Opt.

Proof: Since the set $\mathbf{f}(\mathcal{X}) + \mathbb{R}^p_+$ is convex, there exists a supporting hyperplane at every boundary point. In particular, let $\mathbf{y}^*$ be a properly minimal point on $\min_{\mathbb{R}^p_+\setminus\{\mathbf{0}\}}(\mathcal{Y})$. Then there exists some $\boldsymbol{\lambda} \in \text{int } \mathbb{R}^p_+$ such that the supporting hyperplane is given by $\boldsymbol{\lambda}^\top(\mathbf{y} - \mathbf{y}^*) = 0$. By the definition of a supporting hyperplane,

$$\boldsymbol{\lambda}^\top \mathbf{y} \geq \boldsymbol{\lambda}^\top \mathbf{y}^* \quad \forall \mathbf{y} \in \mathbf{f}(\mathcal{X}),$$

and therefore $\boldsymbol{\lambda}^\top \mathbf{y}^*$ is the minimum for the scalar optimization problem based on the (linear) utility function represented by $\boldsymbol{\lambda}$. The proof for (ii) is similar. ∎

Remark 7.3.2 Theorem 7.3.2 is applicable only when the set $\mathcal{Y} = \mathbf{f}(\mathcal{X})$ is cone-convex. This is a very strong condition, and is not likely to hold, when either $\mathbf{f}$ is not convex, or $\mathcal{X}$ is not convex, or when $\mathcal{X}$ is a discrete set whence $\mathbf{f}(\mathcal{X})$ is also a discrete set and therefore convexity has no meaning.

Now $t(\mathbf{y}^* - \mathbf{b}) - (\mathbf{y} - \mathbf{b}) \in \mathbb{R}^p_+$ by (7.3.1), and $(1-t)(\mathbf{y}^* - \mathbf{b}) \in \text{int } \mathbb{R}^p_+$ since $t < 1$, therefore $\mathbf{y}^* - \mathbf{y} \in \text{int } \mathbb{R}^p_+$ or $\mathbf{y}^* \geq_{\text{int } \mathbb{R}^p_+} \mathbf{y}$, contradicting that $\mathbf{y}^*$ is weakly minimal. ∎

Theorem 7.3.8 (Non-convex scalarization theorem [GW1]) Let $\mathcal{Y} \subset \mathbb{R}^p$ be a $\mathbb{R}^p_+$-bounded subset. Then $\mathbf{y}^* \in \min_{\text{int } \mathbb{R}^p_+}(\mathcal{Y})$ if and only if, for some $\mathbf{a} \in \mathbb{R}^p$ and some $\mathbf{e} \in \text{int } \mathbb{R}^p_+$,

$$\xi_{\mathbf{ea}}(\mathbf{y}^*) = \min\ \xi_{\mathbf{ea}}(\mathcal{Y}).$$

Proof: (Sufficiency) Assume that for some $\mathbf{a} \in \mathbb{R}^p$ and some $\mathbf{e} \in \text{int } \mathbb{R}^p_+$, $\xi_{\mathbf{ea}}(\mathbf{y}^*) = \min \xi_{\mathbf{ea}}(\mathcal{Y})$. If $\mathbf{y}^* \notin \min_{\text{int } \mathbb{R}^p_+}(\mathcal{Y})$, then there exists an $\mathbf{y} \in \mathcal{Y}$ such that $\mathbf{y} \leq_{\text{int } \mathbb{R}^p_+} \mathbf{y}^*$. By the strict monotonicity of the Gerstewitz's function, we have $\xi_{\mathbf{ea}}(\mathbf{y}) < \xi_{\mathbf{ea}}(\mathbf{y}^*)$, which contradicts the assumption that $\xi_{\mathbf{ea}}(\mathbf{y}^*) = \min \xi_{\mathbf{ea}}(\mathcal{Y})$. Hence $\xi_{\mathbf{ea}}(\mathbf{y}^*) > \min \xi_{\mathbf{ea}}(\mathcal{Y})$.

(Necessity) Conversely, assume that $\mathbf{y}^* \in \min_{\text{int } \mathbb{R}^p_+}(\mathcal{Y})$. Thus

$$\mathbf{y}^* \not\geq_{\text{int } \mathbb{R}^p_+} \mathbf{y} \quad \forall \mathbf{y} \in \mathcal{Y}.$$

Note that int $\mathbb{R}^p_+ \neq \emptyset$ and $\mathcal{Y}$ is strongly $\mathbb{R}^p_+$-bounded. Then there exists $\mathbf{b} \in -\text{int } \mathbb{R}^p_+$ such that $\mathcal{Y} \subset \mathbf{b} + \text{int } \mathbb{R}^p_+$ or $\mathcal{Y} - \mathbf{b} \subset \text{int } \mathbb{R}^p_+$. Choosing $\mathbf{e} = \mathbf{y}^* - \mathbf{b} \in \text{int } \mathbb{R}^p_+$ and $\mathbf{a} = 0$, we have, by definition,

$$\xi_{\mathbf{e}0}(\mathbf{y} - \mathbf{b}) = \min\{t \mid \mathbf{y} - \mathbf{b} \in t(\mathbf{y}^* - \mathbf{b}) - \mathbb{R}^p_+\} \geq 1 = \xi_{\mathbf{e}0}(\mathbf{y}^* - \mathbf{b}) \text{ by Lemma 7.3.7.}$$

This is equivalent to $\xi_{\mathbf{eb}}(\mathbf{y}) \geq \xi_{\mathbf{eb}}(\mathbf{y}^*)$ $\forall \mathbf{y}$, or $\xi_{\mathbf{eb}}(\mathbf{y}^*) = \min \xi_{\mathbf{eb}}(\mathcal{Y})$. ∎

Theorem 7.3.8 is a strong result that establishes that finding a weakly minimal solution of VO is equivalent to solving the scalar optimization problem using the special utility function $\xi_{\mathbf{ea}}$, even when the set $\mathcal{Y}$ is non-convex or discrete. While we have established that the weighted Tchebyshev norm is in some theoretical sense ideal for solving multicriteria optimization problems, it should however be noted that this method is not without limitation. The first limitation, of course, is that it may yield weakly minimal solutions along with minimal solutions. Secondly, the scalarized optimization problem is inherently non-smooth, and thus sophisticated non-smooth optimization techniques may be required. In practice, one can possibly get around this by solving a sequence of smooth but only approximate sub-problems using a large but finite value of ρ. Then by increasing ρ it is sometimes possible to converge to some consistent solution, though for a very large value of ρ, numerical instability may occur.

7.4 Optimality Conditions for Multicriteria Optimization

We now discuss some basic optimality conditions for multicriteria optimization where the feasible set is defined by an explicit set of constraints. Consider the following multicriteria optimization problem.

Definition 7.4.1 (Inequality constrained multicriteria optimization problem) Let $\mathbf{f} : \mathbb{R}^n \to \mathbb{R}^p, \quad \mathbf{g} : \mathbb{R}^n \to \mathbb{R}^m$ be differentiable functions.

$$\text{(Problem VO)} \qquad \min_{\mathbf{R}^p_+ \setminus \{\mathbf{0}\}} \mathbf{f}(\mathbf{x})$$
$$\text{subject to} \quad \mathbf{x} \in \mathcal{X} = \{\mathbf{x} \in \mathbb{R}^n \mid \mathbf{g}(\mathbf{x}) \leq_{\mathbf{R}^m_+} \mathbf{0}\}.$$

Definition 7.4.2 (Kuhn-Tucker constraint qualification) A point $\mathbf{x}^* \in \mathcal{X}$ is said to satisfy the *Kuhn-Tucker constraint qualification* if, for any $\mathbf{y} \in \mathbb{R}^n$ such that $\nabla g_j(\mathbf{x}^*)\mathbf{y} \leq 0$ for all $j \in \mathcal{J} = \{i \mid g_i(\mathbf{x}^*) = 0\}$ (the active set), there exists a scalar $\bar{t} > 0$, a function $\boldsymbol{\psi} : [0, \bar{t}] \to \mathbb{R}^n$ that is differentiable at $t = 0$, and a scalar $\alpha > 0$ such that

$$\boldsymbol{\psi}(0) = \mathbf{x}^*, \qquad \mathbf{g}(\boldsymbol{\psi}(t)) \leq_{\mathbf{R}^m_+} \mathbf{0}, \ \forall\, t \in [0, \bar{t}], \qquad \dot{\boldsymbol{\psi}}(0) = \alpha \mathbf{y}.$$

Theorem 7.4.1 (Kuhn-Tucker necessary condition for weakly minimal solutions) Let $\mathbf{x}^*$ be such that the Kuhn-Tucker constraint qualification is satisfied. If $\mathbf{x}^*$ is a weakly minimal solution to the problem VO, then it is necessary that there exist $\boldsymbol{\mu} \in \mathbb{R}^p$ and $\boldsymbol{\lambda} \in \mathbb{R}^m$ such that

(i) $\boldsymbol{\mu}^\top \nabla \mathbf{f}(\mathbf{x}^*) + \boldsymbol{\lambda}^\top \nabla \mathbf{g}(\mathbf{x}^*) = \mathbf{0}^\top$,
(ii) $\boldsymbol{\lambda}^\top \mathbf{g}(\mathbf{x}^*) = 0$,
(iii) $\boldsymbol{\mu} \geq_{\mathbf{R}^p_+ \setminus \{\mathbf{0}\}} \mathbf{0}, \quad \boldsymbol{\lambda} \geq_{\mathbf{R}^m_+} \mathbf{0}$.

Proof: If $\mathbf{x}^*$ is a weakly minimal solution that satisfies the Kuhn-Tucker constraint qualification, then there exists a differentiable arc $\boldsymbol{\psi}(t), t \in [0, \bar{t}]$ such that

$$\boldsymbol{\psi}(0) = \mathbf{x}^*, \qquad \mathbf{g}(\boldsymbol{\psi}(t)) \leq_{\mathbf{R}^m_+} \mathbf{0}, \ \forall\, t \in [0, \bar{t}], \qquad \dot{\boldsymbol{\psi}}(0) = \alpha \mathbf{y},$$

for some $\mathbf{y}$ such that $\nabla g_j(\mathbf{x}^*)\mathbf{y} \leq 0$, for all $j \in \mathcal{J} = \{i \mid g_i(\mathbf{x}^*) = 0\}$. Then

$$f_i(\boldsymbol{\psi}(t)) = f_i(\mathbf{x}^*) + t\alpha \nabla f_i(\mathbf{x}^*)\mathbf{y} + o(t).$$

Then the following system must be empty in order for $\mathbf{x}^*$ to be minimal:

$$\nabla f_i(\mathbf{x}^*)\mathbf{y} < 0, \quad i = 1, 2, \cdots, p,$$
$$\nabla g_j(\mathbf{x}^*)\mathbf{y} \leq 0, \quad \forall\, j \in \mathcal{J} = \{i \mid g_i(\mathbf{x}^*) = 0\},$$

for otherwise $f_i(\boldsymbol{\psi}(t)) < f_i(\mathbf{x}^*)\ \forall i = 1, 2, \cdots, p$, and hence $\mathbf{x}^*$ cannot be weakly minimal. The conclusion that the conditions (i), (ii), and (iii) hold then follows from the Motzkin's theorem of alternative (Theorem 3.4.11) and the fact that the corresponding $\lambda_i = 0$ for an inactive constraint $g_i(\mathbf{x}) < 0$. ∎

Theorem 7.4.2 (Kuhn-Tucker sufficient condition for weak VO) If the functions $\mathbf{f}$ and $\mathbf{g}$ are convex, and if the Kuhn-Tucker conditions (i), (ii) and (iii) of Theorem 7.4.1 hold, then $\mathbf{x}^*$ is a weakly minimal solution to the problem VO.

Proof: Let $\mathbf{x}$ be a feasible solution. Then

$$\begin{aligned}
\boldsymbol{\mu}^{\top}\mathbf{f}(\mathbf{x}) &\geq \boldsymbol{\mu}^{\top}\mathbf{f}(\mathbf{x}) + \boldsymbol{\lambda}^{\top}\mathbf{g}(\mathbf{x}) \quad \text{since } \boldsymbol{\lambda} \geq_{\mathbf{R}^m_+} \mathbf{0},\ \mathbf{g}(\mathbf{x}) \leq_{\mathbf{R}^m_+} \mathbf{0}, \\
&= \boldsymbol{\mu}^{\top}\mathbf{f}(\mathbf{x}^* + (\mathbf{x} - \mathbf{x}^*)) + \boldsymbol{\lambda}^{\top}\mathbf{g}(\mathbf{x}^* + (\mathbf{x} - \mathbf{x}^*)) \\
&\geq \boldsymbol{\mu}^{\top}\mathbf{f}(\mathbf{x}^*) + \boldsymbol{\lambda}^{\top}\mathbf{g}(\mathbf{x}^*) + \left[\boldsymbol{\mu}^{\top}\nabla\mathbf{f}(\mathbf{x}^*) + \boldsymbol{\lambda}^{\top}\nabla\mathbf{g}(\mathbf{x}^*)\right](\mathbf{x} - \mathbf{x}^*) \\
&\qquad\qquad \text{by the convexity of } \mathbf{f} \text{ and } \mathbf{g} \\
&\geq \boldsymbol{\mu}^{\top}\mathbf{f}(\mathbf{x}^*), \quad \text{by (i) and (ii) of the Kuhn-Tucker conditions.}
\end{aligned}$$

Since for $\boldsymbol{\mu} \geq_{\mathbf{R}^p_+\setminus\{\mathbf{0}\}} \mathbf{0}$, the scalarization $\boldsymbol{\mu}^{\top}\mathbf{f}(\mathbf{x})$ constitutes a weak utility function. By Theorem 7.3.1(ii), we conclude that $\mathbf{x}^*$ is a weakly minimal solution to the problem VO. ■

Stronger versions of Theorem 7.4.1 and 7.4.2 can also be obtained for a properly minimal solution in the sense of Kuhn-Tucker, which is defined as follows.

Definition 7.4.3 (Kuhn-Tucker proper efficiency) A solution $\mathbf{x}^*$ of the problem VO is said to be *properly efficient or properly minimal in the sense of Kuhn-Tucker* if it is minimal and there exists no $\mathbf{y} \in \mathbb{R}^n$ such that

$$\begin{aligned}
&\nabla\mathbf{f}(\mathbf{x}^*)\mathbf{y} \leq_{\mathbf{R}^p_+\setminus\{\mathbf{0}\}} \mathbf{0}^{\top} \\
&\nabla g_j(\mathbf{x}^*)\mathbf{y} \leq 0, \quad \forall\, j \in \mathcal{J} = \{i \mid g_i(\mathbf{x}^*) = 0\}.
\end{aligned}$$

In the event that $\mathbf{f}$ and $\mathbf{g}$ are convex, it can be shown that (Kuhn-Tucker) proper efficiency implies the (Geoffrion) proper efficiency as defined in Definition 7.3.2 (iii) (see Theorem 3.1.5 of [SNT1]).

Theorem 7.4.3 (Kuhn-Tucker necessary condition for properly minimal solution) Let $\mathbf{x}^*$ be a properly minimal (in the sense of Kuhn-Tucker) solution to the problem VO, then it is necessary that there exist $\boldsymbol{\mu} \in \mathbb{R}^p$ and $\boldsymbol{\lambda} \in \mathbb{R}^m$ such that

(i) $\boldsymbol{\mu}^{\top}\nabla\mathbf{f}(\mathbf{x}^*) + \boldsymbol{\lambda}^{\top}\nabla\mathbf{g}(\mathbf{x}^*) = \mathbf{0}^{\top}$,
(ii) $\boldsymbol{\lambda}^{\top}\mathbf{g}(\mathbf{x}^*) = 0$,
(iii) $\boldsymbol{\mu} \geq_{\text{int } \mathbf{R}^p_+} \mathbf{0}, \quad \boldsymbol{\lambda} \geq_{\mathbf{R}^m_+} \mathbf{0}$.

Proof: Follows very closely to that of Theorem 7.4.1, except that the Kuhn-Tucker theorem of alternative (Theorem 3.4.12) is used instead of the Motzkin theorem of alternative. ■

Theorem 7.4.4 (Kuhn-Tucker sufficient condition for properly minimal solution) If the functions $\mathbf{f}$ and $\mathbf{g}$ are convex, and if the Kuhn-Tucker conditions (i), (ii) and (iii) of Theorem 7.4.3 hold, then $\mathbf{x}^*$ is a weakly minimal solution to the problem VO.

Proof: Similar to the proof for Theorem 7.4.2. ∎

Hitherto, the Lagrange multiplier used is a vector-valued one. To proceed, it is useful to look at the case of a matrix-valued Lagrange multiplier, which leads to the concept of a vector-valued Lagrangian.

Definition 7.4.4 (Set of positive matrices) Let

$$\mathcal{L} = \{\Lambda \in \mathbb{R}^{p \times m} \mid \Lambda \mathbb{R}^m_+ \subseteq \mathbb{R}^p_+\}.$$

Definition 7.4.5 (Vector-valued Lagrangian) For the problem VO, the vector-valued Lagrangian is a function $\mathbf{L} : \mathbb{R}^n \times \mathcal{L} \to \mathbb{R}^p$ defined by:

$$\mathbf{L}(\mathbf{x}, \Lambda) = \mathbf{f}(\mathbf{x}) + \Lambda \mathbf{g}(\mathbf{x}).$$

Theorem 7.4.5 (Necessary condition for a properly minimal solution) Assume that $\mathbf{f}$ and $\mathbf{g}$ are convex functions. If $\mathbf{x}^*$ is a properly minimal solution for the problem VO, and if the Slater constraint qualification condition is satisfied, then there exists $\Lambda^* \in \mathcal{L}$ such that

$$\begin{aligned} \mathbf{f}(\mathbf{x}^*) &\in \min\nolimits_{\mathbf{R}^p_+ \setminus \{\mathbf{0}\}} \{\mathbf{L}(\mathbf{x}, \Lambda^*) \mid \mathbf{x} \in \mathbb{R}^n\}, \\ \Lambda^* \mathbf{g}(\mathbf{x}^*) &= \mathbf{0}. \end{aligned}$$

Proof: Since $\mathbf{x}^*$ is properly minimal for problem VO, there exists $\boldsymbol{\mu}^* \geq_{\text{int } \mathbf{R}^p_+} \mathbf{0}$ such that

$$(\boldsymbol{\mu}^*)^\top \mathbf{f}(\mathbf{x}^*) \leq (\boldsymbol{\mu}^*)^\top \mathbf{f}(\mathbf{x}) \quad \forall \mathbf{x} \in \mathcal{X}.$$

It is easy to verify that $(\boldsymbol{\mu}^*)^\top \mathbf{f}(\mathbf{x})$ is a convex scalar-valued function. By Theorem 4.3.8, there exists a $\boldsymbol{\lambda}^* \geq_{\mathbf{R}^m_+} \mathbf{0}$ such that

$$(\boldsymbol{\mu}^*)^\top \mathbf{f}(\mathbf{x}^*) + (\boldsymbol{\lambda}^*)^\top \mathbf{g}(\mathbf{x}^*) \leq (\boldsymbol{\mu}^*)^\top \mathbf{f}(\mathbf{x}) + (\boldsymbol{\lambda}^*)^\top \mathbf{g}(\mathbf{x}) \quad \forall \mathbf{x} \in \mathbb{R}^n \tag{7.4.1}$$

and

$$(\boldsymbol{\lambda}^*)^\top \mathbf{g}(\mathbf{x}^*) = 0. \tag{7.4.2}$$

Given that $\boldsymbol{\mu}^* \geq_{\text{int } \mathbf{R}^p_+} \mathbf{0}$ and $\boldsymbol{\lambda}^* \geq_{\mathbf{R}^m_+} \mathbf{0}$, we choose a special matrix multiplier $\Lambda^* = [\lambda_1^* \boldsymbol{\xi}, \lambda_2^* \boldsymbol{\xi}, \cdots, \lambda_m^* \boldsymbol{\xi}]$ such that $\boldsymbol{\xi} \geq_{\mathbf{R}^p_+} \mathbf{0}$ and $\boldsymbol{\xi}^\top \boldsymbol{\mu}^* = 1$. Consequently,

$$\Lambda^* \in \mathcal{L},$$

$$(\Lambda^*)^\top \boldsymbol{\mu}^* = \boldsymbol{\lambda}^*, \tag{7.4.3}$$

$$\text{and} \quad \Lambda^* \mathbf{g}(\mathbf{x}^*) = \mathbf{0}. \tag{7.4.4}$$

(7.4.4) yields the second necessary condition. If

$$\mathbf{f}(\mathbf{x}^*) \notin \min\nolimits_{\mathbf{R}^p_+ \setminus \{\mathbf{0}\}} \{\mathbf{f}(\mathbf{x}) + \Lambda^* \mathbf{g}(\mathbf{x}) \mid \mathbf{x} \in \mathbb{R}^n\},$$

then there exists $\mathbf{x} \in \mathbb{R}^n$ such that

$$\mathbf{f}(\mathbf{x}^*) \geq_p \mathbf{f}(\mathbf{x}) + \boldsymbol{\Lambda}^* \mathbf{g}(\mathbf{x}). \tag{7.4.5}$$

By (7.4.3) and (7.4.4), we have

$$\begin{aligned}(\boldsymbol{\mu}^*)^\top \mathbf{f}(\mathbf{x}^*) + (\boldsymbol{\mu}^*)^\top \boldsymbol{\Lambda}^* \mathbf{g}(\mathbf{x}^*) &= (\boldsymbol{\mu}^*)^\top \mathbf{f}(\mathbf{x}^*) \\ &\geq (\boldsymbol{\mu}^*)^\top \mathbf{f}(\mathbf{x}) + (\boldsymbol{\mu}^*)^\top \boldsymbol{\Lambda}^* \mathbf{g}(\mathbf{x}) \\ &= (\boldsymbol{\mu}^*)^\top \mathbf{f}(\mathbf{x}) + (\boldsymbol{\lambda}^*)^\top \mathbf{g}(\mathbf{x}).\end{aligned}$$

This is a contradiction to (7.4.1). So $\mathbf{f}(\mathbf{x}^*) \in \min_{\mathbf{R}^p_+ \setminus \{\mathbf{0}\}} \{\mathbf{f}(\mathbf{x}) + \Lambda^* \mathbf{g}(\mathbf{x}) \mid \mathbf{x} \in \mathbb{R}^n\}$.■

Definition 7.4.6 (Vector-valued Saddle Point) For the problem VO, A pair $(\mathbf{x}^*, \boldsymbol{\Lambda}^*) \in \mathbb{R}^n \times \mathcal{L}$ is said to be a *saddle point* of $\mathbf{L}$ if

$$\mathbf{L}(\mathbf{x}^*, \boldsymbol{\Lambda}^*) \in \min_{\mathbf{R}^p_+ \setminus \{\mathbf{0}\}} \{\mathbf{L}(\mathbf{x}, \boldsymbol{\Lambda}^*) \mid \mathbf{x} \in \mathbb{R}^n\} \cap \max_{\mathbf{R}^p_+ \setminus \{\mathbf{0}\}} \{\mathbf{L}(\mathbf{x}^*, \boldsymbol{\Lambda}) \mid \boldsymbol{\Lambda} \in \mathcal{L}\}.$$

Lemma 7.4.6 $(\mathbf{x}^*, \boldsymbol{\Lambda}^*) \in \mathbb{R}^n \times \mathcal{L}$ is a saddle point for problem VO if and only if
(i) $\mathbf{L}(\mathbf{x}^*, \boldsymbol{\Lambda}^*) \in \min_{\mathbf{R}^p_+ \setminus \{\mathbf{0}\}} \{\mathbf{L}(\mathbf{x}, \boldsymbol{\Lambda}^*) \mid \mathbf{x} \in \mathbb{R}^n\}$,
(ii) $\mathbf{g}(\mathbf{x}^*) \leq_{\mathbf{R}^m_+} \mathbf{0}$,
(iii) $\boldsymbol{\Lambda}^* \mathbf{g}(\mathbf{x}) = \mathbf{0}$.

Proof: (Necessity) The proof of (i) follows directly from the definition of a saddle point. For (ii), we note that since $\mathbf{L}(\mathbf{x}^*, \boldsymbol{\Lambda}^*) \in \max_{\mathbf{R}^p_+ \setminus \{\mathbf{0}\}} \{\mathbf{L}(\mathbf{x}^*, \boldsymbol{\Lambda}) \mid \boldsymbol{\Lambda} \in \mathcal{L}\}$ by definition, then

$$\mathbf{f}(\mathbf{x}^*) + \boldsymbol{\Lambda}^* \mathbf{g}(\mathbf{x}^*) \not\leq_{\mathbf{R}^p_+ \setminus \{\mathbf{0}\}} \mathbf{f}(\mathbf{x}^*) + \boldsymbol{\Lambda} \mathbf{g}(\mathbf{x}^*) \quad \forall \boldsymbol{\Lambda} \in \mathcal{L}.$$

This implies that

$$(\boldsymbol{\Lambda} - \boldsymbol{\Lambda}^*) \mathbf{g}(\mathbf{x}^*) \not\geq_{\mathbf{R}^p_+ \setminus \{\mathbf{0}\}} \mathbf{0} \quad \forall \boldsymbol{\Lambda} \in \mathcal{L}, \tag{7.4.6}$$

or, for some $\boldsymbol{\mu}^* \geq_{\mathbf{R}^p_+} \mathbf{0}$,

$$(\boldsymbol{\mu}^*)^\top (\boldsymbol{\Lambda} - \boldsymbol{\Lambda}^*) \mathbf{g}(\mathbf{x}^*) \leq 0 \quad \forall \boldsymbol{\Lambda} \in \mathcal{L}. \tag{7.4.7}$$

If (ii) does not hold, i.e., if $\mathbf{g}(\mathbf{x}^*) \not\leq_{\mathbf{R}^m_+} \mathbf{0}$, then there exists $\boldsymbol{\lambda}^* \geq_{\mathbf{R}^m_+} \mathbf{0}$ such that $(\boldsymbol{\lambda}^*)^\top \mathbf{g}(\mathbf{x}^*) > 0$. For any $t > 0$, we can select some $\boldsymbol{\Lambda} \in \mathcal{L}$ such that

$$(\boldsymbol{\mu}^*)^\top \boldsymbol{\Lambda} = t (\boldsymbol{\lambda}^*)^\top.$$

Consequently,

$$(\boldsymbol{\mu}^*)^\top \boldsymbol{\Lambda} \mathbf{g}(\mathbf{x}^*) - (\boldsymbol{\mu}^*)^\top \boldsymbol{\Lambda}^* \mathbf{g}(\mathbf{x}^*) = t (\boldsymbol{\lambda}^*)^\top \mathbf{g}(\mathbf{x}^*) - (\boldsymbol{\mu}^*)^\top \boldsymbol{\Lambda}^* \mathbf{g}(\mathbf{x}^*) > 0$$

since t can be made arbitrarily large. This contradicts (7.4.7). So (ii) must hold.

By definition of $\mathcal{L}$,

$$\boldsymbol{\Lambda}^*\mathbf{g}(\mathbf{x}^*) \leq_{\mathbf{R}^p_+} \mathbf{0} \quad \forall \boldsymbol{\Lambda}^* \in \mathcal{L}. \tag{7.4.8}$$

Let $\boldsymbol{\Lambda} = \mathbf{0} \in \mathcal{L}$ in (7.4.6), we have

$$\boldsymbol{\Lambda}^*\mathbf{g}(\mathbf{x}^*) \not\leq_{\mathbf{R}^p_+\setminus\{\mathbf{0}\}} \mathbf{0}. \tag{7.4.9}$$

(7.4.8) and (7.4.9) together imply that $\boldsymbol{\Lambda}^*\mathbf{g}(\mathbf{x}^*) = \mathbf{0}$, which is (iii).

(Sufficiency) We need only to prove that (ii) and (iii) imply

$$\mathbf{L}(\mathbf{x}^*, \boldsymbol{\Lambda}^*) \in \max_{\mathbf{R}^p_+\setminus\{\mathbf{0}\}}\{\mathbf{L}(\mathbf{x}^*, \boldsymbol{\Lambda}) \mid \boldsymbol{\Lambda} \in \mathcal{L}\}.$$

Since $\mathbf{g}(\mathbf{x}^*) \leq_{\mathbf{R}^m_+} \mathbf{0}$, we have

$$\boldsymbol{\Lambda}\mathbf{g}(\mathbf{x}^*) \leq_{\mathbf{R}^p_+} \mathbf{0} \quad \forall \boldsymbol{\Lambda} \in \mathcal{L}.$$

This implies that

$$\max_{\mathbf{R}^p_+\setminus\{\mathbf{0}\}}\{\boldsymbol{\Lambda}\mathbf{g}(\mathbf{x}^*) \mid \boldsymbol{\Lambda} \in \mathcal{L}\} = \{\mathbf{0}\},$$

and hence,

$$\max_{\mathbf{R}^p_+\setminus\{\mathbf{0}\}}\{\mathbf{L}(\mathbf{x}^*, \boldsymbol{\Lambda}) \mid \boldsymbol{\Lambda} \in \mathcal{L}\} = \max_{\mathbf{R}^p_+\setminus\{\mathbf{0}\}}\{\mathbf{f}(\mathbf{x}^*) + \boldsymbol{\Lambda}\mathbf{g}(\mathbf{x}^*) \mid \boldsymbol{\Lambda} \in \mathcal{L}\} = \{\mathbf{f}(\mathbf{x}^*)\}.$$

Since $\boldsymbol{\Lambda}^*\mathbf{g}(\mathbf{x}^*) = \mathbf{0}$ by (ii), we have

$$\mathbf{L}(\mathbf{x}^*, \boldsymbol{\Lambda}^*) = \mathbf{f}(\mathbf{x}^*) \in \max_{\mathbf{R}^p_+\setminus\{\mathbf{0}\}}\{\mathbf{L}(\mathbf{x}^*, \boldsymbol{\Lambda}) \mid \boldsymbol{\Lambda} \in \mathcal{L}\}. \tag{7.4.10}$$

■

The following is a generalization of the saddle point sufficient condition (Theorem 4.2.2) for scalar optimization to multicriteria optimization.

Theorem 7.4.7 (Saddle point sufficient condition) If $(\mathbf{x}^*, \boldsymbol{\Lambda}^*) \in \mathbb{R}^n \times \mathcal{L}$ is a saddle point, then $\mathbf{x}^*$ is a minimal solution to the problem VO.

Proof: Suppose $\mathbf{x}^*$ is not minimal, then there exists some feasible $\mathbf{x} \in \mathbb{R}^n$ such that $\mathbf{g}(\mathbf{x}) \leq_{\mathbf{R}^m_+} \mathbf{0}$ and $\mathbf{f}(\mathbf{x}) \leq_{\mathbf{R}^p_+\setminus\{\mathbf{0}\}} \mathbf{f}(\mathbf{x}^*)$. Since $\boldsymbol{\Lambda}^* \in \mathcal{L}$,

$$\boldsymbol{\Lambda}^*\mathbf{g}(\mathbf{x}) \leq_{\mathbf{R}^p_+} \mathbf{0}.$$

Noting that $\boldsymbol{\Lambda}^*\mathbf{g}(\mathbf{x}^*) = \mathbf{0}$ from Lemma 7.4.6, we have,

$$\mathbf{L}(\mathbf{x}, \boldsymbol{\Lambda}^*) = \mathbf{f}(\mathbf{x}) + \boldsymbol{\Lambda}^*\mathbf{g}(\mathbf{x}) \leq_{\mathbf{R}^p_+} \mathbf{f}(\mathbf{x}) \leq_{\mathbf{R}^p_+\setminus\{\mathbf{0}\}} \mathbf{f}(\mathbf{x}^*) + \boldsymbol{\Lambda}^*\mathbf{g}(\mathbf{x}^*) = \mathbf{L}(\mathbf{x}^*, \boldsymbol{\Lambda}^*).$$

This contradicts that

$$\mathbf{L}(\mathbf{x}^*, \boldsymbol{\Lambda}^*) \in \min_{\mathbf{R}^p_+\setminus\{\mathbf{0}\}}\{\mathbf{L}(\mathbf{x}, \boldsymbol{\Lambda}^*) \mid \mathbf{x} \in \mathbb{R}^n\}.$$

■

In the presence of convexity, some sort of converse result to the above saddle point condition also holds. The following result can be regarded as a generalization of Theorem 4.2.8 for scalar optimization to multicriteria optimization.

Theorem 7.4.8 (Saddle point necessary condition) Assume that $\mathbf{f}$ and $\mathbf{g}$ are convex. If $\mathbf{x}^*$ is a properly minimal solution for the problem VO, and if the Slater constraint qualification condition is satisfied, then there exists $\boldsymbol{\Lambda}^* \in \mathcal{L}$ such that $(\mathbf{x}^*, \boldsymbol{\Lambda}^*) \in \mathbb{R}^n \times \mathcal{L}$ is a saddle point for $\mathbf{L}(\mathbf{x}, \boldsymbol{\Lambda})$.

Proof: Follows from Theorem 7.4.5 and Lemma 7.4.6. ■

CHAPTER 8

DUALITY IN MULTICRITERIA OPTIMIZATION

Unlike the case of scalar optimization, the subject of multicriteria optimization is characterized by far more theories than solution methods/algorithms. This is understandable since the computation of a solution set is clearly much harder than the computation of a solution point. In particular, the theory of duality in multicriteria optimization has attracted the attention of many researchers. Notwithstanding a more complex notion of optimality, it turns out that most of the duality results in scalar optimization can be extended and generalized to the multicriteria case, albeit in non-trivial ways.

We begin with the simplest case of duality arising from a multicriteria linear program in Section 8.1. This result can be manifested in many different forms. The version that we have chosen to present is due to [CGY3] which will be used directly in the analysis of gap functions in Section 8.5. In Section 8.2, we present a conjugate duality theory due to [SNT]. In Section 8.3, we present a convex duality result due to [TS1] that is a direct generalization of the Lagrangian duality in Section 4.3. In Section 8.4, we present a case of symmetric and self dual optimization problems. In Section 8.5, we generalize the concepts of Wolfe duality and gap function of Section 4.5 to convex multicriteria optimization problems. In Section 8.6, we conclude the chapter by presenting a duality theory based on the multicriteria convex composite formulation in [JY1]. This provides a unified framework for multicriteria fractional programming (see [C7], [CC2]), minimax optimization (see [JA3]) and generalized convex optimization problems (see [WM2]). This is a generalization of the duality theory for scalar convex composite optimization problems which has been extensively studied recently, see [J1], [JY2], and [Y2]. Other related references are found in [B1],[B2],[C3],[dG1],[J2],[J4],[K1],[L3],[M2],[R6],[TS2].

8.1 Duality in Linear Multicriteria Optimization

Duality of multicriteria linear programs was first studied by Gale et al [GKT1]. Subsequently, more duality results were derived by Iserman [I1], [I2], and [I3], and Nakayama [N2] for different formulations of the problem. In this section, we have chosen to present a version of multicriteria LP duality due to [CGY3] for the reason that the underlying formulation of this LP problem is directly relevant to the analysis of gap functions in Section 8.5. For comparison, we present similar results taken from [SNT1] and [I1] and deduce the proof as a corollary of the result due to [CGY3].

Definition 8.1.1 (Primal and dual multicriteria linear programs) Consider the primal problem

$$\begin{aligned}&\text{(Primal VLP Problem)}\\&\min_{\mathbf{R}^p_+\setminus\{0\}}\mathbf{Cx}\\&\text{subject to } \mathbf{x}\in\mathbb{R}^n, A\mathbf{x}\geq_{\mathbf{R}^m_+}\mathbf{b},\end{aligned}$$

where $\mathbf{x}\in\mathbb{R}^n$, $\mathbf{C}\in\mathbb{R}^{p\times n}$, $\mathbf{A}\in\mathbb{R}^{m\times n}$, and $\mathbf{b}\in\mathbb{R}^m$ and its dual problem

$$\begin{aligned}&\text{(Dual VLP Problem)}\\&\max_{\mathbf{R}^p_+\setminus\{0\}}\boldsymbol{\Lambda}\mathbf{b}\\&\text{subject to } \boldsymbol{\mu}^\top\boldsymbol{\Lambda}\geq_{\mathbf{R}^m_+}\mathbf{0}^\top \text{ and } \mathbf{A}^\top\boldsymbol{\Lambda}^\top\boldsymbol{\mu}=\mathbf{C}^\top\boldsymbol{\mu}\\&\qquad\text{for some } \boldsymbol{\mu}\geq_{\text{int }\mathbf{R}^p_+}\mathbf{0}.\end{aligned}$$

Let

$$\mathcal{X}=\{\mathbf{x}\in\mathbb{R}^n\mid\mathbf{Ax}\geq_{\mathbf{R}^m_+}\mathbf{b}\}$$

and

$$\mathcal{L}=\{\boldsymbol{\Lambda}\in\mathbb{R}^{p\times m}\mid\exists\boldsymbol{\mu}\geq_{\text{int }\mathbf{R}^p_+}\mathbf{0}\ \text{such that}\ \boldsymbol{\mu}^\top\boldsymbol{\Lambda}\geq_{\mathbf{R}^m_+}\mathbf{0}^\top\ \text{and}\ \mathbf{A}^\top\boldsymbol{\Lambda}^\top\boldsymbol{\mu}=\mathbf{C}^\top\boldsymbol{\mu}\}.$$

Different versions of primal and dual multicriteria linear programs differ in the way the linear constraints of the primal are being included. Nevertheless each can be reduced to the above, and vice versa, by some suitable transformation. The following duality result and its proof share some similarity to that of [I1] and Theorem 5.1.4 of [SNT1].

Theorem 8.1.1 (Multicriteria linear programming duality)
(i) (Weak Duality) $\forall(\mathbf{x},\boldsymbol{\Lambda})\in\mathcal{X}\times\mathcal{L}$, $\boldsymbol{\Lambda}\mathbf{b}\not\geq_{\mathbf{R}^p_+\setminus\{0\}}\mathbf{Cx}$.
(ii) (Strong Duality) if $\boldsymbol{\Lambda}^*\in\mathcal{L}$ and $\mathbf{x}^*\in\mathcal{X}$ satisfies $\boldsymbol{\Lambda}^*\mathbf{b}=\mathbf{Cx}^*$, then

$$\mathbf{x}^*\in\operatorname{argmin}_{\mathbf{R}^p_+\setminus\{0\}}\{\mathbf{Cx}\mid\mathbf{x}\in\mathcal{X}\}\ \text{ and }\ \boldsymbol{\Lambda}^*\in\max_{\mathbf{R}^p_+\setminus\{0\}}\{\boldsymbol{\Lambda}\mathbf{b}\mid\boldsymbol{\Lambda}^*\in\mathcal{L}\}.$$

(iii) (Set equality of efficient frontiers)

$$\min_{\mathbf{R}^p_+\setminus\{0\}}\{\mathbf{Cx}\mid\mathbf{x}\in\mathcal{X}\}=\max_{\mathbf{R}^p_+\setminus\{0\}}\{\boldsymbol{\Lambda}\mathbf{b}\mid\boldsymbol{\Lambda}\in\mathcal{L}\}.$$

Proof: (i) Suppose to the contrary, there exists some $\mathbf{x} \in \mathcal{X}$ and $\Lambda \in \mathcal{L}$ such that

$$\Lambda \mathbf{b} \geq_{\mathbf{R}^p_+ \backslash \{0\}} \mathbf{C}\mathbf{x}. \tag{8.1.1}$$

Then since $\Lambda \in \mathcal{L}$, there exists $\boldsymbol{\mu} \geq_{\text{int } \mathbf{R}^p_+} \mathbf{0}$ such that

$$\boldsymbol{\mu}^\top \Lambda \geq_{\mathbf{R}^m_+} \mathbf{0}^\top \tag{8.1.2}$$

and

$$\boldsymbol{\mu}^\top \Lambda \mathbf{A} = \boldsymbol{\mu}^\top \mathbf{C}. \tag{8.1.3}$$

Then equations (8.1.2), (8.1.3) and the feasibility of $\mathbf{x}$ together imply that

$$\boldsymbol{\mu}^\top \mathbf{C}\mathbf{x} = \boldsymbol{\mu}^\top \Lambda \mathbf{A}\mathbf{x} \geq \boldsymbol{\mu}^\top \Lambda \mathbf{b}. \tag{8.1.4}$$

But from (8.1.1) and the fact that $\boldsymbol{\mu} \geq_{\text{int } \mathbf{R}^p_+} \mathbf{0}$, we have

$$\boldsymbol{\mu}^\top \Lambda \mathbf{b} > \boldsymbol{\mu}^\top \mathbf{C}\mathbf{x}, \tag{8.1.5}$$

which is a contradiction to (8.1.4). Thus (i) holds.

(ii) Suppose, to the contrary, that

$$\mathbf{x}^* \notin \operatorname{argmin}_{\mathbf{R}^p_+ \backslash \{0\}} \{\mathbf{C}\mathbf{x} \mid \mathbf{x} \in \mathcal{X}\},$$

then there exists $\mathbf{x} \in \mathcal{X}$ such that

$$\mathbf{C}\mathbf{x} \leq_{\mathbf{R}^p_+ \backslash \{0\}} \mathbf{C}\mathbf{x}^* = \Lambda^* \mathbf{b},$$

which contradicts the weak duality (i).

Similarly, if to the contrary, that

$$\Lambda^* \notin \operatorname{argmax}_{\mathbf{R}^p_+ \backslash \{0\}} \{\Lambda \mathbf{b} \mid \boldsymbol{\lambda} \in \mathcal{L}\},$$

then there exists $\Lambda \in \mathcal{L}$ such that,

$$\Lambda \mathbf{b} \geq_{\mathbf{R}^p_+ \backslash \{0\}} \Lambda^* \mathbf{b} = \mathbf{C}\mathbf{x}^*,$$

which contradicts the weak duality (i).

(iii) First, we prove the inclusion that

$$\min_{\mathbf{R}^p_+ \backslash \{0\}} \{\mathbf{C}\mathbf{x} \mid \mathbf{x} \in \mathcal{X}\} \subseteq \max_{\mathbf{R}^p_+ \backslash \{0\}} \{\Lambda \mathbf{b} \mid \Lambda \in \mathcal{L}\}.$$

Let $\mathbf{x}^*$ be a minimal solution to the primal, i.e.,

$$\mathbf{C}\mathbf{x}^* \in \min_{\mathbf{R}^p_+ \backslash \{0\}} \{\mathbf{C}\mathbf{x} \mid \mathbf{x} \in \mathcal{X}\}.$$

Theorem 7.3.2 (i) asserts that there exists $\boldsymbol{\mu} \geq_{\text{int } \mathbf{R}^p_+} \mathbf{0}$ such that

$$\boldsymbol{\mu}^\top \mathbf{C}\mathbf{x}^* = \min\{\boldsymbol{\mu}^\top \mathbf{C}\mathbf{x} \mid \mathbf{x} \in \mathcal{X}\}. \tag{8.1.6}$$

Consider the scalar linear program defined by the right hand side of (8.1.6) and its duality. To facilitate the use of standard linear programming duality result as stated in Section 3.2, we cast the problem in the following equivalent standard form:

$$\min\ \tilde{\mathbf{c}}^\top \tilde{\mathbf{x}} \quad \text{subject to} \quad \tilde{\mathbf{A}}\tilde{\mathbf{x}} = \mathbf{b},\ \tilde{\mathbf{x}}^\top \geq_{\mathbf{R}^{2n+m}} \mathbf{0},$$

where

$$\tilde{\mathbf{x}}^\top = \begin{bmatrix} \mathbf{x}^1 \\ \mathbf{x}^2 \\ \mathbf{y} \end{bmatrix}, \quad \tilde{\mathbf{c}}^\top = \boldsymbol{\mu}^\top [\mathbf{C}, -\mathbf{C}, \mathbf{0}], \quad \tilde{\mathbf{A}} = [\mathbf{A}, -\mathbf{A}, -\mathbf{I}].$$

Let $\tilde{\mathbf{B}}$ be the basis of $\tilde{\mathbf{A}}$ corresponding to the optimal (basic feasible) solution, and $\tilde{\mathbf{c}}^\top_{\tilde{\mathbf{B}}}$ be the corresponding (sub-) cost vector of $\tilde{\mathbf{c}}^\top$. From Theorem 3.1.3, the optimality condition of the linear program is given by

$$\tilde{\mathbf{c}}^\top - \tilde{\mathbf{c}}^\top_{\tilde{\mathbf{B}}} \tilde{\mathbf{B}}^{-1}\tilde{\mathbf{A}} \geq_{\mathbf{R}^{2n+m}\setminus\{0\}} \mathbf{0}^\top,$$

or, expressing $\tilde{\mathbf{c}}^\top_{\tilde{\mathbf{B}}}$ as $\tilde{\mathbf{c}}^\top_{\tilde{\mathbf{B}}} = \boldsymbol{\mu}^\top \mathbf{C}_{\tilde{\mathbf{B}}}$, (where $\mathbf{C}_{\tilde{\mathbf{B}}}$ is a particular submatrix of $\mathbf{C}$,) we have:

$$\boldsymbol{\mu}^\top \mathbf{C} - \boldsymbol{\mu}^\top \mathbf{C}_{\tilde{\mathbf{B}}} \tilde{\mathbf{B}}^{-1}\mathbf{A} \geq_{\mathbf{R}^n_+} \mathbf{0}^\top \tag{8.1.7}$$

$$-\boldsymbol{\mu}^\top \mathbf{C} + \boldsymbol{\mu}^\top \mathbf{C}_{\tilde{\mathbf{B}}} \tilde{\mathbf{B}}^{-1}\mathbf{A} \geq_{\mathbf{R}^n_+} \mathbf{0}^\top \tag{8.1.8}$$

$$\boldsymbol{\mu}^\top \mathbf{C}_{\tilde{\mathbf{B}}} \tilde{\mathbf{B}}^{-1} \geq_{\mathbf{R}^m_+} \mathbf{0}^\top. \tag{8.1.9}$$

(8.1.7) and (8.1.8) together imply that

$$\boldsymbol{\mu}^\top (\mathbf{C} - \mathbf{C}_{\tilde{\mathbf{B}}} \tilde{\mathbf{B}}^{-1}\mathbf{A}) = \mathbf{0}^\top. \tag{8.1.10}$$

Let

$$\boldsymbol{\Lambda}^* = \mathbf{C}_{\tilde{\mathbf{B}}} \tilde{\mathbf{B}}^{-1},$$

then (8.1.10) reduces to

$$\boldsymbol{\mu}^\top (\mathbf{C} - \boldsymbol{\Lambda}^* \mathbf{A}) = \mathbf{0}^\top, \tag{8.1.11}$$

and (8.1.9) reduces to

$$\boldsymbol{\mu}^\top \boldsymbol{\Lambda}^* \geq_{\mathbf{R}^m_+} \mathbf{0}^\top. \tag{8.1.12}$$

Therefore $\Lambda^* \in \mathcal{L}$. Furthermore,

$$\boldsymbol{\Lambda}^* \mathbf{b} = \mathbf{C}_{\tilde{\mathbf{B}}} \tilde{\mathbf{B}}^{-1}\mathbf{b} = \mathbf{C}_{\tilde{\mathbf{B}}} \mathbf{x}^0_{\tilde{\mathbf{B}}} = \mathbf{C}\mathbf{x}^*,$$

where $\mathbf{x}^*$ is the optimal solution to the scalar linear program (8.1.6), and $\mathbf{x}^0_{\tilde{\mathbf{B}}}$ is its basic component. Thus $\boldsymbol{\Lambda}^*$ is a maximal solution to the dual VLP by (ii), and hence

$$\min_{\mathbf{R}^p_+\setminus\{0\}}\{\mathbf{C}\mathbf{x} \mid \mathbf{x} \in \mathcal{X}\} \subseteq \max_{\mathbf{R}^p_+\setminus\{0\}}\{\boldsymbol{\Lambda}\mathbf{b} \mid \boldsymbol{\Lambda} \in \mathcal{L}\}.$$

Conversely, we assume that $\boldsymbol{\Lambda}^*$ is a maximal solution to the dual VLP, then,

$$\forall \boldsymbol{\mu} \geq_{\text{int } \mathbf{R}_+^p} \mathbf{0} \quad \nexists \boldsymbol{\Lambda} \in \mathcal{L} \text{ such that } \boldsymbol{\mu}^\top \boldsymbol{\Lambda} \mathbf{b} > \boldsymbol{\mu}^\top \boldsymbol{\Lambda}^* \mathbf{b}.$$

Let $\boldsymbol{\lambda} = \boldsymbol{\Lambda}^\top \boldsymbol{\mu}$. Then it follows that

$$\nexists \boldsymbol{\lambda} \geq_{\mathbf{R}_+^m} \mathbf{0} \quad \nexists \boldsymbol{\mu} \geq_{\text{int } \mathbf{R}_+^p} \mathbf{0} \text{ such that } \boldsymbol{\lambda}^\top \mathbf{b} > \boldsymbol{\mu}^\top \boldsymbol{\Lambda}^* \mathbf{b} \text{ and } \mathbf{A}^\top \boldsymbol{\lambda} = \mathbf{C}^\top \boldsymbol{\mu}. \quad (8.1.13)$$

In other words, the system

$$\{(\boldsymbol{\lambda}, \boldsymbol{\mu}) \geq_{\mathbf{R}_+^{p-m}} \mathbf{0}^\top \mid \mathbf{A}^\top \boldsymbol{\lambda} = \mathbf{C}^\top \boldsymbol{\mu} \text{ and } \boldsymbol{\lambda}^\top \mathbf{b} > \boldsymbol{\mu}^\top \boldsymbol{\Lambda}^* \mathbf{b}\}$$

is infeasible. By Gale's Transposition Theorem (Theorem 3.4.9), this is true if and only if the system

$$\{(\mathbf{x}^*, \eta) \in \mathbb{R}^n \times \mathbb{R} \mid \mathbf{A}\mathbf{x}^* \geq_{\mathbf{R}_+^m} \eta \mathbf{b},\ \mathbf{C}\mathbf{x}^* \leq_{\mathbf{R}_+^p} \eta \boldsymbol{\Lambda}^* \mathbf{b},\ \eta > 0\}$$

is non-empty. Without loss of generality, let $\eta = 1$, then $\mathbf{x}^*$ satisfies $\mathbf{A}\mathbf{x} \geq_{\mathbf{R}_+^m} \mathbf{b}$ and hence is feasible for the primal VLP. Furthermore,

$$\mathbf{C}\mathbf{x}^* \leq_{\mathbf{R}_+^p} \boldsymbol{\Lambda}^* \mathbf{b}. \quad (8.1.14)$$

(8.1.14) together with (i) imply that

$$\mathbf{C}\mathbf{x}^* = \boldsymbol{\Lambda}^* \mathbf{b}. \quad (8.1.15)$$

Thus by strong duality (ii),

$$\boldsymbol{\Lambda}^* \mathbf{b} \in \min_{\mathbf{R}_+^p \setminus \{0\}} \{\mathbf{C}\mathbf{x} \mid \mathbf{x} \in \mathcal{X}\},$$

and hence,

$$\max_{\mathbf{R}_+^p \setminus \{0\}} \{\boldsymbol{\Lambda} \mathbf{b} \mid \boldsymbol{\Lambda} \in \mathcal{L}\} \subseteq \min_{\mathbf{R}_+^p \setminus \{0\}} \{\mathbf{C}\mathbf{x} \mid \mathbf{x} \in \mathcal{X}\}.$$

∎

For comparison, we present another variant of the primal and dual VLP. The following duality result is due to Sawaragi et. al. [SNT1], where the ordering relationships for the primal are induced by some closed pointed convex polyhedral cones $\mathcal{C}, \mathcal{M}, \mathcal{Q}$ in $\mathbb{R}^p, \mathbb{R}^n$, and $\mathbb{R}^m$ respectively, while the ordering relationships for the dual are induced by the *positive polar cones* of $\mathcal{C}, \mathcal{M}, \mathcal{Q}$ respectively. The positive polar cone of a cone $\mathcal{M}$ is defined by:

$$\mathcal{M}^0 = \{\mathbf{y} \in \mathbb{R}^n \mid \mathbf{x}^\top \mathbf{y} \geq 0 \quad \forall \mathbf{x} \in \mathcal{M}\}.$$

If $\mathcal{M} = \mathbb{R}_+^n$ then $\mathcal{M}^0 = \mathcal{M}$. In the following result, we specialize the result to a case where the ordering cone is the positive orthant, and as such the ordering cone for the dual is also the positive orthant.

Definition 8.1.2 (Primal and dual vector linear programs, Sawaragi's version)

(Primal VLP Problem)

$\min_{\mathbb{R}^p_+\setminus\{0\}} \mathbf{C}_s\mathbf{x}$

subject to $\mathbf{x} \geq_{\mathbb{R}^n_+} \mathbf{0},\ \mathbf{A}_s\mathbf{x} \geq_{\mathbb{R}^m_+} \mathbf{b}_s$,

where $\mathbf{C}_s \in \mathbb{R}^{p\times n},\ \mathbf{A}_s \in \mathbb{R}^{m\times n}, \mathbf{b}_s \in \mathbb{R}^m$,

(Dual VLP Problem)

$\max_{\mathbb{R}^p_+\setminus\{0\}} \Lambda_s\mathbf{b}_s$

subject to $\boldsymbol{\mu}^\top \boldsymbol{\Lambda} \geq_{\mathbb{R}^m_+} \mathbf{0}^\top$ and

$\mathbf{A}_s^\top \boldsymbol{\Lambda}_s^\top \boldsymbol{\mu} \leq_{\mathbb{R}^p_+} \mathbf{C}_s^\top \boldsymbol{\mu}$

for some $\boldsymbol{\mu} \geq_{\text{int } \mathbb{R}^p_+} \mathbf{0}$.

Let

$$\mathcal{X}_s = \{\mathbf{x} \in \mathbb{R}^n \mid \mathbf{x} \geq_{\mathbb{R}^n_+} \mathbf{0},\ \mathbf{A}_s\mathbf{x} \geq_{\mathbb{R}^m_+} \mathbf{b}_s\ \},$$

and

$$\mathcal{L}_s = \{\boldsymbol{\Lambda}_s \in \mathbb{R}^{p\times m} \mid \exists \boldsymbol{\mu} \geq_{\text{int } \mathbb{R}^p_+} \mathbf{0} \text{ such that } \boldsymbol{\mu}^\top \boldsymbol{\Lambda} \geq_{\mathbb{R}^m_+} \mathbf{0}^\top \text{ and } \mathbf{A}_s^\top \boldsymbol{\Lambda}_s^\top \boldsymbol{\mu} \leq_{\mathbb{R}^p_+} \mathbf{C}_s^\top \boldsymbol{\mu}\ \}.$$

Theorem 8.1.2 (Multicriteria linear programming duality, Sawaragi's version)

(i) (Weak Duality) $\forall(\mathbf{x}, \boldsymbol{\Lambda}) \in \mathcal{X}_s \times \mathcal{L}_s,\ \ \boldsymbol{\Lambda}_s\mathbf{b}_s \not\geq_{\mathbb{R}^p_+\setminus\{0\}} \mathbf{C}_s\mathbf{x}$.

(ii) (Strong Duality) if $\boldsymbol{\Lambda}_s^* \in \mathcal{L}_s$ and $\mathbf{x}^* \in \mathcal{X}_s$ satisfies $\boldsymbol{\Lambda}_s^*\mathbf{b}_s = \mathbf{C}_s\mathbf{x}^*$, then

$$\mathbf{x}^* \in \text{argmin}_{\mathbb{R}^p_+\setminus\{0\}}\{\mathbf{C}_s\mathbf{x} \mid \mathbf{x} \in \mathcal{X}_s\} \text{ and } \boldsymbol{\Lambda}_s^* \in \text{argmax}_{\mathbb{R}^p_+\setminus\{0\}}\{\boldsymbol{\Lambda}\mathbf{b}_s \mid \boldsymbol{\Lambda}_s \in \mathcal{L}_s\}.$$

(iii) (Set equality of efficient frontiers)

$$\min_{\mathbb{R}^p_+\setminus\{0\}}\{\mathbf{C}_s\mathbf{x} \mid \mathbf{x} \in \mathcal{X}_s\} = \max_{\mathbb{R}^p_+\setminus\{0\}}\{\boldsymbol{\Lambda}_s\mathbf{b}_s \mid \boldsymbol{\Lambda}_s \in \mathcal{L}_s\}.$$

Proof: We shall prove this as a corollary to Theorem 8.1.1 by examining a special case of the primal and dual multicriteria linear program in Definition 8.1.1. Let

$$\mathbf{A} = \begin{bmatrix} \mathbf{A}_s \\ \mathbf{I} \end{bmatrix}, \qquad \mathbf{b} = \begin{bmatrix} \mathbf{b}_s \\ \mathbf{0} \end{bmatrix}.$$

Then $\mathcal{X}$ is reduced to $\mathcal{X}_s$. Let

$$\begin{aligned}
\mathcal{L} &= \{[\boldsymbol{\Lambda}_s, \boldsymbol{\Lambda}'] \in \mathbb{R}^{p\times(m+n)} \mid \exists \boldsymbol{\mu} \geq_{\text{int } \mathbb{R}^p_+} \mathbf{0} \text{ such that } \boldsymbol{\mu}^\top[\boldsymbol{\Lambda}_s, \boldsymbol{\Lambda}'] \geq_{\mathbb{R}^{m+n}_+} \mathbf{0}^\top \\
&\qquad \text{and } [\mathbf{A}_s^\top, \mathbf{I}] \begin{bmatrix} \boldsymbol{\Lambda}_s \\ \boldsymbol{\Lambda}' \end{bmatrix} \boldsymbol{\mu} = \mathbf{C}_s^\top \boldsymbol{\mu}\ \} \\
&= \{[\boldsymbol{\Lambda}_s, \boldsymbol{\Lambda}'] \in \mathbb{R}^{p\times(m+n)} \mid \exists \boldsymbol{\mu} \geq_{\text{int } \mathbb{R}^p_+} \mathbf{0} \text{ such that } \boldsymbol{\mu}^\top \boldsymbol{\Lambda}_s \geq_{\mathbb{R}^m_+} \mathbf{0}^\top, \\
&\qquad \boldsymbol{\mu}^\top \boldsymbol{\Lambda}' \geq_{\mathbb{R}^n_+} \mathbf{0}^\top \text{ and } \mathbf{A}_s^\top \boldsymbol{\Lambda}_s^\top \boldsymbol{\mu} + (\boldsymbol{\Lambda}')^\top \boldsymbol{\mu} = \mathbf{C}_s^\top \boldsymbol{\mu}\} \\
&= \{[\boldsymbol{\Lambda}_s, \boldsymbol{\Lambda}'] \in \mathbb{R}^{p\times(m+n)} \mid \exists \boldsymbol{\mu} \geq_{\text{int } \mathbb{R}^p_+} \mathbf{0} \text{ such that } \boldsymbol{\mu}^\top \boldsymbol{\Lambda}_s \geq_{\mathbb{R}^m_+} \mathbf{0}, \\
&\qquad \text{and } \mathbf{A}_s^\top \boldsymbol{\Lambda}_s^\top \boldsymbol{\mu} \leq_{\mathbb{R}^p_+} \mathbf{C}_s^\top \boldsymbol{\mu}\}.
\end{aligned}$$

The dual problem of Definition 6.1.1 becomes

$$\max_{\mathbf{R}^p_+\setminus\{0\}}\{[\Lambda_s,\Lambda']\begin{bmatrix}\mathbf{b}_s\\ \mathbf{0}\end{bmatrix} \mid [\Lambda_s,\Lambda']\in\mathcal{L}\} = \max_{\mathbf{R}^p_+\setminus\{0\}}\{\Lambda_s\mathbf{b}_s \mid \Lambda_s\in\mathcal{L}_s\}$$

which is the dual VLP problem as defined in Definition 8.1.3. The conclusion follows from Theorem 8.1.1. ■

Clearly Theorem 8.1.1 can also be conversely proved as a corollary of Theorem 8.1.3.

8.2 Conjugate Duality in Convex Multicriteria Optimization

In this section, we present a generalization of the conjugate duality theory for scalar optimization (see Section 4.1 and [G2]). Although the theory is established for an unconstrained multicriteria optimization problem, it is only easy to allow for constraints implicitly by defining the underlying objective function to assume an abstract infinity whenever the explicit constraints are violated.

In Chapter Seven, we have discussed the generalization of conjugate functions to vector-valued functions. As it turns out, this is still insufficient to cope with the conjugate duality theory for multicriteria optimization as most of the functions involved are now set-valued. We begin this section by further defining the tools required for these set-valued functions.

Definition 8.2.1 Let $\mathbf{h} : \mathbb{R}^n \to 2^{\bar{\mathbf{R}}^p}$ be a set-valued function, and $\mathbf{y} \in \mathbf{h}(\mathbf{x})$.

(i) The conjugate dual (or Fenchel transform) of $\mathbf{h}$, is a set-valued function denoted by $\mathbf{h}^* : \mathbb{R}^{p\times n} \to 2^{\bar{\mathbf{R}}^p}$ and is defined as follows:

$$\mathbf{h}^*(\mathbf{U}) = \max_{\mathbf{R}^p_+\setminus\{0\}} \cup_{\mathbf{x}\in\mathbf{R}^n} [\mathbf{U}\mathbf{x} - \mathbf{h}(\mathbf{x})], \qquad \mathbf{U} \in \mathbb{R}^{p\times n}.$$

(ii) The biconjugate of $\mathbf{h}$, or the conjugate of $\mathbf{h}^*$, is a set-valued function denoted by $\mathbf{h}^{**} : \mathbb{R}^n \to 2^{\bar{\mathbf{R}}^p}$ and is defined as follows:

$$\mathbf{h}^{**}(\mathbf{x}) = \max_{\mathbf{R}^p_+\setminus\{0\}} \cup_{\mathbf{U}\in\mathbf{R}^{p\times n}} [\mathbf{U}\mathbf{x} - \mathbf{h}^*(\mathbf{U})], \qquad \mathbf{x} \in \mathbb{R}^n.$$

(iii) $\mathbf{U}$ is said to be a *subgradient* of the set-valued function $\mathbf{h}$ at $(\mathbf{x};\mathbf{y})$ if

$$\mathbf{y} - \mathbf{U}\mathbf{x} \in \min_{\mathbf{R}^p_+\setminus\{0\}} \cup_{\mathbf{x}'\in\mathbf{R}^n} [\mathbf{h}(\mathbf{x}') - \mathbf{U}\mathbf{x}'].$$

(iv) The set of all subgradients of $\mathbf{h}$ at $(\mathbf{x};\mathbf{y})$ is denoted by $\partial\mathbf{h}(\mathbf{x};\mathbf{y})$, the *subdifferential* of $\mathbf{h}$ at $(\mathbf{x};\mathbf{y})$.

(v) $\mathbf{h}$ is said to be *subdifferentiable* at $\mathbf{x}$ if $\partial\mathbf{h}(\mathbf{x};\mathbf{y}) \neq \emptyset \quad \forall \mathbf{y} \in \mathbf{h}(\mathbf{x})$.

Let $\mathbf{f} : \mathbb{R}^n \to \bar{\mathbb{R}}^p$ be an extended vector-valued function. An unconstrained multicriteria optimization problem is as follows:

$$\text{(Unconstrained Multicriteria Optimization Problem UP)}$$
$$\min_{\mathbf{R}^p_+\setminus\{\mathbf{0}\}}\mathbf{f}(\mathbf{x}), \text{ subject to } \mathbf{x} \in \mathbb{R}^n.$$

To build up a conjugate duality theory for the above problem, we embed $\mathbf{f}$ into a family of perturbed functions. Let $\psi : \mathbb{R}^n \times \mathbb{R}^m \to \bar{\mathbb{R}}^p$ be another vector-valued function such that

$$\mathbf{f}(\mathbf{x}) = \psi(\mathbf{x}, \mathbf{0}) \quad \forall \mathbf{x} \in \mathbb{R}^n. \tag{8.2.1}$$

The perturbation function is a set-valued function $\mathbf{p} : \mathbb{R}^m \to 2^{\bar{\mathbf{R}}^p}$ defined as:

$$\text{(Perturbation Function)} \qquad \mathbf{p}(\mathbf{y}) = \min_{\mathbf{R}^p_+\setminus\{\mathbf{0}\}}\{\psi(\mathbf{x},\mathbf{y}) \mid \mathbf{x} \in \mathbb{R}^n\}. \tag{8.2.2}$$

Clearly $\mathbf{p}(\mathbf{0}) = \min_{\mathbf{R}^p_+\setminus\{\mathbf{0}\}}\mathbf{f}(\mathbb{R}^n)$ is the minimal frontier for the problem UP.

The problem UP may now be stated as the primal of a pair of dual optimization problems.

$$\text{(Primal Problem P)} \qquad \min_{\mathbf{R}^p_+\setminus\{\mathbf{0}\}}\{\psi(\mathbf{x},\mathbf{0}) \mid \mathbf{x} \in \mathbb{R}^n\}.$$

The conjugate dual of ψ, denoted as $\psi^* : \mathbb{R}^{p\times n} \times \mathbb{R}^{p\times m} \to 2^{\bar{\mathbf{R}}^p}$ is a set-valued function defined in the usual manner:

$$\psi^*(\mathbf{U},\mathbf{V}) = \max_{\mathbf{R}^p_+\setminus\{\mathbf{0}\}} \cup_{\mathbf{x}\in\mathbf{R}^n,\mathbf{y}\in\mathbf{R}^m} [\mathbf{U}\mathbf{x} + \mathbf{V}\mathbf{y} - \psi(\mathbf{x},\mathbf{y})]. \tag{8.2.3}$$

The conjugate dual optimization problem is now defined as follows:

$$\text{(Dual Problem D)} \qquad \max_{\mathbf{R}^p_+\setminus\{\mathbf{0}\}} -\psi^*(\mathbf{0},\mathbf{V}), \text{ subject to } \mathbf{V} \in \mathbb{R}^{p\times n}.$$

Since $-\psi^*$ is a set-valued function, Problem D is not an ordinary multicriteria optimization problem. To avoid any possible confusion, a more accurate statement of the dual problem is given as follows:

(Restated Dual Problem D): Find: $\mathbf{V}^* \in \mathbb{R}^{p\times m}$ such that

$$-\psi^*(\mathbf{0},\mathbf{V}^*) \cap \max_{\mathbf{R}^p_+\setminus\{\mathbf{0}\}} \cup_{\mathbf{V}} -\psi^*(\mathbf{0},\mathbf{V}) \neq \emptyset. \tag{8.2.4}$$

The first result is a generalization of Lemma 4.1.4.

Theorem 8.2.1 (Weak Duality)

$$\forall \mathbf{x} \in \mathbb{R}^n, \forall \mathbf{V} \in \mathbb{R}^{p\times m} \qquad \psi(\mathbf{x},\mathbf{0}) \notin -\psi^*(\mathbf{0},\mathbf{V}) - \mathbb{R}^p_+ \setminus \{\mathbf{0}\}. \tag{8.2.5}$$

Proof: Let $\boldsymbol{\xi} = \boldsymbol{\psi}(\mathbf{x}, \mathbf{0})$ and $\boldsymbol{\xi}' \in \boldsymbol{\psi}^*(\mathbf{0}, \mathbf{V})$. By definition of $\boldsymbol{\psi}^*(\mathbf{0}, \mathbf{V})$,

$$\boldsymbol{\xi}' \not\leq_{\mathbf{R}^p_+\backslash\{\mathbf{0}\}} \mathbf{V}\mathbf{y} - \boldsymbol{\psi}(\mathbf{x}, \mathbf{y}) \quad \forall \mathbf{x}, \mathbf{y}.$$

In particular when $\mathbf{y} = \mathbf{0}$,

$$\boldsymbol{\xi}' \not\leq_{\mathbf{R}^p_+\backslash\{\mathbf{0}\}} -\boldsymbol{\psi}(\mathbf{x}, \mathbf{0}) = -\boldsymbol{\xi},$$

or

$$\boldsymbol{\xi} + \boldsymbol{\xi}' \not\leq_{\mathbf{R}^p_+\backslash\{\mathbf{0}\}} \mathbf{0}.$$

∎

Corollary 8.2.2 $\forall \boldsymbol{\xi} \in \min_{\mathbf{R}^p_+\backslash\{\mathbf{0}\}} \boldsymbol{\psi}(\mathbf{x}, \mathbf{0})$ and $\forall \boldsymbol{\xi}' \in \max_{\mathbf{R}^p_+\backslash\{\mathbf{0}\}} \cup_{\mathbf{V}} -\boldsymbol{\psi}^*(\mathbf{0}, \mathbf{V})$,

$$\boldsymbol{\xi} \not\leq_{\mathbf{R}^p_+\backslash\{\mathbf{0}\}} \boldsymbol{\xi}'.$$

Definition 8.2.2 (i) The (set-valued) perturbation function

$$\mathbf{p}(\mathbf{y}) = \min_{\mathbf{R}^p_+\backslash\{\mathbf{0}\}} \boldsymbol{\psi}(\mathbf{R}^n, \mathbf{y})$$

is said to be *externally stable* if

$$\{\boldsymbol{\psi}(\mathbf{x}, \mathbf{y}) \in \mathbb{R}^p \mid \mathbf{x} \in \mathbb{R}^n\} \subseteq \mathbf{p}(\mathbf{y}) + \mathbb{R}^p_+ = \min_{\mathbf{R}^p_+\backslash\{\mathbf{0}\}} \boldsymbol{\psi}(\mathbf{R}^n, \mathbf{y}) + \mathbb{R}^p_+ \quad \forall \mathbf{y} \in \mathbb{R}^m.$$

(ii) Let $\mathbf{h} : \mathbb{R}^n \to 2^{\bar{\mathbb{R}}^p}$ be a set-valued function. We say that the set-valued function $\min_{\mathbf{R}^p_+\backslash\{\mathbf{0}\}} \mathbf{h}(\mathbf{x})$ is *externally stable* if

$$\mathbf{h}(\mathbf{x}) \subseteq \min_{\mathbf{R}^p_+\backslash\{\mathbf{0}\}} \mathbf{h}(\mathbf{x}) + \mathbb{R}^p_+, \quad \forall \mathbf{x} \in \mathbb{R}^n.$$

(iii) Similarly, the set-valued function $\max_{\mathbf{R}^p_+\backslash\{\mathbf{0}\}} \mathbf{h}(\mathbf{x})$ is said to be *externally stable* if

$$\mathbf{h}(\mathbf{x}) \subseteq \max_{\mathbf{R}^p_+\backslash\{\mathbf{0}\}} \mathbf{h}(\mathbf{x}) - \mathbb{R}^p_+, \quad \forall \mathbf{x} \in \mathbb{R}^n.$$

(iv) Given a set $\mathcal{B} \subseteq \mathbb{R}^p$, $\min_{\mathbf{R}^p_+\backslash\{\mathbf{0}\}}(\mathcal{B})$ is said to be *externally stable* if $\mathcal{B} \subseteq \min_{\mathbf{R}^p_+\backslash\{\mathbf{0}\}}(\mathcal{B}) + \mathbb{R}^p_+$. Similarly, $\max_{\mathbf{R}^p_+\backslash\{\mathbf{0}\}}(\mathcal{B})$ is said to be *externally stable* if $\mathcal{B} \subseteq \max_{\mathbf{R}^p_+\backslash\{\mathbf{0}\}}(\mathcal{B}) - \mathbb{R}^p_+$.

The following result is a generalization of Lemma 4.1.5.

Lemma 8.2.3 If $\boldsymbol{\psi}$ is convex on $\mathbb{R}^n \times \mathbb{R}^m$ and the perturbation function $\mathbf{p}(\mathbf{y})$ is externally stable, then the perturbation function is type II convex.

Proof: Refer to Lemma 6.1.2 of [SNT1].

∎

Lemma 8.2.4 Let $\mathbf{h} : \mathbb{R}^n \to 2^{\bar{\mathbb{R}}^p}$ be a set-valued function, and assuming that $\max_{\mathbf{R}^p_+\setminus\{0\}} -\mathbf{h}(\mathbf{x})$ is externally stable. Then

$$\mathbf{h}^*(\mathbf{U}) = \max_{\mathbf{R}^p_+\setminus\{0\}} \cup_{\mathbf{x}} [\mathbf{U}\mathbf{x} - \min_{\mathbf{R}^p_+\setminus\{0\}}\mathbf{h}(\mathbf{x})].$$

Proof: Since $\max_{\mathbf{R}^p_+\setminus\{0\}} -\mathbf{h}(\mathbf{x})$ is externally stable, we have

$$-\mathbf{h}(\mathbf{x}) - \mathbb{R}^p_+ = \max_{\mathbf{R}^p_+\setminus\{0\}} -\mathbf{h}(\mathbf{x}) - \mathbb{R}^p_+ \quad \forall\mathbf{x},$$

or

$$\mathbf{U}\mathbf{x} - \mathbf{h}(\mathbf{x}) - \mathbb{R}^p_+ = \mathbf{U}\mathbf{x} - \min_{\mathbf{R}^p_+\setminus\{0\}}\mathbf{h}(\mathbf{x}) - \mathbb{R}^p_+ \quad \forall\mathbf{x},$$

or

$$\cup_{\mathbf{x}}[\mathbf{U}\mathbf{x} - \mathbf{h}(\mathbf{x})] - \mathbb{R}^p_+ = \cup_{\mathbf{x}}[\mathbf{U}\mathbf{x} - \min_{\mathbf{R}^p_+\setminus\{0\}}\mathbf{h}(\mathbf{x})] - \mathbb{R}^p_+.$$

Taking max on both sides, we have,

$$\mathbf{h}^*(\mathbf{U}) = \max_{\mathbf{R}^p_+\setminus\{0\}} \cup_{\mathbf{x}} [\mathbf{U}\mathbf{x} - \min_{\mathbf{R}^p_+\setminus\{0\}}\mathbf{h}(\mathbf{x})].$$

■

The following is a generalization of Lemma 4.1.6.

Theorem 8.2.5 Assuming that the perturbation function $\mathbf{p}$ is externally stable, then

$$\mathbf{p}^*(\mathbf{V}) = \psi^*(\mathbf{0}, \mathbf{V}) \quad \forall\mathbf{V} \in \mathbb{R}^{p\times m}.$$

Proof:

$$\begin{aligned}\mathbf{p}^*(\mathbf{V}) &= \max_{\mathbf{R}^p_+\setminus\{0\}} \cup_{\mathbf{y}} [\mathbf{V}\mathbf{y} - \mathbf{p}(\mathbf{y})] \\ &= \max_{\mathbf{R}^p_+\setminus\{0\}} \cup_{\mathbf{y}} [\mathbf{0}\mathbf{x} + \mathbf{V}\mathbf{y} - \min_{\mathbf{R}^p_+\setminus\{0\}}\psi(\mathbb{R}^n, \mathbf{y})] \\ &= \psi^*(\mathbf{0}, \mathbf{V}) \qquad \text{by Lemma 8.2.4.}\end{aligned}$$

■

Remark 8.2.1

(i) If $p = 1$, $p(y) = \min \psi(\mathbb{R}^n, y)$ is always externally stable.

(ii) Note that without the external stability assumption, the above result becomes considerably weaker, and it is only possible to establish that (See Lemma 6.1.3 of [SNT1]):

$$\mathbf{p}^*(\mathbf{V}) \subseteq \psi^*(\mathbf{0}, \mathbf{V}) \quad \forall\mathbf{V} \in \mathbb{R}^{p\times m}.$$

Henceforth we shall assume throughout that $\mathbf{p}$ is externally stable.

Corollary 8.2.6

$$\max_{\mathbf{R}^p_+\setminus\{0\}} \cup_{\mathbf{V}} -\psi^*(\mathbf{0}, \mathbf{V}) = \max_{\mathbf{R}^p_+\setminus\{0\}} \cup_{\mathbf{V}} -\mathbf{p}^*(\mathbf{V}) = \mathbf{p}^{**}(\mathbf{0}).$$

Proof:

$$\begin{aligned}\mathbf{p}^{**}(\mathbf{0}) &= \max_{\mathbf{R}^p_+\setminus\{\mathbf{0}\}} \cup_{\mathbf{V}} [\mathbf{V0} - \mathbf{p}^*(\mathbf{V})] \\ &= \max_{\mathbf{R}^p_+\setminus\{\mathbf{0}\}} \cup_{\mathbf{V}} -\psi^*(\mathbf{0}, \mathbf{V}) \qquad \text{by Theorem 8.2.5.}\end{aligned}$$

■

Remark 8.2.2 As a result of the definition of $\mathbf{p}$ and Corollary 8.2.6, the minimal frontier of the primal is just $\mathbf{p}(\mathbf{0})$ while the maximal frontier of the dual is $\mathbf{p}^{**}(\mathbf{0})$.

Lemma 8.2.7 (Generalized Young's Inequality) Let $\mathbf{h} : \mathbb{R}^n \to 2^{\bar{\mathbf{R}}^p}$ be a set-valued function. If $\mathbf{y} \in \mathbf{h}(\mathbf{x})$ and $\mathbf{y}' \in \mathbf{h}^*(\mathbf{U})$, then

$$\mathbf{y} + \mathbf{y}' \not\leq_{\mathbf{R}^p_+\setminus\{\mathbf{0}\}} \mathbf{Ux}.$$

Proof: Assume that $\mathbf{y} \in \mathbf{h}(\mathbf{x})$ and $\mathbf{y}' \in \mathbf{h}^*(\mathbf{U})$. Then

$$\mathbf{y}' \in \max_{\mathbf{R}^p_+\setminus\{\mathbf{0}\}} \cup_{\mathbf{x}'} [\mathbf{Ux}' - \mathbf{h}(\mathbf{x}')].$$

In particular, let $\mathbf{x}' = \mathbf{x}$ and $\mathbf{y} \in \mathbf{h}(\mathbf{x})$, then

$$\mathbf{y}' \not\leq_{\mathbf{R}^p_+\setminus\{\mathbf{0}\}} \mathbf{Ux} - \mathbf{y}.$$

■

Lemma 8.2.8 Let $\mathbf{h} : \mathbb{R}^n \to 2^{\bar{\mathbf{R}}^p}$ be a set-valued function, and $\mathbf{y} \in \mathbf{h}(\mathbf{x})$. Then

$$\partial\mathbf{h}(\mathbf{x};\mathbf{y}) \neq \emptyset \quad \text{if and only if} \quad \mathbf{y} \in \mathbf{h}^{**}(\mathbf{x}).$$

In other word, $\mathbf{h}$ is subdifferentiable at $\mathbf{x}$ if and only if $\mathbf{h}(\mathbf{x}) \subseteq \mathbf{h}^{**}(\mathbf{x})$.

Proof: By definition of $\mathbf{h}^*$, $\partial\mathbf{h}(\mathbf{x};\mathbf{y}) \neq \emptyset$ if and only if

$$\exists \mathbf{U} \in \mathbb{R}^{p\times n} \ \text{ such that } \mathbf{Ux} - \mathbf{y} \in \mathbf{h}^*(\mathbf{U}). \tag{8.2.4}$$

Hence if

$$\mathbf{y} \in \mathbf{h}^{**}(\mathbf{x}) = \max_{\mathbf{R}^p_+\setminus\{\mathbf{0}\}} \cup_{\mathbf{U}} [\mathbf{Ux} - \mathbf{h}^*(\mathbf{U})],$$

then

$$\mathbf{y} \in \mathbf{h}^{**}(\mathbf{x}) = \mathbf{Ux} - \mathbf{h}^*(\mathbf{U}) \quad \text{for some } \mathbf{U} \in \mathbb{R}^{p\times n},$$

or

$$\mathbf{Ux} - \mathbf{y} \in \mathbf{h}^*(\mathbf{U}) \quad \text{for some } \mathbf{U} \in \mathbb{R}^{p\times n}.$$

Thus by (8.2.4), we have $\partial\mathbf{h}(\mathbf{x};\mathbf{y}) \neq \emptyset$.

Conversely, if $\partial\mathbf{h}(\mathbf{x};\mathbf{y}) \neq \emptyset$, then $\mathbf{y} \in \mathbf{Ux} - \mathbf{h}^*(\mathbf{U})$ for some $\mathbf{U}$. Pick any $\mathbf{y}' \in \mathbf{Ux} - \mathbf{h}^*(\mathbf{U})$ or $\mathbf{Ux} - \mathbf{y}' \in \mathbf{h}^*(\mathbf{U})$. By Definition 8.2.1 (i) and the fact that $\mathbf{y} \in \mathbf{h}(\mathbf{x})$, we have

$$\mathbf{y} + \mathbf{Ux} - \mathbf{y}' \not\leq_{\mathbf{R}^p_+\setminus\{\mathbf{0}\}} \mathbf{Ux},$$

or $\mathbf{y} \not\leq_{\mathbf{R}^p_+\backslash\{0\}} \mathbf{y}'$. Since $\mathbf{y}' \in \mathbf{Ux} - \mathbf{h}^*(\mathbf{U})$ with any $\mathbf{U}$. Therefore

$$\mathbf{y} \in \max_{\mathbf{R}^p_+\backslash\{0\}} \cup_{\mathbf{U}} [\mathbf{Ux} - \mathbf{h}^*(\mathbf{U})] = \mathbf{h}^{**}(\mathbf{x}).$$

■

Definition 8.2.3 The primal problem P is said to be *stable* if the perturbation function $\mathbf{p}$ is subdifferentiable at $\mathbf{y} = \mathbf{0}$.

Theorem 8.2.9 (Strong Duality)

(i) The primal problem P is stable if and only if for each solution $\mathbf{x}^*$ of the primal problem P, there exists a solution $\mathbf{V}^*$ for the dual problem D such that

$$\psi(\mathbf{x}^*, \mathbf{0}) \in -\psi^*(\mathbf{0}, \mathbf{V}^*). \tag{8.2.5}$$

(ii) Conversely, if $\mathbf{x}^*$ and $\mathbf{V}^*$ satisfy (8.2.5), then $\mathbf{x}^*$ is a solution of P and $\mathbf{V}^*$ is a solution of D.

Proof: (i) The primal problem P is stable if and only if

$$\partial\mathbf{p}(\mathbf{0}; \mathbf{z}) \neq \emptyset \quad \forall \mathbf{z} \in \mathbf{p}(\mathbf{0}) = \min_{\mathbf{R}^p_+\backslash\{0\}} \psi(\mathbb{R}^n, \mathbf{0}),$$

if and only if, by Lemma 8.2.8 and Theorem 8.2.5,

$$\begin{aligned} \mathbf{z} &\subseteq \mathbf{p}^{**}(\mathbf{0}) \\ &= \max_{\mathbf{R}^p_+\backslash\{0\}} \cup_{\mathbf{V}} [\mathbf{V0} - \mathbf{p}^*(\mathbf{V})] \\ &= \max_{\mathbf{R}^p_+\backslash\{0\}} \cup_{\mathbf{V}} -\psi^*(\mathbf{0}, \mathbf{V}), \end{aligned}$$

if and only if,

$$\min_{\mathbf{R}^p_+\backslash\{0\}} \psi(\mathbb{R}^n, \mathbf{0}) \subseteq \max_{\mathbf{R}^p_+\backslash\{0\}} \cup_{\mathbf{V}} -\psi^*(\mathbf{0}, \mathbf{V}).$$

Thus for each solution $\mathbf{x}^*$ of the primal problem P and for each solution $\mathbf{V}^*$ of the dual problem D, we have

$$\psi(\mathbf{x}^*, \mathbf{0}) \in -\psi^*(\mathbf{0}, \mathbf{V}^*).$$

(ii) Follows from the definitions. ■

Note that the conjugate duality can be specialized to the Lagrangian duality as discussed in Section 8.3. Note also that the above conjugate duality theory is based on the type I conjugacy, i.e., the conjugate function is a function of a matrix variable. Another version of the conjugate duality theory based on the type II conjugacy can be similarly derived.

8.3 Lagrangian Duality in Convex Multicriteria Optimization

In this section, we consider a Lagrangian type duality theory for a vector generalization of the scalar duality theory discussed in Section 4.3. In the most general case, a complete duality theory can be established for multicriteria optimization problems based on an arbitrary cone ordering. For our present purpose, however, we shall sacrifice some generality for clarity and discuss the duality under the assumption that all vector ordering relationships are induced by the positive orthant.

Let $\mathcal{X} \subset \mathbb{R}^n$ be a nonempty compact and convex set in $\mathbb{R}^n$, and let $\mathbf{f}$ and $\mathbf{g}$ be $\mathbb{R}^p_+$-convex and $\mathbb{R}^m_+$-convex mappings respectively.

Definition 8.3.1 Consider the primal convex multicriteria program.

$$\text{(Primal CVO Problem)} \qquad \min_{\mathbb{R}^p_+ \setminus \{\mathbf{0}\}} \mathbf{f}(\mathbf{x}), \text{ subject to } \mathbf{x} \in \mathcal{X}_0$$

where $\mathcal{X}_0 = \{\mathbf{x} \in \mathcal{X} \mid \mathbf{g}(\mathbf{x}) \leq_{\mathbb{R}^m_+} \mathbf{0}\}$. Let $\mathcal{L} \subset \mathbb{R}^{p \times m}$ denote the family of all positive matrices: $\mathcal{L} = \{\Lambda \in \mathbb{R}^{p\times m} \mid \Lambda \mathbb{R}^m_+ \subseteq \mathbb{R}^p_+\}$, and $\mathbf{L} : \mathbb{R}^n \times \mathcal{L} \to \mathbb{R}^p$ denote the vector-valued Lagrangian:

$$\mathbf{L}(\mathbf{x}, \Lambda) = \mathbf{f}(\mathbf{x}) + \Lambda \mathbf{g}(\mathbf{x}).$$

Definition 8.3.2 (Lagrangian dual function) Define the set-valued Lagrangian dual function $\Phi : \mathcal{L} \to 2^{\mathbb{R}^p}$ (or simply dual function) by:

$$\Phi(\Lambda) = \min_{\mathbb{R}^p_+ \setminus \{\mathbf{0}\}} \{\mathbf{L}(\mathbf{x}, \Lambda) \mid \mathbf{x} \in \mathcal{X}\}. \tag{8.3.1}$$

In the case $p = 1$, $L(\mathbf{x}, \boldsymbol{\lambda}) = f(\mathbf{x}) + \boldsymbol{\lambda}\mathbf{g}(\mathbf{x})$ is reduced to the ordinary Lagrangian function (4.2.4) and $\phi(\boldsymbol{\lambda})$ is reduced to the ordinary Lagrangian dual function for convex programming (see (4.3.6)):

$$\phi(\boldsymbol{\lambda}) = \inf_{\mathbf{x} \in \mathcal{X}} \{f(\mathbf{x}) + \boldsymbol{\lambda}^\top \mathbf{g}(\mathbf{x})\}, \tag{8.3.2}$$

Definition 8.3.3 (Dual convex multicriteria optimization problem) The dual problem for CVO is defined as

$$\text{(Dual CVO Problem)} \qquad \max_{\mathbb{R}^p_+ \setminus \{\mathbf{0}\}} \Phi(\Lambda), \text{ subject to } \Lambda \in \mathcal{L}.$$

The following result is a generalization of Theorem 4.3.1.

Theorem 8.3.1 (Type II concavity of the dual function) The dual function Φ is type II $\mathbb{R}^p_+$-concave.

Proof: Since $\mathcal{X}$ is compact,

$$\min_{\mathbb{R}^p_+ \setminus \{\mathbf{0}\}} \left\{ t\{\mathbf{f}(\mathbf{x}) + \Lambda^1 \mathbf{g}(\mathbf{x})\} + (1-t)\{\mathbf{f}(\mathbf{x}) + \Lambda^2 \mathbf{g}(\mathbf{x})\} \mid \mathbf{x} \in \mathcal{X} \right\}$$

is externally stable. For $t \in (0,1), \Lambda^1, \Lambda^2 \in \mathcal{L}$, we have

$$\begin{aligned}
&\Phi(t\Lambda^1 + (1-t)\Lambda^2) \\
&= \min_{\mathbf{R}^p_+\setminus\{\mathbf{0}\}}\{\mathbf{f}(\mathbf{x}) + (t\Lambda^1 + (1-t)\Lambda^2)\mathbf{g}(\mathbf{x}) \mid \mathbf{x} \in \mathcal{X}\} \\
&= \min_{\mathbf{R}^p_+\setminus\{\mathbf{0}\}}\{t(\mathbf{f}(\mathbf{x}) + \Lambda^1\mathbf{g}(\mathbf{x})) + (1-t)(\mathbf{f}(\mathbf{x}) + \Lambda^2\mathbf{g}(\mathbf{x})) \mid \mathbf{x} \in \mathcal{X}\} \\
&\subseteq \{t(\mathbf{f}(\mathbf{x}) + \Lambda^1\mathbf{g}(\mathbf{x})) + (1-t)(\mathbf{f}(\mathbf{x}) + \Lambda^2\mathbf{g}(\mathbf{x})) \mid \mathbf{x} \in \mathcal{X}\} \\
&\subseteq t\{\mathbf{f}(\mathbf{x}) + \Lambda^1\mathbf{g}(\mathbf{x}) \mid \mathbf{x} \in \mathcal{X}\} + (1-t)\{\mathbf{f}(\mathbf{x}) + \Lambda^2\mathbf{g}(\mathbf{x}) \mid \mathbf{x} \in \mathcal{X}\} \\
&\subseteq \min_{\mathbf{R}^p_+\setminus\{\mathbf{0}\}}\{t\{\mathbf{f}(\mathbf{x}) + \Lambda^1\mathbf{g}(\mathbf{x})\} + (1-t)\{\mathbf{f}(\mathbf{x}) + \Lambda^2\mathbf{g}(\mathbf{x})\} \mid \mathbf{x} \in \mathcal{X}\} + \mathbb{R}^p_+ \\
&\subseteq \min_{\mathbf{R}^p_+\setminus\{\mathbf{0}\}}\{t\{\mathbf{f}(\mathbf{x}) + \Lambda^1\mathbf{g}(\mathbf{x})\} \mid \mathbf{x} \in \mathcal{X}\} \\
&\qquad + \min_{\mathbf{R}^p_+\setminus\{\mathbf{0}\}}\{(1-t)\{\mathbf{f}(\mathbf{x}) + \Lambda^2\mathbf{g}(\mathbf{x})\} \mid \mathbf{x} \in \mathcal{X}\} + \mathbb{R}^p_+ \\
&\qquad\qquad \text{by Lemma 7.1.4} \\
&= t\min_{\mathbf{R}^p_+\setminus\{\mathbf{0}\}}\{\mathbf{f}(\mathbf{x}) + \Lambda^1\mathbf{g}(\mathbf{x}) \mid \mathbf{x} \in \mathcal{X}\} + \\
&\qquad (1-t)\min_{\mathbf{R}^p_+\setminus\{\mathbf{0}\}}\{\mathbf{f}(\mathbf{x}) + \Lambda^2\mathbf{g}(\mathbf{x}) \mid \mathbf{x} \in \mathcal{X}\} + \mathbb{R}^p_+ \\
&= t\,\Phi(\Lambda^1) + (1-t)\Phi(\Lambda^2) + \mathbb{R}^p_+
\end{aligned}$$

∎

Theorem 8.3.2 (Weak Duality) If $\mathbf{x} \in \mathcal{X}_0$ is feasible for the primal CVO, and $\Lambda \in \mathcal{L}$ is feasible for the dual CVO, then

$$\zeta \not\geq_{\mathbf{R}^p_+\setminus\{\mathbf{0}\}} \mathbf{f}(\mathbf{x}) \quad \forall\, \zeta \in \Phi(\Lambda).$$

Proof: For any feasible $\mathbf{x} \in \mathcal{X}_0$ and feasible $\Lambda \in \mathcal{L}$,

$$\Lambda\mathbf{g}(\mathbf{x}) \leq_{\mathbf{R}^p_+\setminus\{\mathbf{0}\}} \mathbf{0}. \tag{8.3.3}$$

For any $\zeta \in \Phi(\Lambda)$,

$$\zeta \not\geq_{\mathbf{R}^p_+\setminus\{\mathbf{0}\}} \mathbf{f}(\mathbf{x}) + \Lambda\mathbf{g}(\mathbf{x}), \quad \forall \mathbf{x} \in \mathcal{X}_0. \tag{8.3.4}$$

The conclusion thus follows from (8.3.3), (8.3.4) and Lemma 7.1.2. ∎

Theorem 8.3.3 (Strong Duality) If $\mathbf{x}^* \in \mathcal{X}_0$, $\Lambda^* \in \mathcal{L}$ and $\mathbf{f}(\mathbf{x}^*) \in \Phi(\Lambda^*)$. Then $\mathbf{f}(\mathbf{x}^*)$ is simultaneously a minimal point to the primal CVO problem and a maximal point to the dual CVO problem.

Proof: If $\mathbf{f}(\mathbf{x}^*)$ is not minimal for the primal CVO problem, i.e.,

$$\mathbf{f}(\mathbf{x}^*) \notin \min_{\mathbf{R}^p_+\setminus\{\mathbf{0}\}}\{\mathbf{f}(\mathbf{x}) \mid \mathbf{x} \in \mathcal{X}_0\},$$

then there exists $\mathbf{x} \in \mathcal{X}_0$ such that

$$\mathbf{f}(\mathbf{x}^*) \geq_{\mathbf{R}^p_+\setminus\{\mathbf{0}\}} \mathbf{f}(\mathbf{x}). \tag{8.3.5}$$

$\mathbf{g}(\mathbf{x}^*) \leq_{\mathbf{R}^m_+} \mathbf{0}$ and $\boldsymbol{\Lambda}^* \in \mathcal{L}$ together imply that

$$\boldsymbol{\Lambda}^*\mathbf{g}(\mathbf{x}^*) \leq_{\mathbf{R}^p_+} \mathbf{0}. \tag{8.3.6}$$

(8.3.5) and (8.3.6) together imply that

$$\mathbf{L}(\mathbf{x}, \boldsymbol{\Lambda}^*) = \mathbf{f}(\mathbf{x}) + \boldsymbol{\Lambda}^*\mathbf{g}(\mathbf{x}) \leq_{\mathbf{R}^p_+} \mathbf{f}(\mathbf{x}) \leq_{\mathbf{R}^p_+\setminus\{0\}} \mathbf{f}(\mathbf{x}^*).$$

This contradicts the premise that

$$\mathbf{f}(\mathbf{x}^*) \in \Phi(\boldsymbol{\Lambda}^*) = \min_{\mathbf{R}^p_+\setminus\{0\}}\{\mathbf{L}(\mathbf{x}, \boldsymbol{\Lambda}^*) \mid \mathbf{x} \in \mathcal{X}_0\}.$$

Hence $\mathbf{f}(\mathbf{x}^*)$ is minimal for the primal CVO problem.

If $\mathbf{f}(\mathbf{x}^*)$ is not maximal for the dual CVO problem, i.e.,

$$\mathbf{f}(\mathbf{x}^*) \notin \max_{\mathbf{R}^p_+\setminus\{0\}}\{\Phi(\boldsymbol{\Lambda}) \mid \boldsymbol{\Lambda} \in \mathcal{L}\},$$

then there exists $\boldsymbol{\zeta} \in \cup_{\boldsymbol{\Lambda}\in\mathcal{L}}\Phi(\boldsymbol{\Lambda})$ such that $\boldsymbol{\zeta} \geq_{\mathbf{R}^p_+\setminus\{0\}} \mathbf{f}(\mathbf{x}^*)$. Let $\boldsymbol{\Lambda}^0 \in \mathcal{L}$ be such that $\boldsymbol{\zeta} \in \Phi(\boldsymbol{\Lambda}^0)$. Since $\boldsymbol{\Lambda}^0\mathbf{g}(\mathbf{x}^*) \leq_{\mathbf{R}^p_+} \mathbf{0}$, this implies that

$$\boldsymbol{\zeta} \geq_{\mathbf{R}^p_+\setminus\{0\}} \mathbf{L}(\mathbf{x}^*, \boldsymbol{\Lambda}^0) = \mathbf{f}(\mathbf{x}^*) + \boldsymbol{\Lambda}^0\mathbf{g}(\mathbf{x}^*) = \mathbf{L}(\mathbf{x}^*, \boldsymbol{\Lambda}^0).$$

This contradicts that

$$\boldsymbol{\zeta} \in \Phi(\boldsymbol{\Lambda}^0) = \min_{\mathbf{R}^p_+\setminus\{0\}}\{\mathbf{L}(\mathbf{x}, \boldsymbol{\Lambda}) \mid \mathbf{x} \in \mathcal{X}\}.$$

Hence $\mathbf{f}(\mathbf{x}^*)$ is maximal for the dual CVO problem. ∎

Corollary 8.3.4 If $\mathbf{f}(\mathbf{x}^*)$ is a properly minimal point to the primal CVO problem and if the Slater's constraint qualification is satisfied, then $\mathbf{f}(\mathbf{x}^*)$ is a maximal point to the dual CVO problem.

Proof: By Theorem 7.4.5, there exists $\boldsymbol{\Lambda}^* \in \mathcal{L}$ such that $\mathbf{f}(\mathbf{x}^*) \in \Phi(\boldsymbol{\Lambda}^*)$. The conclusion thus follows from Theorem 8.3.3. ∎

8.4 A Case of Symmetric and Self Duality

A special case of duality in multicriteria optimization warrants some special attention in view of the nice symmetric properties involved.

Definition 8.4.1 (Symmetry and self dual problems)

(i) For a given primal optimization problem, its dual optimization problem is said to be *symmetric* if the dual of the dual is the original primal optimization problem.

(ii) A primal optimization problem is said to be *self-dual* if, when recast in the form of the primal, its dual is the same as the primal problem.

As an example in scalar optimization, the primal and dual linear program as cast in (3.2.1)-(3.2.6) is symmetric, as is shown in Theorem 3.2.1. However, it is not self-dual. A special formulation of a pair of primal and dual multicriteria optimization problems is given in [WM1], and these exhibit both the symmetric and self-dual properties. Another example is given in [KYK1].

A slight deviation from our convention will simplify the notation considerably. Partial derivatives of functions of vector variables will, for the purpose of this section only, be assumed to be column vectors. Let $h : \mathbb{R}^n \times \mathbb{R}^m \to \mathbb{R}$ be some scalar-valued function of two vector variables, the first is denoted as $\mathbf{x}$ or $\mathbf{u} \in \mathbb{R}^n$ and the second is denoted as $\mathbf{y}$ or $\mathbf{v} \in \mathbb{R}^m$. Let $h_{\mathbf{x}}(\mathbf{x}, \mathbf{y}) \in \mathbb{R}^n$, and $h_{\mathbf{y}}(\mathbf{x}, \mathbf{y}) \in \mathbb{R}^m$ (both as a column vector) denote the first order partial derivatives of h with respect to the corresponding variable, $h_{\mathbf{xx}}(\mathbf{x}, \mathbf{y}) \in \mathbb{R}^{n \times n}$, $h_{\mathbf{xy}}(\mathbf{x}, \mathbf{y}) \in \mathbb{R}^{m \times n}$, and $h_{\mathbf{yy}}(\mathbf{x}, \mathbf{y}) \in \mathbb{R}^{m \times m}$ denote the second order partial derivatives of h with respect to the corresponding variable.

Definition 8.4.2 (Skew symmetric function) If $m = n$, i.e., the two variables are of the same dimension, then h is said to be *skew-symmetric* if $h(\mathbf{x}, \mathbf{y}) = -h(\mathbf{y}, \mathbf{x})$. A vector-valued function of the two variables is said to be *skew-symmetric* if each component is skew-symmetric.

Definition 8.4.3 The pair of primal and dual problems of interest are defined as follows. Let $\mathbf{f} = (f_1, \cdots, f_p)^\top : \mathbb{R}^n \times \mathbb{R}^m \to \mathbb{R}^p$. Let $\Lambda^+ = \{\boldsymbol{\lambda} \in \text{int}\mathbb{R}^p_+ \mid \sum \lambda_i = 1\}$, and $\mathbf{1} = (1, \cdots, 1)^\top \in \mathbb{R}^p$.

(Primal problem P)

$$\min_{\mathbf{R}^p_+ \setminus \{\mathbf{0}\}} \mathbf{f}(\mathbf{x}, \mathbf{y}) - [\mathbf{y}^\top (\boldsymbol{\lambda}^\top \mathbf{f})_{\mathbf{y}}(\mathbf{x}, \mathbf{y})]\mathbf{1}$$

$$\text{subject to } (\boldsymbol{\lambda}^\top \mathbf{f})_{\mathbf{y}}(\mathbf{x}, \mathbf{y}) \leq_{\mathbf{R}^m_+} \mathbf{0} \tag{8.4.1a}$$

$$\mathbf{x} \geq_{\mathbf{R}^n_+} \mathbf{0}, \tag{8.4.1b}$$

$$\boldsymbol{\lambda} \in \Lambda^+. \tag{8.4.1c}$$

(Dual problem D)

$$\max_{\mathbf{R}^p_+ \setminus \{\mathbf{0}\}} \mathbf{f}(\mathbf{u}, \mathbf{v}) - [\mathbf{u}^\top (\boldsymbol{\lambda}^\top \mathbf{f})_{\mathbf{u}}(\mathbf{u}, \mathbf{v})]\mathbf{1}$$

$$\text{subject to } (\boldsymbol{\lambda}^\top \mathbf{f})_{\mathbf{u}}(\mathbf{u}, \mathbf{v}) \geq_{\mathbf{R}^n_+} \mathbf{0} \tag{8.4.1d}$$

$$\mathbf{v} \geq_{\mathbf{R}^m_+} \mathbf{0}, \tag{8.4.1e}$$

$$\boldsymbol{\lambda} \in \Lambda^+. \tag{8.4.1f}$$

Theorem 8.4.1 Problems P and D are a pair of symmetric dual problems.

Proof: Follows trivially from the definition. ∎

Lemma 8.4.2 Let $\boldsymbol{\xi} \in \mathbb{R}^p$ and $\boldsymbol{\lambda} \in \Lambda^+$. Then, $\boldsymbol{\lambda}^\top \boldsymbol{\xi} \geq \alpha \;\Rightarrow\; \boldsymbol{\xi} \not\leq_{\mathbb{R}^p_+ \setminus \{0\}} \alpha \mathbf{1}$.

Proof: If $\boldsymbol{\xi} \leq_{\mathbb{R}^p_+ \setminus \{0\}} \alpha \mathbf{1}$, then $\boldsymbol{\lambda}^\top \boldsymbol{\xi} < \alpha \boldsymbol{\lambda}^\top \mathbf{1} = \alpha$. ∎

Assumption 8.4.1 $\mathbf{f}(\cdot, \mathbf{y})$ is $\mathbb{R}^p_+$-convex for a fixed $\mathbf{y}$ and $\mathbf{f}(\mathbf{x}, \cdot)$ is $\mathbb{R}^p_+$-concave for a fixed $\mathbf{x}$.

Theorem 8.4.3 (Weak Duality) Assume that Assumption 8.4.1 holds. Let $(\mathbf{x}, \mathbf{y}, \boldsymbol{\lambda})$ be feasible for P and $(\mathbf{u}, \mathbf{v}, \boldsymbol{\lambda})$ be feasible for D. Then

$$\mathbf{f}(\mathbf{x}, \mathbf{y}) - [\mathbf{y}^\top (\boldsymbol{\lambda}^\top \mathbf{f})_{\mathbf{y}}(\mathbf{x}, \mathbf{y})]\mathbf{1} \not\leq_{\mathbb{R}^p_+ \setminus \{0\}} \mathbf{f}(\mathbf{u}, \mathbf{v}) - [\mathbf{u}^\top (\boldsymbol{\lambda}^\top \mathbf{f})_{\mathbf{u}}(\mathbf{u}, \mathbf{v})]\mathbf{1}.$$

Proof: By the convexity of $\mathbf{f}(\cdot, \mathbf{y})$ (hence $\boldsymbol{\lambda}^\top \mathbf{f}(\cdot, \mathbf{y})$) and the concavity of $\mathbf{f}(\mathbf{x}, \cdot)$ (hence $\boldsymbol{\lambda}^\top \mathbf{f}(\mathbf{x}, \cdot)$), we have,

$$\boldsymbol{\lambda}^\top \mathbf{f}(\mathbf{x}, \mathbf{v}) - \boldsymbol{\lambda}^\top \mathbf{f}(\mathbf{u}, \mathbf{v}) \geq (\mathbf{x} - \mathbf{u})^\top (\boldsymbol{\lambda}^\top \mathbf{f})_{\mathbf{u}}(\mathbf{u}, \mathbf{v}), \tag{8.4.2}$$

$$\boldsymbol{\lambda}^\top \mathbf{f}(\mathbf{x}, \mathbf{v}) - \boldsymbol{\lambda}^\top \mathbf{f}(\mathbf{x}, \mathbf{y}) \leq (\mathbf{v} - \mathbf{y})^\top (\boldsymbol{\lambda}^\top \mathbf{f})_{\mathbf{y}}(\mathbf{x}, \mathbf{y}). \tag{8.4.3}$$

Subtracting (8.4.3) from (8.4.2)and rearranging gives

$$\begin{aligned} &[\boldsymbol{\lambda}^\top \mathbf{f}(\mathbf{x}, \mathbf{y}) - \mathbf{y}^\top (\boldsymbol{\lambda}^\top \mathbf{f})_{\mathbf{y}}(\mathbf{x}, \mathbf{y})] - [\boldsymbol{\lambda}^\top \mathbf{f}(\mathbf{u}, \mathbf{v}) - \mathbf{u}^\top (\boldsymbol{\lambda}^\top \mathbf{f})_{\mathbf{u}}(\mathbf{u}, \mathbf{v})] \\ &\qquad \geq \mathbf{x}^\top (\boldsymbol{\lambda}^\top \mathbf{f})_{\mathbf{u}}(\mathbf{u}, \mathbf{v}) - \mathbf{v}^\top (\boldsymbol{\lambda}^\top \mathbf{f})_{\mathbf{y}}(\mathbf{x}, \mathbf{y}) \\ &\qquad \geq 0, \qquad \text{by (8.4.1b,c,e,f).} \end{aligned} \tag{8.4.4}$$

By Lemma 8.4.2, this implies that

$$\mathbf{f}(\mathbf{x}, \mathbf{y}) - [\mathbf{y}^\top (\boldsymbol{\lambda}^\top \mathbf{f})_{\mathbf{y}}(\mathbf{x}, \mathbf{y})]\mathbf{1} \not\leq_{\mathbb{R}^p_+ \setminus \{0\}} \mathbf{f}(\mathbf{u}, \mathbf{v}) - [\mathbf{u}^\top (\boldsymbol{\lambda}^\top \mathbf{f})_{\mathbf{u}}(\mathbf{u}, \mathbf{v})]\mathbf{1}.$$

∎

Theorem 8.4.4 (Strong Duality) Let Assumption 8.4.1 hold. Let $(\mathbf{x}^0, \mathbf{y}^0, \boldsymbol{\lambda}^0)$ be a properly minimal solution of P and fix $\boldsymbol{\lambda} = \boldsymbol{\lambda}^0$ in D. Let $((\boldsymbol{\lambda}^0)^\top \mathbf{f})_{\mathbf{yy}}(\mathbf{x}^0, \mathbf{y}^0)$ be positive definite or negative definite and let the set

$$\{(f_{1\mathbf{y}}(\mathbf{x}^0, \mathbf{y}^0)), (f_{2\mathbf{y}}(\mathbf{x}^0, \mathbf{y}^0)), \cdots, (f_{p\mathbf{y}}(\mathbf{x}^0, \mathbf{y}^0))\}$$

be linearly independent. Then $(\mathbf{x}^0, \mathbf{y}^0)$ is a properly maximal solution of (D).

Proof: Since $(\mathbf{x}^0, \mathbf{y}^0, \boldsymbol{\lambda}^0)$ is a properly minimal solution of P, it is also a weakly minimal solution. Hence from [C6], there exists $\boldsymbol{\tau} \in \mathbb{R}^p, \mathbf{r} \in \mathbb{R}^m, \mathbf{s} \in \mathbb{R}^n, \mathbf{t} \in \mathbb{R}^p$, not all zero such that the following Fritz-John conditions are satisfied at $(\mathbf{x}^0, \mathbf{y}^0, \boldsymbol{\lambda}^0)$:

$$(\boldsymbol{\tau}^\top \mathbf{f})_{\mathbf{x}}(\mathbf{x}^0, \mathbf{y}^0) + ((\boldsymbol{\lambda}^0)^\top \mathbf{f})_{\mathbf{yx}}(\mathbf{x}^0, \mathbf{y}^0)(\mathbf{r} - (\boldsymbol{\tau}^\top \mathbf{1})\mathbf{y}^0) - \mathbf{s} = 0, \quad (8.4.5)$$

$$\mathbf{f}_{\mathbf{y}}(\mathbf{x}^0, \mathbf{y}^0)(\boldsymbol{\tau} - (\boldsymbol{\tau}^\top \mathbf{1})\boldsymbol{\lambda}^0) + ((\boldsymbol{\lambda}^0)^\top \mathbf{f})_{\mathbf{yy}}(\mathbf{x}^0, \mathbf{y}^0)(\mathbf{r} - (\boldsymbol{\tau}^\top \mathbf{1})\mathbf{y}^0) = 0, \quad (8.4.6)$$

$$\mathbf{f}_{\mathbf{y}}(\mathbf{x}^0, \mathbf{y}^0)^\top (\mathbf{r} - (\boldsymbol{\tau}^\top \mathbf{1})\mathbf{y}^0) - \mathbf{t} = \mathbf{0}, \quad (8.4.7)$$

$$\mathbf{r}^\top ((\boldsymbol{\lambda}^0)^\top \mathbf{f})_{\mathbf{y}}(\mathbf{x}^0, \mathbf{y}^0) = \mathbf{0}, \quad (8.4.8)$$

$$\mathbf{s}^\top \mathbf{x}^0 = \mathbf{0}, \quad (8.4.9)$$

$$\mathbf{t}^\top \boldsymbol{\lambda}^0 = \mathbf{0}, \quad (8.4.10)$$

$$\boldsymbol{\tau} \geqq_{\mathbf{R}^p_+} \mathbf{0}, \ \mathbf{r} \geqq_{\mathbf{R}^m_+} \mathbf{0}, \ \mathbf{s} \geqq_{\mathbf{R}^n_+} \mathbf{0}, \ \mathbf{t} \geqq_{\mathbf{R}^p_+} \mathbf{0}, \quad (8.4.11)$$

$$(\boldsymbol{\tau}, \mathbf{r}, \mathbf{s}, \mathbf{t}) \neq \mathbf{0}. \quad (8.4.12)$$

Since $\boldsymbol{\lambda}^0 \in \Lambda^+$ and $\mathbf{t} \geqq_{\mathbf{R}^p_+} \mathbf{0}$, then (8.4.10) implies that

$$\mathbf{t} = \mathbf{0} \quad (8.4.13)$$

It then follows from (8.4.7) and (8.4.13) that

$$\mathbf{f}_{\mathbf{y}}(\mathbf{x}^0, \mathbf{y}^0)^\top (\mathbf{r} - (\boldsymbol{\tau}^\top \mathbf{1})\mathbf{y}^0) = \mathbf{0}. \quad (8.4.14)$$

Multiplying (8.4.6) by $(\mathbf{r} - (\boldsymbol{\tau}^\top \mathbf{1})\mathbf{y}^0)^\top$ and applying (8.4.14) gives

$$(\mathbf{r} - (\boldsymbol{\tau}^\top \mathbf{1})\mathbf{y}^0)((\boldsymbol{\lambda}^0)^\top \mathbf{f})_{\mathbf{yy}}(\mathbf{x}^0, \mathbf{y}^0)(\mathbf{r} - (\boldsymbol{\tau}^\top \mathbf{1})\mathbf{y}^0) = 0. \quad (8.4.15)$$

Since $((\boldsymbol{\lambda}^0)^\top \mathbf{f})_{\mathbf{yy}}(\mathbf{x}^0, \mathbf{y}^0)$ is either positive definite or negative definite, we obtain

$$\mathbf{r} = (\boldsymbol{\tau}^\top \mathbf{1})\mathbf{y}^0. \quad (8.4.16)$$

(8.4.6) and (8.4.16) imply that

$$\mathbf{f}_{\mathbf{y}}(\mathbf{x}^0, \mathbf{y}^0)(\boldsymbol{\tau} - (\boldsymbol{\tau}^\top \mathbf{1})\boldsymbol{\lambda}^0) = \mathbf{0}. \quad (8.4.17)$$

Since, by the assumption of the theorem, the set

$$\{(f_{1\mathbf{y}}(\mathbf{x}^0, \mathbf{y}^0)), (f_{2\mathbf{y}}(\mathbf{x}^0, \mathbf{y}^0)), \cdots, (f_{p\mathbf{y}}(\mathbf{x}^0, \mathbf{y}^0))\}$$

is linearly independent, then (8.4.17) implies that

$$\boldsymbol{\tau} = (\boldsymbol{\tau}^\top \mathbf{1})\boldsymbol{\lambda}^0. \quad (8.4.18)$$

If

$$\boldsymbol{\tau} = \mathbf{0}, \quad (8.4.19)$$

then by (8.4.16),

$$\mathbf{r} = \mathbf{0}, \quad (8.4.20)$$

and by (8.4.5),

$$\mathbf{s} = \mathbf{0}. \tag{8.4.21}$$

(8.4.19)-(8.4.20) together with (8.4.13) together imply $(\boldsymbol{\tau}, \mathbf{r}, \mathbf{s}, \mathbf{t}) = \mathbf{0}$ which contradicts (8.2.12). Hence $\boldsymbol{\tau} \neq \mathbf{0}$ and

$$\boldsymbol{\tau}^\top \mathbf{1} > 0. \tag{8.4.22}$$

It follows from (8.4.22) and (8.4.18) that

$$\boldsymbol{\tau} \geq_{\text{int } \mathbf{R}^p_+} \mathbf{0}. \tag{8.4.23}$$

By (8.4.11), (8.4.16) and (8.4.22),

$$\mathbf{y}^0 \geq_{\mathbf{R}^m_+} \mathbf{0}. \tag{8.4.24}$$

Combining (8.4.5), (8.4.16) and (8.4.18) we have

$$(\boldsymbol{\tau}^\top \mathbf{1})\ ((\boldsymbol{\lambda}^0)^\top \mathbf{f})_{\mathbf{x}}(\mathbf{x}^0, \mathbf{y}^0) = \mathbf{s}. \tag{8.4.25}$$

Since $\mathbf{s} \geq_{\mathbf{R}^n_+} \mathbf{0}$, (8.4.22) implies that

$$((\boldsymbol{\lambda}^0)^\top \mathbf{f})_{\mathbf{x}}(\mathbf{x}^0, \mathbf{y}^0) \geqq_{\mathbf{R}^n_+} \mathbf{0}. \tag{8.4.26}$$

Hence $(\mathbf{x}^0, \mathbf{y}^0)$ is feasible for D.

Next, we show that the objective functions are equal. Pre-multiplying (8.4.5) by $\mathbf{x}^0$, and noting (8.4.9) and (8.4.16), we have:

$$(\mathbf{x}^0)^\top ((\boldsymbol{\lambda}^0)^\top \mathbf{f})_{\mathbf{x}}(\mathbf{x}^0, \mathbf{y}^0) = 0. \tag{8.4.27}$$

By (8.4.8), (8.4.16), we have

$$(\boldsymbol{\tau}^\top \mathbf{1})(\mathbf{y}^0)^\top ((\boldsymbol{\lambda}^0)^\top \mathbf{f})_{\mathbf{y}}(\mathbf{x}^0, \mathbf{y}^0) = 0. \tag{8.4.28}$$

It follows from (8.4.22), that

$$(\mathbf{y}^0)^\top ((\boldsymbol{\lambda}^0)^\top \mathbf{f})_{\mathbf{y}}(\mathbf{x}^0, \mathbf{y}^0) = 0. \tag{8.4.29}$$

The objectives of the primal and dual are therefore equal by virtue of (8.4.27) and (8.4.29).

If $(\mathbf{x}^0, \mathbf{y}^0)$ were feasible for D, with the equality of objective functions, but were not maximal, then there exists $(\bar{\mathbf{u}}, \bar{\mathbf{v}})$ such that

$$\mathbf{f}(\bar{\mathbf{u}}, \bar{\mathbf{v}}) - [\bar{\mathbf{u}}^\top ((\boldsymbol{\lambda}^0)^\top \mathbf{f})_{\mathbf{x}}(\bar{\mathbf{u}}, \bar{\mathbf{v}})]\mathbf{1} \geq_{\mathbf{R}^p_+ \setminus \{\mathbf{0}\}} \mathbf{f}(\mathbf{x}^0, \mathbf{y}^0) - [(\mathbf{x}^0)^\top ((\boldsymbol{\lambda}^0)^\top \mathbf{f})_{\mathbf{x}}(\mathbf{x}^0, \mathbf{y}^0)]\mathbf{1}.$$

Since $(\mathbf{x}^0)^\top ((\boldsymbol{\lambda}^0)^\top \mathbf{f})_{\mathbf{x}}(\mathbf{x}^0, \mathbf{y}^0) = 0 = (\mathbf{y}^0)^\top ((\boldsymbol{\lambda}^0)^\top \mathbf{f})_{\mathbf{y}}(\mathbf{x}^0, \mathbf{y}^0)$, it follows that

$$\mathbf{f}(\bar{\mathbf{u}}, \bar{\mathbf{v}}) - [\bar{\mathbf{u}}^\top ((\boldsymbol{\lambda}^0)^\top \mathbf{f})_{\mathbf{x}}(\bar{\mathbf{u}}, \bar{\mathbf{v}})]\mathbf{1} \geq_{\mathbf{R}^p_+ \setminus \{\mathbf{0}\}} \mathbf{f}(\mathbf{x}^0, \mathbf{y}^0) - [(\mathbf{y}^0)^\top ((\boldsymbol{\lambda}^0)^\top \mathbf{f})_{\mathbf{y}}(\mathbf{x}^0, \mathbf{y}^0)]\mathbf{1},$$

which contradicts the weak duality of Theorem 8.4.3.

If $(\mathbf{x}^0, \mathbf{y}^0)$ were maximal, but were not properly maximal, then for some feasible $(\bar{\mathbf{u}}, \bar{\mathbf{v}})$ and some i,

$$f_i(\bar{\mathbf{u}}, \bar{\mathbf{v}}) - [\bar{\mathbf{u}}^\top((\boldsymbol{\lambda}^0)^\top \mathbf{f})_{\mathbf{x}}(\bar{\mathbf{u}}, \bar{\mathbf{v}})] - \{f_i(\mathbf{x}^0, \mathbf{y}^0) - [(\mathbf{x}^0)^\top((\boldsymbol{\lambda}^0)^\top \mathbf{f})_{\mathbf{x}}(\mathbf{x}^0, \mathbf{y}^0)]\} > M,$$

for any $M > 0$. Since $(\mathbf{x}^0)^\top((\boldsymbol{\lambda}^0)^\top \mathbf{f})_{\mathbf{x}}(\mathbf{x}^0, \mathbf{y}^0) = 0 = (\mathbf{y}^0)^\top((\boldsymbol{\lambda}^0)^\top \mathbf{f})_{\mathbf{y}}(\mathbf{x}^0, \mathbf{y}^0)$ it follows that

$$f_i(\bar{\mathbf{u}}, \bar{\mathbf{v}}) - [\bar{\mathbf{u}}^\top((\boldsymbol{\lambda}^0)^\top \mathbf{f})_{\mathbf{x}}(\bar{\mathbf{u}}, \bar{\mathbf{v}})] - \{f_i(\mathbf{x}^0, \mathbf{y}^0) - [(\mathbf{y}^0)^\top((\boldsymbol{\lambda}^0)^\top \mathbf{f})_{\mathbf{y}}(\mathbf{x}^0, \mathbf{y}^0)]\} > M,$$

and so,

$$\begin{aligned}(\boldsymbol{\lambda}^0)^\top &\{\mathbf{f}(\bar{\mathbf{u}}, \bar{\mathbf{v}}) - [\bar{\mathbf{u}}^\top((\boldsymbol{\lambda}^0)^\top \mathbf{f})_{\mathbf{x}}(\bar{\mathbf{u}}, \bar{\mathbf{v}})]\mathbf{1}\} \\ &> (\boldsymbol{\lambda}^0)^\top\{\mathbf{f}(\mathbf{x}^0, \mathbf{y}^0) - [(\mathbf{x}^0)^\top((\boldsymbol{\lambda}^0)^\top \mathbf{f})_{\mathbf{y}}(\mathbf{x}^0, \mathbf{y}^0)]\mathbf{1}\},\end{aligned}$$

which contradicts the weak duality of Theorem 8.4.3. ∎

Theorem 8.4.5 (Self Duality) If $\mathbf{f}$ is skew-symmetric, then the primal problem is self-dual. If P and D are dual, and $(\mathbf{x}^0, \mathbf{y}^0, \boldsymbol{\lambda}^0)$ is a joint optimal solution, then so is $(\mathbf{y}^0, \mathbf{x}^0, \boldsymbol{\lambda}^0)$ and furthermore,

$$\mathbf{f}(\mathbf{x}^0, \mathbf{y}^0) - [(\mathbf{y}^0)^\top((\boldsymbol{\lambda}^0)^\top \mathbf{f})_{\mathbf{y}}(\mathbf{x}^0, \mathbf{y}^0)]\mathbf{1} = \mathbf{f}(\mathbf{x}^0, \mathbf{y}^0) = 0.$$

Proof: Since $\mathbf{f}$ is skew-symmetric, $(\boldsymbol{\lambda}^\top \mathbf{f})_{\mathbf{x}}(\mathbf{u}, \mathbf{v}) = -(\boldsymbol{\lambda}^\top \mathbf{f})_{\mathbf{y}}(\mathbf{v}, \mathbf{u})$ and the dual problem D can be rewritten as:

$$\begin{aligned}\text{(Dual problem D')} \qquad & \min_{\mathbf{R}^p_+\setminus\{0\}} \mathbf{f}(\mathbf{v}, \mathbf{u}) - [\mathbf{u}^\top(\boldsymbol{\lambda}^\top \mathbf{f})_{\mathbf{y}}(\mathbf{v}, \mathbf{u})]\mathbf{1} \\ & \text{subject to } -(\boldsymbol{\lambda}^\top \mathbf{f})_{\mathbf{y}}(\mathbf{v}, \mathbf{u}) \geq_{\mathbf{R}^m_+} \mathbf{0} \\ & \mathbf{v} \geq_{\mathbf{R}^n_+} \mathbf{0}, \\ & \boldsymbol{\lambda} \in \Lambda^+,\end{aligned}$$

which is identical to the primal problem P. Consequently, $(\mathbf{x}^0, \mathbf{y}^0, \boldsymbol{\lambda}^0)$ is optimal for D if and only if $(\mathbf{y}^0, \mathbf{x}^0, \boldsymbol{\lambda}^0)$ is optimal for P.

If P and D are dual to each other and $(\mathbf{x}^0, \mathbf{y}^0, \boldsymbol{\lambda}^0)$ is jointly optimal, then

$$-(\mathbf{y}^0)^\top((\boldsymbol{\lambda}^0)^\top \mathbf{f})_{\mathbf{y}}(\mathbf{x}^0, \mathbf{y}^0) = -(\mathbf{x}^0)^\top((\boldsymbol{\lambda}^0)^\top \mathbf{f})_{\mathbf{x}}(\mathbf{x}^0, \mathbf{y}^0). \tag{8.4.30}$$

Since $-(\mathbf{y}^0)^\top((\boldsymbol{\lambda}^0)^\top \mathbf{f})_{\mathbf{y}}(\mathbf{x}^0, \mathbf{y}^0) \geq 0$ by virtue of (8.4.1a) and (8.4.1e), and

$$-(\mathbf{x}^0)^\top((\boldsymbol{\lambda}^0)^\top \mathbf{f})_{\mathbf{x}}(\mathbf{x}^0, \mathbf{y}^0) \leq 0$$

by virtue of (8.4.1d) and (8.4.1b), it follows from (8.4.30) that

$$-(\mathbf{y}^0)^\top((\boldsymbol{\lambda}^0)^\top \mathbf{f})_{\mathbf{y}}(\mathbf{x}^0, \mathbf{y}^0) = 0.$$

In a similar way, we can show that $-(\mathbf{x}^0)^\top((\boldsymbol{\lambda}^0)^\top \mathbf{f})_{\mathbf{y}}(\mathbf{y}^0, \mathbf{x}^0) = 0$. Thus

$$
\begin{aligned}
\mathbf{f}(\mathbf{x}^0,\mathbf{y}^0) - [(\mathbf{y}^0)^\top((\boldsymbol{\lambda}^0)^\top\mathbf{f})_{\mathbf{y}}(\mathbf{x}^0,\mathbf{y}^0)]\mathbf{1} &= \mathbf{f}(\mathbf{y}^0,\mathbf{x}^0) - [(\mathbf{x}^0)^\top((\boldsymbol{\lambda}^0)^\top\mathbf{f})_{\mathbf{y}}(\mathbf{y}^0,\mathbf{x}^0)]\mathbf{1} \\
&= \mathbf{f}(\mathbf{x}^0,\mathbf{y}^0) \\
&= \mathbf{f}(\mathbf{y}^0,\mathbf{x}^0) \\
&= -\mathbf{f}(\mathbf{x}^0,\mathbf{y}^0).
\end{aligned}
$$

Therefore

$$
\mathbf{f}(\mathbf{x}^0,\mathbf{y}^0) - [(\mathbf{y}^0)^\top((\boldsymbol{\lambda}^0)^\top\mathbf{f})_{\mathbf{y}}(\mathbf{x}^0,\mathbf{y}^0)]\mathbf{1} = \mathbf{f}(\mathbf{x}^0,\mathbf{y}^0) = 0.
$$

∎

8.5 Gap Functions for Convex Multicriteria Optimization

The notion of gap function for convex (scalar) optimization was previously discussed in Section 4.5, where the meaning of "gap" was interpreted as the difference between the (scalar) cost function and the maximum of its Wolfe dual. It turns out that this interpretation can be generalized, in a non-trivial manner, to the case of convex multicriteria optimization problems, provided that the notion of set-valued gap function and the convexity of set-valued functions are appropriately defined. The materials from this section are taken from [CGY3].

Definition 8.5.1 Consider the following multicriteria optimization problem:

$$
\text{Problem } \mathrm{P}_\ell \qquad \min_{\mathbb{R}^p_+\setminus\{\mathbf{0}\}}\mathbf{f}(\mathbf{x}), \text{ subject to } \mathbf{x} \in \mathcal{X},
$$

where $\mathbf{f} : \mathbb{R}^n \to \mathbb{R}^p$ is $\mathbb{R}^p_+$-convex, $\mathcal{X} = \{\mathbf{x} \in \mathbb{R}^n | \mathbf{b} - \mathbf{A}\mathbf{x} \leq_{\mathbb{R}^m_+} \mathbf{0}\}$ is a convex polyhedral feasible set, $\mathbf{A} \in \mathbb{R}^{m\times n}$, $\mathbf{b} \in \mathbb{R}^m$.

Definition 8.5.2 (Defining properties of a gap function) A set-valued function $\gamma : \mathbb{R}^n \to 2^{\mathbb{R}^p}$ is said to be a *gap function* for problem P_ℓ if
(i) $\gamma(\mathbf{x}) \not\leq_{\mathbb{R}^p_+\setminus\{\mathbf{0}\}} \mathbf{0}$,
(ii) $\mathbf{0} \in \gamma(\mathbf{x}^*)$ if and only if $\mathbf{x}^*$ solves P_ℓ, i.e., $\mathbf{x}^* \in \operatorname{argmin}_{\mathbb{R}^p_+\setminus\{\mathbf{0}\}}(\mathbf{f},\mathcal{X})$.

Definition 8.5.3 For problem P_ℓ, define a set-valued function $\gamma : \mathbb{R}^n \to 2^{\mathbb{R}^p}$ by

$$
\gamma(\mathbf{x}) = \max_{\mathbb{R}^p_+\setminus\{\mathbf{0}\}}\{\nabla\mathbf{f}(\mathbf{x})(\mathbf{x}-\mathbf{y}) \mid \mathbf{y} \in \mathcal{X}\}.
$$

Theorem 8.5.1 If $\mathbf{f}$ is $\mathbb{R}^p_+$-concave, then γ (as given in Definition 8.5.3) is a gap function.

Proof: (i) Let $\mathbf{y} = \mathbf{x}$, then $\nabla \mathbf{f}(\mathbf{x})(\mathbf{x} - \mathbf{y}) = \mathbf{0}$ implies that

$$\max_{\mathbf{R}^p_+ \setminus \{0\}} \{\nabla \mathbf{f}(\mathbf{x})(\mathbf{x} - \mathbf{y}) \mid \mathbf{y} \in \mathcal{X}\} \not\leq_{\mathbf{R}^p_+ \setminus \{0\}} \mathbf{0}.$$

(ii) Suppose $\mathbf{x}^*$ solves P_ℓ, and $\nabla \mathbf{f}(\mathbf{x}^*)(\mathbf{y} - \mathbf{x}^*) \leq_{\mathbf{R}^p_+ \setminus \{0\}} \mathbf{0}$ for some $\mathbf{y} \in \mathcal{X}$. Since $\mathcal{X}$ is convex and $\mathbf{f}$ is $\mathbb{R}^p_+$-concave, we have

$$\mathbf{f}(\mathbf{y}) \leq_{\mathbf{R}^p_+} \mathbf{f}(\mathbf{x}^*) + \nabla \mathbf{f}(\mathbf{x}^*)(\mathbf{y} - \mathbf{x}^*).$$

We have

$$\mathbf{f}(\mathbf{y}) \leq_{\mathbf{R}^p_+ \setminus \{0\}} \mathbf{f}(\mathbf{x}^*),$$

contradicting that $\mathbf{x}^*$ solves P_ℓ. Thus $\mathbf{x}^*$ solves P_ℓ implies that

$$\nabla \mathbf{f}(\mathbf{x}^*)(\mathbf{y} - \mathbf{x}^*) \not\leq_{\mathbf{R}^p_+ \setminus \{0\}} \mathbf{0} \ \forall \mathbf{y} \in \mathcal{X}.$$

In particular,

$$\nabla \mathbf{f}(\mathbf{x}^*)(\mathbf{y} - \mathbf{x}^*) \not\leq_{\mathbf{R}^p_+ \setminus \{0\}} \nabla \mathbf{f}(\mathbf{x}^*)(\mathbf{x}^* - \mathbf{x}^*) = \mathbf{0}, \ \forall \mathbf{y} \in \mathcal{X}.$$

So

$$\mathbf{0} \in \gamma(\mathbf{x}^*) = \max_{\mathbf{R}^p_+ \setminus \{0\}} \{\nabla \mathbf{f}(\mathbf{x}^*)(\mathbf{y} - \mathbf{x}^*) \mid \mathbf{y} \in \mathcal{X}\}.$$

Conversely, if $\mathbf{x}^*$ does not solve P_ℓ. Then there exists $\mathbf{y} \in \mathcal{X}$ such that $\mathbf{f}(\mathbf{y}) \leq_{\mathbf{R}^p_+ \setminus \{0\}} \mathbf{f}(\mathbf{x}^*)$, i.e., $\mathbf{f}(\mathbf{y}) - \mathbf{f}(\mathbf{x}^*) \leq_{\mathbf{R}^p_+ \setminus \{0\}} \mathbf{0}$. Since $\mathbf{f}$ is convex, $\mathbf{f}(\mathbf{y}) - \mathbf{f}(\mathbf{x}^*) \geq_{\mathbf{R}^p_+} \nabla \mathbf{f}(\mathbf{x}^*)(\mathbf{y} - \mathbf{x}^*)$. Thus $\nabla \mathbf{f}(\mathbf{x}^*)(\mathbf{y} - \mathbf{x}^*) \leq_{\mathbf{R}^p_+ \setminus \{0\}} \mathbf{0} = \nabla \mathbf{f}(\mathbf{x}^*)(\mathbf{x}^* - \mathbf{x}^*)$, implying that $\mathbf{0} \notin \gamma(\mathbf{x}^*)$. ∎

We may now interpret the gap function in terms of the Wolfe dual of P_ℓ. As in Section 8.3, we define the vector-valued Lagrangian of problem P_ℓ to be:

$$\mathbf{L}(\mathbf{x}, \boldsymbol{\Lambda}) = \mathbf{f}(\mathbf{x}) + \boldsymbol{\Lambda}(\mathbf{b} - \mathbf{A}\mathbf{x})$$

where $\boldsymbol{\Lambda} \in \mathbb{R}^{p \times m}$ is the matrix-valued Lagrange multiplier. Define the set-valued function $\mathbf{d} : \mathbb{R}^n \to 2^{\mathbf{R}^p}$ to be

$$\mathbf{d}(\mathbf{x}) = \max_{\mathbf{R}^p_+ \setminus \{0\}} \{\mathbf{L}(\mathbf{x}, \boldsymbol{\Lambda}) \mid \boldsymbol{\Lambda} \in \mathcal{L}\}$$

where

$$\begin{aligned} \mathcal{L} = \{\boldsymbol{\Lambda} \in \mathbb{R}^{p \times m} \mid \exists \boldsymbol{\mu} \geq_{\text{int } \mathbf{R}^p_+} \mathbf{0}, \text{ such that } \boldsymbol{\mu}^\top \boldsymbol{\Lambda} \geq_{\mathbf{R}^m_+} \mathbf{0}^\top, \text{ and} \\ \boldsymbol{\mu}^\top \nabla_x \mathbf{L}(\mathbf{x}, \boldsymbol{\Lambda}) \geq_{\mathbf{R}^n_+} \mathbf{0}^\top \} \\ = \{\boldsymbol{\Lambda} \in \mathbb{R}^{p \times m} \mid \exists \boldsymbol{\mu} \geq_{\text{int } \mathbf{R}^p_+} \mathbf{0}, \text{ such that } \boldsymbol{\mu}^\top \boldsymbol{\Lambda} \geq_{\mathbf{R}^m_+} \mathbf{0}^\top, \text{ and} \\ \boldsymbol{\mu}^\top (\nabla \mathbf{f}(\mathbf{x}) - \boldsymbol{\Lambda}\mathbf{A}) \geq_{\mathbf{R}^n_+} \mathbf{0}^\top \}. \end{aligned}$$

Theorem 8.5.2 (Interpretation of gap functions)

$$\gamma(\mathbf{x}) = \mathbf{f}(\mathbf{x}) - \mathbf{d}(\mathbf{x}).$$

Proof: Write

$$\mathbf{d}(\mathbf{x}) = \mathbf{f}(\mathbf{x}) + \max_{\mathbf{R}^p_+\setminus\{0\}}\{\Lambda(\mathbf{b} - \mathbf{A}\mathbf{x}) \mid \Lambda \in \mathcal{L}\}.$$

By part (iii) of the multicriteria linear programming duality Theorem 8.1.1, we have

$$\begin{aligned}
&\max_{\mathbf{R}^p_+\setminus\{0\}}\{\Lambda(\mathbf{b} - \mathbf{A}\mathbf{x}) \mid \Lambda \in \mathcal{L}\}\\
&= \min_{\mathbf{R}^p_+\setminus\{0\}}\{\nabla\mathbf{f}(\mathbf{x})\boldsymbol{\xi} \mid \mathbf{A}\mathbf{y} \geq_{\mathbf{R}^m_+} \mathbf{b}\}\\
&= \min_{\mathbf{R}^p_+\setminus\{0\}}\{\nabla\mathbf{f}(\mathbf{x})(\mathbf{y} - \mathbf{x}) \mid \mathbf{A}\mathbf{y} \geq_{\mathbf{R}^m_+} \mathbf{b}\} \quad (\text{ Note: let } \boldsymbol{\xi} = \mathbf{y} - \mathbf{x}\ )\\
&= -\max_{\mathbf{R}^p_+\setminus\{0\}}\{\nabla\mathbf{f}(\mathbf{x})(\mathbf{x} - \mathbf{y}) \mid \mathbf{y} \in \mathcal{X}\} = -\gamma(\mathbf{x}).
\end{aligned}$$

Hence $\mathbf{d}(\mathbf{x}) = \mathbf{f}(\mathbf{x}) - \gamma(\mathbf{x})$, and the proof is complete. ∎

From Definition 8.5.3, we may rewrite the gap function as:

$$\gamma(\mathbf{x}) = \nabla\mathbf{f}(\mathbf{x})\mathbf{x} - \min_{\mathbf{R}^p_+\setminus\{0\}}\{\nabla\mathbf{f}(\mathbf{x})\mathbf{y} \mid \mathbf{y} \in \mathcal{X}\} = \nabla\mathbf{f}(\mathbf{x})\mathbf{x} - \mathbf{W}(\nabla\mathbf{f}(\mathbf{x})) \quad (8.5.1)$$

where $\mathbf{W} : \mathbb{R}^{p\times n} \to 2^{\mathbf{R}^p}$ is a set-valued function defined by

$$\mathbf{W}(\mathbf{Z}) = \min_{\mathbf{R}^p_+\setminus\{0\}}\{\mathbf{Z}\mathbf{y} \mid \mathbf{y} \in \mathcal{X}\},$$

and $\mathcal{X} = \{\mathbf{x} \mid \mathbf{A}\mathbf{x} \geq_{\mathbf{R}^m_+} \mathbf{b}\ \}$.

The following theorem is related to Theorem 7.2.5, which states that the Fenchel conjugate of a set-valued function is type I convex.

Theorem 8.5.3 The set-valued function $\mathbf{W} : \mathbb{R}^{p\times n} \to 2^{\mathbf{R}^p}$ is type II concave.

Proof: Let $\mathbf{x} \in \mathcal{X}$ and $\mathbf{Z}^1, \mathbf{Z}^2 \in \mathbb{R}^{p\times n}$. Then

$$(t\mathbf{Z}^1 + (1-t)\mathbf{Z}^2)\mathbf{x} = t\mathbf{Z}^1\mathbf{x} + (1-t)\mathbf{Z}^2\mathbf{x}.$$

Since

$$\begin{aligned}
&\mathbf{Z}^1\mathbf{x} \in \min_{\mathbf{R}^p_+\setminus\{0\}}\{\mathbf{Z}^1\mathbf{y} \mid \mathbf{y} \in \mathcal{X}\} + \mathbb{R}^p_+,\\
\text{and}\quad &\mathbf{Z}^2\mathbf{x} \in \min_{\mathbf{R}^p_+\setminus\{0\}}\{\mathbf{Z}^2\mathbf{y} \mid \mathbf{y} \in \mathcal{X}\} + \mathbb{R}^p_+,
\end{aligned}$$

we have,

$$\begin{aligned}
&t\mathbf{Z}^1\mathbf{x} + (1-t)\mathbf{Z}^2\mathbf{x}\\
&\in t\min_{\mathbf{R}^p_+\setminus\{0\}}\{\mathbf{Z}^1\mathbf{y} \mid \mathbf{y} \in \mathcal{X}\} + (1-t)\min_{\mathbf{R}^p_+\setminus\{0\}}\{\mathbf{Z}^2\mathbf{y} \mid \mathbf{y} \in \mathcal{X}\} + \mathbb{R}^p_+ \quad \forall \mathbf{x} \in \mathcal{X}.
\end{aligned}$$

Hence

$$(t\mathbf{Z}^1 + (1-t)\mathbf{Z}^2)\mathcal{X} \subseteq t\mathbf{W}(\mathbf{Z}^1) + (1-t)\mathbf{W}(\mathbf{Z}^2) + \mathbb{R}^p_+.$$

But

$$\min_{\mathbb{R}^p_+\setminus\{0\}}\{(t\mathbf{Z}^1 + (1-t)\mathbf{Z}^2)\mathbf{y} \mid \mathbf{y} \in \mathcal{X}\} \subseteq (t\mathbf{Z}^1 + (1-t)\mathbf{Z}^2)\mathcal{X},$$

we conclude that

$$\mathbf{W}(t\mathbf{Z}^1 + (1-t)\mathbf{Z}^2) \subseteq t\mathbf{W}(\mathbf{Z}^1) + (1-t)\mathbf{W}(\mathbf{Z}^2) + \mathbb{R}^p_+,$$

i.e., $\mathbf{W}$ is type II concave. ∎

In the scalar case, the concavity of $\nabla\mathbf{f}$ is sufficient for the vector-valued gap function to be convex (see Section 4.5). This is no longer the case for the current set-valued gap function due to the various intricate convexity relations for set-valued functions. The following is a slightly more restricted sufficient condition for the gap function to be convex.

Theorem 8.5.4 If
(i) the (vector-valued) function $\nabla\mathbf{f}(\cdot)(\cdot) : \mathcal{X} \to \mathbb{R}^p$ is convex and
(ii) the set-valued function $\nabla\mathbf{f}(\cdot)\mathcal{X} : \mathcal{X} \to 2^{\mathbf{R}^p}$ is type II concave,
then the gap function $\gamma : \mathbb{R}^n \to 2^{\mathbf{R}^p}$ is type I convex.

Proof: By definition, the gap function is equivalent to

$$\gamma(\mathbf{x}) = \nabla\mathbf{f}(\mathbf{x})\mathbf{x} - \min_{\mathbf{R}^p_+\setminus\{0\}}\{\nabla\mathbf{f}(\mathbf{x})\mathbf{y} \mid \mathbf{y} \in \mathcal{X}\}. \tag{8.5.2}$$

The first term is convex by Assumption (i).
We need to show that the second term $\min_{\mathbf{R}^p_+\setminus\{0\}}\{\nabla\mathbf{f}(\mathbf{x})\mathbf{y} \mid \mathbf{y} \in \mathcal{X}\}$ is type II concave, and hence its negative is type I convex. Now

$$\min_{\mathbf{R}^p_+\setminus\{0\}}\{\nabla\mathbf{f}(\mathbf{x})\mathbf{y} \mid \mathbf{y} \in \mathcal{X}\} = \min_{\mathbf{R}^p_+\setminus\{0\}}\nabla\mathbf{f}(\mathbf{x})\mathcal{X} \subseteq \nabla\mathbf{f}(\mathbf{x})\mathcal{X}.$$

So, given $\mathbf{x}^1, \mathbf{x}^2 \in \mathcal{X}$,

$$\begin{aligned}
&\min_{\mathbf{R}^p_+\setminus\{0\}}\{\nabla\mathbf{f}(t\mathbf{x}^1 + (1-t)\mathbf{x}^2)\mathbf{y} \mid \mathbf{y} \in \mathcal{X}\} \\
&\qquad \subseteq \nabla\mathbf{f}(t\mathbf{x}^1 + (1-t)\mathbf{x}^2)\mathcal{X} \\
&\qquad \subseteq t\nabla\mathbf{f}(\mathbf{x}^1)\mathcal{X} + (1-t)\nabla\mathbf{f}(\mathbf{x}^2)\mathcal{X} + \mathbb{R}^p_+ \quad \text{by Assumption (ii)} \\
&\qquad \subseteq t\,\min_{\mathbf{R}^p_+\setminus\{0\}}\nabla\mathbf{f}(\mathbf{x}^1)\mathcal{X} + (1-t)\min_{\mathbf{R}^p_+\setminus\{0\}}\nabla\mathbf{f}(\mathbf{x}^2)\mathcal{X} + \mathbb{R}^p_+,
\end{aligned}$$

i.e., $\min_{\mathbf{R}^p_+\setminus\{0\}}\{\nabla\mathbf{f}(\mathbf{x})\mathbf{y} \mid \mathbf{y} \in \mathcal{X}\}$ is type II concave, and hence the gap function γ is type I convex. ∎

Definition 8.5.4 The matrix-valued function $\nabla\mathbf{f} : \mathcal{X} \to \mathbb{R}^{p\times n}$ is said to be *monotone* if for any $\mathbf{x}^1, \mathbf{x}^2 \in \mathcal{X}$,

$$(\nabla\mathbf{f}(\mathbf{x}^1) - \nabla\mathbf{f}(\mathbf{x}^2))(\mathbf{x}^1 - \mathbf{x}^2) \geq_{\mathbf{R}^p_+} \mathbf{0}.$$

Lemma 8.5.5 If $\nabla \mathbf{f} : \mathcal{X} \to \mathbb{R}^{p\times n}$ is affine and monotone, i.e., $\nabla \mathbf{f}(\alpha \mathbf{x}^1 + \beta \mathbf{x}^2) = \alpha \nabla \mathbf{f}(\mathbf{x}^1) + \beta \nabla \mathbf{f}(\mathbf{x}^2)$, $\forall \mathbf{x}^1, \mathbf{x}^2 \in \mathcal{X}$, $\alpha, \beta \in \mathbb{R}$, $\alpha + \beta = 1$, then the function $\nabla \mathbf{f}(\cdot)(\cdot) : \mathcal{X} \to \mathbb{R}^p$ is $\mathbb{R}^p_+$-convex.

Proof: Given $t \in (0,1)$, $\mathbf{x}^1, \mathbf{x}^2 \in \mathcal{X}$,

$$\begin{aligned}
&\nabla \mathbf{f}(t\mathbf{x}^1 + (1-t)\mathbf{x}^2)(t\mathbf{x}^1 + (1-t)\mathbf{x}^2) - t\nabla \mathbf{f}(\mathbf{x}^1)\mathbf{x}^1 - (1-t)\nabla \mathbf{f}(\mathbf{x}^2)\mathbf{x}^2 \\
&= t^2 \nabla \mathbf{f}(\mathbf{x}^1)\mathbf{x}^1 + (1-t)^2 \nabla \mathbf{f}(\mathbf{x}^2)\mathbf{x}^2 + t(1-t)(\nabla \mathbf{f}(\mathbf{x}^1)\mathbf{x}^2 + \nabla \mathbf{f}(\mathbf{x}^2)\mathbf{x}^1) \\
&\qquad - t\nabla \mathbf{f}(\mathbf{x}^1)\mathbf{x}^1 - (1-t)\nabla \mathbf{f}(\mathbf{x}^2)\mathbf{x}^2 \\
&= -t(1-t)(\nabla \mathbf{f}(\mathbf{x}^1) - \nabla \mathbf{f}(\mathbf{x}^2))(\mathbf{x}^1 - \mathbf{x}^2) \leq_{\mathbb{R}^p_+} \mathbf{0}.
\end{aligned}$$

■

Lemma 8.5.6 If $\mathbf{W} : \mathbb{R}^{p\times n} \to 2^{\mathbb{R}^p}$ is type II concave, and $\nabla \mathbf{f} : \mathcal{X} \to \mathbb{R}^{p\times n}$ is affine, then the composite $\mathbf{W} \circ \nabla \mathbf{f} : \mathcal{X} \to 2^{R^p}$ is type II concave.

Proof: Given $\mathbf{x}^1, \mathbf{x}^2 \in \mathcal{X}$ and $t \in (0,1)$,

$$\begin{aligned}
\mathbf{W} \circ \nabla \mathbf{f}(t\mathbf{x}^1 + (1-t)\mathbf{x}^2) &= \mathbf{W}(\nabla \mathbf{f}(t\mathbf{x}^1 + (1-t)\mathbf{x}^2)) \\
&= \mathbf{W}(t\nabla \mathbf{f}(\mathbf{x}^1) + (1-t)\nabla \mathbf{f}(\mathbf{x}^2)) \\
&\subseteq t\mathbf{W}(\nabla \mathbf{f}(\mathbf{x}^1)) + (1-t)\mathbf{W}(\nabla \mathbf{f}(\mathbf{x}^2)) + \mathbb{R}^p_+ \\
&= t(\mathbf{W} \circ \nabla \mathbf{f})(\mathbf{x}^1) + (1-t)(\mathbf{W} \circ \nabla \mathbf{f})(\mathbf{x}^2) + \mathbb{R}^p_+.
\end{aligned}$$

■

Theorem 8.5.7 If $\nabla \mathbf{f} : \mathbb{R}^n \to \mathbb{R}^{p\times n}$ is affine and monotone, then the gap function γ is type I convex.

Proof: In (8.5.1), the first term is convex if $\nabla \mathbf{f}$ is affine and monotone by Lemma 8.5.5. The second term is the composition of a type II concave function (by Theorem 8.5.3) with an affine function, and is type II concave by Lemma 8.5.6. ■

8.6 Duality in Convex Composite Multicriteria Optimization

Consider the following convex composite multicriteria optimization problem P

$$\text{(Problem P)} \quad \min_{\mathbb{R}^p_+ \setminus \{0\}} (f_1(\mathbf{F}_1(\mathbf{x})), ..., f_p(\mathbf{F}_p(\mathbf{x})))$$
$$\text{subject to } \mathbf{x} \in \mathcal{X}, \quad g_j(\mathbf{G}_j(\mathbf{x})) \leq 0, \ j = 1, 2, \cdots, m,$$

where $\mathcal{X}$ is a convex subset of $\mathbb{R}^n$, $f_i, i = 1, 2, \cdots, p$, $g_j, j = 1, 2, \cdots, m$, are convex functions on $\mathbb{R}^l$, and $\mathbf{F}_j$ and $\mathbf{G}_j$ are locally Lipschitz and differentiable functions from $\mathbb{R}^n$ to $\mathbb{R}^l$ with (Jacobian) derivatives $\mathbf{F}'_j(\cdot)$ and $\mathbf{G}'_j(\cdot)$ respectively. It is known that the composite programming model problem provides a unified framework for studying Lagrangian conditions and the convergence behaviour of various algorithms. The main reference of this section is [JY1] and [JY2], see also [BP1],[CC2],[Y2],[J3],[JK1].

Two examples of the convex composite multicriteria optimization problem are as follows.

Example 8.6.1 Define $\mathbf{F}_i, \mathbf{G}_j : \mathbb{R}^n \to \mathbb{R}^{p+m}$ by

$$\mathbf{F}_i(\mathbf{x}) = (0, 0, \cdots, l_i(\mathbf{x}), 0, \cdots, 0)^\top, \ i = 1, 2, \cdots, p,$$

$$\mathbf{G}_j(\mathbf{x}) = (0, 0, \cdots, h_j(\mathbf{x}), 0, \cdots, 0)^\top, \ j = 1, 2...., m,$$

where $l_i(\mathbf{x})$ and $h_j(\mathbf{x})$ are locally Lipschitz and differentiable functions on $\mathbb{R}^n$. Define $f_i, g_j : \mathbb{R}^{p+m} \to \mathbb{R}$ by

$$f_i(\mathbf{z}) = z_i, \ i = 1, 2, \cdots, p,$$

$$g_j(\mathbf{z}) = z_{p+j}, \ j = 1, 2, \cdots, m.$$

Then the problem P reduces to

$$\min_{\mathbb{R}^p_+ \setminus \{0\}} (l_1(\mathbf{x}), ..., l_p(\mathbf{x}))$$
$$\text{subject to } \mathbf{x} \in \mathcal{X}, \ h_j(\mathbf{x}) \leq 0, \ j = 1, 2, \cdots, m.$$

This is a standard multicriteria optimization problem.

Example 8.6.2 Consider the vector approximation problem:

$$\min_{\mathbb{R}^p_+ \setminus \{0\}} (\|\mathbf{F}_1(\mathbf{x})\|, \cdots, \|\mathbf{F}_p(\mathbf{x})\|)^\top$$
$$\text{subject to } \mathbf{x} \in \mathcal{X} \subseteq \mathbb{R}^n \qquad ,$$

where $\|\cdot\|$ is a norm function in $\mathbb{R}^m$, and for each $i = 1, 2, \cdots, p, \mathbf{F}_i : \mathbb{R}^n \to \mathbb{R}^m$ is a differentiable (error) function. This problem is also a special case of P, where for each $i = 1, 2, \cdots, p,$ $f_i : \mathbb{R}^n \to \mathbb{R}$ is given by $f_i(\mathbf{z}) = \|\mathbf{z}\|$ and $\mathbf{G}_j : \mathbb{R}^n \to \mathbb{R}^n$ and $g_j : \mathbb{R}^n \to \mathbb{R}$ are given by $\mathbf{G}_j(\mathbf{x}) = \mathbf{x}$ and $g_j(\mathbf{z}) = 0$.

Consider the following dual problem of P:

(Problem D) $\max_{\mathbb{R}^p_+\setminus\{0\}}(f_1(\mathbf{F}_1(\mathbf{u})), \cdots, f_p(\mathbf{F}_p(\mathbf{u})))^\top$

$$\text{subject to } \mathbf{0} \in \sum_{i=1}^{p} \tau_i \partial f_i(\mathbf{F}_i(\mathbf{u}))\mathbf{F}_i'(\mathbf{u}) + \sum_{j=1}^{m} \lambda_j \partial g_j(\mathbf{G}_j(\mathbf{u}))\mathbf{G}_j'(\mathbf{u}) - (\mathcal{X} - \mathbf{u})^*,$$
$$\lambda_j g_j(\mathbf{G}_j(\mathbf{u})) \geq 0,\ j = 1, 2, \cdots, m,$$
$$\mathbf{u} \in \mathcal{X},\ \boldsymbol{\tau} \in \text{int}\mathbb{R}^p,\ \boldsymbol{\lambda} \in \mathbb{R}^m_+,$$

where $\mathcal{X}^*$ is the polar cone of $\mathcal{X}$ defined by

$$\mathcal{X}^* = \{\mathbf{y} \in \mathbb{R}^n \mid \mathbf{y}^\top \mathbf{x} \geq 0 \quad \forall \mathbf{x} \in \mathcal{X}\}.$$

Note that, unlike the case of Wolfe dual (see Section 4.5), the primal and the dual problems here have the same form of objective functions. This is known as Mond-Weir type dual problems [WM2]. We examine duality properties for the convex composite problem that include the corresponding results for convex problems and other related generalized convex problems.

Assumption 8.6.1 The Generalized Representation Condition (GRC) holds for P if for primal feasible $\mathbf{x}$ and for dual feasible $(\mathbf{u}, \boldsymbol{\tau}, \boldsymbol{\lambda})$, there exist $\alpha_i(\mathbf{x}, \mathbf{u}) > 0$, $\beta_j(\mathbf{x}, \mathbf{u}) > 0$ and $\boldsymbol{\eta}(\mathbf{x}, \mathbf{u}) \in \mathcal{X} - \mathbf{u}$ such that

$$\mathbf{F}_i(\mathbf{x}) - \mathbf{F}_i(\mathbf{u}) = \alpha_i(\mathbf{x}, \mathbf{u})\mathbf{F}_i'(\mathbf{x})\boldsymbol{\eta}(\mathbf{x}, \mathbf{u}), \quad i = 1, \cdots, p$$

$$\mathbf{G}_j(\mathbf{x}) - \mathbf{G}_j(\mathbf{u}) = \beta_j(\mathbf{x}, \mathbf{u})\mathbf{G}_j'(\mathbf{x})\boldsymbol{\eta}(\mathbf{x}, \mathbf{u}) \quad j = 1, \cdots, m.$$

Theorem 8.6.1 (Weak Duality) Let $\mathbf{x}$ be feasible for P and let $(\mathbf{u}, \boldsymbol{\tau}, \boldsymbol{\lambda})$ be feasible for D, and assume that Assumption 8.6.1 holds. Then

$$(f_1(\mathbf{F}_1(\mathbf{x})), \cdots, f_p(\mathbf{F}_p(\mathbf{x})))^\top \not\leq_{\mathbb{R}^p_+\setminus\{0\}} (f_1(\mathbf{F}_1(\mathbf{u})), \cdots, f_p(\mathbf{F}_p(\mathbf{u})))^\top.$$

Proof: Since $(\mathbf{u}, \boldsymbol{\tau}, \boldsymbol{\lambda})$ is feasible for D,there exist $\mathbf{v}^i \in \partial f_i(\mathbf{F}_i(\mathbf{u})), \mathbf{w}^j \in \partial g_j(\mathbf{G}_j(\mathbf{u}))$ for $i = 1, 2, \cdots, p,\ j = 1, 2, \cdots, m$ satisfying $\lambda_j g_j(\mathbf{G}_j(\mathbf{u})) \geq 0$, for $j = 1, 2, \cdots, m$, and

$$\sum_{i=1}^{p} \tau_i (\mathbf{v}^i)^\top \mathbf{F}_i'(\mathbf{u}) + \sum_{j=1}^{m} \lambda_j (\mathbf{w}^j)^\top \mathbf{G}_j'(\mathbf{u}) \in (\mathcal{X} - \mathbf{u})^*.$$

Suppose that $\mathbf{x} \neq \mathbf{u}$ and

$$(f_1(\mathbf{F}_1(\mathbf{x})), \cdots, f_p(\mathbf{F}_p(\mathbf{x})))^\top \leq_{\mathbb{R}^p_+\setminus\{0\}} (f_1(\mathbf{F}_1(\mathbf{u})), \cdots, f_p(\mathbf{F}_p(\mathbf{u})))^\top.$$

Then

$$\sum_{i=1}^{p} \frac{\tau_i}{\alpha_i(\mathbf{x}, \mathbf{u})}(f_i(\mathbf{F}_i(\mathbf{x})) - f_i(\mathbf{F}_i(\mathbf{u}))) < 0,$$

since $\tau_i/\alpha_i(\mathbf{x},\mathbf{u}) > 0$. Now, by the convexity of f_i and by Assumption 8.6.1, we have

$$\sum_{i=1}^{p} \tau_i(\mathbf{v}^i)^\top F_i'(\mathbf{u})\eta(\mathbf{x},\mathbf{u}) < 0.$$

From the feasibility conditions, we have

$$\lambda_j g_j(\mathbf{G}_j(\mathbf{x})) \le 0, \quad \lambda_j g_j(\mathbf{G}_j(\mathbf{u})) \ge 0,$$

and so

$$\sum_{j=1}^{m} \frac{\lambda_j}{\beta_j(\mathbf{x},\mathbf{u})}(g_j(\mathbf{G}_j(\mathbf{x})) - g_j(\mathbf{G}_j(\mathbf{u})) \le 0.$$

Similarly, by the convexity of g_j, the positivity of $\beta_j(\mathbf{x},\mathbf{u})$, and by Assumption 8.6.1,

$$\sum_{j=1}^{m} \lambda_j(\mathbf{w}^j)^\top \mathbf{G}_j'(\mathbf{u})\eta(\mathbf{x},\mathbf{u}) \le 0.$$

Hence,

$$\left[\sum_{i=1}^{p} \tau_i(\mathbf{v}^i)^\top \mathbf{F}_i'(\mathbf{u}) + \sum_{j=1}^{m} \lambda_j(\mathbf{w}^j)^\top \mathbf{G}_j'(\mathbf{u})\right] \eta(\mathbf{x},\mathbf{u}) < 0.$$

This is a contradiction. The proof is completed by noting that when $\mathbf{x} = \mathbf{u}$, the conclusion trivially holds. ∎

Definition 8.6.1 We say that *the generalized Slater constraint qualification* holds at $\mathbf{a}$ for P if

$$\exists \mathbf{x}_0 \in \text{cone}(\mathcal{X} - \mathbf{a}), \qquad \mathbf{v}^\top G_j'(\mathbf{x})\mathbf{x}_0 < 0, \quad \forall \mathbf{v} \in \partial g_j(\mathbf{G}_j(\mathbf{x})), \ \forall j \in \mathcal{J}(\mathbf{x}),$$

where $\mathcal{J}(\mathbf{x}) = \{j | g_j(\mathbf{G}_j(\mathbf{x})) = 0, \ j = 1, \cdots, m\}$, and $\text{cone}(\mathcal{X} - \mathbf{a}) = \{\lambda\mathbf{x} \mid \mathbf{x} \in \mathcal{X} - \mathbf{a}, \ \lambda \ge 0\}$.

Theorem 8.6.2 (Strong Duality) For the problem P, assume that the generalized Slater constraint qualification holds and that Assumption 8.6.1 holds at each feasible point of P and D. If $\mathbf{x}$ is a properly minimal solution for P then there exist $\boldsymbol{\tau} \in \mathbb{R}^p, \tau_j > 0, \boldsymbol{\lambda} \in R^m, \lambda_j \ge 0$ such that $(\mathbf{x}, \boldsymbol{\tau}, \boldsymbol{\lambda})$ is a properly maximal solution for D and the objective values of P and D at these points are equal.

Proof: It follows from Theorem 2.1 of [JY1] that there exist $\boldsymbol{\tau} \in \mathbb{R}^p, \tau_j > 0, \boldsymbol{\lambda} \in \mathbb{R}^m, \lambda_j \ge 0$ such that

$$\mathbf{0} \in \sum_{i=1}^{p} \tau_i \partial f_i(\mathbf{F}_i(\mathbf{x}))\mathbf{F}_i'(\mathbf{x}) + \sum_{j=1}^{m} \lambda_j \partial g_j(\mathbf{G}_j(\mathbf{x}))\mathbf{G}_j'(\mathbf{x}) - (\mathcal{X} - \mathbf{x})^*,$$

$$\lambda_j g_j(\mathbf{G}_j(\mathbf{x})) = 0, j = 1, \cdots, m.$$

Then $(\mathbf{x}, \boldsymbol{\tau}, \boldsymbol{\lambda})$ is a feasible solution for D.

From the weak duality Theorem 8.6.1, the point $(\mathbf{x}, \boldsymbol{\tau}, \boldsymbol{\lambda})$ is a maximal solution for D. We shall prove that $(\mathbf{x}, \boldsymbol{\tau}, \boldsymbol{\lambda})$ is a properly maximal solution for D by contradiction. Suppose that there exists $(\mathbf{x}^*, \boldsymbol{\tau}^*, \boldsymbol{\lambda}^*)$ feasible for D satisfying, for some i,

$$f_i(\mathbf{F}_i(\mathbf{x}^*)) - f_i(\mathbf{F}_i(\mathbf{x})) > M[f_j(\mathbf{F}_j(\mathbf{x})) - f_j(\mathbf{F}_j(\mathbf{x}^*))],$$

for any $M > 0$ and all j satisfying $f_j(\mathbf{F}_j(\mathbf{x})) > f_j(\mathbf{F}_j(\mathbf{x}^*))$. Let $\mathcal{A} = \{j \in I \mid f_j(\mathbf{F}_j(\mathbf{x}) > f_j(\mathbf{F}_j(\mathbf{x}^*))\}$, where $I = \{1, 2, \cdots, p\}$. Let $\mathcal{B} = I \setminus (\mathcal{A} \cup \{i\})$. Choose $M > 0$ such that

$$M/|\mathcal{A}| > \tau_j/\tau_i, \qquad j \in \mathcal{A},$$

where $|\mathcal{A}|$ denotes the cardinality of $\mathcal{A}$. Then,

$$\tau_i(f_i(\mathbf{F}_i(\mathbf{x}^*)) - f_i(\mathbf{F}_i(\mathbf{x}))) > \sum_{k\in\mathcal{A}}(f_k(\mathbf{F}_k(\mathbf{x})) - f_k(F_k(\mathbf{x}^*))),$$

since $f_k(\mathbf{F}_k(\mathbf{x})) - f_k(\mathbf{F}_k(\mathbf{x}^*)) > 0$, for $k \in \mathcal{A}$. So,

$$\begin{aligned}\sum_{i=1}^{p} \tau_i f_i(\mathbf{F}_i(\mathbf{x})) &= \tau_i f_i(\mathbf{F}_i(\mathbf{x})) + \sum_{k\in\mathcal{A}} \tau_k f_k(\mathbf{F}_k(\mathbf{x})) + \sum_{k\in\mathcal{B}} f_k(\mathbf{F}_k(\mathbf{x})) \\ &< \tau_i f_i(\mathbf{F}_i(\mathbf{x})) + \sum_{k\in\mathcal{A}} \tau_k f_k(\mathbf{F}_k(\mathbf{x})) + \sum_{k\in\mathcal{B}} f_k(\mathbf{F}_k(\mathbf{x})) \\ &= \sum_{i=1}^{p} \tau_i^* f_i(\mathbf{F}_i(\mathbf{x}^*)).\end{aligned}$$

This contradicts the weak duality. Hence, $(\mathbf{x}, \boldsymbol{\tau}, \boldsymbol{\lambda})$ is a properly maximal solution for D. ∎

As a special case, the following multicriteria pseudolinear programming problem (PLP) is considered in [CC2]:

$$\begin{aligned}&\text{(Problem PLP)} \quad \min_{\mathbf{R}^p_+\setminus\{0\}}(l_1(\mathbf{x}), ..., l_p(\mathbf{x}))^\top \\ &\qquad\qquad\qquad \text{subject to } h_j(\mathbf{x}) - b_j \le 0,\ j = 1, 2, \cdots, m, \\ &\qquad\qquad\qquad\qquad \mathbf{x} \in \mathbb{R}^n,\end{aligned}$$

where $l_j : \mathbb{R}^n \to \mathbb{R}$ and $h_j : \mathbb{R}^n \to \mathbb{R}$ are differentiable and pseudolinear, i.e., pseudoconvex and pseudoconcave [CC2], and $b_j \in \mathbb{R}, j = 1, 2, \cdots, m$. It should be noted that a real-valued function $h : \mathbb{R}^n \to \mathbb{R}$ is pseudolinear if and only if for each $\mathbf{x}, \mathbf{y} \in \mathbb{R}^n$, there exists a real constant $\alpha(\mathbf{x}, \mathbf{y}) > 0$ such that

$$h(\mathbf{y}) = h(\mathbf{x}) + \alpha(\mathbf{x}, \mathbf{y})h'(\mathbf{x})(\mathbf{y} - \mathbf{x}).$$

Moreover, any fractional linear functions of the form $(\mathbf{a}^T\mathbf{x} + b)/(\mathbf{c}^T\mathbf{x} + d)$ on $\mathbb{R}^n$ is pseudolinear, where $\mathbf{a}, \mathbf{c} \in \mathbb{R}^n, b, d \in \mathbb{R}$. Define $\mathbf{F}_i, \mathbf{G}_j : \mathbb{R}^n \to \mathbb{R}^{p+m}$ by

$$\mathbf{F}_i(\mathbf{x}) = (0, 0, \cdots, l_i(\mathbf{x}), 0, \cdots, 0)^\top,\ i = 1, 2, \cdots, p,$$

$$\mathbf{G}_j(\mathbf{x}) = (0, 0, \cdots, h_j(\mathbf{x}) - b_j, 0, \cdots, 0)^\top,\ j = 1, 2, \cdots, m.$$

Define $f_i, g_j : \mathbb{R}^{p+m} \to \mathbb{R}$ by

$$f_i(\mathbf{z}) = z_i, i = 1, 2, \cdots, p,$$

$$g_j(\mathbf{z}) = z_{p+j}, j = 1, 2, \cdots, m.$$

Then, we can rewrite PLP as the convex composite multicriteria problem. The dual problem of PLP is as follows:

$$\text{(Problem DLP)}\quad \max_{\mathbf{R}^p_+ \setminus \{0\}} (l_1(\mathbf{u}), \cdots, l_p(\mathbf{u}))^\top$$

$$\text{such that}\quad \sum_{i=1}^{p} \tau_i l_i'(\mathbf{u}) + \sum_{j=1}^{m} \lambda_j h_j'(\mathbf{u}) = 0,$$

$$\lambda_j h_j(\mathbf{u}) = 0, \quad j = 1, 2, \cdots, m,$$

$$\boldsymbol{\tau} \in \text{int}\mathbb{R}^p_+, \quad \boldsymbol{\lambda} \in \mathbb{R}^m_+.$$

Theorem 8.6.3 (Weak duality) Let $\mathbf{x}$ be feasible for PLP, and $(\mathbf{u}, \boldsymbol{\tau}, \boldsymbol{\lambda})$ be feasible for problem DLP, then

$$(l_1(\mathbf{x}), \cdots, l_p(\mathbf{x})) \not\leq_{\mathbf{R}^p_+ \setminus \{0\}} (l_1(\mathbf{u}), \cdots, l_p(\mathbf{u})).$$

Proof: Follows from Theorem 8.6.1. ∎

Theorem 8.6.4 (Strong duality) For the problem PLP, assume that the generalized Slater constraint qualification holds at each feasible point of P and D. If $\mathbf{x}$ is a properly minimal solution for PLP then there exist $\boldsymbol{\tau} \in \mathbb{R}^p, \tau_j > 0, \boldsymbol{\lambda} \in \mathbb{R}^m, \lambda_j \geq 0$ such that $(\mathbf{x}, \boldsymbol{\tau}, \boldsymbol{\lambda})$ is a properly maximal solution for DLP and the objective values of PLP and DLP at these points are equal.

Proof: Follows from Theorem 8.6.2. ∎

CHAPTER 9

DUALITY IN VECTOR VARIATIONAL INEQUALITIES

The concept of vector variational inequalities (VVI) was introduced by Giannessi [G6] in 1980 as a generalization of scalar variational inequalities. Since then the subject has found many applications, particularly in the areas of multicriteria optimization (VO) and vector equilibria (VE) problems (see [G9]). This follows the idea discussed in Chapter Seven that most real world problems are concerned with not just a single criterion, but a number of criteria has to be considered simultaneously. In fact some early work in multicriteria optimization (see [J1], [L2]) had led to some forms of vector variational inequalities, albeit being called different names. More specifically, VVI often turns out to be a necessary optimality condition for some VO problems. Under certain convexity assumptions, it may also turn out to be a sufficient optimality condition. Further works relating VO and VVI can be found in [CY2], [CC1] and [Y1]. Applications of VVI in vector equilibria problems are studied in [CY1], [YG1] and [YG3].

Because of its immense theoretical interest, the subject of VVI has been generalized and extended in many different directions in recent years. Some notable ones are:

- Generalized Vector Variational Inequalities (GVVI), see, for example [C4].
- Extended Vector Variational Inequalities (EVVI), see, for example [Y1].
- Quasi-vector Variational Inequalities (QVVI), see, for example, [CL1], [CGY5].
- Pre-vector Variational Inequalities (PVVI) or Vector Variational-like Inequalities, see, for example, [L5], [Y5], [Y6].
- Vector Variational Inequalities for Set-valued Mappings, see, [LLKC1].
- Vector Variational Inequalities for Fuzzy Mappings, see, for example, [LKLC1].
- Vector Complementarity Problems, see, for example, [CY2], [Y7].
- Relationships with multicriteria optimization, see, for example, [CY2], [L10], [Y3].

Much of the literature on VVI is centered around existence results, which are beyond the scope of this book. There are very few results in terms of the solution of VVI as this is a highly non-trivial problem involving the computation of a set. A recent result is found in [GY7]. As the subject is fairly new and there is yet a serious book on the subject, we shall devote Section 9.1 to an introduction to the subject. Scalarization of VVI will be discussed in Section 9.2. The first duality result in VVI due to [Y1] will be the subject matter of Section 9.3. In [CGY1], the duality of VVI is further investigated and the concept of a gap function is generalized for VVI. This will be presented in Section 9.4. A generalization of the Giannessi's gap function as discussed in Chapter Six will be presented in Section 9.5. Lastly, a solution method is studied in Section 9.6 because of its relevance to gap functions. Although most of the results on VVI in the literature are derived in the framework of infinite dimensional abstract space like Banach or Hausdorff spaces, we shall present the results only in finite dimensional vector spaces, in order to be consistent with the rest of this book.

The vector ordering relationships in Sections 9.3, 9.4 and 9.6 are induced by a closed and convex cone $C \subset \mathbb{R}^p$ as in Definition 7.1.1. In Sections 9.1, 9.2 and 9.5 however, the ordering cone is assumed to be $\mathbb{R}^p_+$.

9.1 Introduction to Vector Variational Inequalities

Definition 9.1.1 Let $\mathbf{F} : \mathbb{R}^n \to \mathbb{R}^{n \times p}$ be a matrix-valued continuous function and $\mathcal{K} \subseteq \mathbb{R}^n$ be a closed and convex ground set. The *vector variational inequality problem* (VVI) is defined as follows:

$$\text{(Problem VVI)} \qquad \text{Find } \mathbf{x}^0 \in \mathcal{K} \text{ such that}$$
$$\mathbf{F}(\mathbf{x}^0)^\top(\mathbf{x} - \mathbf{x}^0) \not\leq_{\mathbb{R}^p_+ \setminus \{\mathbf{0}\}} \mathbf{0}, \quad \forall \mathbf{x} \in \mathcal{K}. \tag{9.1.1}$$

A weaker version of VVI, called the *weak vector variational inequality (WVVI)*, is defined as follows. This is concerned with a closed set.

$$\text{(Problem WVVI)} \qquad \text{Find } \mathbf{x}^0 \in \mathcal{K} \text{ such that}$$
$$\mathbf{F}(\mathbf{x}^0)^\top(\mathbf{x} - \mathbf{x}^0) \not\leq_{\text{int } \mathbb{R}^p_+} \mathbf{0}, \quad \forall \mathbf{x} \in \mathcal{K}. \tag{9.1.2}$$

Obviously, a solution to VVI is also a solution to WVVI, but not the converse. Furthermore, the solution of a VVI or a WVVI is often non-unique, and a complete solution requires the computation of the set of all solutions. This is much harder to do than to compute the point solution of a variational inequality.

We shall illustrate the significance of VVI and WVVI via a couple of well-known examples.

Example 9.1.1 (Multicriteria optimization) Consider the problem VO as introduced in Chapter Seven.

$$\text{(Problem VO)} \qquad \min_{\mathbf{R}^p_+\setminus\{\mathbf{0}\}}\mathbf{f}(\mathbf{x}), \text{ subject to } \mathbf{x}\in\mathcal{K}. \tag{9.1.3}$$

The optimality conditions in terms of VVI was obtained by [CY2]. Let $\mathbf{f}$ be assumed to be Gateaux differentiable.

Theorem 9.1.1 If $\mathbf{x}$ is a weakly minimal solution of problem VO, then $\mathbf{x}$ solves the WVVI (9.1.2) with $\mathbf{F}(\mathbf{x})^\top = \nabla\mathbf{f}(\mathbf{x})$. If $\mathbf{f}$ is convex and $\mathbf{x}$ solves the WVVI (9.1.2) with $\mathbf{F}(\mathbf{x})^\top = \nabla\mathbf{f}(\mathbf{x})$, then $\mathbf{x}$ is a weakly minimal solution to the problem VO.

Proof: If $\mathbf{x}$ is a weakly minimal solution of VO, then by definition,

$$\mathbf{f}(\mathbf{x}) \not\geq_{\text{int }\mathbf{R}^p_+} \mathbf{f}(\mathbf{x}+t(\mathbf{y}-\mathbf{x})) \quad \forall t\in(0,1), \mathbf{y}\in\mathcal{K},$$

implying that

$$\frac{\mathbf{f}(\mathbf{x}+t(\mathbf{y}-\mathbf{x}))-\mathbf{f}(\mathbf{x})}{t} \not\leq_{\text{int }\mathbf{R}^p_+} \mathbf{0} \quad \forall t\in(0,1), \mathbf{y}\in\mathcal{K}.$$

Taking the limit $t\downarrow 0$, we have

$$\nabla\mathbf{f}(\mathbf{x})(\mathbf{y}-\mathbf{x}) \not\leq_{\text{int }\mathbf{R}^p_+} \mathbf{0} \quad \forall \mathbf{y}\in\mathcal{K}.$$

Conversely, assume that $\mathbf{x}$ solves WVVI, and $\mathbf{f}$ is $\mathbb{R}^p_+$-convex. By the $\mathbb{R}^p_+$-convexity of $\mathbf{f}$, we have

$$\mathbf{f}(\mathbf{y}) \geq_{\mathbf{R}^p_+} \mathbf{f}(\mathbf{x}) + \nabla\mathbf{f}(\mathbf{x})(\mathbf{y}-\mathbf{x}), \quad \forall \mathbf{y}\in\mathcal{K}.$$

Since

$$\nabla\mathbf{f}(\mathbf{x})(\mathbf{y}-\mathbf{x}) \not\leq_{\text{int }\mathbf{R}^p_+} \mathbf{0}, \quad \forall \mathbf{y}\in\mathcal{K},$$

we have

$$\mathbf{f}(\mathbf{y}) \not\leq_{\text{int }\mathbf{R}^p_+} \mathbf{f}(\mathbf{x}), \quad \forall \mathbf{y}\in\mathcal{K},$$

implying that $\mathbf{x}$ is a weakly minimal solution to problem VO. ∎

It is also straightforward to prove that if $\mathbf{f}$ is $\mathbb{R}^p_+$-convex, and if $\mathbf{x}$ solves the VVI (9.1.1) with $\mathbf{F}(\mathbf{x})^\top = \nabla\mathbf{f}(\mathbf{x})$, then $\mathbf{x}$ is a minimal solution of VO. However, it should be noted that *a minimal solution of VO may not be a solution of VVI.* As a simple example, let $n=1, p=2, \mathbf{f}(x) = (x, x^2+1)^\top$, and $\mathcal{K}=[-1,0]$. It is easy to check that every $x\in[-1,1]$ is minimal. But for $x=0$ and $y=-1$, we have

$$\nabla\mathbf{f}(x)(y-x) = \begin{pmatrix}-1\\0\end{pmatrix} \leq_{\mathbf{R}^2_+} \mathbf{0},$$

implying that $x=0$ is not a solution of the corresponding VVI.

Example 9.1.2 (Vector Equilibria) This example is based on a traffic equilibrium model with a single criterion taken from [M5]. The generalization to a multicriteria consideration by using VVI was studied in [CY1], [GY6] and [YG1].

We shall introduce the necessary notation by summarizing the (scalar) traffic equilibria problem. Consider a transportation network $\mathcal{G} = (\mathcal{N}, \mathcal{A})$ where $\mathcal{N}$ denotes the set of nodes and $\mathcal{A}$ denotes the set of arcs. Because the interest is more in paths than arcs, we shall use a slightly different set of notations from that of Chapter Two. Let $\mathcal{I}$ be a set of origin-destination (O-D) pairs and P_i, $i \in \mathcal{I}$ denotes the set of available paths joining O-D pair i. For a given path $k \in P_i$, let h_k denote the traffic flow on this path and $\mathbf{h} = [h_k] \in \mathbb{R}^N$, $N = \sum_{i \in \mathcal{I}} |P_i|$. A (path) flow vector $\mathbf{h}$ induces a flow v_a on each arc $a \in \mathcal{A}$ given by:

$$v_a = \sum_{i \in \mathcal{I}} \sum_{k \in P_i} \delta_{ak} h_k$$

where $\Delta = [\delta_{ak}] \in \mathbb{R}^{|\mathcal{A}| \times N}$ is the arc path incidence matrix with $\delta_{ak} = 1$ if arc a belongs to path k and 0 otherwise. Let $\mathbf{v} = [v_a] \in \mathbb{R}^{|\mathcal{A}|}$ be the vector of arc flow. Succinctly $\mathbf{v} = \Delta \mathbf{h}$. We shall assume that the demand of traffic flow is fixed for each O-D pair, i.e., $\sum_{k \in P_i} h_k = d_i$, where d_i is a given demand for each O-D pair i. In the scalar cost case, the demand d_i can be easily generalized to be a function of the minimum cost (see [M5]). Unfortunately to do this for vector costs will surely incur complicated set-valued notations. For simplicity, we shall assume a fixed demand for the moment. A flow $\mathbf{h} \geq_{\mathbf{R}^N_+} \mathbf{0}$ satisfying the demand is called a *feasible flow*. Let $\mathcal{H} = \{\mathbf{h} \mid \mathbf{h} \geq_{\mathbf{R}^N_+} \mathbf{0}, \sum_{k \in P_i} h_k = d_i \ \forall i \in \mathcal{I}\}$. $\mathcal{H}$ is clearly a closed and convex set. Let $\mathbf{t}_a(\mathbf{v})$ be a (vector) cost on arc a (usually the delay) and is in general a function of all the arc flows, and $\mathbf{t}(\mathbf{v}) = [\mathbf{t}_a(\mathbf{v})]$. If $\mathbf{t}_a$ is only a function of v_a, we say that the cost is *separable*. If $\partial t_a / \partial v_{a'} = \partial t_{a'} / \partial v_a$, we say that the cost is *integrable*. This nomenclature comes about because if a vector-valued function is integrable, then there exists a scalar-valued function whose gradient yields the vector-valued function, see [OR1]. Note that a separable cost is a special case of an integrable cost. As it turns out, if the cost is integrable, then the subsequent equilibrium problem can be cast in the form of an optimization problem, see [M5].

The cost along a path τ_k is assumed to be the sum of all the arc costs along this path, thus

$$\tau_k(\mathbf{h}) = \sum_{a \in A} \delta_{ak} t_a(\mathbf{v}).$$

Let $\boldsymbol{\tau} = [\tau_k] \in \mathbb{R}^N$. Succinctly $\boldsymbol{\tau}(\mathbf{h}) = \Delta^\top \mathbf{t}(\mathbf{v})$. Given a flow vector $\mathbf{h}$, the minimum cost for an O-D pair i is defined by

$$u_i = \min_{k \in P_i} \tau_k(\mathbf{h}).$$

The Wardrop's user principle [W3] is a behavioral principle which asserts that, at equilibrium, users only choose minimum cost paths to travel on, i.e., a flow $\mathbf{h} \in \mathcal{H}$ is said to be in *Wardrop equilibrium* if

• **Wardrop equilibrium principle:** $\forall i \in \mathcal{I}, \forall k \in P_i, \tau_k(\mathbf{h}) > u_i \Longrightarrow h_k = 0.$

Amongst several possible ways of stating the conditions for Wardrop equilibrium, the most popular one seems to be in the form of a (scalar) variational inequality (VI):

$$\text{(Problem VI)} \qquad \text{Find } \mathbf{h} \in \mathcal{H}, \text{ s.t. } \tau(\mathbf{h})^\top(\bar{\mathbf{h}} - \mathbf{h}) \geq 0, \ \forall \bar{\mathbf{h}} \in \mathcal{H}.$$

Existence and uniqueness for traffic equilibria can be established under easily satisfiable conditions such as the continuity and strict monotonicity of the function τ.

The assumption that users choose their path based on a single criterion may be unreasonably restrictive. In reality, users may choose their path based on several criteria, for example, time delay and (monetary) cost, amongst others. In general, these costs are often conflicting. We generalize the scalar costs $t_a(\mathbf{h})$ and $\tau_k(\mathbf{h})$ to vector costs $\mathbf{t}_a(\mathbf{h})$ and $\tau_k(\mathbf{h}) \in \mathbb{R}^p$. Let $\mathbf{T}(\mathbf{h})$ be an $p \times N$ matrix with columns given by $\tau_k(\mathbf{h})$. In [CY1], the following vector equilibrium principle is proposed. Essentially it asserts that, at equilibrium, users choose only minimal or efficient paths to travel on.

• **Vector equilibrium principle:** A flow $\mathbf{h} \in \mathcal{H}$ is said to be in *vector equilibrium* if:

$$\forall i \in \mathcal{I}, \forall k, \bar{k} \in P_i, \tau_k(\mathbf{h}) - \tau_{\bar{k}}(\mathbf{h}) \geq_{\mathbf{R}^p_+ \setminus \{0\}} \mathbf{0} \Longrightarrow h_k = 0.$$

A weaker form of vector equilibrium which turns out to be rather important is defined as follows.

• **Weak vector equilibrium principle:** A flow $\mathbf{h} \in \mathcal{H}$ is said to be in *weak vector equilibrium* if:

$$\forall i \in \mathcal{I}, \forall k, \bar{k} \in P_i, \tau_k(\mathbf{h}) - \tau_{\bar{k}}(\mathbf{h}) \geq_{\text{int } \mathbf{R}^p_+} \mathbf{0} \Longrightarrow h_k = 0.$$

The latter is called *weak* since a flow $\mathbf{h}$ satisfying $\tau_k(\mathbf{h}) - \tau_{\bar{k}}(\mathbf{h}) \geq_{\text{int } \mathbf{R}^p_+} \mathbf{0}$ and $h_k > 0$ (i.e., not weak vector equilibrium) also satisfies $\tau_k(\mathbf{h}) - \tau_{\bar{k}}(\mathbf{h}) \geq_{\mathbf{R}^p_+ \setminus \{0\}} \mathbf{0}$ and $h_k > 0$ (i.e., not vector equilibrium).

The following sufficient condition for a flow $\mathbf{h}$ to be in vector equilibrium is due to [YG1].

Theorem 9.1.2 (Sufficient condition for vector equilibria) $\mathbf{h} \in \mathcal{H}$ is in vector equilibrium if $\mathbf{h}$ solves the VVI:

$$\text{Find } \mathbf{h} \text{ such that} \quad \mathbf{T}(\mathbf{h})(\bar{\mathbf{h}} - \mathbf{h}) \not\leq_{\mathbf{R}^p_+ \setminus \{0\}} \mathbf{0}, \ \forall \bar{\mathbf{h}} \in \mathcal{H}. \tag{9.1.4}$$

Proof: Let $\mathbf{h}$ satisfy (9.1.4), Choose $\bar{\mathbf{h}}$ to be such that

$$\bar{h}_j = \begin{cases} h_j & \text{if } j \neq k \text{ or } \bar{k}; \\ 0 & \text{if } j = k; \\ h_k + h_{\bar{k}} & \text{if } j = \bar{k}. \end{cases}$$

Clearly $\bar{\mathbf{h}} \in \mathcal{H}$ since $\forall i \in \mathcal{I}$, $\sum_{j\in P_i} h_j = \sum_{j\in P_i} \bar{h}_j = d_i$. Now

$$\begin{aligned}\mathbf{T}(\mathbf{h})(\bar{\mathbf{h}} - \mathbf{h}) &= \sum_{i\in I}\sum_{j\in P_i} (\bar{h}_j - h_j)\tau_j(\mathbf{h}) \\ &= (\bar{h}_k - h_k)\tau_k(\mathbf{h}) + (\bar{h}_{\bar{k}} - h_{\bar{k}})\tau_{\bar{k}}(\mathbf{h}) \\ &= h_k(\tau_{\bar{k}}(\mathbf{h}) - \tau_k(\mathbf{h})) \not\leq_{\mathbf{R}^p_+\setminus\{\mathbf{0}\}} \mathbf{0}. \end{aligned} \tag{9.1.5}$$

If

$$\tau_k(\mathbf{h}) - \tau_{\bar{k}}(\mathbf{h}) \geq_{\mathbf{R}^p_+\setminus\{\mathbf{0}\}} \mathbf{0}, \tag{9.1.6}$$

then (9.1.5) and (9.1.6) together imply that $h_k = 0$. ∎

A similar sufficient condition can be established for a weak vector equilibrium:

Theorem 9.1.3 $\mathbf{h} \in \mathcal{H}$ is in weak vector equilibrium if h solves the WVVI:

$$\text{Find } \mathbf{h} \text{ such that} \qquad \mathbf{T}(\mathbf{h})(\bar{\mathbf{h}} - \mathbf{h}) \not\leq_{\text{int } \mathbf{R}^p_+} \mathbf{0},\ \forall \bar{\mathbf{h}} \in \mathcal{H}.$$

Proof: The proof follows exactly as in Theorem 9.1.2 but with $\not\leq_{\mathbf{R}^p_+\setminus\{\mathbf{0}\}}$ in (9.1.5) replaced by $\not\leq_{\text{int } \mathbf{R}^p_+}$ and $\geq_{\mathbf{R}^p_+\setminus\{\mathbf{0}\}}$ in (9.1.6) replaced by $\geq_{\text{int } \mathbf{R}^p_+}$. ∎

9.2 Scalarization of Vector Variational Inequalities

In contrast to multicriteria optimization, the solution of vector variational inequalities is a much less mature subject. In Section 7.3, we discuss in detail how solutions to multicriteria optimization problems can be found by scalarization or composing the vector cost with a utility function. Unfortunately, such scalarization techniques in general do not work for VVI. There are very few results on solving VVI until recently. The first such result is found in [GY7] where the solution of WVVI is reduced to finding the level sets of some related scalar function. As this is related to the notion of gap functions for VVI, we shall defer the discussion of this result to Section 9.6. The main reference for this section is [GY8]. The ordering cone for this section is assumed to be $\mathbb{R}^p_+$.

If a certain integrability assumption can be made about the underlying (matrix valued) function, it is possible to apply a scalarization technique to relate the problem of solving a WVVI to a (scalar) variational inequality or to a multicriteria optimization problem [GY8]. These results are discussed as follows.

Definition 9.2.1 Let $\boldsymbol{\lambda} \in \mathbb{R}^p_+ \setminus \{\mathbf{0}\}$. The *scalar variational inequality* for the vector variational inequality as defined in Definition 9.1.1 is defined by:

$$\text{(Problem VI}(\boldsymbol{\lambda}))\qquad \text{Find } \mathbf{x} \in \mathcal{K}, \text{such that } \boldsymbol{\lambda}^\top \mathbf{F}(\mathbf{x})^\top(\mathbf{y}-\mathbf{x}) \geq 0,\ \forall \mathbf{y} \in \mathcal{K},$$

where $\mathcal{K}, \mathbf{F}$ are as given in Definition 9.1.1.

We first present a straightforward relationship between WVVI and VI($\boldsymbol{\lambda}$).

Theorem 9.2.1 (Sufficient condition for WVVI in terms of a scalar VI) If there exists $\boldsymbol{\lambda} \in \mathbb{R}^p_+ \setminus \{\mathbf{0}\}$ such that $\mathbf{x} \in \mathcal{K}$ is a solution of VI($\boldsymbol{\lambda}$), then $\mathbf{x}$ is a solution of the WVVI (9.1.2).

Proof: If not, there exists $\mathbf{y} \in \mathcal{K}$ such that

$$\mathbf{F}(\mathbf{x})^\top(\mathbf{y}-\mathbf{x}) \leq_{\text{int } \mathbf{R}^p_+} \mathbf{0}.$$

Then for all $\boldsymbol{\lambda} \in \mathbb{R}^p_+ \setminus \{\mathbf{0}\}$, we have

$$\boldsymbol{\lambda}^\top \mathbf{F}(\mathbf{x})(\mathbf{y}-\mathbf{x}) < 0.$$

Hence $\mathbf{x} \in \mathcal{K}$ is not a solution of VI($\boldsymbol{\lambda}$). ∎

Unlike the above sufficient condition, a necessary condition for the solution of WVVI is much harder to obtain. Several non-trivial conditions are required which are to be discussed as follows.

Definition 9.2.2 A (cost) matrix function $\mathbf{F} : \mathbb{R}^n \to \mathbf{R}^{n \times p}$ is said to be *integrable* if,

$$\frac{\partial F_{ik}(\mathbf{x})}{\partial x_j} = \frac{\partial F_{jk}(\mathbf{x})}{\partial x_i},\quad \forall\ i,j = 1,\cdots,n,\ \forall\ k = 1,\cdots,p.$$

Thus a cost matrix is integrable if the Jacobian of each column of the matrix is symmetric. This nomenclature is a generalization of integrable field in physics. The following is a well-known extension of Green's theorem to functions in $\mathbb{R}^n$.

Theorem 9.2.2 If $\mathbf{F}(\mathbf{x})$ is integrable, then there exists $\mathbf{f} : \mathbb{R}^n \to \mathbb{R}^p$, with $\mathbf{f}(\mathbf{x}) = (f_1(\mathbf{x}),\cdots,f_p(\mathbf{x}))^\top$, denoted by

$$f_k(\mathbf{x}) = \oint^{\mathbf{x}} \mathbf{F}^k(\mathbf{y})d\mathbf{y},\ k = 1,\cdots,p,$$

such that

$$\mathbf{F}(\mathbf{x})^\top = \nabla \mathbf{f}(\mathbf{x}).$$

Proof: See Theorem 4.1.6 of [OR1]. ∎

Definition 9.2.3 Let $\mathbf{f} : \mathcal{K} \to \mathbb{R}^p$ be a differentiable vector-valued function. $\mathbf{f}$ is said to be int$\mathbb{R}^p_+$-*pseudoconvex* (or weakly pseudoconvex) on $\mathcal{K}$ if

$$\forall \mathbf{x}, \mathbf{y} \in \mathcal{K}, \quad \nabla \mathbf{f}(\mathbf{x})(\mathbf{y}-\mathbf{x}) \not\leq_{\text{int } \mathbf{R}^p_+} \mathbf{0} \text{ implies } \mathbf{f}(\mathbf{y}) \not\leq_{\text{int } \mathbf{R}^p_+} \mathbf{f}(\mathbf{x}),$$

$\mathbf{f}$ is said to be $\mathbb{R}^p_+ \setminus \{\mathbf{0}\}$-*pseudoconvex* (or strongly pseudoconvex) on $\mathcal{K}$ if

$$\forall \mathbf{x}, \mathbf{y} \in \mathcal{K}, \quad \nabla \mathbf{f}(\mathbf{x})(\mathbf{y}-\mathbf{x}) \not\leq_{\mathbf{R}^p_+ \setminus \{\mathbf{0}\}} \mathbf{0} \text{ implies } \mathbf{f}(\mathbf{y}) \not\leq_{\mathbf{R}^p_+ \setminus \{\mathbf{0}\}} \mathbf{f}(\mathbf{x}).$$

Clearly strong pseudoconvexity implies weak pseudoconvexity, but not vice versa.

Theorem 9.2.3 If each component of $\mathbf{f}$ is pseudoconvex on $\mathcal{K}$, then $\mathbf{f}$ is int$\mathbb{R}^p_+$-pseudoconvex on $\mathcal{K}$. If in addition, at least one component of $\mathbf{f}$ is strictly pseudoconvex on $\mathcal{K}$, then $\mathbf{f}$ is $\mathbb{R}^p_+ \setminus \{\mathbf{0}\}$-pseudoconvex on $\mathcal{K}$.

Proof: Let $\nabla \mathbf{f}(\mathbf{x})(\mathbf{y}-\mathbf{x}) \not\leq_{\text{int } \mathbf{R}^p_+} \mathbf{0}$. Then there exists i, such that $\nabla f_i(\mathbf{x})(\mathbf{y}-\mathbf{x}) \geq 0$. By the pseudoconvexity of f_i, $f_i(\mathbf{y}) \geq f_i(\mathbf{x})$. Thus $\mathbf{f}(\mathbf{y}) \not\leq_{\text{int } \mathbf{R}^p_+} \mathbf{f}(\mathbf{x})$. Then $\mathbf{f}$ is int$\mathbb{R}^p_+$-pseudoconvex on $\mathcal{K}$. If one component (j say) of $\mathbf{f}$ is strictly pseudoconvex, then by the strict pseudoconvexity of f_j, $f_j(\mathbf{y}) > f_j(\mathbf{x})$. Thus $\mathbf{f}(\mathbf{y}) \not\leq_{\mathbf{R}^p_+ \setminus \{\mathbf{0}\}} \mathbf{f}(\mathbf{x})$. Then $\mathbf{f}$ is $\mathbb{R}^p_+ \setminus \{\mathbf{0}\}$-pseudoconvex. ∎

Theorem 9.2.4 If $\mathbf{f}$ is int$\mathbb{R}^p_+$-pseudoconvex and $\mathbf{x}$ solves the following WVVI:

$$\text{Find } \mathbf{x} \in \mathcal{K}, \quad \nabla \mathbf{f}(\mathbf{x})(\mathbf{y}-\mathbf{x}) \not\leq_{\text{int } \mathbf{R}^p_+} \mathbf{0}, \quad \forall \mathbf{y} \in \mathcal{K}$$

then $\mathbf{x}$ solves the problem WVO as defined in Definition 7.3.3, with $\mathcal{K}$ as the feasible set. If $\mathbf{f}$ is $\mathbb{R}^p_+ \setminus \{\mathbf{0}\}$ -pseudoconvex and $\mathbf{x}$ solves the following VVI

$$\text{Find } \mathbf{x} \in \mathcal{K}, \quad \nabla \mathbf{f}(\mathbf{x})(\mathbf{y}-\mathbf{x}) \not\leq_{\mathbf{R}^p_+ \setminus \{\mathbf{0}\}} \mathbf{0}, \quad \forall \mathbf{y} \in \mathcal{K},$$

then $\mathbf{x}$ solves the problem VO as defined in Definition 7.3.1 with $\mathcal{K}$ as the feasible set.

Proof: Follows from Definition 9.2.3. ∎

Let $\boldsymbol{\lambda} \in \mathbb{R}^p_+ \setminus \{\mathbf{0}\}$. Consider the scalar optimization problem

$$\text{(Problem P}(\boldsymbol{\lambda})) \qquad \text{Min } \boldsymbol{\lambda}^\top \mathbf{f}(\mathbf{x}) \\ \text{subject to } \mathbf{x} \in \mathcal{K},$$

where $\mathbf{f}$ and $\mathcal{K}$ are as given in Definition 9.1.4. From Section 7.3, If the set $\mathbf{f}(\mathcal{K})+\mathbb{R}^p_+$ is convex, then a weakly minimal solution of WVO also solves $\text{P}(\boldsymbol{\lambda})$ (Theorem 7.3.2).

Lemma 9.2.5 If $\mathbf{x}$ is a globally minimum solution of $\text{P}(\boldsymbol{\lambda})$, then $\mathbf{x}$ is a solution of $\text{VI}(\boldsymbol{\lambda})$ with $\mathbf{F}(\mathbf{x})^\top = \nabla \mathbf{f}(\mathbf{x})$. If, in addition, that $\mathbf{f}$ is $\mathbb{R}^p_+$-convex, then the converse also holds, i.e., $\mathbf{x}$ is a solution of $\text{VI}(\boldsymbol{\lambda})$ implies that $\mathbf{x}$ is also a global minimum solution of $\text{P}(\boldsymbol{\lambda})$.

Proof: Assume that $\mathbf{x}$ is a global minimum solution of $\mathrm{P}(\boldsymbol{\lambda})$. Since $\mathcal{K}$ is a convex set and $\mathbf{f}$ is differentiable, $\mathbf{x}$ is a solution of the variational inequality

$$\nabla(\boldsymbol{\lambda}^\top \mathbf{f})(\mathbf{x})(\mathbf{y}-\mathbf{x}) \geq 0, \ \forall \mathbf{y} \in \mathcal{K}.$$

It is clear that $\nabla(\boldsymbol{\lambda}^\top \mathbf{f})(\mathbf{x}) = \boldsymbol{\lambda}^\top \nabla \mathbf{f}(\mathbf{x})$. Thus $\mathbf{x}$ is a solution of $\mathrm{VI}(\boldsymbol{\lambda})$ with $\mathbf{F}(\mathbf{x})^\top = \nabla \mathbf{f}(\mathbf{x})$.

The second part of the result follows from Theorem 1.5.1 and the argument that $\mathbf{f}$ is $\mathbb{R}^p_+$-convex if and only if $\boldsymbol{\lambda}^\top \mathbf{f}$ is convex for all $\boldsymbol{\lambda} \in \mathbb{R}^p_+$. ■

Remark 9.2.1 Note that the result of the first part in Lemma 9.2.5 is not true if $\mathbf{x}$ is only a local minimum solution.

Theorem 9.2.6 (Necessary condition for WVVI in terms of a scalar VI) Assume that $\mathbf{F}(\mathbf{x})$ is integrable, i.e., there exists $\mathbf{f}$ such that $\mathbf{F}(\mathbf{x})^\top = \nabla \mathbf{f}(\mathbf{x})$. If $\mathbf{f}$ is $\mathrm{int}\mathbb{R}^p_+$-pseudoconvex, $\mathbf{f}(\mathcal{K})$ is $\mathbb{R}^p_+$-bounded and $\mathbf{x} \in \mathcal{K}$ is a solution of WVVI, then there exists $\boldsymbol{\lambda} \in \mathbb{R}^p_+ \setminus \{\mathbf{0}\}$ such that $\mathbf{x}$ is a solution of $\mathrm{VI}(\boldsymbol{\lambda})$.

Proof: Following Theorem 9.2.4, $\mathbf{x}$ is a weakly minimal solution of WVO. By Theorem 7.3.8, there exists $\mathbf{e} \in \mathrm{int}\mathbb{R}^p_+, \mathbf{a} \in \mathbb{R}^p$ such that $\mathbf{x}$ is a solution of the following optimization problem:

$$\begin{array}{c} \text{Min } \ \xi_{\mathbf{ea}}(\mathbf{f}(\mathbf{x})) \\ \text{subject to } \mathbf{x} \in \mathcal{K}, \end{array}$$

where the Gerstewitz function $\boldsymbol{\xi}_{\mathbf{ea}}$ is as defined in Definition 7.3.5. From a corollary on page 52 of [C5],

$$\mathbf{0} \in \partial^\circ(\xi_{\mathbf{ea}} \circ \mathbf{f})(\mathbf{x}) + \mathcal{N}(\mathcal{K}, \mathbf{x}),$$

where $\mathcal{N}(\mathcal{K}, \mathbf{x})$ is the normal cone defined by $\mathcal{N}(\mathcal{K}, \mathbf{x}) = \{\mathbf{x}^* \in \mathbb{R}^n \mid (\mathbf{z}-\mathbf{x})^\top \mathbf{x}^* \leq 0 \ \ \forall \mathbf{z} \in \mathcal{K}\}$. Then $\mathbf{x}$ satisfies the following condition:

$$\exists \mathbf{z} \in \partial^\circ(\xi_{\mathbf{ea}} \circ \mathbf{f})(\mathbf{x}) \text{ such that } \mathbf{z}^\top(\mathbf{y}-\mathbf{x}) \geq 0, \ \forall \mathbf{y} \in \mathcal{K},$$

where $\partial^\circ g(\mathbf{x})$ is the Clarke generalized subgradient [C5]. Note that

$$(\xi_{\mathbf{ea}} \circ \mathbf{f})(\mathbf{x}) = \max_{1 \leq i \leq p} \left\{ \frac{f_i(\mathbf{x}) - f_i(\mathbf{a})}{e_i} \right\}.$$

Thus

$$\partial^\circ(\xi_{\mathbf{ea}} \circ \mathbf{f})(\mathbf{x}) = \left\{ \sum_{i=1}^p \frac{t_i}{e_i} \nabla f_i(\mathbf{x}) \mid \sum_{i=1}^p t_i = 1, t_i \geq 0, t_i = 0, i \notin I(\mathbf{x}) \right\},$$

where $I(\mathbf{x})$ is the set of index that achieves the maximum in (9.1.7). Thus $\mathbf{x}$ is a solution of the variational inequality problem:

$$\left(\sum_{i=1}^p \frac{t_i}{e_i} \nabla f_i(\mathbf{x}) \right) (\mathbf{y}-\mathbf{x}) \geq 0, \quad \forall \mathbf{y} \in \mathcal{K},$$

where for some t_i such that $\sum_{i=1}^{p} t_i = 1, t_i \geq 0, t_i = 0, i \notin \mathcal{I}(\mathbf{x})$. Let

$$\lambda_i = \frac{t_i}{e_i}, \quad i = 1, \cdots, p.$$

Then $\boldsymbol{\lambda} \in \mathbb{R}_+^p \setminus \{\mathbf{0}\}$, and the proof is complete. ∎

The above necessary and sufficient conditions for weak vector variational inequalities can be extended to vector variational inequalities as follows.

Theorem 9.2.7 (Sufficient condition for VVI in terms of a scalar VI) If there exists $\boldsymbol{\lambda} \in \text{int}\mathbb{R}_+^p$ such that $\mathbf{x} \in \mathcal{K}$ is a solution of VI($\boldsymbol{\lambda}$), then $\mathbf{x}$ is a solution of VVI.

Proof: If not, there exists $\mathbf{y} \in \mathcal{K}$ such that

$$\mathbf{F}(\mathbf{x})^\top (\mathbf{y} - \mathbf{x}) \leq_{\mathbb{R}_+^p \setminus \{0\}} \mathbf{0}.$$

Since $\boldsymbol{\lambda} \in \text{int}\mathbb{R}_+^p$, we have

$$\boldsymbol{\lambda}^\top \mathbf{F}(\mathbf{x})^\top (\mathbf{y} - \mathbf{x}) < 0.$$

Hence $\mathbf{x} \in \mathcal{K}$ is not a solution of VI($\boldsymbol{\lambda}$). ∎

Theorem 9.2.8 (Necessary condition for VVI in terms of a scalar VI) Assume that $\mathbf{F}(\mathbf{x})$ is integrable, i.e., there exists $\mathbf{f}$ such that $\mathbf{F}(\mathbf{x})^\top = \nabla \mathbf{f}(\mathbf{x})$. If $\mathbf{x} \in \mathcal{K}$ is a solution of VVI, and $\mathbf{f}$ is $\mathbb{R}_+^p \setminus \{\mathbf{0}\}$-pseudoconvex, then there exists $\boldsymbol{\lambda} \in \mathbb{R}_+^p \setminus \{\mathbf{0}\}$ such that $\mathbf{x}$ is a solution of VI($\boldsymbol{\lambda}$).

Proof: Let $\mathbf{x} \in \mathcal{K}$ be a solution of VVI. Since $\mathbf{f}$ is $\mathbb{R}_+^p$-pseudoconvex, it follows from part two of Theorem 9.2.4 that $\mathbf{x}$ solves VO. The rest of the proof is similar to that of Theorem 9.2.6 and is omitted. ∎

9.3 Duality in Extended Vector Variational Inequalities

Like variational inequalities, duality plays a very important role in the analysis of vector variational inequalities. In this section, we study the duality of extended vector variational inequalities and establish their equivalences with VVI. Two examples are used to illustrate the role of VVI duality in multicriteria optimization and vector approximation problems. The main reference for this section is [Y1].

Like in the scalar case, the duality of VVI is more appropriately analysed in the extended case. For this section, the underlying vector ordering relationships are induced by a closed and convex cone $\mathcal{C} \subset \mathbb{R}^p$.

Definition 9.3.1 (Extended vector variational inequality) Let $\mathbf{F} : \mathbb{R}^n \to \mathbb{R}^{n\times p}$ and $\mathbf{f} : \mathbb{R}^n \to \overline{\mathbb{R}}^p$ (see Remark 9.3.1). The extended vector variational inequality (EVVI) is defined as:

$$\text{(Problem EVVI)} \qquad \text{Find a point } \mathbf{x}^0 \in \mathbb{R}^n \text{ such that}$$
$$\mathbf{F}(\mathbf{x}^0)^\top(\mathbf{x}-\mathbf{x}^0) \not\leq_{\mathcal{C}\setminus\{\mathbf{0}\}} \mathbf{f}(\mathbf{x}^0) - \mathbf{f}(\mathbf{x}), \quad \forall \mathbf{x} \in \mathbb{R}^n. \tag{9.3.1}$$

Remark 9.3.1 Note that in order to be consistent with the definition of Fenchel Transform (Definition 7.2.1), $\mathbf{f}$ is assumed to be a function from $\mathbb{R}^n$ to $\mathbb{R}^p$. Thus, in general, VVI is not a special case of EVVI. However, if we append an abstract "∞" to $\mathbb{R}^p$, written as $\overline{\mathbb{R}}^p = \mathbb{R}^p \cup \{\infty\}$ (see [B4]), then the EVVI (9.3.1) includes the VVI (9.1.1) as a special case where $\mathbf{f}$ is just the following indicator function for the ground set $\mathcal{K}$:

$$\mathbf{f}(\mathbf{x}) = \begin{cases} \mathbf{0}, & \text{if } \mathbf{x} \in \mathcal{K}; \\ \infty, & \text{if } \mathbf{x} \notin \mathcal{K}. \end{cases}$$

We make the following assumptions.

Assumption 9.3.1
(i) $\mathbf{F}$ is one-to-one (injective),
(ii) $\mathbf{f}$ is continuous, and
(iii) $\mathbf{f}^*(\mathbf{U}) \neq \emptyset, \forall \mathbf{U} \in \mathbb{R}^{p\times n}$.

Under Assumption 9.3.1, we may define the adjoint function $\mathbf{F}^\dagger : \mathbb{R}^{p\times n} \longrightarrow \mathbb{R}^n$ as follows:

$$\mathbf{F}^\dagger(\mathbf{U}) = -\mathbf{F}^{-1}(-\mathbf{U}), \quad \forall \mathbf{U} \in \text{dom}(\mathbf{F}^\dagger) = -\text{Range}(\mathbf{F}) \subset \mathbb{R}^{p\times n}.$$

If $\mathbf{F}$ is linear, then $\mathbf{F}^\dagger = \mathbf{F}^{-1}$.

Definition 9.3.2 (Dual extended vector variational inequality) The dual extended vector variational inequality (DEVVI) is defined by:

$$\text{(Problem DEVVI)} \qquad \text{Find } \mathbf{U}^0 \in \text{dom}(\mathbf{F}^\dagger) \quad \text{such that}$$
$$(\mathbf{U}-\mathbf{U}^0)\mathbf{F}^\dagger(\mathbf{U}^0) \not\leq_{\mathcal{C}\setminus\{\mathbf{0}\}} \mathbf{f}^*(\mathbf{U}^0) - \mathbf{f}^*(\mathbf{U}), \quad \forall \mathbf{U} \in \mathbb{R}^{p\times n} \tag{9.3.2}$$

where the (set-valued) function

$$\mathbf{f}^*(\mathbf{U}) = \max\nolimits_{\mathcal{C}\setminus\{\mathbf{0}\}}\{\mathbf{U}\mathbf{x} - \mathbf{f}(\mathbf{x}) \mid \mathbf{x} \in \mathbb{R}^n\}$$

is the type I Fenchel transform of $\mathbf{f}$.

Theorem 9.3.1 Under Assumption 9.3.1, we have

(i) If $\mathbf{x}^0$ is a solution of EVVI, then $\mathbf{U}^0 = -\mathbf{F}(\mathbf{x}^0)^\top$ is a solution of the DEVVI.

(ii) In addition to Assumption 9.3.1, we assume that the closed convex cone C is connected, i.e., $C \cup (-C) = \mathbb{R}^p$, and $\partial \mathbf{f}(\mathbf{x}^0) \neq \emptyset$, where $\mathbf{x}^0 = -\mathbf{F}^\dagger(\mathbf{U}^0)$. If $\mathbf{U}^0$ solves DEVVI, then $\mathbf{x}^0$ solves EVVI.

(iii) In both cases (i) and (ii), the following relation holds

$$\mathbf{U}^0\mathbf{x}^0 \in \mathbf{f}(\mathbf{x}^0) + \mathbf{f}^*(\mathbf{U}^0).$$

Proof: (i) Let $\mathbf{x}^0$ be a solution of EVVI, then

$$\mathbf{F}(\mathbf{x}^0)^\top(\mathbf{x} - \mathbf{x}^0) \not\leq_{C\setminus\{\mathbf{0}\}} \mathbf{f}(\mathbf{x}^0) - \mathbf{f}(\mathbf{x}), \quad \forall \mathbf{x} \in \mathbb{R}^n,$$

or,

$$-\mathbf{F}(\mathbf{x}^0)^\top\mathbf{x}^0 - \mathbf{f}(\mathbf{x}^0) \not\leq_{C\setminus\{\mathbf{0}\}} -\mathbf{F}(\mathbf{x}^0)^\top\mathbf{x} - \mathbf{f}(\mathbf{x}), \quad \forall \mathbf{x} \in \mathbb{R}^n,$$

or,

$$-\mathbf{F}(\mathbf{x}^0)^\top\mathbf{x}^0 - \mathbf{f}(\mathbf{x}^0) \in \mathbf{f}^*(-\mathbf{F}(\mathbf{x}^0)), \tag{9.3.3}$$

or,

$$-\mathbf{F}(\mathbf{x}^0)^\top\mathbf{x}^0 - \mathbf{f}(\mathbf{x}^0) - \mathbf{f}^*(\mathbf{U}) \subseteq \mathbf{f}^*(-\mathbf{F}(\mathbf{x}^0)) - \mathbf{f}^*(\mathbf{U}), \quad \forall \mathbf{U} \in \mathbb{R}^{p\times n}. \tag{9.3.4}$$

If $\mathbf{U}^0 = -\mathbf{F}(\mathbf{x}^0)^\top$ is not a solution of DEVVI, then there exists $\mathbf{U} \in -\text{Range}(\mathbf{F})$ such that

$$(\mathbf{U} - \mathbf{U}^0)\mathbf{F}^\dagger(\mathbf{U}^0) \leq_{C\setminus\{\mathbf{0}\}} \mathbf{f}^*(\mathbf{U}^0) - \mathbf{f}^*(\mathbf{U}).$$

It follows from (9.3.4) that

$$(\mathbf{U} - \mathbf{U}^0)\mathbf{F}^\dagger(\mathbf{U}^0) \leq_{C\setminus\{\mathbf{0}\}} -\mathbf{F}(\mathbf{x}^0)^\top\mathbf{x}^0 - \mathbf{f}(\mathbf{x}^0) - \mathbf{f}^*(\mathbf{U}),$$

or,

$$-\mathbf{U}\mathbf{x}^0 \leq_{C\setminus\{\mathbf{0}\}} -\mathbf{f}(\mathbf{x}^0) - \mathbf{f}^*(\mathbf{U}).$$

This is a contradiction to the generalized Young's inequality of Theorem 7.2.1. Thus $\mathbf{U}^0$ is a solution of DEVVI.

(ii) Let $\mathbf{U}^0$ be a solution of DEVVI. Let $\mathbf{x}^0 = -\mathbf{F}^\dagger(\mathbf{U}^0)$ and $\mathbf{U}^0 = -\mathbf{F}(\mathbf{x}^0)^\top$. Then

$$(\mathbf{U} - \mathbf{U}^0)\mathbf{F}^\dagger(\mathbf{U}^0) \not\leq_{C\setminus\{\mathbf{0}\}} \mathbf{f}^*(\mathbf{U}^0) - \mathbf{f}^*(\mathbf{U}), \quad \forall \mathbf{U} \in \mathbb{R}^{p\times n},$$

$$(\mathbf{U} + \mathbf{F}(\mathbf{x}^0)^\top)(-\mathbf{x}^0) \not\leq_{C\setminus\{\mathbf{0}\}} \mathbf{f}^*(-\mathbf{F}(\mathbf{x}^0)) - \mathbf{f}^*(\mathbf{U}),$$

$$-\mathbf{F}(\mathbf{x}^0)^\top\mathbf{x}^0 - \mathbf{U}\mathbf{x}^0 + \mathbf{f}^*(\mathbf{U}) \not\leq_{C\setminus\{\mathbf{0}\}} \mathbf{f}^*(-\mathbf{F}(\mathbf{x}^0)). \tag{9.3.5}$$

Since $\partial \mathbf{f}(\mathbf{x}^0) \neq \emptyset$, let $\mathbf{U} \in \partial \mathbf{f}(\mathbf{x}^0)$. Then by Lemma 7.2.2,

$$\mathbf{U}\mathbf{x}^0 - \mathbf{f}(\mathbf{x}^0) \in \mathbf{f}^*(\mathbf{U}). \tag{9.3.6}$$

It follows from (9.3.5) and (9.3.6) that

$$-\mathbf{F}(\mathbf{x}^0)^\top\mathbf{x}^0 - \mathbf{f}(\mathbf{x}^0) \not\leq_{C\setminus\{\mathbf{0}\}} \mathbf{f}^*(-\mathbf{F}(\mathbf{x}^0)).$$

From the definition of $\mathbf{f}^*$ and the assumption that C is connected, we have,

$$-\mathbf{F}(\mathbf{x}^0)^\top \mathbf{x}^0 - \mathbf{f}(\mathbf{x}^0) \in \mathbf{f}^*(-\mathbf{F}(\mathbf{x}^0)). \tag{9.3.7}$$

If $\mathbf{x}^0 = -\mathbf{F}^\dagger(\mathbf{U}^0)$ is not a solution of EVVI, then there exists $\mathbf{x} \in \mathbb{R}^n$ such that

$$\mathbf{F}(\mathbf{x}^0)^\top(\mathbf{x} - \mathbf{x}^0) \leq_{C\setminus\{\mathbf{0}\}} \mathbf{f}(\mathbf{x}^0) - \mathbf{f}(\mathbf{x}),$$

or

$$-\mathbf{F}(\mathbf{x}^0)^\top \mathbf{x}^0 - \mathbf{f}(\mathbf{x}^0) \leq_{C\setminus\{\mathbf{0}\}} -\mathbf{F}(\mathbf{x}^0)^\top \mathbf{x} - \mathbf{f}(\mathbf{x}).$$

Then

$$-\mathbf{F}(\mathbf{x}^0)^\top \mathbf{x}^0 - \mathbf{f}(\mathbf{x}^0) \notin \mathbf{f}^*(-\mathbf{F}(\mathbf{x}^0)),$$

a contradiction. Thus $\mathbf{x}^0 = -\mathbf{F}^\dagger(\mathbf{U}^0)$ is a solution of EVVI.

(iii) From (9.3.3), (9.3.6) and the fact that $\mathbf{U}^0 = -\mathbf{F}(\mathbf{x}^0)^\top$, we have

$$\mathbf{U}^0\mathbf{x}^0 \in \mathbf{f}(\mathbf{x}^0) + \mathbf{f}^*(\mathbf{U}^0).$$

■

Remark 9.3.2 Note that the interior of the closed convex cone C does not need to be assumed nonempty. Note also that in order for Theorem 9.3.1 (ii) to hold, the restrictive assumption that C is connected is required. This is certainly not the case when the cone is assumed to be $\mathbb{R}^p_+$. Moreover, note that since the condition $\text{int}C \neq \emptyset$ is not required, Theorem 9.3.1 includes the Mosco's result as a special case (see Theorem 6.1.1).

The weak vector variational inequality WVVI and its dual are defined as follows:

Definition 9.3.3 (Weak extended vector variational inequality WEVVI)

(Problem WVVI) Find $\mathbf{x}^0 \in \mathbb{R}^n$ such that

$$\mathbf{F}(\mathbf{x}^0)^\top(\mathbf{x} - \mathbf{x}^0) \not\leq_{\text{int }C} \mathbf{f}(\mathbf{x}^0) - \mathbf{f}(\mathbf{x}), \quad \forall \mathbf{x} \in \mathbb{R}^n.$$

Definition 9.3.4 (Dual weak extended vector variational inequality DW-EVVI)

(Problem DWVVI) Find $\mathbf{U}^0 \in \text{dom}(\mathbf{F}^\dagger)$ such that

$$(\mathbf{U} - \mathbf{U}^0)\mathbf{F}^\dagger(\mathbf{U}^0) \not\leq_{\text{int }C} \mathbf{f}^*_w(\mathbf{U}^0) - \mathbf{f}^*_w(\mathbf{U}), \quad \forall \mathbf{U} \in \mathbb{R}^{p\times n} \tag{9.3.8}$$

where the (set-valued) function

$$\mathbf{f}^*_w(\mathbf{U}) = \max\nolimits_{\text{int}C}\{\mathbf{U}\mathbf{x} - \mathbf{f}(\mathbf{x}) \mid \mathbf{x} \in \mathbb{R}^n\} \tag{9.3.9}$$

is the weak Fenchel transform of $\mathbf{f}$, see Definition 7.2.1.

Theorem 9.3.2 Under Assumption 9.3.1, we have

(i) If $\mathbf{x}^0$ solves WEVVI, then $\mathbf{U}^0 = -\mathbf{F}(\mathbf{x}^0)^\top$ solves DWEVVI.

(ii) In addition to Assumption 9.3.1, we assume that the closed convex cone $\mathcal{C}$ is connected, i.e., $\mathcal{C} \cup (-\mathcal{C}) = \mathbb{R}^p$, and $\partial \mathbf{f}(\mathbf{x}^0) \neq \emptyset$, where $\mathbf{x}^0 = -\mathbf{F}^\dagger(\mathbf{U}^0)$. If $\mathbf{U}^0$ solves DWEVVI, then $\mathbf{x}^0$ solves WEVVI.

(iii) In both cases (i) and (ii), the following relation holds

$$\mathbf{U}^0\mathbf{x}^0 \in \mathbf{f}(\mathbf{x}^0) + \mathbf{f}_w^*(\mathbf{U}^0).$$

Proof: The proof follows essentially the same lines as in the proof of Theorem 9.3.1, with $\not\leq_{\mathcal{C}\setminus\{0\}}$ replaced by $\not<_{\text{int } \mathcal{C}}$ and $\mathbf{f}^*(\mathbf{U})$ replaced by $\mathbf{f}_w^*(\mathbf{U})$. ∎

In the following, two examples are given to illustrate the applications of a dual weak vector variational inequality.

Example 9.3.1 Consider the vector approximation problem:

(Vector Approximation Problem)

$$\min_{\text{int } \mathcal{C}} \left(\|\mathbf{a}^1 - \mathbf{x}\|^2, \cdots, \|\mathbf{a}^p - \mathbf{x}\|^2\right), \text{subject to } \mathbf{x} \in \mathcal{K}$$

where $\mathcal{K} \subseteq \mathbb{R}^n$ is a closed and convex set, $\mathbf{a}^i$, $i = 1, \cdots, p$ are fixed vectors in $\mathbb{R}^n$. We say that $\mathbf{x}^0 \in \mathcal{K}$ is a weakly minimal approximation to all the $\mathbf{a}^i$ if

$$\left(\|\mathbf{a}^1 - \mathbf{x}\|^2, \cdots, \|\mathbf{a}^p - \mathbf{x}\|^2\right)^\top \not\leq_{\text{int } \mathcal{C}} \left(\|\mathbf{a}^1 - \mathbf{x}^0\|^2, \cdots, \|\mathbf{a}^p - \mathbf{x}^0\|^2\right)^\top, \ \forall \mathbf{x} \in \mathcal{K},$$

or $\forall \mathbf{x} \in \mathcal{K}$

$$\left((\mathbf{x}^0)^\top(\mathbf{x} - \mathbf{x}^0), \cdots, (\mathbf{x}^0)^\top(\mathbf{x} - \mathbf{x}^0)\right)^\top \not\leq_{\text{int } \mathcal{C}} \left((\mathbf{a}^1)^\top(\mathbf{x} - \mathbf{x}^0), \cdots, (\mathbf{a}^p)^\top(\mathbf{x} - \mathbf{x}^0)\right)^\top$$

This is a weak extended vector variational inequality (WEVVI) with

$$\mathbf{F}(\mathbf{x}) = [\mathbf{x}, \mathbf{x}, \cdots, \mathbf{x}] \in \mathbb{R}^{n\times p}, \quad \mathbf{f}(\mathbf{x}) = -\begin{bmatrix} (\mathbf{a}^1)^\top \mathbf{x} \\ (\mathbf{a}^2)^\top \mathbf{x} \\ \vdots \\ (\mathbf{a}^p)^\top \mathbf{x} \end{bmatrix} \in \mathbb{R}^p.$$

Since $\mathbf{F}^\dagger(\mathbf{U}^0) = -\mathbf{x}^0$, $\mathbf{U}^0$ is a solution of the dual weak extended vector variational inequality. It should be easy to verify that $\mathbf{U}^0 = -\mathbf{F}(\mathbf{x}^0)^\top$ satisfies the DWEVVI:

$$\left(\mathbf{U} - \mathbf{U}^0\right)(-\mathbf{x}^0) \not\leq_{\text{int } \mathcal{C}} \mathbf{f}_w^*(\mathbf{U}^0) - \mathbf{f}_w^*(\mathbf{U}), \quad \forall \mathbf{U} \in \mathbb{R}^{p\times n}.$$

Example 9.3.2 Consider the multicriteria optimization problem VO:

$$\text{(Problem VO)} \qquad \min_{\text{int } \mathcal{C}} \ \mathbf{h}(\mathbf{x}), \ \text{subject to } \mathbf{x} \in \mathbb{R}^n$$

where $\mathbf{h} : \mathbb{R}^n \longrightarrow \mathbb{R}^p$ is a differentiable vector-valued function. Let $\mathbf{x}^0$ be a weakly minimal solution of problem VO. Then by Theorem 9.1.1, $\mathbf{x}^0$ is a solution of the following weak vector variational inequality:

$$\nabla \mathbf{h}(\mathbf{x}^0)(\mathbf{x} - \mathbf{x}^0) \not\leq_{\text{int } C} \mathbf{0}, \quad \forall \mathbf{x} \in \mathbb{R}^n. \tag{9.3.10}$$

From Theorem 9.3.1, $\mathbf{U}^0 = -\nabla \mathbf{h}(\mathbf{x}^0)^\top$ satisfies the dual weak extended vector variational inequality:

$$(\mathbf{U} - \mathbf{U}^0)(-\mathbf{x}^0) \not\leq_{\text{int } C} \boldsymbol{\sigma}_w(\mathbf{U}^0) - \boldsymbol{\sigma}_w(\mathbf{U}), \quad \forall \mathbf{U} \in \mathbb{R}^{p\times n}, \tag{9.3.11}$$

where $\boldsymbol{\sigma}_w(\mathbf{U}) = \max_{\text{int } C}\{\mathbf{U}\mathbf{x} \mid \mathbf{x} \in \mathbb{R}^n\}$. This is the dual weak extended vector variational inequality of (9.3.10) if we let $\mathbf{F} = \nabla \mathbf{h}$ and $-\mathbf{x}^0 = \mathbf{F}^\dagger(\mathbf{U}^0)$.

Note that the problem VO can be expressed as

$$\min_{\text{int } C} \boldsymbol{\phi}(\mathbf{x}, \mathbf{0}), \qquad \text{subject to } \mathbf{x} \in \mathbb{R}^n$$

where $\boldsymbol{\phi} : \mathbb{R}^n \times \mathbb{R}^{n\times p} \longrightarrow \mathbb{R}^p$ is the perturbation function satisfying

$$\boldsymbol{\phi}(\mathbf{x}, \mathbf{0}) = \mathbf{h}(\mathbf{x}), \quad \forall \mathbf{x} \in \mathbb{R}^n.$$

Let $\boldsymbol{\phi}_w^*(\mathbf{x}, \boldsymbol{\Gamma})$ be the weak Fenchel transform of $\boldsymbol{\phi}(\mathbf{x}, \boldsymbol{\Gamma})$. Then we may construct the dual problem DVO of VO as follows:

$$\text{(Problem DVO)} \qquad \max_{\text{int } C} \; -\boldsymbol{\phi}_w^*(0, \boldsymbol{\Gamma}), \quad \text{subject to } \boldsymbol{\Gamma} \in \mathbb{R}^{n\times p}.$$

It is easy to verify that the following weak duality relation of VO and DVO holds. For any $\mathbf{x} \in \mathbb{R}^n$, $\boldsymbol{\Gamma} \in \mathbb{R}^{n\times p}$,

$$\boldsymbol{\phi}(\mathbf{x}, \mathbf{0}) \notin -\boldsymbol{\phi}_w^*(\mathbf{0}, \boldsymbol{\Gamma}) - \text{int } C. \tag{9.3.12}$$

The problem VO is said to be *weakly stable* if the set-valued mapping $\boldsymbol{\zeta} : \mathbb{R}^n \to 2^{\mathbb{R}^p}$

$$\boldsymbol{\zeta}(\mathbf{U}) = -\max_{\text{int } C}\{-\boldsymbol{\phi}(\mathbf{x}, \mathbf{U}) \mid \mathbf{x} \in \mathbb{R}^n\} \tag{9.3.13}$$

has a weak subgradient at $\mathbf{U} = \mathbf{0}$. From [SNT1], if VO is weakly stable, then there exists a solution $\boldsymbol{\Gamma}^0 \in \mathbb{R}^{n\times p}$ of DVO satisfying

$$(\mathbf{0}, \boldsymbol{\Gamma}^0) \in \partial_w \boldsymbol{\phi}(\mathbf{x}^0, \mathbf{0}). \tag{9.3.14}$$

Assuming that C is connected, then from the weak vector variational inequality (9.3.1),

$$-\nabla \mathbf{h}(\mathbf{x}^0)\mathbf{x} \leq_{C\setminus\{0\}} \mathbf{0}, \quad \forall \mathbf{x} \in \mathbb{R}^n. \tag{9.3.15}$$

Hence $(\mathbf{U}^0, \boldsymbol{\Gamma}^0) \in \partial_w \boldsymbol{\phi}(\mathbf{x}^0, \mathbf{0})$. We have proved:

Theorem 9.3.3. Assume that the problem VO is weakly stable and C is connected. If $\mathbf{x}^0$ is a solution of VO, then there exists $\boldsymbol{\Gamma}^0 \in \mathbb{R}^{n\times p}$ such that $\mathbf{U}^0 = -\nabla \mathbf{h}(\mathbf{x}^0)$ is a solution to the weak dual vector variational inequality (9.3.7) and $\boldsymbol{\Gamma}^0$ is a solution of DVO that satisfies the inclusion

$$(\mathbf{U}^0, \boldsymbol{\Gamma}^0) \in \partial_w \boldsymbol{\phi}(\mathbf{x}^0, \mathbf{0}). \tag{9.3.16}$$

9.4 Gap Functions for Extended Vector Variational Inequalities

In Chapter Six, we studied the gap function of a (scalar) VI and found a number of interesting properties relating to the duality of VI. We may now extend the concept of scalar gap functions to the vector case. This effort is a non-trivial one since many of the functions involved will now be set-valued, and the concept of convexity for set-valued functions has some twists and turns (see Section 7.1). The main reference for this section is [CGY1].

Let all vector ordering relationships be induced by a closed and convex cone $C \subset \mathbb{R}^p$. The set-valued gap function for the EVVI (9.3.1) is defined as follows:

Definition 9.4.1 A set-valued mapping $\gamma_0 : \mathbb{R}^n \to 2^{\mathbb{R}^p}$ is said to be a *gap function* for the EVVI (9.3.1) if and only if

1. $\mathbf{0} \in \gamma_0(\mathbf{x})$ if and only if $\mathbf{x}$ solves the EVVI;
2. $\mathbf{0} \not\geq_{C\setminus\{\mathbf{0}\}} \gamma_0(\mathbf{y}),\ \forall \mathbf{y} \in \mathbb{R}^n$.

The following is a generalization of the scalar gap function introduced in Section 6.2.1.

Definition 9.4.2 Let $\gamma : \mathbb{R}^n \to 2^{\mathbb{R}^p}$ be such that:

$$\gamma(\mathbf{x}) = \max_{C\setminus\{\mathbf{0}\}} \{\mathbf{F}(\mathbf{x})^\top(\mathbf{x} - \mathbf{y}) + \mathbf{f}(\mathbf{x}) - \mathbf{f}(\mathbf{y}) \mid \mathbf{y} \in \mathbb{R}^n\}.$$

It follows immediately from this definition that

$$\gamma(\mathbf{x}) = \mathbf{f}(\mathbf{x}) + \mathbf{f}^*(-\mathbf{F}(\mathbf{x})) + \mathbf{F}(\mathbf{x})^\top\mathbf{x}, \tag{9.4.1}$$

where $\mathbf{f}^*$ is the Fenchel transform of $\mathbf{f}$. We have a simple proof that γ is a gap function for EVVI, and the meaning of "gap" is now apparent.

Theorem 9.4.1 γ is a gap function for the problem EVVI.

Proof: We need to verify the defining properties of a gap function. First, from Theorem 9.3.1, $\mathbf{x}$ solves EVVI implies that $\mathbf{U} = -\mathbf{F}(\mathbf{x})^\top$ solves DEVVI. Then part (iii) of Theorem 9.3.1 together with (9.4.1) imply that $\mathbf{0} \in \gamma(\mathbf{x})$.

Conversely, if $\mathbf{x}$ does not solve EVVI, then there exists some $\mathbf{y} \in \mathbb{R}^n$ such that:

$$\mathbf{F}(\mathbf{x})^\top(\mathbf{x} - \mathbf{y}) - \mathbf{f}(\mathbf{y}) + \mathbf{f}(\mathbf{x}) \geq_{C\setminus\{\mathbf{0}\}} \mathbf{0},$$

hence $\mathbf{0} \notin \gamma(\mathbf{x})$.

Furthermore, taking $\mathbf{y} = \mathbf{x}$ in $\mathbf{F}(\mathbf{x})^\top(\mathbf{x} - \mathbf{y}) - \mathbf{f}(\mathbf{y}) + \mathbf{f}(\mathbf{x})$, we have

$$\mathbf{F}(\mathbf{x})^\top(\mathbf{x} - \mathbf{x}) - \mathbf{f}(\mathbf{x}) + \mathbf{f}(\mathbf{x}) = \mathbf{0}.$$

Thus $\mathbf{0} \not\geq_{C\setminus\{\mathbf{0}\}} \gamma(\mathbf{x}),\ \forall \mathbf{x} \in \mathbb{R}^n$. Alternatively $\gamma(\mathbf{x}) \not\leq_{C\setminus\{\mathbf{0}\}} \mathbf{0}\ \ \forall \mathbf{x}$ follows directly from (9.4.1) and the Young's inequality of Theorem 7.2.1. ∎

The above EVVI is trivially extendable to the weak case. We shall include this as follows for completeness.

Definition 9.4.3 (Weak extended vector variational inequality WEVVI) Assume that int $C \neq \emptyset$. Find $\mathbf{x} \in \mathbb{R}^n$ such that

$$\mathbf{F}(\mathbf{x})^\top(\mathbf{y}-\mathbf{x}) - \mathbf{f}(\mathbf{x}) + \mathbf{f}(\mathbf{y}) \not\leq_{\text{int } C} \mathbf{0}, \quad \forall \mathbf{y} \in \mathbb{R}^n,$$

where $\mathbf{F} : \mathbb{R}^n \to \mathbb{R}^{n\times p}$ is assumed to be injective, and $\mathbf{f} : \mathbb{R}^n \to \mathbb{R}^p$ is assumed to be lower-semi-continuous and convex.

Definition 9.4.4 A set-valued mapping $\gamma'_w : \mathbb{R}^n \to 2^{\mathbb{R}^p}$ is said to be a *gap function* of the WEVVI if and only if

1. $\mathbf{0} \in \gamma_w(\mathbf{x})$ if and only if $\mathbf{x}$ solves WEVVI;
2. $\mathbf{0} \not\geq_{\text{int } C} \gamma_w(\mathbf{y}), \ \forall \mathbf{y} \in \mathbb{R}^n$.

Definition 9.4.5 (Weak gap function) Define the set-valued function $\gamma_w : \mathbb{R}^n \to 2^{\mathbb{R}^p}$:

$$\gamma_w(\mathbf{x}) = \max_{\text{int } C}\{\mathbf{F}(\mathbf{x})^\top(\mathbf{x}-\mathbf{y}) + \mathbf{f}(\mathbf{x}) - \mathbf{f}(\mathbf{y}) \mid \mathbf{y} \in \mathbb{R}^n\}. \tag{9.4.2}$$

Theorem 9.4.2 The set-valued function $\gamma_w : \mathbb{R}^n \to 2^{\mathbb{R}^p}$ is a gap function for WEVVI.

Proof: Follows similar arguments as in the proof of Theorem 9.4.1, with $\not\leq_{C\setminus\{\mathbf{0}\}}$ and $\leq_{C\setminus\{\mathbf{0}\}}$ replaced by $\not\leq_{\text{int } C}$ and $\leq_{\text{int } C}$ respectively. ∎

Under further assumptions, the gap function for both the extended vector variational inequality and weak extended vector variational inequality can be shown to be convex. We shall prove it only for the extended vector variational inequality defined in Definition 9.3.1.

Lemma 9.4.3 Let $\mathbf{F} : \mathbb{R}^n \to \mathbb{R}^{n\times p}$ be a matrix-valued function. If $\mathbf{F}$ is affine and monotone (see Definitions 7.1.10 and 7.1.11), then the function $\mathbf{F}(\cdot)^\top(\cdot) : \mathbb{R}^n \to \mathbb{R}^p$ is type I convex.

Proof: Given $t \in (0,1), \mathbf{x}^1, \mathbf{x}^2 \in \mathbb{R}^n$,

$$\begin{aligned}
&\mathbf{F}(t\mathbf{x}^1 + (1-t)\mathbf{x}^2)^\top(t\mathbf{x}^1 + (1-t)\mathbf{x}^2) - t\mathbf{F}(\mathbf{x}^1)^\top\mathbf{x}^1 - (1-t)\mathbf{F}(\mathbf{x}^2)^\top\mathbf{x}^2 \\
&\quad = t^2\mathbf{F}(\mathbf{x}^1)^\top\mathbf{x}^1 + (1-t)^2\mathbf{F}(\mathbf{x}^2)^\top\mathbf{x}^2 + t(1-t)(\mathbf{F}(\mathbf{x}^1)^\top\mathbf{x}^2 + \mathbf{F}(\mathbf{x}^2)^\top\mathbf{x}^1) \\
&\qquad - t\mathbf{F}(\mathbf{x}^1)^\top\mathbf{x}^1 - (1-t)\mathbf{F}(\mathbf{x}^2)^\top\mathbf{x}^2 \\
&\quad = -t(1-t)\mathbf{F}(\mathbf{x}^1 - \mathbf{x}^2)^\top(\mathbf{x}^1 - \mathbf{x}^2) \leq_{C\setminus\{\mathbf{0}\}} \mathbf{0} \quad \text{by the monotonicity of } \mathbf{F}.
\end{aligned}$$

∎

Lemma 9.4.4 If $\mathbf{f}^* : \mathbb{R}^{n\times p} \to 2^{\mathbb{R}^p}$ is type I convex, and $\mathbf{F} : \mathbb{R}^n \to \mathbb{R}^{n\times p}$ is affine, then the composite $\mathbf{f}^* \circ \mathbf{F} : \mathbb{R}^n \to 2^{\mathbb{R}^p}$ is type I convex.

Proof: Given $\mathbf{x}^1, \mathbf{x}^2 \in \mathbb{R}^n$ and $t \in (0,1)$,

$$\begin{aligned}
\mathbf{f}^* \circ \mathbf{F}(t\mathbf{x}^1 + (1-t)\mathbf{x}^2) &= \mathbf{f}^*(\mathbf{F}(t\mathbf{x}^1 + (1-t)\mathbf{x}^2)) \\
&= \mathbf{f}^*(t\mathbf{F}(\mathbf{x}^1) + (1-t)\mathbf{F}(\mathbf{x}^2)) \\
&\subseteq t\mathbf{f}^*(\mathbf{F}(\mathbf{x}^1)) + (1-t)\mathbf{f}^*(\mathbf{F}(\mathbf{x}^2)) - \mathbb{R}^p_+ \\
&= t\mathbf{f}^* \circ \mathbf{F}(\mathbf{x}^1) + (1-t)\mathbf{f}^* \circ \mathbf{F}(\mathbf{x}^2) - \mathbb{R}^p_+.
\end{aligned}$$

∎

Theorem 9.4.5 Consider the problem EVVI. If $\mathbf{F}$ is affine and monotone, and $\mathbf{f} : \mathbb{R}^n \to \mathbb{R}^p$ is convex, then the gap function γ is type I convex.

Proof: Recall from (9.4.1) that the gap function $\gamma(\mathbf{x})$ can be rewritten as,

$$\gamma(\mathbf{x}) = \mathbf{f}^* \circ (-\mathbf{F})(\mathbf{x}) + \mathbf{F}(\mathbf{x})^\top \mathbf{x} + \mathbf{f}(\mathbf{x}),$$

where the (type I) Fenchel transform $\mathbf{f}^*$ of $\mathbf{f}$ is type I convex by Theorem 7.2.5. Since $\mathbf{F}$ is affine, so is $-\mathbf{F}$. By Lemma 9.4.4, $\mathbf{f}^* \circ (-\mathbf{F})$ is type I convex. By Lemma 9.4.3, $\mathbf{F}(\cdot)^\top(\cdot)$ is type I convex, and hence γ is type I convex. ∎

Next we study the convexity of the weak gap function for WVVI. We have previously established the type I convexity of the Fenchel transform of a vector-valued function in Theorem 7.2.5. A similar result for the weak Fenchel transform is as follows.

Lemma 9.4.6 Let $\mathbf{f} : \mathbb{R}^n \to \mathbb{R}^p$ be convex. The weak Fenchel transform of $\mathbf{f}$ is type I convex.

Proof: Recall in Definition 7.2.1 that the Weak Fenchel transform of a vector-valued function $\mathbf{f}$ is a set-valued function $\mathbf{f}^*_w : \mathbb{R}^{p\times n} \to 2^{\mathbb{R}^p}$ such that

$$\mathbf{f}^*_w(\mathbf{U}) = \max\nolimits_{\mathrm{int}\ C} \{\mathbf{U}\mathbf{x} - \mathbf{f}(\mathbf{x}) \mid \mathbf{x} \in \mathbb{R}^n\}.$$

We have, $\forall t \in (0,1),\ \mathbf{U}^1, \mathbf{U}^2 \in \mathbb{R}^{n\times p}$,

$$\begin{aligned}
\mathbf{f}^*_w(t\mathbf{U}^1 + (1-t)\mathbf{U}^2) &= \max\nolimits_{\mathrm{int}\ C} \{(t\mathbf{U}^1 + (1-t)\mathbf{U}^2)\mathbf{x} - \mathbf{f}(\mathbf{x}) \mid \mathbf{x} \in \mathbb{R}^n\} \\
&\subseteq \{(t\mathbf{U}^1 + (1-t)\mathbf{U}^2)\mathbf{x} - \mathbf{f}(\mathbf{x}) \mid \mathbf{x} \in \mathbb{R}^n\} \\
&= \{t((\mathbf{U}^1)\mathbf{x} - \mathbf{f}(\mathbf{x})) + (1-t)((\mathbf{U}^2)\mathbf{x} - \mathbf{f}(\mathbf{x})) \mid \mathbf{x} \in \mathbb{R}^n\} \\
&\subseteq t\{(\mathbf{U}^1)\mathbf{x} - \mathbf{f}(\mathbf{x}) \mid \mathbf{x} \in \mathbb{R}^n\} + (1-t)\{(\mathbf{U}^2)\mathbf{x} - \mathbf{f}(\mathbf{x}) \mid \mathbf{x} \in \mathbb{R}^n\} \\
&\subseteq t \max\nolimits_{\mathrm{int}\ C} \{(\mathbf{U}^1)\mathbf{x} - \mathbf{f}(\mathbf{x}) \mid \mathbf{x} \in \mathbb{R}^n\} - \mathbb{R}^p_+ \\
&\qquad + (1-t)\max\nolimits_{\mathrm{int}\ C} \{(\mathbf{U}^2)\mathbf{x} - \mathbf{f}(\mathbf{x}) \mid \mathbf{x} \in \mathbb{R}^n\} - \mathbb{R}^p_+ \\
&= t\mathbf{f}^*_w(\mathbf{U}^1) + (1-t)\mathbf{f}^*_w(\mathbf{U}^2) - \mathbb{R}^p_+.
\end{aligned}$$

Thus $\mathbf{f}^*_w$ is type I convex. ∎

Lemma 9.4.7 If $\mathbf{f}_w^* : \mathbb{R}^{n\times p} \to 2^{\mathbb{R}^p}$ is type I convex, and $\mathbf{F} : \mathbb{R}^n \to \mathbb{R}^{n\times p}$ is affine, then the composite $\mathbf{f}_w^* \circ \mathbf{F} : \mathbb{R}^n \to 2^{\mathbb{R}^p}$ is type I convex.

Proof: The proof is similar to that of Lemma 9.4.4 and is omitted. ∎

Theorem 9.4.8 Consider the problem WEVVI. If $\mathbf{F}$ is affine and monotone, and $\mathbf{f} : \mathbb{R}^n \to \mathbb{R}^p$ is convex, then the gap function γ_w is type I convex.

Proof: It is clear that from the definition (9.4.2) of the weak gap function, $\gamma_w(\mathbf{x})$ can be rewritten as,

$$\gamma_w(\mathbf{x}) = \mathbf{f}_w^* \circ (-\mathbf{F})(\mathbf{x}) + \mathbf{F}(\mathbf{x})^\top \mathbf{x} + \mathbf{f}(\mathbf{x}),$$

where the Weak Fenchel transform $\mathbf{f}_w^*$ of $\mathbf{f}$ is type I convex by Lemma 9.4.6. Since $\mathbf{F}$ is affine, so is $-\mathbf{F}$. By Lemma 9.4.7, $\mathbf{f}_w^* \circ (-\mathbf{F})$ is type I convex. By Lemma 9.4.3, $\mathbf{F}(\cdot)^\top(\cdot)$ is type I convex, and hence γ_w is type I convex. ∎

9.5 Generalized Gap Functions

The gap function for EVVI discussed in the previous section can be considered a generalization of the Auslender's gap function for scalar EVI. This does not incorporate the constraints defining the ground set explicitly. In this section, we study a generalization of the gap function for scalar VI previously discussed in Section 7.2. This generalization to VVI allows the constraints to be incorporated explicitly into the gap function. For this section, we assumed that all vector ordering relationships are induced by the positive orthant.

Let $\mathbf{F} : \mathbb{R}^n \to \mathbb{R}^{n\times p}$ be a matrix-valued continuous function, and the convex polyhedral ground set $\mathcal{K}$ be defined by

$$\mathcal{K} = \{\mathbf{x} \in \mathbb{R}^n \mid \mathbf{A}\mathbf{x} - \mathbf{b} \geq_{\mathbb{R}^m_+} \mathbf{0}\},$$

where $\mathbf{A} \in \mathbb{R}^{m\times n}$ and $\mathbf{b} \in \mathbb{R}^m$. The problem VVI is restated below for easy reference.

(Problem VVI) Find the set of points $\mathbf{x}^0 \in \mathcal{K}$ such that

$$\mathbf{F}(\mathbf{x}^0)^\top(\mathbf{x} - \mathbf{x}^0) \not\leq_{\mathbb{R}^p_+\setminus\{\mathbf{0}\}} \mathbf{0}, \quad \forall \mathbf{x} \in \mathcal{K}. \tag{9.5.1}$$

Definition 9.5.1 (Generalized gap function) The *generalized gap function* $\gamma_G : \mathcal{K} \to 2^{\mathbb{R}^p}$ is a set-valued function defined by:

$$\gamma_G(\mathbf{x}) = \min_{\mathbf{R}^m_+ \setminus \{\mathbf{0}\}} \cup_{\boldsymbol{\Gamma} \in \mathcal{L}_0} \Phi(\boldsymbol{\Gamma}), \quad \forall \mathbf{x} \in \mathcal{K},$$

where

$$\mathcal{L}_0 = \{\boldsymbol{\Gamma} \in \mathbb{R}^{p \times m} \mid \exists \boldsymbol{\mu} \in \text{int } \mathbb{R}^p_+ \text{ such that } \boldsymbol{\mu}^\top \boldsymbol{\Gamma} \geq_{\mathbf{R}^m_+} \mathbf{0}^\top\},$$

$$\Phi(\boldsymbol{\Gamma}) = \max_{\mathbf{R}^p_+ \setminus \{\mathbf{0}\}} [\mathbf{F}(\mathbf{x})^\top (\mathbf{x} - \mathbb{R}^n) + \boldsymbol{\Gamma}(\mathbf{A}\,\mathbb{R}^n - \mathbf{b})],$$

$$[\mathbf{F}(\mathbf{x})^\top (\mathbf{x} - \mathbb{R}^n) + \boldsymbol{\Gamma}(\mathbf{A}\mathbb{R}^n - \mathbf{b})] = \cup_{\mathbf{y} \in \mathbf{R}^n} [\mathbf{F}(\mathbf{x})^\top (\mathbf{x} - \mathbf{y}) + \boldsymbol{\Gamma}(\mathbf{A}\mathbf{y} - \mathbf{b})].$$

As in the scalar case where the generalized gap function reduces to the Auslender's gap function, we shall show that, in the event that the ground set $\mathcal{K}$ is polyhedral, then the generalized gap function for VVI reduces to the gap function of the previous section.

Lemma 9.5.1 Consider the problem

$$\max_{\mathbf{R}^m_+ \setminus \{\mathbf{0}\}} \mathbf{B}\mathbf{y} \qquad \text{subject to } \mathbf{y} \in \mathbb{R}^n,$$

where $\mathbf{B}$ is an $p \times n$ matrix. If there exists $\boldsymbol{\mu} \in \text{int}\mathbb{R}^p_+$ such that $\boldsymbol{\mu}^\top \mathbf{B} = \mathbf{0}^\top$, then every $\mathbf{y} \in \mathbb{R}^n$ is a minimal solution.

Proof: The problem is a convex multicriteria optimization problem. By the assumption, every $\mathbf{y} \in \mathbb{R}^n$ is an optimal solution of the scalarized problem:

$$\max \boldsymbol{\mu}^\top \mathbf{B}\mathbf{y}, \quad \text{subject to } \mathbf{y} \in \mathbb{R}^n.$$

It follows from $\boldsymbol{\mu} \in \text{int}\mathbb{R}^p_+$ and the scalarization Theorem 7.3.1 that every $\mathbf{y} \in \mathbb{R}^n$ is a minimal solution. ■

As part of the analysis, the *primal and dual multicriteria linear programs* of Definition 8.1.1 are restated below.

$$\text{(Primal VLP)} \qquad \min_{\mathbf{R}^p_+ \setminus \{\mathbf{0}\}} \mathbf{Q}\Lambda$$

where $\mathbf{Q}\Lambda = \{\mathbf{Q}\mathbf{x} \mid \mathbf{x} \in \mathbb{R}^n, \ \mathbf{A}\mathbf{x} \geq_{\mathbf{R}^m_+} \mathbf{b}\ \}$, $\mathbf{Q} \in \mathbb{R}^{p \times n}$, $\mathbf{A} \in \mathbb{R}^{m \times n}, \mathbf{b} \in \mathbb{R}^m$, and

$$\text{(Dual VLP)} \qquad \max_{\mathbf{R}^p_+ \setminus \{\mathbf{0}\}} \mathcal{L}\mathbf{b}$$

where $\mathcal{L}\mathbf{b} = \{\boldsymbol{\Gamma}\mathbf{b} \mid \boldsymbol{\Gamma} \in \mathbb{R}^{p \times m}$ such that $\exists \boldsymbol{\mu} \in \text{int}\mathbb{R}^p_+$ satisfying

$$\boldsymbol{\mu}^\top \boldsymbol{\Gamma} \geq_{\mathbf{R}^m_+} \mathbf{0}^\top \text{ and } \mathbf{A}^\top \boldsymbol{\Gamma}^\top \boldsymbol{\mu} = \mathbf{Q}^\top \boldsymbol{\mu}\ \}.$$

The following duality result was previously established in Theorem 8.1.1

Theorem 9.5.2 (Multicriteria linear programming duality)
(i) (Weak Duality) $\forall(\mathbf{x},\boldsymbol{\Gamma}) \in \boldsymbol{\Lambda} \times \mathcal{L}$, $\boldsymbol{\Gamma}\mathbf{b} \not\geq_{\mathbf{R}^p_+\setminus\{0\}} \mathbf{Qx}$.
(ii) (Strong Duality) If $\boldsymbol{\Gamma}^0 \in \mathcal{L}$ and $\mathbf{x}^0 \in \boldsymbol{\Lambda}$ satisfies $\boldsymbol{\Gamma}^0\mathbf{b} = \mathbf{Qx}^0$, then

$$\mathbf{x}^0 \in \text{argmin}_{\mathbf{R}^p_+\setminus\{0\}} \mathbf{Q}\boldsymbol{\Lambda} \text{ and } \boldsymbol{\Gamma}^0 \in \text{argmax}_{\mathbf{R}^p_+\setminus\{0\}} \mathcal{L}\mathbf{b}.$$

(iii) (Set equality of efficient frontiers)

$$\min_{\mathbf{R}^p_+\setminus\{0\}} \mathbf{Q}\boldsymbol{\Lambda} = \max_{\mathbf{R}^p_+\setminus\{0\}} \mathcal{L}\mathbf{b}.$$

The gap function of Definition 9.4.2 is defined for EVVI. In the event that the function $\mathbf{f}$ is identically zero, it reduces to a special case that can be considered a generalization of the Auslender's gap function for a scalar VI.

$$\gamma_A(\mathbf{x}) = \max_{\mathbf{R}^p_+\setminus\{0\}} \cup \{\mathbf{F}(\mathbf{x})^\top(\mathbf{x}-\mathbf{y}) \mid \mathbf{y} \in \mathcal{K}\}.$$

Theorem 9.5.3 $\gamma_G(\mathbf{x}) = \gamma_A(\mathbf{x})$ and $\gamma_G(\mathbf{x})$ is type I convex.

Proof: Note that

$$\begin{aligned}\gamma_G(\mathbf{x}) &= \min_{\mathbf{R}^p_+\setminus\{0\}} \cup_{\boldsymbol{\Gamma}\in\mathcal{L}_0} \boldsymbol{\Phi}(\boldsymbol{\Gamma}) \\ &= \mathbf{F}(\mathbf{x})^\top\mathbf{x} + \min_{\mathbf{R}^p_+\setminus\{0\}} \cup_{\boldsymbol{\Gamma}\in\mathcal{L}_0} \boldsymbol{\Phi}_1(\boldsymbol{\Gamma}),\end{aligned}$$

where $\boldsymbol{\Phi}$ is as defined in Definition 9.5.1, and

$$\boldsymbol{\Phi}_1(\boldsymbol{\Gamma}) = \min_{\mathbf{R}^p_+\setminus\{0\}} [-\boldsymbol{\Gamma}\mathbf{b} + \boldsymbol{\Phi}_2(\boldsymbol{\Gamma})],$$

$$\boldsymbol{\Phi}_2(\boldsymbol{\Gamma}) = \max_{\mathbf{R}^p_+\setminus\{0\}} \cup_{\mathbf{y}\in\mathbf{R}^n} (\boldsymbol{\Gamma}\mathbf{A} - \mathbf{F}(\mathbf{x})^\top)\mathbf{y}.$$

It follows from Lemma 9.5.1 that

$$\gamma_G(\mathbf{x}) = \mathbf{F}(\mathbf{x})^\top\mathbf{x} + \min_{\mathbf{R}^p_+\setminus\{0\}} \cup_{\boldsymbol{\Gamma}\in\mathcal{L}_1} [-\boldsymbol{\Gamma}\,\mathbf{b}],$$

where

$\mathcal{L}_1 =$
$\{\boldsymbol{\Gamma} \in \mathbb{R}^{p\times m} \mid \exists\boldsymbol{\mu} \in \text{int}\mathbb{R}^p_+ \text{ such that } \boldsymbol{\mu}^\top\boldsymbol{\Gamma} \geq_{\mathbf{R}^m_+} 0^\top \text{ and } \mathbf{A}^\top\boldsymbol{\Gamma}^\top\boldsymbol{\mu} = \mathbf{F}(\mathbf{x})^\top\boldsymbol{\mu}\ \}.$

Then

$$\begin{aligned}\gamma_G(\mathbf{x}) &= \mathbf{F}(\mathbf{x})^\top\mathbf{x} + \min_{\mathbf{R}^p_+\setminus\{0\}} \cup_{\boldsymbol{\Gamma}\in\mathcal{L}_1} -\boldsymbol{\Gamma}\mathbf{b} \\ &= \mathbf{F}(\mathbf{x})^\top\mathbf{x} - \max_{\mathbf{R}^p_+\setminus\{0\}} \cup_{\boldsymbol{\Gamma}\in\mathcal{L}_1} \boldsymbol{\Gamma}\mathbf{b}.\end{aligned}$$

Applying the multicriteria linear programming duality result of Theorem 9.5.2, it follows that

$$\begin{aligned}\gamma_G(\mathbf{x}) &= \mathbf{F}(\mathbf{x})^\top\mathbf{x} - \min_{\mathbf{R}^p_+\setminus\{0\}} \cup \{\mathbf{F}(\mathbf{x})^\top\mathbf{y} \mid \mathbf{y} \in \mathcal{K}\} \\ &= \max_{\mathbf{R}^p_+\setminus\{0\}} \cup \{\mathbf{F}(\mathbf{x})^\top(\mathbf{x}-\mathbf{y}) \mid \mathbf{y} \in \mathcal{K}\} \\ &= \gamma_A(\mathbf{x}).\end{aligned}$$

Furthermore, it follows from Theorem 9.4.5 that $\gamma_G(\mathbf{x})$ is type I convex. ∎

9.6 A Solution Method and a Related Gap Function

While the study of solution methods for solving variational inequalities is a fairly mature subject, (see [HP1] and [N1]), the same cannot be said about vector variational inequalities, in spite of its theoretical advances. In particular, the closedness issue of the solution set of a VVI is still not entirely clear. Like in multicriteria optimization, the solution of VVI entails the computation of sets, which makes it difficult in all but trivial cases. Nevertheless, the case of weak vector variational inequalities is slightly simpler, and a solution method for WVVI has been proposed in [GY6]. This result warrants some discussion here as it can be shown to be related to a particular scalar-valued gap function. The starting point of analysis is a necessary and sufficient optimality condition for the solution of a given WVVI. We show that solving the WVVI essentially reduces to the computation of level sets of some related scalar-valued function.

Let $C \subset \mathbb{R}^p$ be a closed, convex and pointed cone with a non-empty interior $\text{int}\, C$. The cone C induces the usual vector ordering relationships in $\mathbb{R}^p$. Let $\mathcal{K} \subset \mathbb{R}^n$ be a compact convex set and $\mathbf{F} : \mathbb{R}^n \to \mathbb{R}^{n \times p}$. Consider the following weak vector variational inequality:

$$\text{(Problem WVVI)} \qquad \text{Find } \mathbf{x} \in \mathcal{K}, \text{ s.t. } \mathbf{F}(\mathbf{x})^\top(\mathbf{y}-\mathbf{x}) \not\leq_{\text{int}\, C} 0, \ \ \forall \mathbf{y} \in \mathcal{K}. \quad (9.6.1)$$

The corresponding vector variational inequality problem is defined by:

$$\text{(Problem VVI)} \qquad \text{Find } \mathbf{x} \in \mathcal{K}, \text{ s.t. } \mathbf{F}(\mathbf{x})^\top(\mathbf{y}-\mathbf{x}) \not\leq_{C\setminus\{\mathbf{0}\}} 0, \ \ \forall \mathbf{y} \in \mathcal{K}. \quad (9.6.2)$$

Recall from Chapter Seven that given $\mathbf{e} \in \text{int}\, C$ and $\mathbf{a} \in \mathbb{R}^p$, the Gerstewitz function $\xi_{\mathbf{ea}} : \mathbb{R}^p \to \mathbb{R}$ is defined by [GW1]:

$$\xi_{\mathbf{ea}}(\mathbf{y}) = \min\{t \in \mathbb{R} \mid \mathbf{y} \in \mathbf{a} + t\mathbf{e} - C\} \quad \text{for} \quad \mathbf{y} \in \mathbb{R}^p. \quad (9.6.3)$$

Lemma 9.6.1 Let $\mathbf{e} \in \text{int}\, C$. Then $\mathbf{y} \notin -\text{int}\, C$ if and only if $\xi_{\mathbf{e0}}(\mathbf{y}) \geq 0$.

Proof: Let $\mathbf{y} \notin -\text{int}\, C$ and $\mathbf{y} \in t\mathbf{e} - C$. Then $t \geq 0$. Otherwise $t < 0$ implies that $t\mathbf{e} \subset -\text{int}\, C$, thus $\mathbf{y} \in t\mathbf{e} - C \in -\text{int}\, C$.

Let $\xi_{\mathbf{e0}}(\mathbf{y}) \geq 0$. Then $\mathbf{y} \notin -\text{int } \mathcal{C}$. Otherwise there exists $t < 0$ such that $\mathbf{y} \in t\mathbf{e} - \mathcal{C}$, a contradiction to $\xi_{\mathbf{e0}}(\mathbf{y}) \geq 0$. ■

Theorem 9.6.2 (Necessary and sufficient condition for solution of WVVI) Let $\mathbf{e} \in \text{ int } \mathcal{C}$. Then $\mathbf{x} \in \mathcal{K}$ solves WVVI if and only if

$$g(\mathbf{x}) = \min_{\mathbf{y} \in \mathcal{K}} \xi_{\mathbf{e0}}(\mathbf{F}(\mathbf{x})^\top(\mathbf{y} - \mathbf{x})) \geq 0. \tag{9.6.4}$$

Proof: Assume that $\mathbf{x} \in \mathcal{K}$ solves the problem WVVI and let $\mathbf{e} \in \text{int } \mathcal{C}$, then,

$$\begin{aligned}
& \mathbf{F}(\mathbf{x})^\top(\mathbf{y} - \mathbf{x}) \not\leq_{\text{int } \mathcal{C}} \mathbf{0} \quad \forall \mathbf{y} \in \mathcal{K} & \\
\Leftrightarrow\ & \mathbf{F}(\mathbf{x})^\top(\mathbf{y} - \mathbf{x}) \notin -\text{int } \mathcal{C} \quad \forall \mathbf{y} \in \mathcal{K} & \text{by definition} \\
\Leftrightarrow\ & \xi_{\mathbf{e0}}(\mathbf{F}(\mathbf{x})^\top(\mathbf{y} - \mathbf{x})) \geq 0 \quad \forall \mathbf{y} \in \mathcal{K}, & \text{by Lemma 9.6.1} \\
\Leftrightarrow\ & \min_{\mathbf{y} \in \mathcal{K}} \xi_{\mathbf{e0}}(\mathbf{F}(\mathbf{x})^\top(\mathbf{y} - \mathbf{x})) \geq 0. &
\end{aligned}$$

■

By virtue of Theorem 9.6.2, the problem of solving WVVI essentially reduces to the following problem:

(Problem P_1) Find all $\mathbf{x} \in \mathcal{K}$ such that $g(\mathbf{x}) \geq 0$.

In the special case where $\mathcal{C} = \mathbb{R}^p_+$, the Gerstewitz function may be expressed in the following equivalent form

$$\xi_{\mathbf{ea}}(\mathbf{y}) = \max_{1 \leq i \leq p} \frac{y_i - a_i}{e_i}. \tag{9.6.5}$$

Corollary 9.6.3 Let $\mathcal{C} = \mathbb{R}^p_+$ and let

$$h(\mathbf{x}) = \min_{\mathbf{y} \in K} \max_{1 \leq i \leq p} \{F_i(\mathbf{x})(\mathbf{y} - \mathbf{x})\}, \tag{9.6.6}$$

where $F_i(\mathbf{x})$ be the ith row of the matrix $\mathbf{F}(\mathbf{x})$. Then $\mathbf{x} \in \mathcal{K}$ solves WVVI problem if and only if $h(\mathbf{x}) \geq 0$.

Proof: The proof follows from Theorem 9.6.2 and (9.6.5) by letting $\mathbf{a} = \mathbf{0}$ and $\mathbf{e} = (1, \cdots, 1)^\top \in \mathbb{R}^p$. ■

Based on the auxiliary function h or g, a scalar-valued gap function for the WVVI can be readily constructed. Note that this is in contrast with the set-valued gap function discussed in Section 9.4. Let $\gamma : \mathbb{R}^n \to \mathbb{R}$ be defined as:

$$\gamma(\mathbf{x}) = \max\{0, -g(\mathbf{x})\}$$

or

$$\gamma(\mathbf{x}) = \max\{0, -h(\mathbf{x})\} \qquad \text{if } \mathcal{C} = \mathbb{R}^p_+. \tag{9.6.7}$$

Theorem 9.6.4 γ is a (scalar) gap function for WVVI in the sense that

(i)$\gamma(\mathbf{x}) \geq 0 \quad \forall \mathbf{x} \in \mathbb{R}^n$;

(ii)$\gamma(\mathbf{x}) = 0$ if and only if $\mathbf{x}$ solves WVVI.

Proof: (i) holds trivially by construction. (ii) follows from Theorem 9.6.2 and Corollary 9.6.3. ∎

If $C = \mathbb{R}^p_+$, then for a given $\mathbf{x}$, the evaluation of the auxiliary function h requires the solution of a minimax problem, where the cost functions are clearly linear in the decision variable $\mathbf{y}$. Furthermore, in many economics and network equilibrium models, the set $\mathcal{K}$ is often a polyhedral set, in which case, the minimax problem reduces to a trivial linear programming problem:

$$\begin{aligned} g(\mathbf{x}) \quad = \quad & \min_{a,\mathbf{y}} \ a \\ & \text{s.t.} \quad \mathbf{F}(\mathbf{x})^\top \mathbf{y} - \begin{bmatrix} 1 \\ 1 \\ \vdots \\ 1 \end{bmatrix} a \leq_{\mathbf{R}^p_+} \mathbf{F}(\mathbf{x})^\top \mathbf{x} \\ & \mathbf{y} \in \mathcal{K}, \end{aligned}$$

which can be solved very quickly, even if $\mathbf{F}$ is highly nonlinear.

In the case of VVI, the condition becomes significantly weaker.

Theorem 9.6.5 (Necessary condition for solving VVI) Let $\mathbf{e} \in$ int C. Then $\mathbf{x} \in \mathcal{K}$ solves VVI only if $g(\mathbf{x}) \geq 0$, where $g(\mathbf{x})$ is defined by (9.6.4). If $C = \mathbb{R}^p_+$, then $\mathbf{x} \in \mathcal{K}$ solves VVI only if $h(\mathbf{x}) \geq 0$, where $h(\mathbf{x})$ is defined by (9.6.6).

Proof: Note that a solution of VVI is also a solution of WVVI. The result follows from Theorem 9.6.2 and Corollary 9.6.3. ∎

Theorem 9.6.6 (Sufficient condition for solving VVI) Let $h(\mathbf{x})$ be defined by (9.6.6). If $\mathbf{x} \in \mathcal{K}$ and $h(\mathbf{x}) > 0$, then $\mathbf{x} \in \mathcal{K}$ solves VVI.

Proof: If not, there is an $\mathbf{y} \in \mathcal{K}$ such that

$$\mathbf{F}(\mathbf{x})^\top (\mathbf{y} - \mathbf{x}) \leq_{C \setminus \{\mathbf{0}\}} 0.$$

Then

$$\max_{1 \leq i \leq p} \{F_i(\mathbf{x})(\mathbf{y} - \mathbf{x})\} \leq 0,$$

and thus

$$h(\mathbf{x}) = \min_{\mathbf{z} \in \mathcal{K}} \max_{1 \leq i \leq p} \{F_i(\mathbf{x})(\mathbf{z} - \mathbf{x})\} \leq \max_{1 \leq i \leq p} \{F_i(\mathbf{x})(\mathbf{y} - \mathbf{x})\} \leq 0.$$

∎

The following example shows that Theorem 9.6.6 is only a sufficient condition for VVI.

Example 9.6.1 Let $\mathcal{K} = [-1, 0]$ and $\mathcal{C} = \mathbb{R}^2_+$. Consider the VVI defined as follows:

$$(1, 2x)(y - x) \not\leq_{\mathbf{R}^2_+ \setminus \{0\}} \mathbf{0}^\top, \qquad \forall y \in \mathcal{K}. \tag{9.6.8}$$

Then $x = 0$ is a solution of VVI, but $h(0) = 0$. Hence Theorem 9.6.6 is only a sufficient condition for VVI.

Consider also the related WVVI defined as follows:

$$(1, 2x)(y - x) \not\leq_{\text{int } \mathbf{R}^2_+} \mathbf{0}^\top, \qquad \forall y \in \mathcal{K}. \tag{9.6.9}$$

This is related to the multicriteria optimization problem:

$$\begin{aligned} \min\nolimits_{\mathbf{R}^2_+ \setminus \{0\}} \quad & (x, x^2)^\top \\ \text{s.t.} \quad & x \in [-1, 0]. \end{aligned}$$

Note that the minimal solution set, the weakly minimal solution set for this multicriteria optimization problem, the solution set for the VVI (9.6.8) and the solution set for the WVVI (9.6.9) are all the same, and is given by: $\{x : x \in \mathcal{K} = [-1, 0]\}$.

Example 9.6.2 Let $\mathcal{K} = [-5,\ 5]^2$ be the underlying convex set and $\mathcal{C} = \mathbb{R}^2_+$. The problem is to find the set of all $\mathbf{x} = (x_1, x_2)^\top \in \mathcal{K}$ such that

$$\begin{bmatrix} 2x_1 + x_2 - 2 & 5x_1 - 5x_2 - 5 \\ 6x_1 + x_2 - 3 & 4x_1 + x_2 - 1 \end{bmatrix} (\mathbf{y} - \mathbf{x}) \not\leq_{\text{int } \mathbf{R}^2_+} \mathbf{0} \quad \forall \mathbf{y} \in \mathcal{K}. \tag{9.6.10}$$

The auxiliary function $h(\mathbf{x})$ using $\mathbf{e} = (1, 1)$ is then given by

$$h(\mathbf{x}) = \min_{\mathbf{y} \in \mathcal{K}} \max \left\{ \begin{array}{c} (2x_1 + x_2 - 2)(y_1 - x_1) + (5x_1 - 5x - 2 - 5)(y_2 - x_2), \\ (6x_1 + x_2 - 3)(y_1 - x_1) + (4x_1 + x_2 - 1)(y_2 - x_2) \end{array} \right\}.$$

Since $\mathcal{K}$ is polyhedral, each function evaluation of h amounts to solving a linear program, and since it is only a two dimensional problem, it is only easy to compute $h(\mathbf{x})$ for all $\mathbf{x} \in \mathcal{K}$ in a brute force manner. From the contour plot of $h(\mathbf{x})$, it can be concluded that the solution of the WVVI is given by the following set, which is the union of 4 disjoint subsets:

$$\begin{aligned} \mathcal{X}^* &= \{\mathbf{x}^* \mid h(\mathbf{x}^*) \geq 0\ \} \\ &= \{(1, 0)\} \cup \{(1, -3)\} \cup \{(t, -5), t \in [1.5, 3.5]\} \cup \{(t, 5), t \in [-0.5, -1.5]\}. \end{aligned}$$

Note that all the $\mathbf{x}^*$ that solves the WVVI (9.6.10) gives $h(\mathbf{x}^*) = 0$, i.e., there is no $\mathbf{x}$ such that $h(\mathbf{x}) > 0$ and hence the gap function in this case is effectively $\gamma(\mathbf{x}) = -h(\mathbf{x})$.

REFERENCES

[AMO1] Ahuja, R.K., Magnanti, T.L. and and Orlin, J.B., *Network Flows, Theory, Algorithms and Applications*, Prentice Hall, Englewood Cliffs, New Jersey, 1993.

[A1] Auslander, A., *Optimisation: Méthodes Numériques*, Masson, Paris, 1976.

[BGKKT1] Bank, B., Guddat, J., Klatte, D., Kummer, B. and Tammer, K., *Nonlinear Parametric Optimization*, Birhauser Verlag, , Basel, 1983.

[BS1] Bazaraa, M.S. and Shetty, C.M., *Nonlinear Programming Theory and Algorithms*, John Wiley, New York, 1979.

[B1] Bitran, G.R., Duality for nonlinear multi-criteria optimization problems, *Journal of Optimization Theory and Applications*, Vol. 35, 1981, pp. 367-401.

[B2] Brumelle, S., Duality for multiple objective convex programs, *Mathematics of Operations Research*, Vol. 6, 1981, pp. 159-172.

[B3] Barbu, V., *Convexity and Optimization in Banach Spaces*, D. Reidel, Dordrecht, 1986.

[B4] Borwein, J.M., Generic differentiability of order-bounded convex operators, *Journal of the Australian Mathematical Society, ser. B*, Vol. 28, 1986, pp. 22-29.

[BP1] Burke, J.V. and Poliquin, R.A., Optimality conditions for non-finite valued convex composite functions, *Mathematical Programming*, Vol. 57, 1992, pp. 103-120.

[CM1] Castellani, M. and Mastroeni, G., On the duality theory for finite dimensional variational inequalities, in *Variational Inequalities and Network Equilibrium Problems*, Eds., F. Giannessi and A. Maugeri, Plenum Press, New York, 1995.

[CP1] Charnes, A. and Cooper, W.W., Goal programming and multiple objective optimization, *J. Operational Research Society*, Vol.1, 1977, pp. 39-54.

[CC1] Chen, G.Y. and Craven, B.D., A vector variational inequality and optimization over an efficient set, *ZOR-Methods and Models of Operations Research*, Vol. 3, 1990, pp. 1-12.

[CGY1] Chen, G.Y., Goh, C.J. and Yang, X.Q., On gap functions of vector variational inequalities, in *Vector Variational Inequalities and Vector Equilibria*, ed. F. Giannessi, Kluwer Academic Publisher, 2000, pp. 55-72.

[CGY2] Chen, G.Y., Goh, C.J. and Yang, X.Q., On gap functions and duality of variational inequality problems, *Journal of Mathematical Analysis and Applications*, Vol. 214, 1997, pp. 658-673.

[CGY3] Chen, G.Y., Goh, C.J. and Yang, X.Q., The gap function of a convex multicriteria optimization problem, *European Journal of Operational Research,* Vol. 111, 1998, pp. 142-151.

[CGY4] Chen, G.Y., Goh, C.J. and Yang, X.Q., Vector network equilibrium problems and nonlinear scalarization methods, *ZOR – Mathematical Models of Operations Research*, Vol. 49, 1999, pp. 239-253.

[CGY5] Chen, G.Y., Goh, C.J. and Yang, X.Q., Existence of solutions for vector variational inequalities, *Optimization*, to appear.

[CL1] Chen, G.Y. and Li, S.J., Existence of solutions for a generalized quasi-vector variatioanl inequality, *Journal of Optimization Theory and Applications*, Vol. 90, 1996, pp. 331-334.

[CY1] Chen, G.Y. and Yen, N.D., On the variational inequality model for network equilibrium, Research Report, Dipartmento di Matematica, Universita di Pisa, 3.196(724), 1993.

[CY2] Chen, G.Y. and Yang, X.Q., The vector complementarity problem and its equivalences with weak minimal element, *Journal of Mathematical Analysis and Applications*, Vol. 153, 1990, pp. 136-158.

[CY3] Chen, G.Y. and Yang, X.Q., On existence of vector complementarity problems, in *Vector Variational Inequalities and Vector Equilibria*, ed. F. Giannessi, Kluwer Academic Publisher, 2000, pp. 87-95.

[CC2] Chew, K.L. and Choo, E.U., Pseudolinearity and efficiency, *Mathematical Programming*, Vol. 28, 1984, pp. 226-239.

[CH1] Chipot, M, *Variational Inequalities and Flow in Porous Media*, Springer-Verlag, New York, 1984.

[C1] Cameron, N., *Introduction to Linear and Convex Programming,* Australian Mathematical Society Lecture Series, Cambridge University Press, 1985.

[C2] Cesari, L., *Optimization: Theory and Applications*, Springer-Verlag, New York, 1983.

[C3] Craven, B.D., Strong vector minimization and duality, *Z. Angew. Math. Mech.*, Vol. 60, 1980, pp. 1-5.

[C4] Chen, G.Y., Existence of solutions for variational inequality: an extension of Hartma-Stampacchia theorem, *Journal of Optimization Theory and Applications*, Vol.74, 1992, pp. 445-456.

[C5] Clarke, F.H., *Optimization and Nonsmooth Analysis*, SIAM, Philadephia, PA, 1990.

[C6] Craven, B.D., Lagrangian conditions and quasiduality, *Bulletin Australian Mathematical Society,* Vol. 16, 1977, pp. 325-339.

[C7] Choo, E.U., Proper efficiency and linear fractional vector maximum problem, *Operations Research*, Vol. 32, 1984, pp. 216-220.

[C8] Cornes, R., *Duality and Modern Economics*, Cambridge University Press, 1992.

[D1] Dantzig, G.B., *Linear Programming and Extensions*, Princeton University Press, Princeton, 1963.

[D2] Dafermos, S., Traffic equilibrium and variational inequalities, *Transportation Science*, Vol. 14, 1980, pp. 42-54.

[dG1] di Guglielmo, F., Nonconvex duality in multiobjective optimization, *Mathematics of Operations Research*, Vol.2., 1977, pp. 285-291.

[E1] Elsgolts, L., *Differential Equations and the Calculus of Variations*, MIR, Moscow, 1977.

[E2] Ekeland, I., Dual variational methods in non-convex optimization and differential equations, in *Mathematical Theories of Optimziation*, Lecture Notes in Mathematics, 979, Eds. A. Dold and B. Eckmann, 1981.

[ET1] Ekeland, I. and Temam, R., *Convex Analysis and Variational Problems*, North Holland, American Elsevier, 1973.

[FP1] Fang, S.C. and Puthenpura, S., *Linear Optimization and Extensions Theory and Algorithms*, Prentice Hall, Englewood Cliffs, 1993.

[F1] Farkas, J., Über die Theorie der einfachen Ungleichungen, *J. Reine Angew. Math.*, Vol. 124, 1902, pp. 1-24.

[F2] Fenchel, W., On conjugate convex functions, *Canadian Journal of Mathematics*, Vol.1, 1949, pp. 73-77.

[F3] Fulkerson, D.R., An out-of-kilter method for minimal cost flow problems, *SIAM Journal*, Vol. 9, 1961, pp. 18-27.

[F4] Fukushima, M., Equivalent differentiable optimization problems and descent methods for asymmetric variational inequality problem, *Mathematical Programming*, Vol. 53, 1992, pp.99-110.

[FF1] Ford, L.R. and Fulkerson, D.R., Maximal flow through a network, *Canadian Journal of Mathematics*, Vol. 8, 1956, pp. 339-404.

[GKT1] Gale, D., Kuhn, H.W. and Tucker, A.W., Linear programming and the theory of games, in *Activity Analysis of Production and Allocation,* Eds. T.C. Koopmans, Wiley, New York 1951.

[G1] Geoffrion, A.M., Duality in nonlinear programming: a simplified applications oriented development, *SIAM Review*, Vol. 13, 1974, pp. 1-37.

[G2] Gros, C., Generalization of Fenchel's duality theorem for convex vector optimization, *European Journal of Operational Research,* Vol.2, 1978, pp. 368-376.

[G3] Giannessi, F., Theorems of the alternative and optimality conditions, *Journal of Optimization Theory and Applications*, Vol. 42(3), 1984, pp. 331-365.

[G4] Giannessi, F., Separation of sets and gap functions for quasi-variational inequalities, in *Variational Inequalities and Network Equilibrium Problems*, Eds., F. Giannessi and A. Maugeri, Plenum Press, New York, 1995.

[G5] Giannessi, F., On some connections among variational inequalities, combinatorial and continuous optimization, *Annals of Operations Research*, Vol. 58, 1995, pp. 181-200.

[G6] Giannessi, F., Theorems of alternative, quadratic programs and complementarity problems, in *Variational Inequality Complementary Problems*, Eds. R.W. Cottle, F. Giannessi, and J.L. Lions, pp.151-186, Wiley, New York, 1980.

[G7] Geoffrion, A.M., Proper efficiency and the theory of vector maximization, *Journal of Mathematical Analysis and Applications*, Vol. 22, 1968, pp. 618-630.

[G8] Grace, A., *Optimization Toolbox for use with MATLAB user guide,* The Maths Works, Inc, South Natick, MA, 1990.

[G9] Giannessi, F., (Eds), *Vector Variational Inequalities and Vector Equilibria*, Kluwer Academic Publisher, 2000.

[G10] Giannessi, F., On Minty variational principle, in *Trends in Mathematical Programming*, Kluwer, 1997.

[GF1] Gelfand, I.M. and Fomin, S.V., *Calculus of Variations*, Prentice Hall, New York, 1963.

[GW1] Gerth, C. and Weidner, P., Nonconvex separation theorems and some applications in vector optimziation, *Journal of Optimization Theory and Applications*, Vol. 67, 1990, pp. 297-320.

[GY1] Goh, C.J. and Yang, X.Q., A sufficient and necessary condition for nonconvex optimization, *Applied Mathematics Letters*, Vol. 10, 1997, pp. 9-12,

[GY2] Goh, C.J. and Yang, X.Q., A nonlinear Lagrangian theory for non-convex optimization, *Journal of Optimization Theory and Applications*, Vol. 109, 2001, pp. 99-121.

[GY3] Goh, C.J. and Yang, X.Q., On Minkowski metric and weighted Tchebyshev norm in vector optimization, *Optimization*, Vol. 43, 1998, pp. 353-365 .

[GY4] Goh, C.J. and Yang, X.Q., Analytic efficient solution set for multicriteria quadratic programs, *European Journal of Operational Research*, Vol. 92, 1996, pp. 166-181.

[GY5] Goh, C.J. and Yang, X.Q., Comment on convexification of a nonconvex frontier, *Journal of Optimization Theory and Applications*, Vol. 97, 1998, pp. 759-768.

[GY6] Goh, C.J. and Yang, X.Q., Vector equilibrium problem and vector optimization, *European Journal of Operational Research*, Vol. 116, 1999, pp. 615-628.

[GY7] Goh, C.J. and Yang, X.Q., On the solution of vector variational inequalities, *Proceedings of the International Conference on Optimization Theory and Applications*, Perth, July, 1998, pp. 1548-1164.

[GY8] Goh, C.J. and Yang, X.Q., On scalarization methods for vector variational inequalities, in *Vector Variational Inequalities and Vector Equilibria*, eds. F. Giannessi, Kluwer Academic Publisher, 2000, pp. 217-232.

[H1] Haneveld, K., *Duality in Stochastic Linear and Dynamic Programming*, Springer-Verlag, Berlin, 1986.

[H2] Hearn, D.W., The gap function of a convex program, *Operations Research Letters,* Vol. 1, 1982, pp. 67-71.

[H3] Harary, F., *Graph Theory*, Addison Wesley, 1969.

[HL1] Hiriart-Urruty, J.B. and Lemaréchal, C., *Convex Analysis and Minimization Algorithms, Volume I and II*, Springer-Verlag, 1996.

[HP1] Harker, P.T. and Pang, J.S., Finite-dimensional variational inequality and nonlinear complementary problems: A survey of theory, algorithms and applications, *Mathematical Programming*, Vol. 48, 1990, pp. 161-220.

[HY1] Huang, X. and Yang, X.Q., Duality and exact penalization for vector optimization via augmented lagrangian, *Journal of Optimization Theory and Applications*, Vol. 111 (2001).

[HY2] Huang, X.X. and Yang, X.Q., Approximate optimal solutions and nonlinear Lagrangian functions *Journal of Global Optimization*, (accepted).

[I1] Isermann, H., Duality in multi-objective linear programming, in *Multiple Criteria Problem Solving,* Eds. S. Zions, Springer Verlag, Berlin and New York, 1978

[I2] Isermann, H., The relevance of a duality in multi-objective linear programming, in *Multiple Criteria Decision Making,* Eds. M.K. Starr and M. Zeleny, North-Holland Publ., Amsterdam, 1977.

[I3] Isermann, H., On some relations between a dual pair of multiple objective linear programs, *Z. Oper. Res.*, Vol. 22, 1978, pp. 33-41.

[J1] Jahn, J., *Mathematical Vector Optimization in Partially Ordered Linear Spaces,* Peter Lang, Frankfurt, 1986.

[J2] Jahn, J., Duality in vector optimization, *Mathematical Programming*, Vol. 29, 1984, pp. 203-218.

[J3] Jeyakumar, V., Composite nonsmooth programming with Gâteaux differentiability, *SIAM Journal of Optimization*, Vol. 1, 1991, pp. 30-41.

[JK1] Jahn, J. and Krabs, W. Applications of multicriteria optimization in approximation theory, in *Multicriteria Optimization in Engineering and in the Science*, Eds. W. Stadler, Plenum Press, New York, 1988, pp. 49-75.

[JW1] Jeyakumar, V. and Wolkowicz, H., Zero duality gaps in infinite-dimensional programming, *Journal of Optimization Theory and Applications*, Vol. 67, 1990, pp. 87-108.

[JY1] Jeyakumar, V. and Yang, X.Q., Convex composite multi-objective nonsmooth programming, *Mathematical Programming*, Vol. 59, 1993, pp. 325-343.

[JY2] Jeyakumar, V. and Yang, X.Q., Convex composite minimization with $C^{1,1}$ functions, *Journal of Optimization Theory and Applications*, Vol. 86, 1995, pp. 631-648.

[K1] Kawasaki, H., A duality theorem in multiobjective nonlinear programming, *Mathematics of Operations Research*, Vol. 6, 1981, pp. 593-607.

[K2] Kuhn, H.W., The Hungarian method for the assignment problem, *Naval Research Logistic Quarterly*, Vol.2, 1955, pp. 83-97.

[K3] Kambo, N.S., *Mathematical Programming Techniques*, Affiliated East-West Press, New Delhi, 1984.

[K4] Kantorovitch, L.V., Mathematical methods of organizing and planning production, (in Russian), *Publication House of the Leningrad State University*, Leningrad, 1939.

[KH1] Kennington, J.L. and Helgason, R.V., *Algorithms for Network Programming*, Wiley Interscience, 1980.

[KS1] Kinderlehrer, D. and Stampacchia, G., *An Introduction to Variational Inequalities and their Applications*, Academic Press, New York, 1980.

[KT1] Kuhn, H.W., and Tucker, A.W., *Nonlinear Programming, Proceeding of the second Berkeley Symposium on Mathematical Statistics and Probability*, Berkeley, 1951, pp.481-492.

[KW1] Kall, P. and Wallace, S.W., *Stochastic Programming*, Wiley-Interscience, Chichester, 1994.

[KYK1] Kim, D.S., Yun, Y.B. and Kuk, H., Second-order symmetric and self duality in multiobjective programming, *Applied Mathematics Letters* Vol. 10, 1997, pp. 17-22.

[L1] Luengerger, D.G., *Linear and Nonlinear Programming, Second edition*, Addition Wesley Publishing, 1989.

[L2] Luc, D.T., *Theory of Vector Optimization*, Springer-Verlag, Berlin, 1989.

[L3] Luc, D.T., Duality theory in multiobjective programming, *Journal of Optimization Theory and Applications*, Vol. 43, 1984, pp. 557-582.

[L4] Lawler, E.L., *Combinatorial Optimization: Networks and Matroids*, Holt, Rinehart and Winston, 1976.

[L5] Lin, L.J., Pre-vector variational inequalities, *Bulletin of the Australian Mathematical Society*, Vol. 53, 1996, pp. 63-70.

[L6] Li, D., Zero duality gap for a class of nonconvex optimization problem, *Journal of Optimization Theory and Applications*, Vol. 85(2), 1995, pp. 309-324.

[L7] Loridan, P., Necesary conditions for ϵ-optimality, *Mathematical Programming Study*, Vol. 19, 1982, pp. 140-152.

[L8] Loridan, P., ϵ-solution in vector minimization problems, *Journal of Optimziation Theory and Applications*, Vol. 43, 1984, pp. 265-276.

[L9] Li, D., Convexification of a Noninferior Frontier, *Journal of Optimization Theory and Applications*, Vol. 88, 1996, pp. 177-196.

[L10] Lee, G.M., On relations between vector variational inequality and vector optimization problem, in *Progress in Optimization II: contributions from Australasia*, Kluwer, eds. Yang, X.Q. et al, pp. 167-179.

[LKLC1] Lee, G.M., Kim, D.S., Lee, B.S. and Cho, S.J., Generalized vector variational inequalities and fuzzy extension, *Applied Mathematics Letters*, Vol. 6, 1993, pp. 47-51.

[LLKC1] Lee, G.M., Lee, B.S., Kim, D.S. and Chen, G.Y., On vector variatioanl inequalities for multifunctions, *Indian Journal of Pure Mathematics and Applied Mathematics*, Vol. 28, 1997, pp. 633-639.

[M1] Mosco, Dual variational inequalities, *Journal of Mathematical Analysis and Applications*, Vol. 40, 1972, pp. 202-206.

[M2] Mazzoleni, P., Duality and reciprocity for vector programming, *European Journal of Operational Research,* Vol. 10, 1980, pp. 42-50.

[M3] Minty, G.J., Monotone networks, *Proceeding Royal Society of London, A*, Vol. 257, 1960, pp. 194-212.

[M4] Magnanti, T.L., Fenchel and Lagrange duality are equivalent, *Mathematical Programming,* Vol. 10, 1974, pp. 253-258.

[M5] Magnanti, T.L., Models and algorithms for predicting urban traffic equilibria, *Transportation planning Models*, Eds. M. Florian, Elsevier Science Publisher, 1984.

[M6] Mangasarian, O.L., Second and higher order duality in nonlinear programming, *Journal of Mathematical Analysis and Applications*, Vol. 51, 1975, pp. 607-620.

[M7] Mond, B., Second order duality for nonlinear programs, *Opsearch*, Vol. 11, 1974, pp.90-99.

[MP1] Mangasarian, O.L. and Ponstein, J., Minimax and duality in nonlinear programming, *Journal of Mathematical Analysis and Applications*, Vol. 17, 1967, pp. 504-518.

[N1] Nagurney, A., *Network Economics: A Variational Inequality Approach*, Kluwer Academic Publichers, London, 1993.

[N2] Nakayama, H., Duality in linear vector optimization, *IIASA Working Paper* WP, 1984, pp. 84-86.

[NW1] Nemhauser G.L. and Wolsey, L.A., *Integer and Combinatorial Optimization*, John Wiley and Sons, New York, 1988.

[OR1] Ortega, J.M. and Rheinboldt, W.C., *Iterative Solution of Nonlinear Equations in Several Variables*, Acadmic Press, New York, 1970.

[P1] Polyak, B.T., Minimization of nonsmooth functionals, *U.S.S.R. Computational Mathematics and Mathematical Physics,* Vol. 9, 1969, pp. 14-29.

[R1] Rockafellar, R.T., *Convex Analysis*, Princeton University Press, 1970.

[R2] Rockafellar, R.T., *Conjugate Duality and Optimization*, Regional Conference Series in Applied Mathematics 16, SIAM, Philadelphia, 1974.

[R3] Rockafellar, R.T., *Network Flows and Monotropic Optimization*, Wiley Interscience, 1984.

[R5] Royden, H.L., *Real Analysis, third edition*, Macmillan Publishing, 1988.

[R6] Rosinger, E.E., Duality and alternative in multiobjective optimization, *Proceeding of American Mathematical Society*, Vol. 64, 1977, pp. 307-312.

[R7] Rockafellar, R.T., Monotropic programming: descent algorithms and duality, in *Nonlinear Programming 4*, eds. O.L. Mangasarian et al., Academic Press, 1981, pp. 327-366.

[R8] Rockafellar, R.T., Duality in optimal control, in *Mathematical Control Theory*, eds. W.A. Coppel, Lecture Notes in Mathematics, No. 680, Springer -Verlag, 1978, pp. 219-257.

[R9] Rockafellar, R.T., Solving a nonlinear programming by way of a dual problem, *Symposia Mathematica*, Vol. XIX, 1976, pp. 135-160.

[RW1] Rockafellar, R.T. and Wets, R., *Variational Analysis*, Springer, 1998.

[RGY1] Rubinov, A., Glover, B.M. and Yang, X.Q., Extended Lagrange and penalty functions in continuous optimization, *Optimization*, Vol. 46, 1999, pp. 327-351.

[RGY2] Rubinov, A., Glover, B.M. and Yang, X.Q., Decreasing functions with applications to penalization, *SIAM Journal on Optimization*, Vol. 10, 1999, pp. 289-313.

[S1] Stampacchia, G., Forms bilinearies coercitives sur les ensembles convexes, *C.R. Acad. Sc., Paris* Vol. 258, 1964, pp. 4413-4416.

[S2] Singer, I., A general theory of dual optimization problems, *Journal of Mathematical Analysis and Applications*, Vol. 116, 1986, pp. 77-130.

[S3] Singer, I., Surrogate dual problems and surrogate Lagrangians, *Journal of Mathematical Analysis and Applications*, Vol. 95, 1984, pp. 31-71.

[S4] Singer, I., *Abstract Convex Analysis*, Wiley-Interscience Publication, New York, 1997.

[SNT1] Sawaragi, Y., Nakayama, H. and Tanino, T., *Theory of Multi-Objective Optimization*, Academic Press, New York, 1985.

[T1] Tucker, A.W., Dual systems of homogeneous relations, in *Linear Inequalities and Related Systems*, Eds. H.W. Kuhn and A.W. Tucker, *Annals of Mathematical Studies*, Vol. 38, 1956, pp. 3-18.

[T2] Taha, H.A., *Operations Research, fourth edition*, Macmillan, New York, 1987.

[TS1] Tanino, T. and Sawaragi, Y., Duality theory in multiobjective programming, *Journal of Optimization Theory and Applications*, Vol. 27, 1979, pp. 509-529.

[TS2] Tanino, T. and Sawaragi, Y., Conjugate maps and duality in multiobjective programming, *Journal of Optimization Theory and Applications*, Vol. 31, 1980, pp. 473-499.

[vT1] Van Tiel, J., *Convex Analysis, An Introductory Text*, John Wiley and Sons, Chichester, 1984.

[W1] Walk, M., *Theory of Duality in Mathematical Programming*, Springer-Verlag, 1989.

[W2] Wolfe, P., A Duality theorem for nonlinear programming, *Quarterly of Applied Mathematics*, Vol. 19, 1961, pp.239-244.

[W3] Wardrop, J., Some theoretical aspects of road traffic research, *Proceeding of the Institute of Civil Engineers, Part II,*, Vol. 1, 1952, pp. 325-378.

[WM1] Weir, T. and Mond, B., Symmetric and self duality in multiple objective programming, *Asia-Pacific Journal of Operations Research*, Vol. 5, 1988, pp. 124-133.

[WM2] Weir, T. and Mond, B., Generalized convexity and duality in multiple objective programming, *Bulletin of the Australian Mathematical Society*, Vol. 39, 1989, pp. 287-299.

[Y1] Yang, X.Q., Vector variational inequality and its duality, *Nonlinear Analysis*, Vol. 21, 1993, pp. 869-877.

[Y2] Yang, X.Q., Second-order global optimality conditions for convex composite optimization, *Mathematical Programming*, Vol 81, 1998, pp. 327-347.

[Y3] Yang, X.Q., On some equivalent conditions of vector variational inequalities, in *Vector Variational Inequalities and Vector Equilibria*, ed. F. Giannessi, Kluwer Academic Publisher, 2000, pp. 423-432.

[Y4] Yu, P.L., *Multiple-Criteria Decision Making*, Plenum Press, New York, 1985.

[Y5] Yang, X.Q., Vector variational inequalities and its duality, *Nonlinear Analysis, Theory, Method and Applications*, Vol. 21, 1993, pp. 867-877.

[Y6] Yang, X.Q., Vector variational inequality and vector pseudolinear optimization, *Journal of Optimization Theory and Applications*, Vol. 95, 1997, pp. 729-734.

[Y7] Yang, X.Q., Vector complementarity and minimal element problems, *Journal of Optimization Theory and Applications*, Vol. 77, No. 3, 1993, pp. 483-495.

[Y8] Yang, X.Q., Generalized convex functions and vector variational inequalities, *Journal of Optimization Theory and Applications*, Vol. 79, No. 3, 1993, pp. 563-580.

[YC1] Yang, X.Q. and Chen, G.Y., A class of nonconvex functions and prevariational inequalities, *Journal of Mathematical Analysis and Applications*, Vol. 169, 1992, pp. 359-373.

[YC2] Yang, X.Q. and Chen, G.Y., On inverse vector variational inequalities, in *Vector Variational Inequalities and Vector Equilibria*, eds. F. Giannessi, Kluwer Academic Publisher, 2000, pp. 433-446.

[YG1] Yang, X.Q. and Goh, C.J., On vector variational inequality. application to vector equilibria, *Journal of Optimization Theory and Applications*, Vol. 95, 1997, pp. 431-443.

[YG2] Yang, X.Q. and Goh, C.J., The nonlinear Lagrangian function for discrete optimization problems, Preprint, Department of Mathematics, The University of Western Australia, 1997.

[YG3] Yang, X.Q., and Goh C.J., Vector variational inequalities, vector equilibrium flow and vector optimization, in *Vector Variational Inequalities and Vector Equilibria*, eds. F. Giannessi, Kluwer Academic Publisher, 2000, pp. 447-465.

[YH1] Yang, X.Q. and Huang, X.X., A nonlinear Lagrangian approach to constrained optimization problems, *SIAM J. Optimization* Vol. 14, 2001, pp. 1119 - 1144.

[YL1] Yang, X.Q. and Li, D., Successive optimization method via parametric monotone composition formulation, Vol. 16, 2000, no. 4, pp. 355-369.

[YT1] Yang, X.Q. and Teo, K.L., Nonlinear Lagrangian functions and applications to semi-infinite programs, Optimization and Numerical Algebra, A Special Issue in Annals of Operations Research, edited by Qi, L. et al., (accepted).

[ZM1] Zhu, D.L. and Marcotte, P., An extended descent framework for variational inequalities, *Journal of Optimization Theory and Applications*, Vol. 80, 1994, pp. 349-366.

[ZT1] Zhou, J.L. and Tits, A.L., User's guide for FFSQP Version 3.5: A Fortran code for solving constrained nonlinear minimax optimization problems, *Technical report TR-92-107r4*, Systems Research Center, University of Maryland, College Park, MD 20742, 1992.

INDEX

G

H

I

J

K

L

T

U

V

W

Y

Z

Milton Keynes UK
Ingram Content Group UK Ltd.
UKHW052019071024
449327UK00027B/2340

9 780367 396152